Lecture Notes in Computer Science 9239

Commenced Publication in 1973
Founding and Former Series Editors:
Gerhard Goos, Juris Hartmanis, and Jan van Leeuwen

More information about this series at http://www.springer.com/series/7409

Christophe Claramunt · Markus Schneider
Raymond Chi-Wing Wong · Li Xiong
Woong-Kee Loh · Cyrus Shahabi
Ki-Joune Li (Eds.)

Advances in Spatial and Temporal Databases

14th International Symposium, SSTD 2015
Hong Kong, China, August 26–28, 2015
Proceedings

Springer

Editors
Christophe Claramunt
Naval Academy Research Institute
Brest naval
France

Markus Schneider
University of Florida
Gainesville, FL
USA

Raymond Chi-Wing Wong
Hong Kong University of Science
 and Technology
Kowloon
Hong Kong SAR

Li Xiong
Emory University
Atlanta
USA

Woong-Kee Loh
Gachon University
Gyeonggi-do
Korea, Republic of (South Korea)

Cyrus Shahabi
University of Southern California
Los Angeles
USA

Ki-Joune Li
Pusan National University
Pusan
Korea, Republic of (South Korea)

ISSN 0302-9743   ISSN 1611-3349 (electronic)
Lecture Notes in Computer Science
ISBN 978-3-319-22362-9   ISBN 978-3-319-22363-6 (eBook)
DOI 10.1007/978-3-319-22363-6

Library of Congress Control Number: 2015945153

LNCS Sublibrary: SL3 – Information Systems and Applications, incl. Internet/Web, and HCI

Springer Cham Heidelberg New York Dordrecht London

Printed on acid-free paper

Springer International Publishing AG Switzerland is part of Springer Science+Business Media
(www.springer.com)

Preface

This volume contains the proceedings of the 14th International Symposium on Spatial and Temporal Databases (SSTD). Included are research contributions in the area of spatial and temporal data management and related computer science domains presented at SSTD 2015 in Hong Kong, China. The symposium brought together, for three days, researchers, practitioners, and developers for the presentation and discussion of current research on concepts, tools, and techniques related to spatial and temporal databases.

SSTD 2015 was the 14th in a series of biannual events. Previous symposia were held in Santa Barbara (1989), Zurich (1991), Singapore (1993), Portland (1995), Berlin (1997), Hong Kong (1999), Los Angeles (2001), Santorini, Greece (2003), Angra dos Reis (2005), Boston (2007), Aalborg (2009), Minneapolis (2011), and Munich (2013). Before 2001, the series was devoted solely to spatial database management, and was called The International Symposium on Spatial Databases. Starting in 2001, the scope was extended in order to also integrate the temporal dimension and accommodate spatial and temporal database management issues, owing to the increasing importance of research that considers spatial and temporal dimensions of data as complementary challenges.

This year the symposium received 64 submissions from authors in 27 different countries, which were reviewed by at least three of the 53 Program Committee members, helped by 89 external reviewers. At the end of a thorough process of reviews and discussions, 24 submissions were accepted for presentation at the symposium. SSTD 2015 also continued several innovative topics that were successfully introduced in previous events. We also formed a Best Paper Committee including eight committee members to select the best paper for SSTD 2015. In addition to the research paper track, the conference hosted a demonstration and vision/challenge track. Demonstrations and vision/challenge papers were solicited by separate calls for papers. While proposals for demonstrations had to illustrate running systems that showcase the applicability of interesting and solid research, the vision/challenge submissions had to discuss novel ideas that are likely to guide research in the near future and/or challenge prevailing assumptions. The submissions to the demo and vision/challenge track (12 demonstration submissions and two vision/challenge papers submissions) were evaluated by dedicated Program Committees, recruited by the demonstrations co-chairs, and eight demos were selected for the conference program.

We were very fortunate to have had two well-accomplished researchers from academia and industry as keynote speakers opening the first two days of the conference: Prof. Dimitris Papadias (The Hong Kong University of Science and Technology) gave a presentation on "Query Processing in Geo-Social Networks" and Jim Steiner (Oracle) talked about "Emerging Geospatial Trends: The Convergence of Technologies." Both are very attractive and timely topics, from the academic and industrial points of view.

The success of SSTD 2015 was the result of a team effort. Special thanks go to many people for their dedication and hard work, in particular to the local organizers, publicity chairs, proceedings chair, and webmasters. Naturally, we owe our gratitude to more people, and in particular we would like to thank the authors, irrespectively of whether their submissions were accepted or not, for supporting the symposium series and for sustaining the high quality of the submissions. Last but most definitely not least, we are very grateful to the members of the Program Committees (and the external reviewers) for their thorough and timely reviews.

Finally, these proceedings reflect the state of the art in the domain of spatiotemporal data management, and as such we believe they form a strong contribution to the related body of research and literature.

June 2015

Christophe Claramunt
Raymond Chi-Wing Wong
Markus Schneider
Li Xiong
Ki-Joune Li
Cyrus Shahabi

Organization

Steering Committee

The SSTD Endowment

Honorary Chair

Kyu-Young Whang   KAIST, South Korea

General Co-chairs

Ki-Joune Li   Pusan National University, South Korea
Cyrus Shahabi   USC, USA

Program Co-chairs

Christophe Claramunt   Naval Academy, France
Raymond Chi-Wing Wong   HKUST, Hong Kong, SAR China
Markus Schneider   University of Florida, USA

Demonstration Chair

Li Xiong   Emory University, USA

Publicity Co-chairs

Jin Soung Yoo   IPFW, USA
Baihua Zheng   SMU, Singapore

Sponsorship Co-chairs

Siva Ravada   Oracle, USA
Erik Hoel   ESRI, USA
Mehdi Sharifzadeh   Google, USA

Proceedings Chair

Woong-Kee Loh   Gachon University, South Korea

Local Arrangements

Hae-Kyong Kang   KRIHS, South Korea

Treasurer

Juhyun Ham   Pusan National University, South Korea

Research Program Committee

Walid Aref	Bart Kuijpers	Matthias Renz
Masatoshi Arikawa	Jae-Gil Lee	Dimitris Sacharidis
Spiridon Bakiras	Dan Lin	Bernhard Seeger
Michela Bertolotto	Cheng Long	Christian Sengstock
Claudio Bettini	Hua Lu	Shashi Shekhar
Michael Böhlen	Nikos Mamoulis	Ryosuke Shibasaki
Reynold Cheng	Yannis Manolopoulos	Yufei Tao
Chi-Yin Chow	Claudia Medeiros	Yannis Theodoridis
Gao Cong	Mohamed Mokbel	Carola Wenk
Maria Luisa Damiani	Kyriakos Mouratidis	Ouri Wolfson
Ugur Demiryurek	Mirco Nanni	Xiaokui Xiao
Ralf Hartmut Güting	Enrico Nardelli	Man Lung Yiu
Yan Huang	Mario Nascimento	Rui Zhang
Seung-Won Hwang	Dimitris Papadias	Dongxiang Zhang
Sergio Ilarri	Spiros Papadimitriou	Wenjie Zhang
Panagiotis Karras	Torben Bach Pedersen	Yu Zheng
Kyoung-Sook Kim	Dieter Pfoser	Xiaofang Zhou
Minsoo Kim	Chiara Renso	

Vision Program Committee

Michael Gertz	Peter Scheuermann	Vassilis Tsotras
Joerg Sander	Timos Sellis	

Demonstration Program Committee

Rui Chen	Apostolos Papadopoulos	Goce Trajcevski
Liyue Fan	Stavros Papadopoulos	Jaideep Vaidya
Shen-Shyang Ho	Cyril Ray	Fusheng Wang
Yaron Kanza	Marcos Vaz Salles	Wendy Hui Wang
Chang-Tien Lu	Ankur Teredesai	Ting Wang
Jun Luo	Kristian Torp	

Best Paper Selection Committee

Michael Böhlen
Christophe Claramunt
Maria Luisa Damiani

Yannis Manolopoulos
Claudia Medeiros
Joerg Sander

Markus Schneider
Raymond Wong

External Reviewers

Ahmed Eldawy
Alessandra Raffaeta
Amr Magdy
Anastasios Gounaris
Andre Santanche
Andreas Zuefle
Andy Yuan Xue
Antonio Corral
Apostolos Papadopoulos
Bin Yang
Brittany TereseFasy
Chris Jonathan
Christian Authmann
Christian Beilschmidt
Chuanfei Xu
Daniele Riboni
Dimitris Tsakalidis
Eleftherios Tiakas
George Skoumas
Han Su
Haozhou Wang
Hoang Vo
Huichu Zhang
Jia-Dong Zhang

Jianqiu Xu
Jianzhong Qi
Jie Bao
Jiexing Li
Johannes Droenner
Kaiqi Zhao
Klaus Arthur Schmid
Kostas Patroumpas
Lefteris Ntaflos
Li Yuhong
Lisi Chen
Long Yuan
Louai Alarabi
Luyi Mo
Lyu Yan
Michael Mattig
Nikos Pelekis
Panos Parchas
Paolo Cintia
Phuc Nguyen
Qing Guo
Ramon Hermoso
Ran Wang
Raquel Trillo Lado

Saeid Hosseini
Sam King
Sandeep Sasidharan
Sergio Mascetti
Shiyu Yang
Shuo Ma
Sibo Wang
Tobias Emrich
Victor Junqiu Wei
Xiang Wang
Xiaojie Lin
Xiaoyang Wang
Xike Xie
Xin Cao
Yang Zhou
Yixiang Fang
Yong Xu
Yu Li
Yu Sun
Zeyi Wen
Zhang Chong
Zhi Liu

Platinum Sponsors

RealtimeTech Co.
Pusan National University
Oracle
ESRI

Silver Sponsors

Gaia3D, Inc.
QBS System Limited
Research Institute of Computers, Information, and Communication, PNU

Bronze Sponsors

LNCS, Springer
KDnuggets
Korea Spatial Information Society
KIISE Database Society of Korea

Contents

Similarity Search and Pattern

Keyword and Pattern

Demonstrations

Reachability Query and Path Query

RICC: Fast Reachability Query Processing on Large Spatiotemporal Datasets

Elena V. Strzheletska$^{(\boxtimes)}$ and Vassilis J. Tsotras

University of California, Riverside, USA
{elenas,tsotras}@cs.ucr.edu

Abstract. Spatiotemporal reachability queries arise naturally when determining how diseases, information, physical items can propagate through a collection of moving objects; such queries are significant for many important domains like epidemiology, public health, security monitoring, surveillance, and social networks. While traditional reachability queries have been studied in graphs extensively, what makes spatiotemporal reachability queries different and challenging is that the associated graph is dynamic and space-time dependent. As the spatiotemporal dataset becomes very large over time, a solution needs to be I/O-efficient. Previous work assumes an 'instant exchange' scenario (where information can be instantly transferred and retransmitted between objects), which may not be the case in many real world applications. In this paper we propose the RICC (Reachability Index Construction by Contraction) approach for processing spatiotemporal reachability queries without the instant exchange assumption. We tested our algorithm on two types of realistic datasets using queries of various temporal lengths and different types (with single and multiple sources and targets). The results of our experiments show that RICC can be efficiently used for answering a wide range of spatiotemporal reachability queries on disk-resident datasets.

1 Introduction

Reachability queries are significant for many important domains such as epidemiology, public health, social networks, security monitoring, and surveillance. The last two application areas involve performing reachability queries on spatiotemporal datasets, which are the main interest of this paper. Such datasets may, for instance, contain information about locations of a set of moving objects collected during some period of time.

Two objects O_i and O_j have a contact at time t_k (denoted as $< O_i, O_j, t_k >$), if they are within some threshold distance d_{cont} from each other at that time instant [24]. During the encounter, the proximity between O_i and O_j gives them an opportunity to exchange physical items or information (perhaps wirelessly), or a virus. As they move through the network, O_i and O_j may encounter other objects, and participate in further exchanges. This pattern permits moving objects to function as couriers, allowing two objects that remain far apart to nonetheless communicate with each other via intermediaries. A spatiotemporal

C. Claramunt et al. (Eds.): SSTD 2015, LNCS 9239, pp. 3–21, 2015.
DOI: 10.1007/978-3-319-22363-6_1

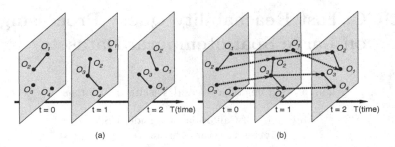

Fig. 1. (a) Positions and contacts between a set of moving objects during the time interval $[0, 2]$; (b) Constructing a supergraph by combining the contact graphs with the object trajectories.

reachability query determines whether two given objects O_S and O_T could have communicated (possibly through other objects), within a given time interval. An example appears in Fig. 1(a), where four moving objects are shown at consecutive time instants. Lines between objects denote contacts at those time instants. For example, objects O_1 and O_2 are in contact at times $t = 0$ and $t = 2$. Note, that objects O_1 and O_3 never contacted each other explicitly, however O_3 is reachable from O_1 within the time interval $[0, 1]$ through object O_2 (O_1 could pass information to O_2 at time $t = 0$, and O_2 could pass it to O_3 at time $t = 1$).

The time to exchange information (or physical items etc.) between objects affects the problem solution and it is application specific. We consider two types of delays that may occur during an exchange: *processing delay* and *transfer delay*. After two objects had a contact, the contacted object may have to spend some time to process the received information (processing delay) before being able to exchange it again; consider for example repackaging the physical item at the receiver object before resending. In other applications, for the transfer of information to occur (transfer delay), two objects are required to stay within the contact distance for some period of time; we call such elongated contact a *meeting*. An example appears if two cars exchange messages through bluetooth and thus have to travel closely together for some time.

Thus we may consider the reachability problem with no delays, one kind of delay (processing or transfer) as well as with both types of delays. To distinguish among the various scenarios we use P to denote the existence of processing delay and T for transfer delay; their absence will be denoted by $\bar{P}$ and $\bar{T}$ respectively. If no delays are present (i.e., $\bar{P}\bar{T}$) the exchange is considered (almost) instantaneous. This scenario (we will call it 'instant exchange') is assumed in [24]. Consider Fig. 1(a) where at time $t = 1$ a chain of contacts occurs: object O_2 contacts O_3, and O_3 contacts O_4. Assuming instantaneous exchanges, at this time instant information can travel from O_2 to its immediate contacts, and at the same time to all the current contacts of its contacts, etc., resulting in object O_4 been reached by O_2 during just one time instant $t = 1$. As another example, consider the case $P\bar{T}$, that is, with processing delay (i.e., an object receiving information at time t may not immediately retransmit it) and no transfer delay

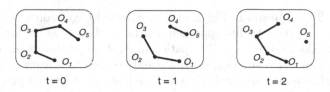

Fig. 2. Contact graphs for a set of moving objects during time interval $[0, 2]$.

(i.e. a simple contact is enough to transfer the information). In Fig. 1(a), at time $t = 1$, object O_2 contacts object O_3, and O_3 contacts O_4, but information from O_2 does not reach O_4 at that time instant.

A trajectory of a moving object O_i is a sequence of pairs (l_j, t_j), where l_j is the location of object O_i at time t_j. We assume that time is discrete, described as a sequence of time instants $(t_1, t_2, ..., t_i, ...)$ and the interval between two consecutive time instants is constant (denoted as Δt). Moreover, each object reports its location at each time instant. We further assume that all contacts between objects are identified by looking at their location records (that is, Δt is small enough that we do not miss any contact between consecutive time instants).

Consider the $P\bar{T}$ reachability scenario: for simplicity we assume that the processing delay is Δt, and after a contact occurs, retransmission starts at the next time instant (our solution can be easily modified to consider the case where the processing delay is a multiple of Δt). The goal of a reachability query $Q: \{O_S, O_T, I\}$, is to determine whether object O_T(target) is reachable from object O_S(source) during time interval $I = [t_s, t_f]$, or in other words if there exists a chain of subsequent contacts $< O_S, O_{i1}, t_1 >$, $< O_{i1}, O_{i2}, t_2 >$, ... ,$< O_{im}, O_T, t_k >$, with $t_s \leq t_1 < t_2... < t_k \leq t_f$. Moreover, if such a chain exists, we would like to find the earliest time instant when O_T was reached (this can have implications on the application: try to control the spread of the disease fast, or identify the shortest time that information traveled through a network).

Note again how the answer to a reachability query depends on the transfer requirements. Consider the example in Fig. 2: here the collection of five moving objects is observed during three time instants. Let $I = [t_0, t_2]$. The answer to the query $\{O_1, O_4, I\}$ under the $\bar{P}\bar{T}$ scenario is $t = 0$. Under the $P\bar{T}$ scenario, the answer is $t = 2$. Another query, $\{O_1, O_5, I\}$ will be answered with $t = 0$ in the first case, however, for the second case, the answer is $t = \infty$. In general, the set of objects, reached by some object O_i during I under the $\bar{P}\bar{T}$ scenario is a superset of the set of objects reached under the $P\bar{T}$ case.

The traditional graph reachability problem examines whether a path exists between two vertices of a static graph, such as a road network. Spatiotemporal reachability is more complex, since even the underlying graph is determined by the time-varying relationships between the positions of objects traversing the road network. Moreover, the contact distance d_{cont} is a parameter, and not an edge of a static graph. One could reduce spatiotemporal reachability into static graph reachability by combining the contact graphs with the object trajectories into a supergraph (by adding an edge connecting two consecutive occurrences of each object). This appears in Fig. 1(b) where dotted edges connect consecutive object positions. However this approach will be inefficient as the supergraph is very large and does not exploit the spatiotemporal properties of the dataset.

There are two naive approaches that could be used to answer a reachability query on a small spatiotemporal dataset. The first approach (no-preprocessing) is to traverse the dataset at query time, from the beginning to the end of the query time interval, collecting all the objects that were reached by the source, and checking whether the target is among the collected objects (in which case the search can be stopped before the end of the interval is reached). If not, the search proceeds, etc. The second approach (precompute-all) is to precompute and store the reachability between every pair of objects for each possible time interval in advance. Both approaches are infeasible for our problem size, since they would require either too much time or space.

Since we consider large sets of moving objects over long periods of time, the trajectory data cannot fit in main memory; hence the solution must be I/O efficient. The first disk-based solution for the spatiotemporal reachability problem with no delays ($\bar{P}\bar{T}$) was recently given by [24]. In this paper, we first develop the RICC (**R**eachability **I**ndex **C**onstruction by **C**ontraction) algorithm for the $P\bar{T}$ reachability problem; we then show how it can be extended to work with no processing but transfer delays ($\bar{P}T$). We also discuss how the PT problem (i.e., with both types of delays) can be solved by a simple modification of $\bar{P}T$.

RICC balances preprocessing time, storage consumption, and query performance time. Its preprocessing consists of several steps: the contact network construction, the reachability network construction, and the contact and reachability index construction. For the reachability network construction, we utilize the *path contraction* idea, introduced in Contraction Hierarchies (CH) [10]. A contraction replaces a path between two nodes of a graph with a (shortcut) edge, which preserves the distance between these nodes. Methods based on CH are currently the fastest known approaches for answering shortest path queries on road networks [9,10]. However, there are two major differences between our problem and computing shortest paths on road networks. CH gains its speed up from creating a hierarchy of nodes on the basis of their importance for the given road network, while in the spatiotemporal reachability problem, there is no preference between the graph nodes. In addition, road networks are typically static graphs, while our environment is dynamic. We thus created our version of path contraction, which decreases the size of the spatiotemporal reachability network, and thus reduces the space search, and consequently the reachability query time.

Figure 3(a) represents the supergraph G_1 constructed on time interval $I = [t_0, t_2)$ for the contact graphs in Fig. 1, under the 'instant exchange' assumption ($\bar{P}\bar{T}$). At time $t = 1$ object O_2 can pass the information to the object O_3, which then can pass it further to O_4 during the same time instant. The supergraph G_1' in Fig. 3(b) is constructed using the same contact graphs but under the 'no instant exchange' assumption. To disallow the 'instant exchange' in G_1', for each pair of contacting objects O_i and O_j at time t_k, we remove edges that represent contacts between them. Next, we connect O_i at time t_k with O_j at

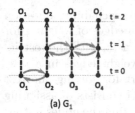 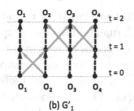 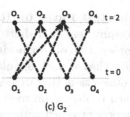

Fig. 3. (a) G_1 is the supergraph under the $\bar{P}\bar{T}$ assumption; (b) DAG G'_1 is the supergraph under the $P\bar{T}$ assumption; (c) the reachability graph G_2 constructed from G'_1 for interval $I = [t_0, t_2)$.

time t_{k+1}, and vice versa. The resulting graph G'_1 satisfies the required condition: in G'_1 at time $t = 1$ object O_2 can pass the information to O_3, but O_3 cannot retransmit it to O_4 at the same time instant. Finally Fig. 3(c) represents the reachability graph G_2, obtained from G'_1 by contracting reachability paths and replacing them with new shortcut edges (and thus G_2 is a much smaller graph than G'_1 while maintaining the same reachability properties).

The rest of the paper is organized as follows: Sect. 2 presents related work while Sect. 3 introduces the RICC algorithm, its index construction and reachability query processing. In Sect. 4, we evaluate the performance of RICC using large spatiotemporal datasets representing objects moving on a real road network (created by the Brinkhoff generator [3]) as well objects moving freely on a 2-dimensional plane (based on the random waypoint model). Finally, Sect. 5 provides conclusions and future work.

2   Related Work

Static Graph Reachability. There are many approaches that have been proposed for the static graph reachability problem and their performance lies between the two naive approaches mentioned in the previous section. They are categorized in [14] as using: (i) transitive closure compression, (ii) hop labeling, and (iii) refined online search. The first category encompasses methods that compute and compress a transitive closure. Examples include interval labeling [1], dual labeling [28], chain decomposition, tree cover, etc. The next category includes hop labeling methods: 2-hop cover [7], 3-hop cover [15] and path-top [4]. For instance, in the 2-hop approach a node u in a graph G is assigned a label, which consists of two sets of nodes: a set L_{in} that contains nodes that can reach u, and a set L_{out} of those nodes that can be reached by u. Then a node v is reachable from u if and only if L_{in} and L_{out} have a non-empty intersection. Representatives from the third category include GRAIL [30], which uses indexing based on randomized multiple interval labeling, and PReaCH [19], that applies the Contraction Hierarchies technique [10] to the reachability problem and utilizes topological levels from GRAIL. GRAIL and PReaCH outperform other reachability methods on large static graphs.

Shortest Paths on Road Networks. In our model, the reachability question is equivalent to a shortest path query in a supergraph with edges of weight 1 for consecutive object positions and edges of weight 0 for contacts, with the restriction that a path should not contain two consecutive 0-weight edges in a row. Contraction Hierarchies [10] represent the state-of-the-art for solving shortest path problems on road networks. The preprocessing of CH consists of assigning an order to each node in the road network, and then contracting the nodes in that order, introducing shortcut edges to preserve the shortest path weight for any two nodes in the graph. A shortest path query is being answered by performing a Dijkstra search in the resulted contracted graph. Nevertheless, directly applying CH would not be efficient for our reachability problem. CH benefits from creating a hierarchy of nodes on the basis of their importance for the given road network, while in the spatiotemporal reachability problem, there is no node preference between the graph nodes. Algorithm PReaCH [19] discussed above, applies CH on the static reachability problem (and thus does not exploit the spatiotemporal properties of data).

Evolving Graphs. Evolving graphs (social, citation, biological networks, etc.) have recently experienced high popularity and received increased interest in the research community. In [17], the DeltaGraph is introduced, an external hierarchical index structure that enables efficient storing and retrieving of historical graph snapshots. For large dynamic graphs, [33] constructs a reachability index, based on a combination of labeling, ordering, and updating techniques. The work in [25] utilizes graph reachability labeling methods to develop techniques for analyzing temporal distance and reachability of temporal graphs. Information, stored in such datasets, is of a different nature, if compared with spatiotemporal data. Our problem is complicated by the need to compute the contacts between the objects, while such contacts are already available in evolving graph applications. In addition, out data has spatial properties, which is usually not the case in the analysis, for example, of social and citation networks.

Spatiotemporal Databases. *Spatiotemporal Access Methods.* There has been a large number of works on spatiotemporal access methods; these typically involve some variation on hierarchical trees [6,8,11,18,23,27,31,32], or some form of a grid-based structure [22,29] or indexing in parametric space [2,5,21]. A recent survey appears in [20]. Nevertheless, existing spatiotemporal indexes typically support traditional range and nearest neighbor queries and not the reachability queries we examine here.

Complex Queries on Spatiotemporal Datasets. Recent work has focused on querying/identifying the behavior of moving objects. Various methods have been developed for determining patterns and similar behavior of a group of objects during a particular time interval. Examples include discovering moving clusters [12,16], flock patterns [26], and convoy queries [13]. Recently, [24] provided the first disk-based solutions for the spatiotemporal reachability problem, namely ReachGrid and ReachGraph. These are indexes on the contact dataset that enable faster query times. In ReachGrid, during query processing only a necessary portion of the contact network which is required for reachability evaluation is constructed and traversed. In ReachGraph, the reachability at different scales is precomputed and then reused at query time. Among the two approaches, ReachGraph is superior (and showed that it also greatly outperforms traditional graph reachability solutions like GRAIL [30]). However, what enables ReachGraph is the assumption that a contact between two objects can be instantaneous, and thus during one time instance, a chain of contacts may occur. Conceptually, this 'instant exchange' assumption, allows ReachGraph to be smaller in size (the new graph uses a single vertex for all objects that could be contacted at a given time instant) and thus reduce query time. On the other hand, ReachGrid does not require the 'instant exchange' assumption and is compared with our proposed methods through experimentation.

3   RICC

We proceed with the description of RICC. First we describe the preprocessing needed to maintain the contact and reachability networks and the indexing used to enable fast query time. Then the query processing algorithm is introduced.

3.1   Preprocessing

We start the preprocessing by dividing the entire time interval covered by the dataset into a number of non-overlapping subintervals, which we call *time blocks*; each of the created time blocks contains the information about the locations of all objects during the corresponding time interval. We call the number of time instants in each time block the *contraction parameter C*. Next, we partition the area covered by the dataset into spatial blocks (or grid cells), such that each cell is inscribed into a square with a side no greater than the contact distance d_{cont}.

For each time block, our algorithm performs several steps: multiple contact graph construction, reachability graph construction, and contact and reachability index construction. During the preprocessing, each time block is read into main memory only once, and all work on a block could be done as soon as the data for this particular block is collected.

Contact Graph Construction. For this step, we need to materialize a contact graph for each time instant. To efficiently find all contacts between the objects during a given time instant, we start with partitioning the set of all objects that are active during this time instant into subsets on the basis of their location, and according to the area partitioning described above. Due to the size of each grid cell, all contacts of object O are located either in the same cell with O, or in adjacent cells. We can start, for example, with the left bottom cell of the grid, find all contacts between the objects in this cell, then all contacts between objects in this cell and objects in all adjacent cells. Further, we move to the next cell and proceed until all cells are visited.

After all contacts are found, a contact graph for this time instant is constructed: each object is represented by a vertex, and each contact between two

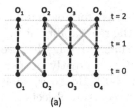 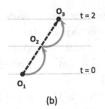 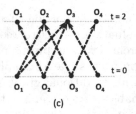

(a)   (b)   (c)

Fig. 4. (a) Supergraph; (b) Path contraction between $O_1^{(0)}$ and $O_3^{(2)}$; (c) Non-trivial reachability graph on interval $I = [t_0, t_2)$ (contraction parameter $C = 2$).

objects - by an edge. Subsequently, when a contact graph is constructed for each time instant of the block, the information is recorded in the file *Contacts* as described later. First, all data about contacts between all the objects during each time instant of a block is collected. The set of the objects is being partitioned on the basis of their location at the first time instant of the block. This time, the size of the grid G (we will call it a grid resolution as in [24]), is much larger, than for the previous partition.(In the Experiments section we describe how to find a good value for G empirically.) Next, objects are sorted according to the order of cells that they belong to. Further, in this order, information about the contacts of each object during the time block, is sequentially written on disk into the file *Contacts*. A record for each object contains its contacts at each time instant of the block in time order. An example of the *Contacts* file appears in Fig. 5.

Reachability Graph Construction. To construct the reachability graph on one time block of the dataset, we start with creating a directed supergraph by collecting contact graphs for each time instant of a block (in time order) and connecting them by introducing an edge for each two consecutive occurrences of each object. Figure 4(a), shows a supergraph, constructed on a time block with contraction parameter $C = 2$ from two contact graphs given in Fig. 1(a). The next step is to contract the reachability graph. Let $O_k^{(i)}$ denote an occurrence of object O_k during an i-th time instant of a block.

Theorem 1. *Let G^s be a supergraph constructed over a time block B. There exists a path in G^s from $O_k^{(0)}$ to $O_l^{(C-1)}$, if and only if, $O_l^{(C-1)}$ is reachable by $O_k^{(0)}$ during B.*

It follows, that to capture all reachability cases during a block, we need to answer, whether there is a path between every pair of vertices $O_k^{(0)}$ and $O_l^{(C-1)}$ in the supergraph constructed for that block. A path non-trivial if $k \neq l$. Next, we consider that any instance of object O_k is reachable from its later instance (there is a trivial path from $O_k^{(i)}$ to $O_k^{(j)}$ for $i \leq j$), and will not record it.

If there is a non-trivial path in G^s between $O_k^{(0)}$ and $O_l^{(C-1)}$, we contract this path, and replace it with an edge. In Fig. 4(a), there is a path between $O_1^{(0)}$ and $O_3^{(2)}$, thus O_3 is reachable from O_1 during this block. This path can be contracted,

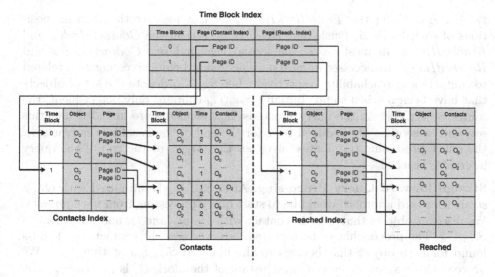

Fig. 5. Two-level index on files Contacts and Reached.

and replaced by a shortcut edge as in Fig. 4(b). We can effectively find all the paths by using multi-source BFS from each object $O_k^{(0)}$ in G^s. Figure 4(c) depicts the final reachability graph. Upon construction of the reachability graph for a given block, all reachability information is written sequentially into file *Reached* in the same object order as for the contact graphs (Fig. 5).

Contact and Reachability Index Construction. To efficiently retrieve information from disk, we use a two-level index, constructed on the files *Contacts* and *Reached*. An example of this index appears in Fig. 5. The first level (*TimeBlockIndex*), is ordered by time block number: each record consists of the time block number, and two pointers to disk pages in the second level indexes, namely the *ContactsIndex* and the *ReachedIndex*. Each record in the *ContactsIndex* is comprised of an object id and a pointer to the page in the file *Contacts*, which contains, which objects and when were contacted by this object during the given time block. Each record in the *ReachedIndex* is composed of object id and a pointer to the page in the file *Reached*, which contains, which objects were reached by this object during the given time block. The order of objects in each page of the *ContactsIndex* and *ReachedIndex* is the same as in *Contacts* and *Reached* respectively. Note that in Fig. 5 with the exception of the Time Block Index, the time block numbers (left columns) are depicted for clarity (i.e., they are not part of the index).

3.2 Query Processing

Consider a query (O_S, O_T, I), where O_S is the source object, O_T is the target object, and time interval $I = [t_s, t_f]$. Before processing this query, we need to identify the time blocks that t_s and t_f belong to. Suppose, $t_s \in B_s$, and

$t_f \in B_{f+1}$. Using the $TimeBlockIndex$, we can identify the starting positions of each block B_i (such that $B_s \leq B_i \leq B_f$) in the $ContactsIndex$ and $ReachedIndex$. In most cases, the second level indexes, $ContactsIndex$ and $ReachedIndex$, are accessed at most once per block, before accessing data related to contacts and reachability respectively. Let $S_{reached}$ denote the set of objects that have been reached so far. Initially, $S_{reached}$ contains only one element, the source object O_S. As the query proceeds, new elements are included into this set, and as soon as O_T is added to it (or the end of the last block is reached), the query processing terminates, as either the target, or the end of the query interval is reached.

Straightforward Query Processing. After $S_{reached}$ is initialized with O_S, a straightforward approach would be to start query processing from file $Contacts$. We discover objects that were in contact with O_S at time t_s, and add them to $S_{reached}$. The process has to be repeated, however now the contacts need to be found for each object that belongs to the updated $S_{reached}$ at time t_{s+1}. We proceed this way until the last time instant of the block B_s is processed. The next step is to find block B_{s+1} in file $Reached$, determine all objects that could be reached by each object from $S_{reached}$, and update $S_{reached}$. The algorithm iterates through these steps in $Reached$ until either B_{f-1}-st block is processed, or the target is reached. Finally, the process returns to file $Contacts$. If O_T has not been reached, the remaining query interval that belongs to block B_f needs to be checked. On the other hand, if O_T was reached during or before B_{f-1}-st block, then the last block, processed in $Reached$ has to be traversed in $Contacts$ once again, to determine the exact time of the contact, when target was reached.

Optimized Query Processing. At the beginning and at the end of the query, when processing information from $Contacts$, new objects are added to $S_{reached}$ at each time instant. This leads to an increase of disk accesses as parts of file $Contacts$ that cover the first and the last blocks may be read multiple times (in the worst case, C times, where C is the contraction parameter). This can be avoided if query processing begins from reading file $ReachedIndex$.

Theorem 2. *Let I and I' be two time intervals such that $I \subseteq I'$. If O_T is reachable from O_S during I, then O_T is reachable from O_S during I' as well. Also, if O_T is not reachable from O_S during I', then O_T is not reachable from O_S during I.*

The optimized query processing algorithm (Algorithm 1) starts from the $ReachedIndex$ (from the page, pointed by the $TimeBlockIndex$), and attempts to find a record for the source object (it will start at B_s and continue until either some record is found, or the end of the interval reached). If such record is found, it points to the page in $Reached$, from where we can determine all objects, that were reached by O_S during the current time block. However, if the current block is the first block of the query, and t_s is not the first time instant of this block, caution is needed, as (according to the theorem above) the set of objects, reached by O_S during B_s is the superset of the set of objects, reached by O_S from t_s

---

Algorithm 1. Reachability query processing

1: **procedure** QUERY PROCESSING(O_S, O_T, I)
2:    $S_{Reached} = \{O_S\}$, $t_{Reached} = \infty$
3:    find B_s and B_f, $B_{cur} = B_s$
4:    $C_{Ind} = readTimeBlockIndex(B_s, B_f)$ ▷ Find position of each B_i in
5:    $R_{Ind} = readTimeBlockIndex(B_s, B_f)$ ▷ *ContactsIndex* and *ReachedIndex*
6:    **while** ($O_T \notin S_{Reached}$ and $B_{cur} \neq B_{f+1}$) **do**
7:       $R_{pageIDs} = \{\emptyset\}$ ▷ $R_{pageIDs}$ - list of pages to be read from *Reached*
8:       **while** ($R_{pageIDs} = \{\emptyset\}$ and $B_{cur} \neq B_{f+1}$) **do**
9:          $R_{pageIDs} = readReachedIndex(R_{ind}, S_{Reached})$
10:          $B_{cur} + +$
11:       $S_{temp} = \{\emptyset\}$ ▷ S_{temp} is the set of objects, reached during the block
12:       $S_{temp} = findReached(R_{PageIDs}, S_{Reached}, B_{cur})$
13:       **if** ($B_{cur} = B_s$ or $B_{cur} = B_f$ or $O_T \in S_{Reached}$) **then**
14:          $C_{pageIDs} = \{\emptyset\}$ ▷ $C_{pageIDs}$ - list of pages to be read from *Contacts*
15:          $C_{pageIDs} = readContactsIndex(C_{ind}, S_{Reached}, S_{temp})$
16:          $S_{new} = filterContacts(C_{PageIDs}, S_{Reached}, S_{temp})$
17:          $S_{Reached} = S_{Reached} \cup S_{new}$
18:          **if** ($O_T \in S_{Reached}$) **then**
19:             update $t_{Reached}$
20:       **else**($S_{Reached} = S_{Reached} \cup S_{temp}$)
21:       $B_{cur} + +$
22:    **return** $t_{Reached}$ ▷ If $t_{Reached} = \infty$, then the target has not been reached

---

to the end of B_s. Hence, we need to traverse *Contacts* to make sure that we filtered all the objects that do not satisfy the time condition (the only time they were reached by the source was before the beginning of the query). After the set $S_{reached}$ is finalized, the algorithm switches to file *Reached* again, and proceeds as in the previous version, with the exception of the last time block. Suppose, we arrived at the end of B_{f-1}, collected all objects that were reached so far, but O_T was not among them. Now, we continue in *Reached*, and record all objects that were reached during B_f. If the target is not one of them, the query processing is completed. However, if O_T was reached during B_f, and t_f is not the last time instant of this block, then (again, it follows from the theorem above) we have to return into *Contacts*, and confirm that the target was reached before the end of the query interval. Although this algorithm may read from *Contacts* at the beginning and/or at the end of the query, just like the straightforward query processing, the major difference is that in this case, we read a time block (or rather its portions) only once, thus minimizing the number of I/Os.

3.3   Reachability with Transfer Delays

We now consider the reachability scenario using transfer delays. This scenario is challenging because transferring information between two objects requires that they remain in contact for some time (we call this a *meeting*). It is thus important to identify when the transfer starts, i.e., when the first contact occurs. To simplify

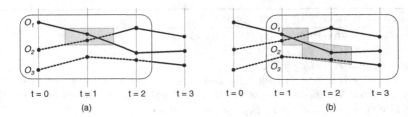

Fig. 6. Reachability with meetings: (a) Sliding window covers time instants t_0, t_1, t_2; (b) Sliding window covers time instants t_1, t_2, t_3. Meeting are shown with gray rectangles.

the problem we first consider no processing delays (i.e. $\bar{PT}$). Let λ_t be the time duration required to transfer information between two objects. We say that two objects O_i and O_j have a meeting if they had been within the threshold distance d_{cont} from each other for at least λ_t. Assuming that object O_i was carrying some information, object O_j will be considered 'reached' itself after λ_t from the start of the meeting (and can thus be able to retransmit this information). Below we show, how RICC can be extended to work with meetings.

To identify when the first contact between two objects occurs, we remove the assumption that contacts can be identified only at reported time instants. The only assumption we need is that between two consecutive location records objects move linearly. Let d_{max} denote the largest distance that can be covered by any object during time Δt. We call two objects O_i and O_j 'candidate contacts' at time t_k, if they are within the distance $d_{cc} = 2d_{max} + d_{cont}$ from each other at that time instant. If two objects are not candidate contacts at t_k, they cannot have a contact during the interval $[t_k, t_{k+1})$.

Again, to find the contacts, we partition the area covered by the dataset into spatial blocks (grid cells). However, this time the size of the cell has to be d_{cc}. During the contact graph construction phase, as opposed to identifying all contacts, we find all candidate contacts at each time instant. Next, for each pair of candidate contacts, we need to verify, whether the contact between them occurred, and if yes, what was the length of their meeting. To perform such a filtering of candidates, we can utilize a sliding time window of size p, i.e., a set of snapshots of all the moving objects taken at p consecutive time instants. For example, for $p = 3$ a window w_i will cover time instants t_i, t_{i+1}, and t_{i+2}. The length of the window p can be calculated as follows: $p = \lceil \frac{\lambda_t}{\Delta t} + 1 \rceil$.

Consider a sliding window that starts at t_k. Since it is assumed that between two consecutive location records objects move linearly, it is easy to verify whether two candidate contacts at t_k had a contact during $[t_k, t_{k+1})$, and find the time of the contact t_c. Now, using a line-sweep algorithm, we find the length of each meeting by "sweeping the line" from t_k to t_{k+p-1}. If we confirm that there was a meeting between O_i (the 'carrier') and O_j, O_j is considered to be 'reached' at $t_c + \lambda_t$. The rest of the preprocessing as well as the query processing is similar to those, described for the RICC.

We should point out that the approach considered above for the $\bar{P}T$ reachability, can be easily modified to solve the PT reachability problem. Object O_j, that was reached at $t_c + \lambda$ in the $\bar{P}T$ scenario, cannot start retransmission immediately under the PT scenario due to the processing delay λ_p. Instead, O_j is ready to start the exchange at $t_c + \lambda_t + \lambda_p$. It is also easy to modify the PT algorithm to solve a general $P\bar{T}$ reachability problem (without the assumption that contacts can be identified only at reported time instants). After all contacts are found, we simply omit the meeting verification portion. As a result, with some modifications, RICC can solve all three problems that involve delays.

4   Experiments

4.1   Dataset Description

We tested the proposed algorithm on two types of realistic datasets. Three of the datasets were created by the Brinkhoff data generator [3], which generates traces of objects, moving on real road networks. For our experiments we chose the San Francisco Bay area road network, which covers an area of about $30000\,km^2$. Three datasets contain the information about $1000, 2000$, and 4000 moving (within the speed limit) vehicles respectively; the location of each vehicle was recorded every 5 seconds and collected during a four month period (a total of $2,040,000$ time instants). Further, we assume that wireless communication is held via the Dedicated Short-Range Communications protocol (DSRC), which can afford contacts for up to $300\,m$. Thus, for the experiments on these datasets $d_{cont} = 300\,m$. We will refer to these sets as the Moving Vehicle datasets (or MV_1, MV_2, and MV_4 for sets of $1000, 2000$, and 4000 objects respectively).

For the second type of datasets, we created our own data generator, which utilizes the popular random waypoint model, frequently used for modeling movements of mobile users. According to this model, each user chooses the direction, speed (between $1.5\,m/s$ and $4\,m/s$), and duration of the next trip, then completes it, after which chooses the parameters for the next trip, and so on. The three generated sets simulate the movements of $10000, 20000$, and 40000 individuals respectively, whose location is recorded every 6 seconds for a period of one month ($432,000$ time instants total), and cover the area of $100\,km^2$ each. These sets will be referred to as Random Waypoint datasets (or RW_1, RW_2, and RW_4 for sets of $10000, 20000$, and 40000 objects respectively). We perform two sets of experiments on these datasets. For the first, we presume the communication over a Bluetooth connection and a contact distance of $d_{cont} = 25\,m$. For the second set of experiments, we assume that the individuals have to transfer a physical item in order for the contact to occur, and set a contact distance to be $d_{cont} = 2\,m$. The size of each dataset is given in Table 1(a).

Since we consider disk-resident datasets, the performance is evaluated using the number of disk accesses (I/Os) for query processing. The ratio of a sequential I/O to a random I/O is system dependent; for our experiments this ratio is $20{:}1$ [24]. In the rest, the total number of I/Os reports the equivalent number of random I/Os (that is, we assume that 20 sequential I/Os are equal to 1 random,

Table 1. (a) Size of datasets and indexes, and (b) System specifications

Dataset	Size of Dataset (GB)	Index Size (GB)	
		RICC	ReachGrid
MV₁	54	17	54
MV₂	107	56	100
MV₄	213	175	194
RW₁	97	31	99
RW₂	194	120	197
RW₄	387	419	392

(a) Size of datasets and indexes

OS	Linux 2.6
Disk Size	3TB, 7200 RPM
CPU	3.3 GHz
RAM	16 GB
Page Size	4096 B

(b) System specifications

Table 2. Parameter optimization on dataset MV_1

		Contraction Parameter (Time instants)			
		20	40	60	80
Grid Resolution (Thousand km)	20	9295	5884	5162	5779
	40	9277	5876	5192	5738
	60	9278	5874	5127	5656
	80	9260	5815	5146	5413

and calculate the total number of I/Os using this ratio). The specifications for the system used for the experiments are given in Table 1(b).

4.2   Parameter Optimization

The query performance of RICC depends on two parameters: the contraction parameter C and the grid resolution G, both of which are dataset dependent. To tune these parameters we used a subset of the dataset (of size 10 %). In general, if data is time-wise homogeneous across a dataset, any portion of it could be used, while if data differs according to some pattern - day/night, rush hour, etc., a sample that reflects the pattern should be created. We tested the performance of RICC using a set of 300 queries (the length of each query was picked uniformly at random between 100 and 500 time instants), and found the pair (C, G), which minimized the number of I/Os. The results of the parameter tuning experiments for dataset MV_1 are shown in Table 2; based on these results for the rest of the experiments involving MV_1 we pick $(C, G) = (60, 60)$ (the values for the other datasets were picked in a similar way).

4.3   Preprocessing and Indexing

Preprocessing Time. The preprocessing time depends on the size of a dataset, as well as on the contraction parameter. During the parameter optimization phase, if there are cases where several pairs of parameters (C, G), give approximately the same query performance, we choose the pair with the smaller contraction parameter C as this leads to less preprocessing.

The preprocessing time for our datasets ranged from 90 min (for the Moving Vehicles, 1000 objects dataset) to 43 h (for the Random Walk, 40000 objects

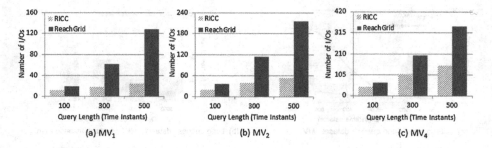

Fig. 7. Query performance evaluation for one-to-one queries; MV datasets

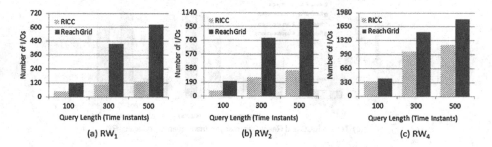

Fig. 8. Query performance evaluation for one-to-one queries; RW datasets

dataset and $d_{cont} = 25$ m). Taking into account the preprocessing speed, as well as the fact, that during the preprocessing each time block of data is read (consequently) into main memory only once, we conclude, that RICC can be applied for processing spatiotemporal data streams.)

Index Size. Fast reachability algorithms often suffer from large index size. The smallest query time is achieved when the transitive closure is precomputed (which however requires space that is quadratic on the graph size). Nevertheless, RICC can achieve very good query performance while its index size is relatively small as it can be seen from Table 1(a). This is because instead of transitive closure we precompute reachability for small portions of the graph.

4.4 Query Processing

For the query processing performance evaluation, we ran different sets of 300 queries on each of the preprocessed datasets. Further we implemented the Reach-Grid for the $P\bar{T}$ reachability, and optimized its parameters as described in [24].

One-to-One Queries. We first consider one-to-one queries $\{O_s, O_t, I\}$, (one source and one target). For the MV and RW (with $d_{cont} = 25$ m) datasets we created three sets of queries, with query lengths of 100, 300, and 500 time instances respectively, and evaluated the performance of RICC and ReachGrid by counting the number of I/Os. The results of these experiments are depicted in Figs. 7 and 8. On all instances, our approach outperforms ReachGrid. This

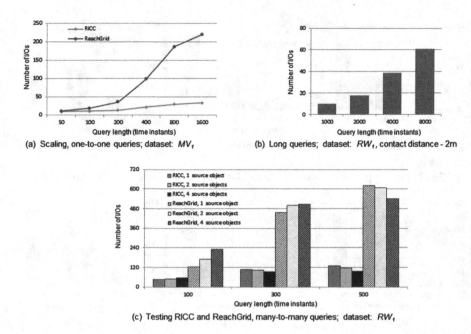

(a) Scaling, one-to-one queries; dataset: MV_1 (b) Long queries; dataset: RW_1, contact distance - 2m

(c) Testing RICC and ReachGrid, many-to-many queries; dataset: RW_1

Fig. 9. (a) Scaling, (b) long interval queries, and (c) multisource queries

improvement is because ReachGrid visits each object in a cell while RICC focuses on precomputed meetings. As the query length increases the number of objects to be checked by ReachGrid increases rapidly. Thus the biggest advantage over ReachGrid (up to 5x improvement) is reached for the longest queries on the smallest datasets (MV_1, RW_1 which have smallest number of meetings).

Scaling. The next set of tests is used to analyze the dependence of the RICC performance on the query length. When starting processing a query we need to retrieve few objects from the disk. If the query specifies a large time interval, more objects become carriers, which in turn (depending on the efficiency of an algorithm) may affect the query performance. We tested our algorithm on the MV_1 dataset, with five sets of queries, with time intervals ranging from 50 to 1600 instants respectively (after 1600 time instants all objects in the MV_1 dataset were reached). As can be seen from Fig. 9(a), while RICC uses a similar number of disk accesses as ReachGrid for the smallest length queries, it achieves much better query performance for the longer ones (up to 6.5 times for the 1600 interval). Further, RICC scales well with the size of the query length.

Many-to-Many Queries. We proceed with the experimental results for many-to-many queries (i.e., queries with several sources and/or several targets). First we note that Single Source Multitarget Queries have the same performance as one-to-one queries. Let $(O_s, \{O_{T_1}, O_{T_2}\}, I)$ be a query with the set of targets $\{O_{T_1}, O_{T_2}\}$. Then the time to answer this query $t = max(t_{Q1}, t_{Q2})$, and $N_{IO} = max(N_{IO}^1, N_{IO}^2)$ (where t_{Q_i} is the time when and if the target t_i was reached (or the end of query interval otherwise), and N_{IO}^i is the number I/Os, needed to answer the query (O_S, O_{T_i}, I).

More interesting are the *Multisource Queries*. In this case if an algorithm strongly utilizes a spatial locality for index construction, its performance should decrease when executing queries with more than one source. In the worst case (when sources are very far from each other), the number of I/Os of a query $(\{O_{S_1}, O_{S_2}\}, O_T, I)$ becomes $N_{IO} = N_{IO_1}^1 + N_{IO_2}^2$. For these experiments we used the MV and RW ($d_{cont} = 25\,m$) datasets; as we can see from Fig. 9(c), with the increase of the number of sources, the gap between the number of I/Os of RICC and ReachGrid, becomes larger.

Long Interval Queries. For the last set of experiments we used the RW_1 dataset with $d_{cont} = 2\,m$. Since the contact distance is much smaller than previously, the average contact degree becomes smaller, which in turn leads to longer average time for two objects to reach each other. We start with queries that are 1000 time instants long. We extended the query length up to 8000 time instants (which for this dataset makes about 95 % objects reachable by the end of the query interval). For these experiments, we were not able to optimize the parameters and complete the preprocessing for ReachGrid, since its query processing was very slow (ReachGrid does not scale well under the given scenario). As it can be seen from Fig. 9(b), RICC can be effectively used for long interval queries as well (it scales almost linear with the query length).

5   Conclusions

We proposed the RICC algorithm for efficient spatiotemporal reachability query processing (without the instant exchange assumption) on large disk-resident datasets. We tested our algorithm on two types of realistic datasets and different types of queries. RICC outperformed the previous known algorithm (Reach-Grid) on all experiments. In addition, our algorithm shows good performance for many-to-many queries, while also scaling well. We are currently examining aggregation-based reachability queries.

References

1. Agrawal, R., Borgida, A., Jagadish, H.V.: Efficient managemet on transitive relationships in large data and knowledge bases. In: ACM SIGMOD, pp. 253–262 (1989)
2. Bakalov, P., Hadjieleftheriou, M., Keogh, E., Tsotras, V.J.: Efficient trajectory joins using symbolic representations. In: MDM, pp. 86–93 (2005)
3. Brinkhoff, T., et al.: Generating traffic data. Data Eng. Bull. **26**(2), 19–25 (2003)
4. Cai, J., Poon, C.K.: Path-hop: efficiently indexing large graphs for reachability queries. In: 19th ACM CIKM, pp. 119–128 (2010)
5. Cai, Y., Ng, R.: Indexing spatio-temporal trajectories with chebyshev polynomials. In: ACM SIGMOD, pp. 599–610 (2004)

6. Chen, S., Ooi, B.C., Tan, K., Nascimento, M.: St2b-tree: a self-tunable spatio-temporal b+-tree index for moving objects. In: ACM SIGMOD, pp. 29–42 (2008)
7. Cohen, E., Halperin, E., Kaplan, H., Zwick, U.: Reachability and distance queries via 2-hop labels. SIAM J. Comput. **32**(5), 1338–1355 (2003)
8. De Almeida, V.T., Güting, R.H.: Indexing the trajectories of moving objects in networks. Geoinformatica **9**(1), 33–60 (2005)
9. Geisberger, R., Rice, M.N., Sanders, P., Tsotras, V.J.: Route planning with flexible edge restrictions. ACM J. Exp. Algorithms **17**(1), 1–20 (2012)
10. Geisberger, R., Sanders, P., Schultes, D., Delling, D.: Contraction hierarchies: faster and simpler hierarchical routing in road networks. In: 7th International Conference on Experimental algorithms (2008)
11. Hadjieleftheriou, M., Kollios, G., Tsotras, V.J., Gunopulos, D.: Efficient indexing of spatiotemporal objects. In: Jensen, C.S., Jeffery, K., Pokorný, J., Šaltenis, S., Bertino, E., Böhm, K., Jarke, M. (eds.) EDBT 2002. LNCS, vol. 2287, pp. 251–268. Springer, Heidelberg (2002)
12. Jensen, C., Lin, D., Ooi, B.: Continuous clustering of moving objects. IEEE TKDE **19**, 1161–1174 (2007)
13. Jeung, H., Yiu, M., Zhou, X., Jensen, C., Shen, H.: Discovery of convoys in trajectory databases. PVLDB **1**, 1068–1080 (2008)
14. Jin, E., Ruan, N., Dey, S., Xu, J.Y.: Scarab: scaling reachability computation on large graphs. In: ACM SIGMOD, pp. 169–180 (2012)
15. Jin, R., Xiang, Y., Ruan, N., Fuhry, D.: 3-hop: a high-compression indexing scheme for reachability query. In: ACM SIGMOD, pp. 813–826 (2009)
16. Kalnis, P., Mamoulis, N., Bakiras, S.: On discovering moving clusters in spatio-temporal data. In: Medeiros, C.B., Egenhofer, M., Bertino, E. (eds.) SSTD 2005. LNCS, vol. 3633, pp. 364–381. Springer, Heidelberg (2005)
17. Khurana, U., Deshpande, A.: Efficient snapshot retrieval over historical graph data. In: IEEE ICDE, pp. 997–1008 (2013)
18. Kollios, G., Gunopulos, D., Tsotras, V.J.: On indexing mobile objects. In: ACM PODS, pp. 261–272 (1999)
19. Merz, F., Sanders, P.: PReaCH: a fast lightweight reachability index using pruning and contraction hierarchies. In: Schulz, A.S., Wagner, D. (eds.) ESA 2014. LNCS, vol. 8737, pp. 701–712. Springer, Heidelberg (2014)
20. Nguyen-Dinh, L.V., Aref, W.G., Mokbel, M.F.: Spatio-temporal access methods: part2 (2003–2010). IEEE Data Eng. Bull. **33**(2), 46–55 (2010)
21. Ni, J., Ravishankar, C.V.: Indexing spatiotemporal trajectories with efficient polynomial approximation. IEEE TKDE **19**, 663–678 (2007)
22. Patel, J.M., Chen, Y., Chakka, V.P.: Stripes: an efficient index for predicted trajectories. In: ACM SIGMOD (2004)
23. Pfoser, D., Jensen, C.S., Theodoridis, Y.: Novel approaches in query processing for moving object trajectories. In: 26th VLDB Conference, pp. 395–406 (2000)
24. Shirani-Mehr, H., Banaei-Kashani, F., Shahabi, C.: Efficient reachability query evaluation in large spatiotemporal contact datasets. PVLDB **5**(9), 848–859 (2012)
25. Tang, J., Musolesi, M., Mascolo, C., Latora, V.: Characterising temporal distance and reachability in mobile and online social networks. ACM SIGCOMM Comput. Commun. Rev. **40**(1), 118–124 (2010)
26. Vieira, M.R., Bakalov, P., Tsotras, V.J.: On-line discovery of flock patterns in spatio-temporal data. In: GIS, pp. 286–295. ACM (2009)
27. Šaltenis, S., Jensen, C.S., Leutenegger, S.T., Lopez, M.A.: Indexing the positions of continuously moving objects. In: ACM SIGMOD, pp. 331–342 (2000)

28. Wang, H., He, H., Yang, J., Yu, P.S., Yu, J.X.: Dual labeling: answering graph reachability queries in constant time. In: ICDE 2006 (2006)
29. Xiong, X., Mokbel, M.F., Aref, W.G.: Lugrid: Update-tolerant grid-based indexing for moving objects. In: MDM (2006)
30. Yildirim, H., Chaoji, V., Zaki, M.J.: GRAIL: scalable reachability index for large graphs. PVLDB **3**, 276–284 (2010)
31. Yiu, M.L., Tao, Y., Mamoulis, N.: The bdual-tree: Indexing moving objects by space filling curves in the dual space. VLDB J. **17**(3), 379–400 (2008)
32. Yufei, T., Papadias, D., Sun, J.: The tpr*-tree: An optimized spatio-temporal access method for predictive queries. In: 29th VLDB Conference (2003)
33. Zhu, A.D., Lin, W., Wang, S., Xiao, X.: Reachability queries on large dynamic graphs: a total order approach. In: ACM SIGMOD pp. 1323–1334 (2014)

COLD. Revisiting Hub Labels on the Database for Large-Scale Graphs

Alexandros Efentakis[1]([✉]), Christodoulos Efstathiades[1,2], and Dieter Pfoser[3]

[1] Research Center "Athena", Marousi, Greece
efentakis@imis.athena-innovation.gr
[2] Knowledge and Database Systems Laboratory,
National Technical University of Athens, Zografou, Greece
cefstathiades@dblab.ece.ntua.gr
[3] Department of Geography and GeoInformation Science,
George Mason University, Fairfax, USA
dpfoser@gmu.edu

Abstract. Shortest-path computation is a well-studied problem in algorithmic theory. An aspect that has only recently attracted attention is the use of databases in combination with graph algorithms to compute distance queries on large graphs. To this end, we propose a novel, efficient, pure-SQL framework for answering exact distance queries on large-scale graphs, implemented entirely on an open-source database system. Our COLD framework (COmpressed Labels on the Database) may answer multiple distance queries (vertex-to-vertex, one-to-many, kNN, RkNN) not handled by previous methods, rendering it a complete solution for a variety of practical applications in large-scale graphs. Experimental results will show that COLD outperforms previous approaches (including popular graph databases) in terms of query time and efficiency, while requiring significantly less storage space than previous methods.

1 Introduction

Answering distance queries on graphs is one of the most well-studied problems on algorithmic theory, mainly due to its wide range of applications. Although a lot of recent research focused exclusively on transportation networks (cf. [9] for the most recent overview) the emergence of social networks has generated massive unweighted graphs of interconnected entities. On such networks, the distance between two vertices is an indication of the closeness of their entities, i.e., for finding users closely related to each other or extracting information about existing communities within the social media users. Although we may always use a breadth first search (BFS) to calculate the distance between any two vertices on such graphs, that approach cannot facilitate fast-enough queries on main memory or be easily adapted to secondary storage solutions.

Moreover, most of the excellent preprocessing techniques available for road networks cannot be adapted to large-scale graphs, such as social or collaboration networks. So far, the most promising approach for this type of graphs builds on the 2-hop labeling or hub labeling (HL) algorithm [12,23], in which we store a

© Springer International Publishing Switzerland 2015
C. Claramunt et al. (Eds.): SSTD 2015, LNCS 9239, pp. 22–39, 2015.
DOI: 10.1007/978-3-319-22363-6_2

two-part label $L(v)$ for every vertex v: a forward label $L_f(v)$ and a backward label $L_b(v)$. These labels are then used to very fast answer vertex-to-vertex shortest-path queries. This technique has been adapted successfully to road networks [2–4,15] and quite recently has also been extended to undirected, unweighted graphs [5,14,25]. The HL method has also been applied for one-to-many, many-to-many and kNN queries in road networks [16,17] and kNN and RkNN queries in the context of social networks in [21].

Although hub labeling is an extremely efficient shortest-path computation method using main memory, there are very few works that try to replicate those algorithms for secondary storage. HLDB [18] stores the calculated hub labels for continental road networks in a commercial database system and translates the typical HL distance query between two vertices to plain SQL commands. More-over, it showed how to efficiently answer kNN queries and k-best via points, again by means of SQL queries. Recently, HopDB [25] proposed a customized solution that utilizes secondary storage also during preprocessing. Unfortunately, both methods have their shortcomings. HLDB has only been tested on road networks and consequently small labels sizes (<100). Its speed would seriously degrade for large-scale graphs due to the much larger label size. HopDB answers only vertex-to-vertex queries and is a customized C++ solution that cannot be used with existing database systems and, hence, has limited practical applicability.

This work presents a database framework that may service multiple distance queries on massive large-scale graphs. Our pure-SQL *COLD* framework (COm-pressed Labels on the Database) can answer multiple exact distance queries (point-to-point, kNN) in addition to RkNN and *one-to-many* queries not han-dled by previous methods, rendering it a complete database solution for a variety of practical massive, large-scale graph problems. Our extensive experimentation will show that COLD outperforms previous solutions, including specialized graph databases, on all aspects (including query performance and memory require-ments), while servicing a larger variety of distance queries. In addition, COLD is implemented using a popular, open-source database engine with no third-party extensions and, thus, our results are easily reproducible by anyone.

The outline of the remainder of this work is as follows. Section 2 presents related work. Section 3 describes the novel COLD framework and its implemen-tation details. Experiments establishing the benefits of COLD are provided in Sect. 4. Finally, Sect. 5 gives conclusions and directions for future work.

2 Related Work

Throughout this work we use undirected, unweighted graphs $G(V, E)$ (where V rep-resents vertices and E arcs). A k-Nearest Neighbor (kNN) query seeks the k-nearest neighbors to an input vertex q. The RkNN query (also referred as the monochro-matic RkNN query), given a query point q and a set of objects P, retrieves all the objects that have q as one of their k-nearest neighbors according to a given dis-tance function $dist()$. In graph networks, $dist(s, t)$ corresponds to the minimum net-work distance between the two objects. Formally RkNN$(q) = \{p \in P : dist(p, q)$

$\leq dist(p, p_k)\}$ where p_k is the k-Nearest Neighbor (kNN) of p. Throughout this work, we assume that objects are located on vertices and we always refer to *snapshot* kNN and RkNN queries on graphs, i.e., objects are not moving. Also, similarly to previous works, the term *object density* D refers to the ratio $|P|/|V|$, where P is a set of objects in the graph and $|V|$ is the total number of vertices. Although, there is extensive literature focusing on kNN and RkNN queries in Euclidean space, since our work focuses on graphs we will only describe related work focusing on the latter.

Regarding road networks and kNN queries, G-tree [33] is a balanced tree structure, constructed by recursively partitioning the road network into subnetworks. Unfortunately, this method cannot scale for continental road networks, since it requires several hours for its preprocessing. Moreover, it requires a *target selection phase* to index which tree-nodes contain objects (requiring few seconds) and thus, cannot be used for moving objects. Recently, the work of [17] expanded the graph-separators CRP algorithm of [13] to handle kNN queries on road networks. Unfortunately, (i) CRP also requires a target selection phase and thus, cannot be applied to moving objects and (ii) it may only perform well for objects near the query location. Hence, this solution is also not optimal. The latest work for kNN queries on road networks is the SALT framework [22] which may be used to answer multiple distance queries on road networks, including *vertex-to-vertex* (v2v), single source (one-to-all, range, one-to-many) and kNN queries. This work expands the graph-separators GRASP algorithms of [20] and the ALT-SIMD adaptation [19] of the ALT algorithm and offers very fast preprocessing time and excellent query times. For kNN queries, SALT does not require a target selection phase and hence it may be used for either static or moving objects.

For RkNN queries on road networks, the work of [30] uses Network Voronoi cells (i.e., the set of vertices and arcs that are closer to the generator object) to answer RkNN queries. This work has only been tested on a relatively small network ($110\,K$ arcs) and all precomputed information is stored in a database. Despite the fact that the preprocessing stage for computing the Network Voronoi cells is quite costly, the queries' executions times range from $1.5\,s$ for $D = 0.05$ and $k = 1$, up to $32\,s$ for $k = 20$, rendering this solution impractical for real-time scenarios. Up until recently, the only work dealing with other graph classes (besides road networks) is [32], although it has only been tested on sparse networks, e.g., road networks, grid networks (max degree 10), p2p graphs (avg degree 4) and a very small, sparse co-authorship graph ($4\,K$ nodes). In this work, the conducted experiments for values of $k > 1$ refer only to road networks, therefore the scalability of this work for denser graphs and larger values of k is questionable. Recently, Borutta et al. [10] extended this work for time-dependent road networks, but presented results were not very encouraging. The larger road network tested had $50\,k$ nodes (queries require more than $1\,s$ for $k = 1$) and for a network of $10\,k$ nodes and $k = 8$, RkNN queries take more than $0.3\,s$ (without even adding the I/O cost). In a nutshell, all existing contributions and methods have not been tested on dense, large-scale graphs, cannot scale for increasing k values and their performance highly depends on the object density D.

Our work builds upon the 2-hop labeling or Hub Labeling (HL) algorithm of [12,23] in which, preprocessing stores at every vertex v a forward $L_f(v)$ and a backward label $L_b(v)$. The forward label $L_f(v)$ is a sequence of pairs $(u, dist(v, u))$, with $u \in V$. Likewise, the backward label $L_b(v)$ contains pairs $(w, dist(w, v))$. Vertices u and w are denoted as the *hubs of v*. The generated labels conform to *the cover property*, i.e., for any s and t, the set $L_f(s) \cap L_b(t)$ must contain at least one hub that is on the shortest $s - t$ path. For undirected graphs $L_b(v) = L_f(v)$. To find the network distance $dist(s, t)$ between two vertices s and t, a HL query must find the hub $v \in L_f(s) \cap L_b(t)$ that minimizes the sum $dist(s, v) + dist(v, t)$. By sorting the pairs in each label by hub, this takes linear time by employing a coordinated sweep over both labels. The HL technique has been successfully adapted for road networks in [2–4,15]. In the case of large-scale graphs, the Pruned Landmark Labeling (PLL) algorithm of [5] *produces a minimal labeling for a specified vertex ordering*. In this work, vertices are ordered by degree, whereas the work of [14] improves the suggested vertex ordering and the storage of the hub labels for maximum compression. The HL method has also been used for one-to-many, many-to-many and kNN queries on road networks in [16] and [17] respectively. Our latest work [21] proposed *ReHub*, a novel main-memory algorithm that extends the Hub Labeling approach to efficiently handle RkNN queries. The main advantage of the *ReHub* algorithm is the separation between its costlier offline phase, which runs only once for a specific set of objects and a very fast online phase which depends on the query vertex q. Still, even the costlier offline phase hardly needs more than $1 s$, whereas the online phase requires usually less than $1 ms$, making *ReHub* the only RkNN algorithm fast enough for real-time applications and big, large-scale graphs.

Regarding secondary-storage solutions, Jiang et al. [25] propose their HopDB algorithm that suggest an efficient HL index construction when the given graphs and the corresponding index are too big to fit into main memory. The work of [1] introduced the HLDB system, which answers distance and kNN queries in road networks entirely within a database by storing the hub labels in database tables and translating the corresponding HL queries to SQL commands. Throughout this work, we will compare our proposed COLD framework to HLDB, since to the best of our knowledge, it is the only framework that may answer exact distance queries entirely within a database. Moreover, within the COLD framework we also adapt our *ReHub* main-memory algorithm into a database context, so that its online phase may be translated to fast and optimized SQL queries.

3 Contribution

This section presents the *COLD* (COmpressed Labels on the Database) database framework. COLD can answer multiple distance queries (vertex-to-vertex, kNN, RkNN and *one-to-many*) for large-scale graphs using SQL commands. Since COLD builds on HLDB [1] and *ReHub* [21], we will follow the notation and running example presented there, for highlighting the necessary concepts and

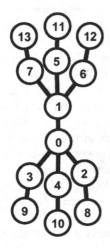

Fig. 1. A sample Graph G

Table 1. The created hub-labels for the sample graph G of Fig. 1

Vertex	Hub Labels (h,d)
0	(0,0)
1	(0,1), (1,0)
2	(0,1), (2,0)
3	(0,1), (3,0)
4	**(0,1), (4,0)**
5	(0,2), (1,1), (5,0)
6	(0,2), (1,1), (6,0)
7	(0,2), (1,1), (7,0)
8	(0,2), (2,1), (8,0)
9	(0,2), (3,1), (9,0)
10	**(0,2), (4,1), (10,0)**
11	(0,3), (1,2), (5,1), (11,0)
12	**(0,3), (1,2), (6,1), (12,0)**
13	(0,3), (1,2), (7,1), (13,0)

challenges for adapting those previous works, (i) in the context of large-scale graphs for [1] and (ii) within the boundaries of a relational database management system (RDBMS) for [21]. To this end, we chose PostgreSQL [29] for our implementation, given that it is a popular, open-source RDBMS. Although we use some PostgreSQL-specific data-types and SQL extensions, we do not use any third-party extensions but only features included in its standard installation.

3.1   Implementation

The COLD framework assumes that we have a correct hub labeling (HL) framework that generates hub-labels for the undirected, unweighted graphs we wish to query. Although COLD will work with any correct HL algorithm, in this work we use the [6] implementation of the PLL algorithm of [5] to generate the necessary labels. To highlight the results of this process, the labels for the undirected, unweighted graph G of Fig. 1 are shown in Table 1. Throughout this work, we will refer to those labels as the *forward labels*. The forward label $L(v)$ for a vertex v is an array of pairs $(u, dist(v, u)$ sorted by hub u. Since our work also focuses on snapshot kNN and RkNN queries, there also some objects $P \in V$ that do not change over time. For our specific running example we assume that $P = \{4, 10, 12\}$ and thus, we highlight the respective entries of Table 1.

Vertex-to-Vertex (v2v) Queries. To find the network distance $dist(s, t)$ between two vertices s and t, a HL query must find the hub $v \in L(s) \cap L(t)$

Table 2. The *forward* table used in HLDB for the sample graph G

v	hub	dist
...	...	...
2	0	1
2	2	0
...	...	...
7	0	2
7	1	1
7	7	0
...	...	...

Table 3. The *forwcold* table used for COLD for the sample graph G

v	hubs	dists
...	...	...
2	$\{0, 2\}$	$\{1, 0\}$
...	...	...
7	$\{0, 1, 7\}$	$\{2, 1, 0\}$
...	...	...

Code 1.1. V2v query for HLDB

```
1 SELECT MIN(n1.dist+n2.dist)
2 FROM forward n1, forward n2
3 WHERE n1.v = s
4 AND n2.v = t
5 AND n1.hub = n2.hub;
```

Code 1.2. V2v query for COLD

```
1 SELECT MIN(n1.d+n2.d) FROM
2 /* Expand hubs, dists arrays */
3 (SELECT UNNEST(hubs) AS hub,
4 UNNEST(dists) AS d
5 FROM forwcold WHERE v = s) n1,
6 (SELECT UNNEST(hubs) AS hub,
7 UNNEST(dists) AS d
8 FROM forwcold WHERE v = t) n2
9 WHERE n1.hub=n2.hub;
```

that minimizes the sum $dist(s, v) + dist(v, t)$. For our sample graph G, the minimum distance between e.g., vertices 2 and 7 is $d(2, 7) = 3$, using the hub 0. To translate this HL query into SQL commands, in HLDB [1] forward labels are stored in a database table denoted *forward* where the labels of vertex v are stored as triples of the form $(v, hub, dist(v, hub))$ (see Table 2). The table *forward* has the combination of (v, hub) as the primary key and is clustered according to those columns, so that "*all rows corresponding to the same label are stored together to minimize random accesses to the database*" [1]. Then we can find the distances between any two vertices s and t by the SQL query of Code 1.1.

Although the HLDB vertex-to-vertex (v2v) query is very simple, there is one major drawback. For such a query, HLDB has to fetch from secondary storage the subset of $|L(s)| + |L(t)|$ rows with common hubs. Although this is practical for road networks where the forward labels have less than 100 hubs per vertex [3], it cannot scale for large-scale graphs where the forward labels have thousand of hubs per vertex. Moreover, on such graphs the *forward* DB table and the corresponding primary key index will become too large, which is also an important disadvantage. To this end, we take advantage of the fact that PostgreSQL features an array data type that allows columns of a DB table to be defined as variable-length arrays. Hence, in COLD we store hubs and distances for a vertex (both ordered by hub) as arrays in two separate columns (i.e., hubs and dists) in a single row. The resulting *forwcold* compressed DB table is shown in Table 3. This approach not only emulates exactly how labels are stored on main-memory for fast v2v queries but also has considerable advantages: (i) The *forwcold* DB table has exactly $|V|$ rows (ii) The *forwcold* DB table has the column v as primary key without needing a composite key. This alone facilitates faster queries. Moreover the size of the corresponding index will be much smaller. In fact, our experimentation will show that the primary-key index for *forwcold* may be $> 4,400\times$ smaller than the index size of HLDB. (iii) For a v2v query, COLD needs to access exactly two rows, regardless of the sizes of $|L(s)|$ and $|L(t)|$. This way, we efficiently minimized the secondary-storage utilization, even working inside a database. The resulting SQL query for COLD is shown in Code 1.2. There we exploit the fact that PostgreSQL *"guarantees that parallel unnesting"* for hubs and distances for each nested query *"will be in sync"*, i.e., each pair (hub, dist) is expanded correctly since for the same v the respective arrays have the same number of elements[1].

Additional Queries Overview. For answering more complex (kNN, RkNN and *one-to-many*) distance queries on a HL framework for a set of objects P, we need to build some additional data structures from the forward labels (for undirected graphs). Then to answer the respective query we only need to combine the forward labels $L(q)$ of query vertex q, with the respective data structure explained in the following. Those data structures are summarized in Table 4.

Table 4. Necessary data structures for the sample graph G, $P = \{4, 10, 12\}$ and *one-to-many*, kNN and RkNN queries

Hub	Backward Labels (to-many) [16]	kNN Backward Labels (k=2) [1]	RkNN Backward Labels (k=1) [21]	Obj	kNN Result (k=1) (Obj., dist) [21]
0	(4,1), (10,2), (12,3)	(4,1), (10,2)	(4,1), (12,3)	4	(10,1)
1	(12,2)	(12,2)	(12,2)		
4	(4,0), (10,1)	(4,0),(10,1)	(4,0), (10,1)	10	(4,1)
6	(12,1)	(12,1)	(12,1)		
10	(10,0)	(10,0)	(10,0)	12	(4,4)
12	(12,0)	(12,0)	(12,0)		

For answering *one-to-many* queries, i.e., calculate distances between a source vertex q and all objects in P, we need to build the *backward labels-to-many* by basically ordering the forward labels of the objects by hub [16] and then by distance for the same hub. For kNN queries we only need to keep at most the k-best pairs (of smallest distances) per hub from the backward labels-to-many to create the kNN *backward labels* [1]. In our specific example, the kNN backward labels for $k = 2$ and hub 0, do not contain the pair $(12, 3)$. Finally, for RkNN queries, we must first calculate the kNN *Results* (i.e., the NN of the object 4 is the object 10 with distance 1) and then we build the RkNN backward labels, based

---

[1] http://stackoverflow.com/a/23838131.

on the observation that *"we need to access those pairs from the backward labels-to-many to a specific object, if and only if those distances are equal or smaller than the distance of the kNN of this object"* [21]. In our specific example, the RkNN backward labels for $k = 1$ and hub 0, do not contain the pair (10,2) since the NN of object 10 (the object 4) is within distance 1. Although for our small graph the differences between the individual data structures seem minimal, for larger graphs those differences become very prominent. This was also showcased by the theoretical analysis provided in [21] which showed that backward labels-to-many will have on average $D \cdot |HL|$ pairs, the kNN backward labels have at most $k \cdot |V|$ pairs and the RkNN backward labels have on average $\varepsilon \cdot D \cdot |HL|$ pairs where ε may be < 0.01 for specific datasets and experimental settings. Moreover, Efentakis et al. [21] have shown how these additional data structures may be constructed from the forward labels in main-memory, requiring less than few seconds, even for the larger tested datasets.

kNN Queries. To translate the HL kNN query into SQL, HLDB stores kNN *backward labels* in a separate DB table denoted *knntab* that stores triples of the form $(hub, dist, obj)$ (see Table 5). The respective table *knntab* has the combination of $(hub, dist, obj)$ as a composite primary key and is clustered according to those columns. Note that in HLDB, we cannot use the combination of $(hub, dist)$ as a primary key, because especially in large scale graphs we will have a lot of distance ties even for k-entries for the same hub. Then we can can answer a kNN query from vertex q by the SQL query of Code 1.3. Again, the kNN HLDB query has the same drawbacks as before, i.e., it has to retrieve $|L(q)|$ rows from *forward* and $k \cdot |L(q)|$ rows from *knntab* tables, for a total of $(k+1) \cdot |L(q)|$ rows retrieved from secondary storage. Moreover in a database, it makes sense to create one large *knntab* table for the maximum value *kmax* of k (e.g., for $k = 16$) that may be serviced by the DB framework and that same table will be used for all kNN queries up to $k = kmax$. In that case, the HLDB framework will have to retrieve $(kmax + 1) \cdot |L(q)|$ rows for every kNN query regardless of the value of k.

To remedy the HLDB drawbacks, COLD creates the *knncold* DB table (Table 6) that has the columns $(hub, dist, objs)$, whereas objects are grouped and ordered per hub and distance (the column *objs* is an array). Although for our sample graph G, the DB tables *knntab* and *knncold* seem identical, COLD's method offers several advantages: (i) We can now use the combination of $(hub, dist)$ as a primary key, which makes the respective index significantly

Table 5. The *knntab* table used in HLDB for the sample graph G, $k = 2$ and $P = \{4, 10, 12\}$

hub	dist	obj
0	1	4
0	2	10
1	2	12
...	...	...

Table 6. The *knntab* table used in COLD for the sample graph G, $k = 2$ and $P = \{4, 10, 12\}$

hub	dist	objs
0	1	{4}
0	2	{10}
1	2	{12}
...	...	...

smaller and faster and (ii) In case of many distance ties (common to large-scale graphs) and one large *knncold* DB table that services all *k*NN queries for values of *k* up to the maximum value *kmax*, we only need to fetch the first *k-objs* entries (i.e., `objs[1:k]`) per hub and dist, which makes the later sorting faster (see Code 1.4).

Code 1.3. *k*NN query for HLDB

```
1 SELECT MIN(n1.dist+n2.dist),
2 n2.obj FROM
3 forward n1, knntab n2
4 WHERE n1.v = q
5 AND n1.hub = n2.hub
6 GROUP BY n2.obj
7 ORDER BY MIN(n1.dist+n2.dist)
8 LIMIT k;
```

Code 1.4. *k*NN query for COLD

```
1 SELECT MIN(n1.d+n2.dist),
2 UNNEST(objs) AS obj FROM
3 (SELECT UNNEST(hubs) AS hub,
4 UNNEST(dists) AS d
5 FROM forwcold WHERE v = q) n1,
6 /* k-entries per hub,dist */
7 (SELECT hub, dist,objs[1:k]
8 FROM knncold) n2
9 WHERE n1.hub=n2.hub
10 GROUP BY obj
11 ORDER BY MIN(n1.d+n2.dist)
12 LIMIT k;
```

One-to-Many Queries. Similar to how COLD handles *k*NN queries, for one-to-many queries, COLD stores the *backward labels-to-many* in a new *objcold* DB table that has an identical format to *knncold*, i.e., it has three columns $(hub, dist, objs)$ whereas objects are grouped and ordered per hub and distance. *Objcold* also uses the combination of $(hub, dist)$ as a primary key. The resulting *one-to-many* query (Code 1.5) is quite similar to COLD's *k*NN query, but (i) it operates on the larger *objcold* DB table (ii) It does not have the ORDER BY ... LIMIT k clause and (iii) We use the entire *objs* array per hub and distance instead of `objs[1:k]`. Note that HLDB cannot possibly support such queries because it will need to retrieve on average $|L(q)|$ rows from the *forward* table and a total of $|L(q)| \cdot D \cdot (|HL|/|V|)$ [21] rows from the corresponding *objlab* table, which will be prohibitively slow for very large datasets.

Table 7. The *knnres* table used in COLD for R*k*NN queries, the sample graph G, $k = 1$ and $P = \{4, 10, 12\}$

obj	dists	objs
4	{1}	{10}
10	{1}	{4}
12	{4}	{4}

R*k*NN Queries. For R*k*NN queries, COLD stores the R*k*NN backward labels in a separate *revcold* DB table that has an identical format to previous *knncold* and *objcold* DB tables, i.e., three columns $(hub, dist, objs)$ where objects are grouped and ordered per hub and distance and the combination of $(hub, dist)$ used as a primary key. COLD also stores the *k*NN *Results*, i.e., the *k*NN of all objects in another *knnres* DB table that has the format $(obj, dists, objs,)$ where obj is the primary key and *objs* and *dists* are arrays (both ordered by distance) (Table 7). Therefore the *k*NN of object p is the `objs[k]` within distance `dists[k]` of the respective row for p. Again it makes sense to build a *knnres* DB table for a max value of *kmax* that may service R*k*NN queries for varying values of *k*. As a result, during the

Code 1.5. *One-to-many* COLD query

```
1 SELECT MIN(n1.d+n2.dist),
2 UNNEST(objs) AS obj FROM
3 (SELECT UNNEST(hubs) AS hub,
4 UNNEST(dists) AS d
5 FROM forwcold
6 WHERE v = q) n1,
7 objcold n2
8 WHERE n1.hub=n2.hub
9 GROUP BY obj;
```

Code 1.6. RkNN query for COLD

```
1 SELECT n3.id2,n3.dist FROM
2 /* n3 subquery is a modified
3    one-many-query to revcold */
4 (SELECT MIN(n1.d+n2.dist) AS d3,
5 UNNEST(objs) AS obj FROM
6 (SELECT UNNEST(hubs) AS hub,
7 UNNEST(dists) AS d
8 FROM forwcold WHERE v = q) n1,
9 revcold n2
10 WHERE n1.hub=n2.hub
11 GROUP BY obj
12 ORDER BY obj,MIN(n1.d+n2.dist)
13 ) n3,
14 /* Join with knnres table */
15 (SELECT obj, dists[k] AS dist
16 FROM knnres) n4
17 WHERE n3.obj=n4.obj
18 AND n3.d3<=n4.dist
19 ORDER BY n3.obj;
```

RkNN COLD query, we will have to use an additional JOIN between the *revcold* and *knnres* DB tables. The resulting query is shown in Code 1.6.

We see that even the more complex RkNN query in COLD requires just a few lines of SQL code that will work on any recent PostgreSQL version without any need of third-party extensions or specialized index structures. In fact, all DB tables in COLD, use only standard B-tree primary key indexes, without any modifications. To satisfy this strict requirement, we effectively compressed the index sizes by grouping rows per vertex (*forcold* table) or object (*knnres* table), or by hub and distance for *knncold, objcold* and *rknncold*. And although we used PostgreSQL specific SQL extensions for expanding the stored arrays, latest versions of other databases (e.g., Oracle) support similar array data-types. Hence, it would be quite easy to port COLD to other database vendors as well.

This section detailed the COLD framework in terms of design and implementation. COLD can answer multiple distance queries (v2v, kNN, RkNN and one-to-many) based on data stored in an off-the-shelf relational database. We also presented the actual queries used and the way the necessary data structures are stored within the database, so that our results are easily reproducible. Although we focused on query efficiency, it is important to note that once we create the *forcold* table, all the adjoining DB tables within COLD may also be created using SQL commands (resulting queries were omitted due to space restrictions). This fact also shows that COLD is truly a pure-SQL framework for servicing multiple distance queries on large-scale graphs. We also provided the necessary theoretical details as to why the COLD framework will outperform existing solutions. This will be further quantified in the following section.

4   Experimental Evaluation

To assess the performance of COLD on various large-scale graphs, we conducted experiments on a workstation with a 4-core Intel i7-4771 processor clocked at 3.5 G Hz and 32 Gb of RAM, running Ubuntu 14.04. We compare our COLD framework with a custom implementation of HLDB in PostgreSQL and with *Neo4j*, a well-known, popular graph database.

We use the same network graphs as our previous work of [21] that are taken from the Stanford Large Network Dataset Collection [26] and the 10th Dimacs Implementation Challenge website [8]. All graphs are undirected, unweighted and strongly connected. We used collaboration graphs (DBLP, Citeseer1, Citeseer2) [24], social networks (Facebook [28], Slashdot1 and Slashdot2 [27]), networks with ground-truth communities (Amazon, Youtube) [31], web graphs (Notre Dame) [7] and location-based social networks (Gowalla) [11]. The graphs' average degree is between 3 and 37 and the PLL algorithm creates $26 - 4,457$ labels per vertex, requiring $0.03 - 5,946\,s$ for the hub labels' construction (see Table 8).

Table 8. Networks graphs statistics

Graph	| V |	| E |	Avg degr	| HL | / | V |	PLL Preproc. Time (s)
Facebook	4,039	88,234	22	26	0.03
NotreDame	325,729	1,090,108	3	55	6
Gowalla	196,591	950,327	5	100	13
Youtube	1,134,890	2,987,624	3	167	123
Slashdot1	77,360	469,180	6	204	11
Slashdot2	82,168	504,230	6	216	13
Citeseer1	268,495	1,156,647	4	408	110
Amazon	334,863	925,872	3	689	230
DBLP	540,486	15,245,729	28	3,628	5,720
Citeseer2	434,102	16,036,720	37	4,457	5,946

COLD and HLDB were implemented in PostgreSQL 9.3.6, 64 bit with reasonable settings (8192 Mb *shared buffers*, 64 Mb *temp buffers*). We also used Neo4j Server v2.1.5. The Neo4j queries were formulated using *Cypher*, Neo4j's declarative query language and we report query times as they were returned by the server. Although Cypher may theoretically facilitate *one-to-many* queries (besides vertex-to-vertex), testing Neo4j with our datasets and the same number of target vertices we tested COLD with, resulted in a "java.lang.Stack OverflowError". Providing the server with additional resources[2] had no positive effect and thus there are no results for *one-to-many* queries and Neo4j.

---

[2] http://neo4j.com/developer/guide-performance-tuning/.

We conducted experiments belonging to four query types: (i) *vertex-to-vertex*, (ii) kNN, (iii) RkNN and (iv) *one-to-many*. For each experiment, we used 10,000 random start vertices, reporting the average running time. Before each experiment, we restart the PostgreSQL and Neo4j servers for clearing their internal cache and we also clear the operating system's cache for accurate benchmarking. All charts are plotted in logarithmic scale.

4.1 Performance on HDD

In our first round of experiments, we ran experiments on an HDD, specifically a SATA3 Seagate Barracude ST3000DM001 7200rpm with 64Mb cache.

Vertex-to-vertex. Fig. 2(a) shows results for vertex-to-vertex (v2v) queries for COLD, HLDB and Neo4j. Results show that COLD is consistently 2 - 20.7× faster than HLDB, with this difference amplified for the Citeseer1, Amazon and Youtube datasets (16.8, 19.1 and 20.7 respectively). Moreover, COLD is also 9 - 143× (for the *Gowalla* dataset) faster than Neo4j, which exhibits stable performance for all datasets, but is slower from both COLD and HLDB. For all datasets, COLD requires less than 9ms for answering v2v queries.

Figure 2(b) shows the difference in memory size for the DB tables *forcold* (COLD) and *forward* (HLDB) and their respective primary-key (PK) indexes. Results show that the size of the PK index in COLD is 3, 600 - 4, 444× smaller than for HLDB (for DBLP and Citeseer2 respectively). As expected, the difference in index sizes is almost identical to the $|HL|/|V|$ ratio, since *forcold* table has $|V|$ rows and *forward* has $|HL|$ rows. Likewise, the corresponding tables are 131 - 188× smaller for COLD. Thus, the techniques used for compressing the forward labels in COLD clearly achieve a considerable reduction in memory size, rendering our proposed framework suitable for real-world scenarios.

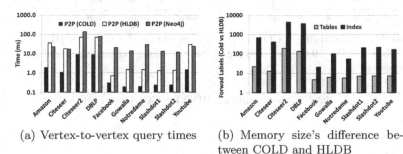

(a) Vertex-to-vertex query times

(b) Memory size's difference between COLD and HLDB

Fig. 2. Experiments on HDD for *vertex-to-vertex*

k**NN.** Fig. 3(a) shows the speedup of COLD compared to HLDB in the case of kNN queries for $D = 0.01$ and $k = \{1, 2, 4, 8, 16\}$. As described in Sect. 3.1, we have created two DB tables for each framework (COLD, HLDB), one for $kmax = 4$ and one for $kmax = 16$. Then the DB table for $kmax = 4$ is used for answering kNN queries for $k = 1$, $k = 2$ and $k = 4$ and the kNN table for $kmax = 16$ is used for answering kNN queries for $k = 8$ and $k = 16$. Results show that for $k = 1$, COLD is 5 - 19× faster for the five largest datasets (Amazon, Citeseer,Citeseer2, DBLP. Youtube) and although this speedup degrades for larger values of k, COLD remains consistently 2 - 10× faster even for $k = 16$. For the smaller datasets, performance between COLD and HLDB is quite similar, with COLD performing better on Facebook and Gowalla, while HLDB performs only marginally better for Slashdot1, Slashdot2 and Notredame. In all cases, COLD answers kNN queries for all datasets in less than $26ms$ even for $k = 16$.

In our second set of kNN experiments, we assess the performance of COLD vs HLDB for varying values of D. For each value for D, we have build separate versions of $knntab$ (HLDB) and $knncold$ (COLD) DB tables for $D \cdot |V|$ objects selected at random from each dataset and $kmax = 4$. Figure 3(b) shows results for $k = 4$ and $D = \{0.001, 0.005, 0.01, 0.05, 0.1\}$. Again, for the five largest datasets COLD is consistently 3.4 - 23.4× faster than HLDB, whereas even for the smaller datasets, COLD is consistently 8.6 - 11.5× faster than HLDB for the largest value of D (for $D = 0.1$). Moreover, COLD may answer kNN queries for $k = 4$ on all datasets and all values of D in less than $14ms$.

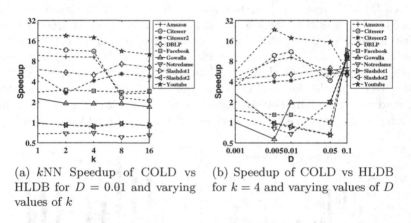

(a) kNN Speedup of COLD vs HLDB for $D = 0.01$ and varying values of k

(b) Speedup of COLD vs HLDB for $k = 4$ and varying values of D

Fig. 3. kNN Experiments on HDD for COLD and HLDB

RkNN. For RkNN experiments, we only report COLD's performance, since there is no other SQL framework that supports these queries. In out first experiment, we report the performance of COLD for $D = 0.01$ and $k = \{1, 2, 4, 8, 16\}$. For all those queries we have built one version of the $knnres$ DB table for $kmax = 16$ (see Sect. 3.1) and 3 separate $revcold$ tables for $kmax = \{1, 4, 16\}$. As expected, for RkNN queries and $k = 1$ we use the $revcold$ table built for

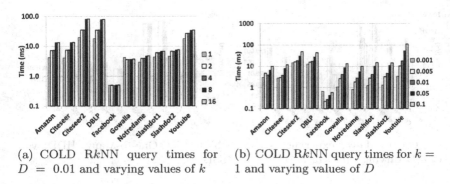

(a) COLD RkNN query times for $D = 0.01$ and varying values of k

(b) COLD RkNN query times for $k = 1$ and varying values of D

Fig. 4. RkNN Experiments on HDD for COLD

$kmax = 1$, for $k = 2$, $k = 4$ we use the *revcold* table built for $kmax = 4$ and for $k = 8$, $k = 16$ we use the *revcold* table built for $kmax = 16$. Figure 4(a) presents the results. In all cases, COLD provides excellent query times that are below $20\,ms$ for $k = 1$ in all datasets and never exceed $82\,ms$ even for $k = 16$.

In our second set of RkNN experiments, we assess the performance of COLD for varying values of D. Figure 4(b) presents results for $k = 1$ (as this is the typical case for RkNN queries) and $D = \{0.001, 0.005, 0.01, 0.05, 0.1\}$. Results show that although COLD's performance degrades for larger values of D, RkNN query times are below $49ms$ for all datasets and values of D, with the exception of Youtube and $D = 0.1$ ($109.3\,ms$). Thus, COLD offers excellent and stable performance in RkNN queries for all all datasets and tested values of k and D.

One-to-Many. Again, COLD is the only SQL framework that supports *one-to-many* queries. Figure 5(a) presents the corresponding results for varying values of D ($D = \{0.001, 0.005, 0.01, 0.05, 0.1\}$). COLD answers such queries in less than a second for all datasets and values of D, except the Citeseer2 and DBLP datasets (those with the highest $|HL|/|V|$ ratio) that require $5601\,ms$ and $4170\,ms$ respectively, for $D = 0.1$. For such high values of D, the *one-to-many* query reaches the complexity of an *one-to-all* query and as expected, it cannot be any faster on a secondary storage device. Note that even specialized graph databases like Neo4j cannot support this type of queries for more than a 1,000 target objects, whereas *COLD answers one-to-many queries to 110,000 target objects in the Youtube dataset in* $401\,ms$ *with a simple SQL query.*

4.2   Performance on SSD

Having established the performance characteristics of COLD in the HDD, in our second round of experiments, we repeat some of the previous experiments, using a SSD to measure the impact of the secondary-storage device type to results. The SSD used is a SATA3 Crucial CT512MX100SSD1 MX100 512 GB 2.5".

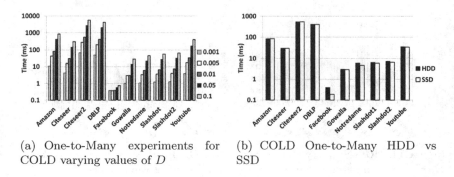

(a) One-to-Many experiments for COLD varying values of D

(b) COLD One-to-Many HDD vs SSD

Fig. 5. One-to-many experiments for COLD

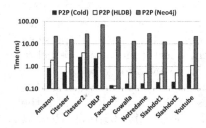

Fig. 6. SSD *vertex-to-vertex*

Vertex-to-vertex. Although the usage of SSD favors HLDB more than COLD (see Fig. 6), COLD is consistently 1.6 - 3.2× faster than HLDB (except Facebook, the smallest of datasets). The SSD has almost no impact on Neo4j and thus, COLD is now 11-171× faster than *Neo4j* on all datasets. Note, than on the SSD, COLD requires less than $0.9\,ms$ for all datasets and v2v queries, except the Citeseer2 and DBLP datasets (those with the highest $|HL|/|V|$ ratio). But even then, vertex-to-vertex queries still require less than $2.6\,ms$ for COLD.

*k*NN. Fig. 7(a) shows the performance speedup of COLD compared to HLDB in the case of *k*NN queries running on the SSD, for $D = 0.01$ and varying value of k. Again, although the SSD lowers the performance gap between COLD and HLDB, COLD is still faster on all datasets (except Facebook). In fact, COLD is 2.6 - 6.75× faster than HLDB for the high $|HL|/|V|$ ratio datasets (Citeseer2, HLDB) requiring less than $24.6\,ms$ even for $k = 16$.

R*k*NN. Fig. 7(b) presents the results of the R*k*NN query time performance on COLD for $D = 0.01$ and varying value of k. Results show that SSD usage accelerates COLD by only 20 % at most, which clearly demonstrates that COLD effectively minimized secondary storage utilization and thus adding a better secondary-storage medium provides minimal benefits for R*k*NN queries.

One-to-Many. Finally, Fig. 5(b) compares *one-to-many* queries on HDD and SSD for COLD. Again, the SSD usage accelerates COLD by only 2- 30 %, which further confirms the optimal secondary storage utilization of COLD.

4.3   Summary

Our experimentation has shown that our proposed COLD framework outperforms previous state-of-the-art HLDB in all performance benchmarks, including

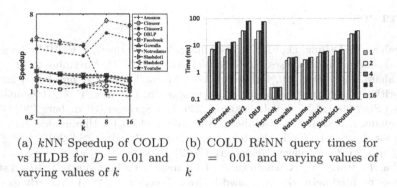

(a) kNN Speedup of COLD vs HLDB for $D = 0.01$ and varying values of k

(b) COLD RkNN query times for $D = 0.01$ and varying values of k

Fig. 7. kNN and RkNN SSD performance

query performance, memory size and scalability. Using HDDs, COLD is $2--21\times$ faster for *vertex-to-vertex* queries and $5--19\times$ faster for kNN queries and the largest datasets. Using SSDs, COLD is $1.6--3.2\times$ faster than HLDB for *vertex-to-vertex* and up to $6.75\times$ faster for kNN queries. COLD also requires up to $4,444\times$ less storage space (indexes) and up to $188\times$ less storage space (DB tables) used for storing forward labels. Even specialized graph databases like Neo4j are outperformed by COLD, which is up to $143\times$ faster. Most importantly COLD may service additional (RkNN, one-to-many) queries, not handled by any other previous secondary-storage solutions, while providing excellent query times and optimal secondary-storage utilization even on standard hard drives.

5 Conclusions

This work presented COLD, a novel SQL framework for answering various exact distance queries for large-scale graphs on a database. Our results showed that COLD outperforms existing solutions (including specialized graph databases) on all levels, including query performance, secondary storage utilization and scalability. Moreover, COLD also answers RkNN and one-to-many queries, not handled by previous methods. This establishes COLD as a competitive database-driven framework for querying large-scale graphs. The paper gives the design and implementation details of COLD using a popular, open-source database system along with the actual SQL queries used in our implementation. This should allow for a simple replication of our results and encourage other researchers to expand the COLD framework towards handling more complex queries and test-cases.

Acknowledgements. This work was partially supported by EU (European Social Fund - ESF) and Greek national funds through the Operational Program "Education and Lifelong Learning" of the National Strategic Reference Framework (NSRF) - Research Funding Program: Thales. Investing in knowledge society through the European Social Fund and the EU/Greece funded KRIPIS Action: MEDA Project. D. Pfoser's work was partially supported by the NGA NURI grant HM02101410004.

References

1. Abraham, I., Delling, D., Fiat, A., Goldberg, A.V., Werneck, R.F.: Hldb: Location-based services in databases. In: SIGSPATIAL GIS. ACM, November 2012
2. Abraham, I., Delling, D., Goldberg, A.V., Werneck, R.F.: A hub-based labeling algorithm for shortest paths in road networks. In: Pardalos, P.M., Rebennack, S. (eds.) SEA 2011. LNCS, vol. 6630, pp. 230–241. Springer, Heidelberg (2011)
3. Abraham, I., Delling, D., Goldberg, A.V., Werneck, R.F.: Hierarchical hub label-ings for shortest paths. In: Epstein, L., Ferragina, P. (eds.) ESA 2012. LNCS, vol. 7501, pp. 24–35. Springer, Heidelberg (2012)
4. Akiba, T., Iwata, Y., Kawarabayashi, K., Kawata, Y.: Fast shortest-path distance queries on road networks by pruned highway labeling. In: 2014 Proceedings of the Sixteenth Workshop on Algorithm Engineering and Experiments, ALENEX 2014, Portland, Oregon, USA, 5 January 2014, pp. 147–154 (2014)
5. Akiba, T., Iwata, Y., Yoshida, Y.: Fast exact shortest-path distance queries on large networks by pruned landmark labeling. In: Proceedings of the ACM SIGMOD International Conference on Management of Data, SIGMOD 2013, New York, USA, pp. 349–360 (2013)
6. Akiba, T., Iwata, Y., Yoshida, Y.: Pruned landmark labeling (2015). https://github.com/iwiwi/pruned-landmark-labeling
7. Albert, R., Jeong, H., Barabási, A.-L.: The diameter of the world wide web, CoRR (1999). http://cond-mat/9907038
8. Bader, D.A., Meyerhenke, H., Sanders, P., Wagner, D. (eds.): Graph Partitioning and Graph Clustering. Contemporary Mathematics, vol. 588. American Mathe-matical Society, Providence (2013)
9. Bast, H., Delling, D., Goldberg, A.V., Müller-Hannemann, M., Pajor, T., Sanders, P., Wagner, D., Werneck, R.F.: Route planning in transportation networks. CoRR, abs/1504.05140 (2015)
10. Borutta, F., Nascimento, M.A., Niedermayer, J., Kröger, P.: Monochromatic rknn queries in time-dependent road networks. In: Proceedings of the Third ACM SIGSPATIAL International Workshop on Mobile Geographic Information Systems, MobiGIS 2014 pp. 26–33, New York, NY, USA. ACM (2014)
11. Cho, E., Myers, S.A., Leskovec, J.: Friendship and mobility: user movement in location-based social networks. In: Proceedings of the 17th ACM SIGKDD Inter-national Conference on Knowledge Discovery and Data Mining, San Diego, CA, USA, 21–24 August 2011, pp. 1082–1090 (2011)
12. Cohen, E., Halperin, E., Kaplan, H., Zwick, U.: Reachability and distance queries via 2-hop labels. In: Proceedings of the Thirteenth Annual ACM-SIAM Sympo-sium on Discrete Algorithms, SODA 2002, pp. 937–946. Society for Industrial and Applied Mathematics, Philadelphia, PA, USA (2002)
13. Delling, D., Goldberg, A.V., Pajor, T., Werneck, R.F.: Customizable route plan-ning. In: Pardalos, P.M., Rebennack, S. (eds.) SEA 2011. LNCS, vol. 6630, pp. 376–387. Springer, Heidelberg (2011)
14. Delling, D., Goldberg, A.V., Pajor, T., Werneck, R.F.: Robust distance queries on massive networks. In: Schulz, A.S., Wagner, D. (eds.) ESA 2014. LNCS, vol. 8737, pp. 321–333. Springer, Heidelberg (2014)
15. Delling, D., Goldberg, A.V., Werneck, R.F.: Hub label compression. In: Demetrescu, C., Marchetti-Spaccamela, A., Bonifaci, V. (eds.) SEA 2013. LNCS, vol. 7933, pp. 18–29. Springer, Heidelberg (2013)

16. Delling, D., Goldberg, A.V., Werneck, R.F.F.: Faster batched shortest paths in road networks. In: ATMOS, pp. 52–63 (2011)
17. Delling, D., Werneck, R.F.: Customizable point-of-interest queries in road networks. In: 21st SIGSPATIAL International Conference on Advances in Geographic Information Systems, SIGSPATIAL 2013, Orlando, FL, USA, 5–8 November 2013, pp. 490–493 (2013)
18. Delling, D., Werneck, R.F.: Better bounds for graph bisection. In: Epstein, L., Ferragina, P. (eds.) ESA 2012. LNCS, vol. 7501, pp. 407–418. Springer, Heidelberg (2012)
19. Efentakis, A., Pfoser, D.: Optimizing landmark-based routing and preprocessing. In: CTS: 6th ACM SIGSPATIAL International Workshop on Computational Transportation Science, 5 November 2013, Orlando, FL, USA, p. 25 (2013)
20. Efentakis, A., Pfoser, D.: GRASP. extending graph separators for the single-source shortest-path problem. In: Schulz, A.S., Wagner, D. (eds.) ESA 2014. LNCS, vol. 8737, pp. 358–370. Springer, Heidelberg (2014)
21. Efentakis, A., Pfoser, D.: ReHub. Extending hub labels for reverse k-nearest neighbor queries on large-scale networks (2015). arXiv preprint http://arXiv:1504.01497
22. Efentakis, A., Pfoser, D., Vassiliou, Y.: SALT. a unified framework for all shortest-path query variants on road networks. In: Bampis, E. (ed.) SEA 2015. LNCS, vol. 9125, pp. 298–311. Springer, Heidelberg (2015)
23. Gavoille, C., Peleg, D., Pérennes, S., Raz, R.: Distance labeling in graphs. In: Proceedings of the Twelfth Annual ACM-SIAM Symposium on Discrete Algorithms, SODA 2001, pp. 210–219. Society for Industrial and Applied Mathematics, Philadelphia, PA, USA (2001)
24. Geisberger, R., Sanders, P., Schultes, D.: Better approximation of betweenness centrality. In: Munro, J.I., Wagner, D. (eds.) ALENEX, pp. 90–100. SIAM (2008)
25. Jiang, M., Fu, A.W., Wong, R.C., Xu, Y.: Hop doubling label indexing for point-to-point distance querying on scale-free networks. PVLDB 7(12), 1203–1214 (2014)
26. Leskovec, J., Krevl, A.: SNAP Datasets: Stanford large network dataset collection, June 2014. http://snap.stanford.edu/data
27. Leskovec, J., Lang, K.J., Dasgupta, A., Mahoney, M.W.: Community structure in large networks: Natural cluster sizes and the absence of large well-defined clusters. Internet Math. 6(1), 29–123 (2009)
28. McAuley, J.J., Leskovec, J.: Learning to discover social circles in ego networks. In: Advances in Neural Information Processing Systems 25: 26th Annual Conference on Neural Information Processing Systems 2012, Proceedings of a meeting held 3–6 December 2012, Lake Tahoe, Nevada, United States, pp. 548–556 (2012)
29. PostgreSQL. The world's most advanced open source database (2015). http://www.postgresql.org/
30. Safar, M., Ibrahimi, D., Taniar, D.: Voronoi-based reverse nearest neighbor query processing on spatial networks. Multimedia Syst. 15(5), 295–308 (2009)
31. Yang, J., Leskovec, J.: Defining and evaluating network communities based on ground-truth. In: 12th IEEE International Conference on Data Mining, ICDM 2012, Brussels, Belgium, 10–13 December 2012, pp. 745–754 (2012)
32. Yiu, M.L., Papadias, D., Mamoulis, N., Tao, Y.: Reverse nearest neighbors in large graphs. IEEE Trans. Knowl. Data Eng. 18(4), 540–553 (2006)
33. Zhong, R., Li, G., Tan, K.-L., Zhou, L.: G-tree: An efficient index for knn search on road networks. In: Proceedings of the 22nd ACM International Conference on Conference on Information Knowledge Management, CIKM 2013, pp. 39–48. ACM, New York, NY, USA (2013)

ParetoPrep: Efficient Lower Bounds for Path Skylines and Fast Path Computation

Michael Shekelyan[2], Gregor Jossé[1]([✉]), and Matthias Schubert[1]

[1] Institute for Informatics, Ludwig-Maximilians-University Munich,
Oettingenstr. 67, 80538 Munich, Germany
{josse,schubert}@dbs.ifi.lmu.de
[2] Faculty of Computer Science, Free University of Bozen-Bolzano, Piazza
Domenicani 3, 39100 Bozen-bolzano, Italy
michael.shekelyan@inf.unibz.it

Abstract. Computing cost-optimal paths in network data is an important task in many application areas like transportation networks, computer networks, or social graphs. In many cases, the cost of an edge can be described by various cost criteria. For example, in a road network possible cost criteria are distance, time, ascent, energy consumption or toll fees. In such a multicriteria network, path optimality can be defined in various ways. In particular, optimality might be defined as a combination of the given cost factors. To avoid finding a suitable combination function, methods like path skyline queries return all potentially optimal paths. To compute alternative paths in larger networks, most efficient algorithms rely on lower bound cost estimations to approximate the remaining costs from an arbitrary node to the specified target. In this paper, we introduce ParetoPrep, a new method for efficient lower bound computation which can be used as a preprocessing step in multiple algorithms for computing path alternatives. ParetoPrep requires less time and visits less nodes in the network than state-of-the-art preprocessing steps. Our experiments show that path skyline and linear path skyline computation can be significantly accelerated by ParetoPrep.

1 Introduction

In recent years, querying network data has become more and more important in many application areas like transportation systems, the world wide web, computer networks, or social graphs. One of the most important tasks in network data is computing cost-optimal paths between nodes. Especially in transportation and computer networks finding cost-optimal paths is essential for optimizing the movement of objects or information. Optimal paths can depend on multiple cost criteria. We refer to such networks, which consider more than one cost criterion, as multicriteria networks. In road networks, for example, possible criteria are travel time, distance, toll fees, environmental hazards, or energy consumption. In computer networks, typical cost criteria are bandwidth, rental cost, and current traffic. To describe the connections of people in a social graph, telephone calls, mails, meetings are aspects which can be transformed into cost criteria.

© Springer International Publishing Switzerland 2015
C. Claramunt et al. (Eds.): SSTD 2015, LNCS 9239, pp. 40–58, 2015.
DOI: 10.1007/978-3-319-22363-6_3

In order to compute a cost-optimal path in a multicriteria network, it is necessary to specify which type of cost should be minimized. This optimization criterion can either be one of the underlying cost criteria or a combination of these criteria, e.g., a weighted sum. Given that any monotone combination of cost values leads to a valid optimality criterion, there is an infinite number of possible cost values. One might be able to anticipate user preferences in some applications, but existing works in this area still rely on the set of skyline paths when recommending suitable path alternatives, e.g., [2,10]. Therefore, the established approach of presenting results to the user is to compute a set of alternative paths which allows the user to evaluate the trade-off between the cost criteria.

The most general set of optimal paths is the set of pareto optimal paths or path skylines [9,13]. This set contains all paths optimal under any monotone combination function. Since this set is typically rather large, newer approaches like the so-called linear path skyline [16,17] restrict the allowed combination function to linear combinations. The difference between the conventional and the linear path skyline is illustrated in Fig. 1.

The state-of-the-art algorithms for computing path skylines employ multicriteria lower bound estimations to approximate the minimum costs which have to be spent to travel from a given node n to a target node t. This way it is possible to compute the best-case cost vector for any path containing n and ending at t. The general idea is analoguos to the concept of the A^* search for shortest path computation: If even the shortest possible path, a straight line with Euclidean distance, is longer than the current best result, the current node can be discarded from the search. Although there are algorithms which do not employ lower bounds (e.g., BRSC in [13]), their use is restricted to small graphs which can be visited entirely, making these algorithms infeasible in most cases.

Considering multiple and arbitrary cost criteria requires lower bound approximations for all cost criteria. One solution is precalcu ating general bounds, as done by a reference node embedding [7,13]. However, these bounds have several drawbacks. First, the approximation quality is often insufficient for the majority of queries. Second, they typically require a lot of memory compared to the size of the network. And finally, in dynamic networks – where edge costs vary over time – the approximation may lose its bounding properties.

Alternatively, [4] conducts a Dijkstra search from the target node to all other nodes in the network w.r.t. each cost criterion. The obtainted lower bounds are query-specific and, thus, exact for the given target. However, this kind of pre-processing has to be done at query time. For d cost criteria this step is performed d times, and for each node d lower bounds have to be stored. Although the preprocessing effort is large, the optimality of the bounds for the given task, often leads to an significant speed-up when performing the actual path computation. In fact, the results of our experiments demonstrate that this method clearly outperforms the use of a reference node embedding for computing path skylines with more than two criteria.

Tackling all of the above problems, we introduce ParetoPrep: A new algorithm for computing all single-criterion shortest paths for arbitrary cost criteria in a multicriteria network. ParetoPrep computes the same optimal lower bounds as [4] for given start and target nodes. However, ParetoPrep requires only a single instead of d graph traversals. Furthermore, the visited part of the graph is significantly smaller than with the approach described in [4]. To show that ParetoPrep is a valid preprocessing step for the path skyline algorithms mentioned above ([9,13,16,17,19]), we show that all nodes contained in any pareto optimal path between the given start and target nodes are visited. Consequently, any node which is not visited by ParetoPrep can be excluded when searching for a cost optimal path w.r.t. any monotone combination function. ParetoPrep reduces processing time, memory consumption, the visited portion of the graph and the number of graph traversals when computing optimal lower bounds. This means, ParetoPrep is a valid and highly efficient preprocessing step for any of the above algorithms, as it yields optimal lower bounds for the computation of all pareto optimal paths.

An additional use of ParetoPrep is to simultaneously compute multiple optimal paths for a given set of cost criteria or combinations of these cost criteria. This is possible, because the computed lower bounds correspond to the exact cost of the optimal path to the target node w.r.t. each of the original cost criteria. Of course, it is possible to consider combined cost criteria (e.g., 0.3· distance +0.6· travel time +0.1· traffic lights) as additional basic cost criteria. Given an arbitrary number of such combined criteria, ParetoPrep computes all cost-optimal paths in a single traversal.

The rest of the paper is organized as follows: In Sect. 2, we present related work in the area of path skylines and lower bound computation in multicriteria networks. Section 3 provides basic notations and concepts. Furthermore, we illustrate the use of point-to-point lower bound computation when computing path skylines. Our new algorithm ParetoPrep is presented in Sect. 4. Additionally, the section contains formal proofs concerning correctness and termination of ParetoPrep. Section 5 describes the results of our experiments, comparing ParetoPrep to state-of-the-art lower bounds for 2 path skyline algorithms. The paper concludes with a summary and directions for future work in Sect. 6.

2   Related Work

In this section we review existing works on (linear) path skylines. We focus not on algorithmical details but the application and use of lower bounds, stressing that performance depends crucially on tight lower bounds. Also, we will survey the state-of-the-art methods for lower bound computation and point out their shortcomings.

In the database community, the task of computing a path skyline between two nodes in a multicriteria network is referred to as path (or route) skyline query in [13]. Synonymously, the path skyline is sometimes referred to as the set of all pareto optimal paths. In Operations Research, the problem is known as the multiobjective shortest path problem. The result set in any case are those paths which have cost vectors that are, mathematically speaking, optimal under

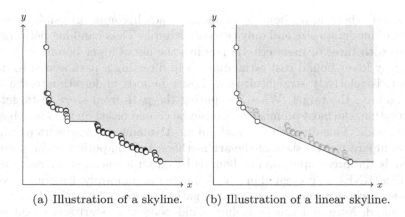

(a) Illustration of a skyline.   (b) Illustration of a linear skyline.

Fig. 1. The linear skyline holds a subset of the elements of the conventional skyline.

some monotone cost function. Surveys on existing solutions to this problem can be found in [5,6,18,20]. Early on, [11] proved that the size of the path skyline may increase exponentially with the number of hops between start and target node, and that the problem therefore is NP-hard. More recently, [15] showed that the number of paths is in practice feasibly low when using strongly correlated cost criteria. Another way of coping with the great number of skyline paths was presented in [17] and extended in [16]. Instead of all paths which are optimal under some monotone cost function, the result set here consists of all paths which are optimal under a linear cost function. This is a restriction of the result set, referred to as linear path skyline. Both sets are visualized in Fig. 1.

Various algorithms have been proposed to solving linear and especially conventional path skyline queries. Going into detail on all methods presented in the above papers would go beyond the scope of this work. We will restrict ourself to noting that the state-of-the-art algorithms either rely on some kind of labeling algorithm or a sequence of target-oriented path searches. Labeling algorithms label nodes with the cost vectors of assembled paths ending at that node. They begin at the start node and follow its outgoing edges. In each iteration all previously assembled paths that may be part of a skyline path are extended. The algorithm terminates once all assembled paths were either extended or pruned. The labeling algorithm for skyline computation that we choose for evaluation purposes is ARSC [13]. Contrary to extending all paths by one edge at a time is the concept of sequentially computing full paths. For such an algorithm to terminate in reasonable time, it is essential to efficiently explore the exponential space of possible results. This means, path searches must be directed, either producing a new result or excluding the existence of results in a particular search direction. LSCH [16] proceeds this way when computing the linear path skyline.

What both algorithmic concepts have in common, is their reliance on lower bound cost estimations. Either to prune branches of the search tree (nonpromising paths which have not yet reached the target) or to direct the search towards the target. Without lower bounds, either approach becomes infeasible even for networks of moderate size and only two cost criteria. Thus handling networks of state-size with three or more criteria requires the use of lower bounds.

The way lower bound cost estimations help directing a path search towards its target, is relatively straightforward. Lower bounds underestimate the cost from a node to the target. When computing the path from start to target, in every iteration, the most promising edge can be chosen based on the lower bound of its end node. This directs the search, in an A*-manner, towards its target.

We compare to two state-of-the-art methods for computing lower bounds. The first is the precomputation of bounds based on a so-called reference node embedding (RNE), as presented in [13]. A number of uniformly distributed reference nodes is chosen, and all cost-optimal paths between all pairs of these nodes are computed. Note that this procedure is independent of start and target node and can therefore be executed prior to any query. Based on the cost differences of these paths, lower bounds are computed. Of course, a reference node embedding only yields optimal lower bounds for some queries. Furthermore, the memory consumption is rather large and any precomputed information has to be checked for validity in case of dynamically changing edge costs.

The other prevalent method ([9,16]) was introduced by Tung and Chew [4]. The authors propose to perform a reversed single-source all-target Dijkstra search for each cost criterion to find the costs of the shortest paths from all other nodes in the graph to the target node. Note that this procedure is not independent of the query input and must be performed at query time. We will refer to this approach as Multi-Dijkstra (MD). Despite the overhead at query time, the superior pruning power in many cases reduces the runtime of path computation to a degree which compensates for the additional effort. A major shortcoming of MD is that it has to process the full graph for every cost criterion. For large graphs, which cannot be held in main memory, this method is unsuitable.

Let us note that in the case of two cost criteria, special properties of (linear) path skylines may be used for further speed-up. We would like to stress that this is only applicable to the case of two cost criteria and that we focus on non-trivial multicriteria networks with a higher number of cost criteria. Hence, we will not go into detail on methods which are limited to the two-dimensional case ([1,11,12,14,17]), but refer the interested reader to these works.

3   Preliminaries

A multicriteria network is represented by a directed weighted graph $G(\mathcal{V}, \mathcal{E}, \mathcal{C})$ comprising a set of nodes $\mathcal{V}$ and a set of directed edges $\mathcal{E} \in \mathcal{V} \times \mathcal{V}$. Each edge $(n, m) \in \mathcal{E}$ is labeled with a cost vector $cost(n, m) \in \mathcal{C} \subset \mathbb{R}^d_+$ which consists of the costs for traversing edge (n, m) w.r.t. each of the d cost criteria. If there exists an edge (n, m) then n and m are neighboring nodes and (n, m) shall be called an outgoing edge of n and an incoming edge of m.

A sequence of consecutive edges connecting two nodes s and t, for instance, $w = ((s, n_1), (n_1, n_2), \ldots, (n_k, t))$, is called a way from s to t. If w does not visit any node twice it is called a path. The cost of a path p for each cost criterion i is the sum of its individual edge costs:

$$cost(p)_i = \sum_{(n,m) \in p} cost(n,m)_i$$

A *monotone cost function* is a function $f : \mathbb{R}^d :\to \mathbb{R}$ for which the following property holds: For cost vectors $a, b \in \mathbb{R}_+^d$ where $a_i \leq b_i$ for some $1 \leq i \leq d$, $f(a) \leq f(b)$ holds.

A cost vector $a \in \mathbb{R}_+^d$ *dominates* a cost vector $b \in \mathbb{R}_+^d$, denoted $a \prec_{\text{dom}} b$, iff a has a smaller cost value than b in at least one dimension i and b does not have a smaller cost value than a in any dimension j:

$$\exists_{i \in \{1,\dots,d\}} : a_i < b_i \ \wedge \ \nexists_{j \in \{1,\dots,d\}} : a_j > b_j \ \Leftrightarrow \ a \prec_{\text{dom}} b$$

Note that there is a unique cost vector associated with each path. We therefore say a path dominates another path, if the cost vector of the former dominates the cost vector of the latter. Note that for any monotone cost function $f(\cdot)$, $a \prec_{\text{dom}} b \Rightarrow f(a) \leq f(b)$.

Paths (more precisely: their cost vectors) which are not dominated by any other path are called nondominated or pareto optimal. The set of nondominated paths includes are all paths which are optimal under some monotone cost function. This set is called path (or route) skyline [3]. In [16,17] this concept was extended in the following way: Instead of computing the paths which are optimal under monotone cost functions, it can be of greater interest to compute the paths which are optimal under linear cost functions. This set is called the linear skyline (see Fig. 1).

3.1 Multicriteria Lower Bounds

The task we examine in this paper is to derive lower bounds $lb(s,t)$ for cost vectors $cost(p)$, where p is a path from node s to node t in a multicriteria network. Formally, a lower bound for two nodes a and b is defined as follows:

Definition 1 (Lower Bound Costs). *Let a and b be nodes, and let $c(\cdot)$ be an arbitrary cost criterion. If for the real value $lb_c(a,b)$ holds $lb_c(a,b) \leq c(p)$ for any path p connecting a and b, then it is called a* lower bound *for c w.r.t. a and b. A vector consisting of lower bounds of all cost criteria w.r.t. a and b is denoted as lb(a, b) and referred to as* lower bound (cost) vector.

A lower bound cost vector contains a lower bound for each of the cost criteria. Thus, for an arbitrary path p connecting nodes a and b, $f(lb(a,b)) < f(cost(p))$ holds for any monotone cost function $f(\cdot)$. Or, in words, the image of a lower bound cost vector (w.r.t. a and b) under a monotone cost function is also a lower bound for the costs of all paths between a and b. The lower bounding property is therefore invariant under monotone cost functions.

We will refer to a lower bound vector $lb(s,t)$ as *optimal* iff $\forall \ 1 \leq i \leq d$: $\exists \ p = ((s,n_1),\dots,(n_k,t))$ where $\text{lb}_i(s,t) = cost(p)_i$.

The bounds computed by Multi-Dijktstra and ParetoPrep are optimal. However, they are query-specific, i.e., only bounds for the cost to reach the target node t are available. Furthermore, since any node is a potential target node, it is not feasible to precompute these bounds, and both methods usually compute bounds at query time. Let us note that the lower bounds provided by a reference node embedding are not specific to a particular query. However, they are usually not optimal. As will be shown in the experiments, for more than two cost criteria and larger distances between start and target, optimal bounds compensate for the additional overhead. In the following, we will explain how lower bounds are employed in algorithms for finding cost optimal and pareto optimal paths in multicriteria networks.

3.2 Lower Bounds in Multicriteria Path Search

Computing an optimal query-specific bound includes computing the cost-optimal path from s to t w.r.t. all given cost criteria. Thus, if all cost criteria or combination functions are known in advance, in order to compute the wanted path alternatives, it suffices to compute optimal lower bounds. In other words, the lower bound method can be used to compute the set of alternatives and no other algorithm is necessary. Of course, the used algorithm has to provide a method to reconstruct these optimal paths as well as to determine their cost. Both algorithms, Multi-Dijkstra and ParetoPrep, provide this possibility. If the desired combination functions are not known in advance, computing (linear) path skylines is recommended, in order to compute all paths potentially of interest.

For computing linear path skylines in general multicriteria networks ($d \geq 2$), LSCH [16] is currently the most efficient algorithm. The method successively constructs a linear skyline as part of a convex hull in the cost space. In each step of the algorithm, LSCH determines a linear combination function and searches the cost-optimal path w.r.t. this monotone combination. If lower bounds are available, these searches can be done using A^*-search. Recall that a lower bound vector induces a lower cost value under all monotone combination functions. Thus, LSCH may exploit a given multicriteria lower bound to accelerate these searches. In this setting, query-specific lower bounds indeed make sense, because the algorithm performs multiple A^*-searches between the same two nodes s and t where each search employs a different cost combination. Note that the bounds are not necessarily optimal anymore when using a combination function to derive the cost values. However, the quality of the bounds usually still outperforms the quality of non-query-specific lower bounds.

Another area of application of multicriteria lower bounds is the computation of path skylines. The most advanced path skyline approaches are label correcting methods like ARSC [13] which we will sketch in the following to explain the necessity of high quality lower bounds. ARSC computes all pareto optimal paths between two nodes s and t in a single graph traversal. To do this efficiently, ARSC employs two ways of pruning the search space. The first is local domination:

Definition 2 (Local Domination). *Let p and q be two paths starting at s and ending at t. Iff $cost(p) \prec_{dom} cost(q)$ holds, we refer to q as dominated by p, denoted by $p \prec_{dom} q$. Correspondingly, q is referred to as nondominated iff $\not\exists\, p = ((s, n_1), \ldots, (n_k, t)) : cost(p) \prec_{dom} cost(q)$.*

For any path $q = ((n_1, n_2), \ldots, (n_{l-1}, n_l))$, any subsequence of edges $p = ((n_i, n_{i+1}), \ldots, (n_{k-1}, n_k))$ is called a subpath of q. As a direct consequence of the following lemma, we may prune any dominated path and all of its possible extensions. This fact constitutes the first domination check referred to as *local domination check*.

Lemma 1 (Local Domination Check). *Any subpath of a nondominated path is nondominated (w.r.t. its start and target node).*

A proof for this lemma can be found in [13]. ARSC maintains a local skyline of paths for each visited node n. Thus, a path is only extended if it is part of local skyline of its end node n. Though local pruning helps to prevent the extension of large amounts of paths, it is not capable of restricting the search to a limited section of the network. To direct this extension towards the target, ARSC employs a second pruning method which relies on lower bounds, being referred to as global domination:

Definition 3 (Global Domination). *Given a start node s and a destination node t, an arbitrary path p is called globally dominated iff it is a subpath of a dominated path q between s and t. Respectively, any subpath p of a nondominated path q between s and t is called globally nondominated.*

Thus, any path p between s and some node n can be excluded from further expansion if there does not exist an extension of p ending an target t which is part of the skyline.

To detect global domination as early as possible, high-quality lower bounds can be used to check for global domination:

Lemma 2 (Global Domination Check). *Let p be a path from node n to node m and q be a path from the start node s to the target node t. If $cost(q) \prec_{dom} lb(s, n) + cost(p) + lb(m, t)$, then p is globally dominated.*

Proof. The cost vector $lb(s, n) + cost(p) + lb(m, t)$ is a lower bound cost of all paths from s to t via p. If this lower bound cost is dominated by the cost of a path from s to t, there is no nondominated path from s to t via p.

ARSC can employ global domination checks as early as the first skyline path is known. This reveals an additional benefit of using query-specific lower bounds. As mentioned before, precomputing optimal bounds includes computing the single-criterion optimal paths for all given cost criteria. These paths are obviously part of the skyline. Thus, when using query-specific bounds, ARSC can use global domination upon initialization.

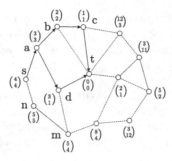

Fig. 2. Exemplary output of ParetoPrep given a start node s and a target node t. The indicated paths $\{s, a, b, c, t\}$ and $\{s, a, d, t\}$ are the shortest paths for the first and the second criterion, respectively. The vectors next to each node are the computed lower bound costs lb of reaching t from the respective nodes.

A final remark on query-specific bounds is that the number of nodes for which a lower bound is required is limited. In general, it is only necessary to provide estimations for all nodes which are part of any pareto optimal path from s to t. This observation is based on the property that the path skyline is the super set of all potentially optimal paths. Thus, any search algorithm reaching a node outside of this area can prune the respective path.

4   Multicriteria Lower Bound Computation

The idea of ParetoPrep is to compute all single-criterion shortest paths between s and t within a single graph traversal. As will be shown later on, there cannot be a node which is part of a skyline path and not visited during this graph traversal. Thus, ParetoPrep is correct in the sense that it visits all nodes required for computing a path skyline.

4.1   ParetoPrep

The goal of a precomputation step like Multi-Dijkstra (MD), the reference node embedding (RNE), or ParetoPrep is to compute the minimum costs from an arbitrary node n to the given target node t for each of the cost criteria. These bounds are computed at or prior to query time and are used by the subsequent path search. ParetoPrep computes all shortest paths for all cost criteria within a single graph traversal. This approach yields major improvements. While ensuring optimal bounds, it is possible to reduce the number of times a node is visited. For example, if the shortest path for two cost criteria ends with the same edge, both shortest paths will be found by processing the same node. The pseudocode of ParetoPrep is provided in Algorithm 1, an exemplary output and execution of the algorithm are shown in Figs. 2 and 3, respectively.

ParetoPrep maintains a set of open nodes *open* and a set $\mathcal{S}$ of paths from s to t. Each visited node n has an entry consisting of two elements $\{lb(n, t), \, succ_i(n)\}$.

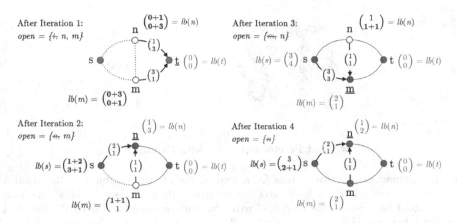

Fig. 3. Exemplary execution of ParetoPrep. Active node of iteration is underlined. After Iteration 2 path through $[s, n_1, t]$ with costs $[3,4]$, after iteration 3 path through $[s, n_1, n_2, t]$ with costs $[3,6]$ is constructed. ParetoPrep terminates upon iteration 4.

The cost vectors $lb(n, t) : \mathcal{V} \to \mathbb{R}^d_+$ are the lower bound costs of all paths from n to t, through which n was previously reached in ParetoPrep. Upon termination of the algorithm, n is reached by all globally nondominated paths from n to t. The edges $succ_i(n) : \mathcal{V} \times \mathbb{N} \to \mathcal{E}$ are the first edges of the currently shortest path from n to t for criterion i. These successor edges are used to reconstruct current shortest paths from s to t. An entry of an unvisited node n is assumed to be $lb(n, t)_i = \infty$, $succ_i(n) = \emptyset$. $lb(t, t)_i$ is always zero because the lower bound cost of reaching t from t are zero.

The first step of the algorithm is the initialization. The open set is created, and the target node t is added to the open set. The second step is node selection. In each iteration, an open node n is selected and removed from the open set. To reduce the number of nodes which have to be visited twice, the nodes closest to t should be visited first. To achieve this, each node is ranked by the linear sum of its cost vector, and the node with the smallest value is selected first. The third step is a check if the selected node has to be extended. If $lb(s, n) + lb(n, t)$ is dominated by the costs of a known path, step 4 and 5 are skipped. The cost vector $lb(s, n)$ is the lower bound cost of all globally nondominated paths from s to n. If no such information is available $lb(s, n) = \mathbf{0}$. The fourth step is the extension of the selected node. The cost of each neighboring node m in dimension i is set to the minumum of $lb(m, t)_i$ and $lb(n, t)_i + cost(m, n)_i$, where (m, n) is an edge from m to n. For each criterion i in which $c(m)_i$ is changed, the i-th predecessor edge $succ_i(m)$ is set to (m, n). If $lb(m, t)$ was changed and m is not the start node s, m is added to the set of open nodes. The fifth step is the construction of paths from s to t. This only happens if $lb(s, t)$ was modified in the previous step. For each modified cost criteria the currently shortest path from s to t is constructed. The paths are constructed by following $succ_i$ of nodes, similarly to how paths are reconstructed in Dijkstra's algorithm. The pseudocode is provided

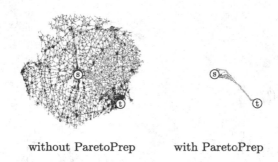

without ParetoPrep   with ParetoPrep

Fig. 4. Comparison of search areas for a routing task from Augsburg (s) to Munich (t) with two cost criteria. The illustration compares thesearch are of ARSC without precomputed bounds to that of ARSC with ParetoPrep-bounds. It is easily observed that the label correcting search considers almost no dominated paths when using the information provided by ParetoPrep.

in Algorithm 2. The sixth step is termination. If after an iteration there are no more open nodes the algorithm terminates, otherwise the algorithm continues with step 2. Upon termination, S contains a shortest path from s to t for each criterion, and lb maps each node which is possibly part of a skyline path from s to t onto its lowerboundcosts of reaching t.

4.2   Correctness and Termination

In this section, we prove that ParetoPrep does indeed visit the portion of the graph which is necessary to compute a path skyline and thus, all nodes which are potentially visited by any cost-optimal path. More precisely, we will show that every node which is part of a nondominated path from start to target is visited (Fig. 4).

Definition 4. *Let $\mathcal{R}(n)$ be all paths through which ParetoPrep reached a node n. For every $n \neq t$, we initialize $\mathcal{R}(n) = \emptyset$. In each iteration, every neighbor m of the selected node n is reached through the edge connecting the two nodes $\mathcal{R}(m) = \mathcal{R}(m) \cup \{p \text{ extended with } (m, n) \mid p \in \mathcal{R}(n)\}$.*

Note that in this case (m, n) becomes the first edge of the path extended with (m, n). This is because ParetoPrep follows incoming edges, moving backward from the target node. For the rest of this section, when we speak of (non)domination, we mean global (non)domination.

Lemma 3. *At the end of each iteration, $lb(n, t)$ equals the minimal costs of all paths through which n was reached, i.e., $lb(n, t)_i = \min_{p \in \mathcal{R}(n)} cost(p)_i$*

Proof. The statement obviously holds for the target node. Now, assume a selected node n is expanded by an edge (m, n). Let i be the index of an arbitrary criterion. The lower bound value of the i-th criterion is set to

```
//step 1: initialization
S ← ∅ and open ← {t}
while open ≠ ∅ do
    //step 2: node selection
    select n with minimal lower bound sum from open and remove from set;
    //step 3: global domination
    if S ≺_dom lb(n,t) + lb(s,n) then
     └ skip step 4 and 5
    //step 4: node expansion
    foreach incoming edge (m,n) of n do
        foreach criterion i do
            if lb(n,t)_i + cost(m,n)_i < lb(m,t)_i then
                lb(m,t)_i ← lb(n,t)_i + cost(m,n)_i
                succ_i(m) ← (m,n)
                open ← open ∪ {m}

    //step 5: path construction
    if lb(s,t) was modified in step 4 then
        foreach modified component i of lb(s,t) do
            p ← constructpath(s,t,succ_i,i)
            S ← S ∪ {p}
            remove paths dominated by p from S
```

Algorithm 1. Pseudocode of ParetoPrep

$\min\{lb(m,t)_i, lb(n,t)_i + cost(m,n)_i\}$ (see Algorithm 1). By the above definition this is exactly the minimal cost of each path through which m is reached. Note that if m was previously unvisited, its cost vector is ∞. Concludingly, the statement holds.

Lemma 4. *If a node is reached by a nondominated path, it is expanded.*

Proof. ParetoPrep expands every node m with two exceptions: (1) if global domination holds in Step 3, or (2) if the m is not added to the *open* set in Step 4 (cf. Algorithm 1). Of course, if m is reached through a nondominated path, global domination does not hold. Hence, the node is expanded, unless $lb(n,t)_i + cost(m,n)_i \geq lb(m,t)_i$ for all of the criteria. If m was previously unvisited, $lb(m,t)_i = \infty$. Therefore, m must have been visited, and if m is reached by a nondominated path, then $lb(n,t)_i + cost(m,n)_i < lb(m,t)_i$ must hold at some point. This means, m is added to the open set and expanded subsequently. This proves the statement.

Lemma 5. *Upon termination, each node n which is part of a nondominated path from s to t was reached. Furthermore, the node n is reached through each nondominated path from n to t.*

Data: s, t, $succ_i$, i
Result: Current shortest path from s to t for criterion i
$m \leftarrow s$
$p \leftarrow$ new empty path
while $m \neq t$ **do**
  $\quad (m, n) \leftarrow succ_i(m)$
  $\quad p \leftarrow p$ extended with (m, n)
  $\quad m \leftarrow n$
return p

Algorithm 2. Pseudocode of ParetoPrep's path construction routine

Proof. Let p be an arbitrary nondominated path from n to t. If no such path exists for n, then n is not contained in any nondominated paths from s to t. Let K be the number of nodes through which p passes. Let $p^{(k)}$ be the k-th node through which p passes. Let $p^{(j,\,k)}$ be a subpath of p which starts at $p^{(j)}$ and ends at $p^{(k)}$. If the claim were incorrect, there would exist some k-th node $p^{(k)}, k < K$, which would not be reached. This implies one of the two cases:

(1) $p^{(k+1)}$ was reached by $p^{(k+1,\,K)}$, but not expanded afterwards
(2) $p^{(k+1)}$ was not reached

Since $p^{(k+1,\,K)}$ is a subpath of the nondominated path p and thereby nondominated itself, it must be expanded by Lemma 4 which contradicts (1). (2) is the inductive shifting of the original statement that $p^{(k)}$ was not reached. This implies two cases, as above. The first one is contradicted as before, the second one is again the inductive shifting. Following the chain of induction, we get $p^{(K)} = t$ was not reached which is contradicted by the empty path starting at t. Concludingly, n is reached by all nondominated paths from n to t.

The above lemmas prove that ParetoPrep is sufficient for path skyline computation, i.e., every node that is possibly part of a skyline path is indeed visited. In addition, let us state precisely, what the values of the lower bound costs are and how they are related to single-criterion shortest paths.

Claim. Let n be a node contained in a nondominated path from s to t. The cost vector $lb(n,t)$ equals the lower bound costs of all nondominated paths from n to t. Furthermore, a shortest path from s to t for each criterion is found.

Proof. The first statement follows directly from Lemmas 3 and 5. If n is contained in a nondominated path from s to t, n is reached through all nondominated paths from n to t. Hence, $lb(n,t)$ equals the lower bound costs of all nondominated paths from n to t. Now, let us investigate the special case of the start node s. The single-criterion shortest paths are obviously a subset of the nondominated paths. Hence, s is reached by all single-criterion shortest paths which are reconstructed in Step 5 of Algorithm 1.

Finally, let us note that ParetoPrep always terminates if executed on finite graphs. This is due to the fact that each previously visited node n is expanded

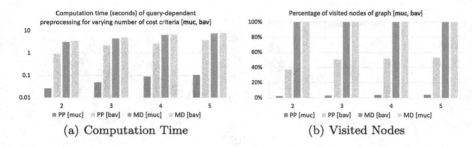

(a) Computation Time (b) Visited Nodes

Fig. 5. Computation time (seconds) and percentage of visited nodes (of all nodes in the graph) of the query-dependent preprocessing steps PP and MD for the settings **muc** and **bav**.

if at least one criterion of $lb(n,t)$ has been lowered. Once a node is reached by the shortest paths for each criterion, it will not be expanded anymore. From a finite number of nodes follows a finite number of paths between nodes which in turn implies that ParetoPrep performs a finite number of iterations.

This concludes our section on the properties and the correctness of ParetoPrep. In the following, we will explore the efficiency and performance of our approach.

5 Experiments

We evaluate ParetoPrep (PP) on settings based on the real world road network of the state of Bavaria, Germany, with 1 023 561 nodes and 2 418 437 edges, extracted from OpenStreetMap[1] (OSM) using the MARiO framework [8]. All experiments were conducted on a work station with an Intel i7 CPU (3.4 GHz) and 32 GB RAM, running Windows 8. Different algorithms are tested on the same randomly generated scenario before comparing results. Runtime evaluations are based on Java's nanotime clock and performed for each algorithm individually.

First, we compare the query-dependent preprocessing steps PP and Multi-Dijkstra (MD) in terms of computation time. Given a start and target pair, we evaluate how long the preprocessing step takes. Second, for PP, MD, and the reference node embedding (RNE), we investigate the quality of the respective bounds, i.e., we evaluate the actual performance gain for different algorithms. For RNE, we select 9 reference nodes on a uniform grid over the map and assume that there is no overhead for loading the embedding. In a first subsetting, we examine how the path skyline algorithm ARSC benefits from the bounds provided by PP, MD and RNE. Given the information of the respective preprocessing step and a start and target pair, we take the runtime of the algorithm as a measure for the quality of the bounds. In a second subsetting, we analyze how the linear path skyline algorithm LSCH benefits from the bounds, again in terms of runtime.

---

[1] http://www.openstreetmap.org/.

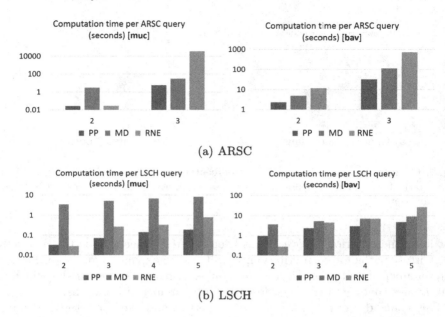

(a) ARSC

(b) LSCH

Fig. 6. Average computation time (seconds) of ARSC and LSCH given the precomputed bounds by the respective methods, evaluated on the **bav** scenario.

Finally, we evaluate PP as a means for single-criterion shortest path computation for given cost criteria. In a network with four criteria, PP can be used to compute the shortest paths w.r.t. given weightings of the criteria or, even simpler, to compute the shortest paths w.r.t. each of the criteria, as produced by four distinct Dijkstra searches. We compare PP to these multiple single-source single-target Dijkstra searches w.r.t. runtime and visited part of the graph. Note that in order to compensate runtime effects in the virtual machine, runtime is measured by performing each task five times and taking the minimum of these runs.

We evaluate the above scenarios on two settings based in Bavaria, Germany. The first one is rather local and set in Munich, capitol of Bavaria, routing from one of the 25 district centers to another, amounting to $\binom{25}{2} = 300$ pairs in total. We refer to this setting as **muc**. The other setting is – relative to the graph – rather global, routing from one of the 5 major cities in Bavaria to all others, amounting to $\binom{5}{2} = 10$ pairs in total. We refer to this setting as **bav**. The cost criteria used are distance (dist), travel time (tt), ascent (asc), penalized travel time (ttpen), and energy expenditure (ener). The basic travel time estimate (tt) assumes travel speeds equal to the speed limits and no delays at crossings. The penalized travel time estimate (ttpen) assumes additional 30 s for each traffic light. The energy expenditure estimate (ener) assumes that 0.2 kWh are lost on friction per kilometer, which is a rough estimate derived from typical battery capacities of electric cars and their respective ranges. For each ascended kilometer 4 kWh are added to the energy usage, which is derived from the increase in potential energy from ascending 1 000 meters with a 1 500 kg vehicle:

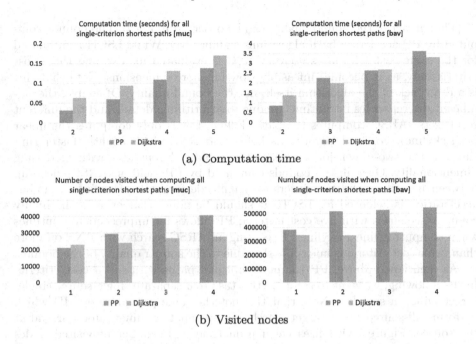

(a) Computation time

(b) Visited nodes

Fig. 7. Computation time (seconds) and visited nodes when computing all k singlecriterion shortest paths with PP and k Dijkstra searches.

$1\,000$ m $\cdot$ $1\,500$ kg $\cdot 9.81\frac{m}{s^2} \cdot \frac{1\text{ kWh}}{3.6\cdot10^6\text{ J}} = 4.0875$ kWh. For each descended kilometer 2 kWh are subtracted from the energy loss and negative energy loss values are corrected to zero. The employed formula for the energy loss in kWh for a road segment from n to m with length $len(n, m)$, ascent $asc(n, m)$ and descent $desc(n, m)$ in kilometers is $\max(0, 0.2 \cdot len(n, m) + 4 \cdot asc(n, m) - 2 \cdot desc(n, m))$, where ascent and descent are derived from OSM data. Let us stress that in order to validate the efficiency of the proposed approach, the cost criteria are not required to be realistcally modeled; we do not claim so for tt, ttpen, ener. We performed queries using the following selection of criteria: 2: dist+tt, 3: dist+tt+asc, 4: dist+tt+asc+ttpen, 5: dist+tt+asc+ttpen+ener.

Figure 5 compares the preprocessing times of the query-dependent methods PP and MD. It shows the computation time in seconds when varying the number of cost criteria for both settings, **muc** and **bav**. We observe that PP always outperforms MD; independent of the number of cost criteria, PP is always around two orders of magnitude faster than MD, hardly ever exceeding 100 ms of computation time. This is especially remarkable, seeing as PP computes optimal bounds. A major reason for this behavior can be observed in the right figure: PP only visits a very restricted part of the graph, yet – as shown before – the necessary part of the graph to compute all pareto optimal path. For the **muc** setting, it only visits at most 6 % of the network and even for examples with large distances (**bav**) between start and target, ParetoPrep needs to visit only about half of all nodes.

The impact of the bound quality can be observed in Fig. 6. The bounds computed by PP accelerate both algorithms significantly. While LSCH is evaluated for the same cost criteria as above, ARSC is evaluated for two and three cost criteria only, as it becomes infeasibly slow for more dimensions. Note that this is a drawback of the algorithm itself, not the bound quality. Of course, the significant discrepancies in runtime are due to algorithmic details and the different result sets (ARSC computes the path skyline, LSCH only computes the linear path skyline). For both algorithms holds that RNE yields slightly faster runtimes in the two-dimensional case, but it rapidly degenerates with increasing dimensionality. Overall, the bounds computed by PP yield an ARSC speed-up between five times and two orders of magnitude for both scenarios. A similar acceleration is achieved for LSCH. It should be noted that even in the highly complex scenarios with five cost criteria, PP allows for unprecendent runtimes when computing linear skylines. Executing an ARSC search with RNE on more than three cost criteria is infeasibly slow due to the subpar quality of the bounds.

As mentioned above, PP computes optimal paths w.r.t. all d cost criteria. In the following, we want to compare this task to d separate single-source single-target Dijkstra searches. Note that this not the same procedure as MD which performs all-source single-target searches and cannot terminate upon arrival at the source. Figure 7 visualizes calculation time and number of visited nodes for both approaches and scenarios. Remarkably, although PP visits significantly more nodes, it is faster than the separate Dijkstra searches for the **muc** scenario. However, when the discrepancy regarding the number of visited nodes becomes too large – as in the **bav** scenario – PP is marginally slower than the Dijkstra approach. Of course, PP visits more nodes than separate Dijkstra searches because it visits all nodes relevant to any skyline path, as proven in Lemma 5. In contrast, the nodes visited by the Dijkstra searches will in general not suffice to build the path skyline. Hence, considering the additional information which PP acquires during its single graph traversal, the overall calculation time is unrivaled.

6   Conclusion

A multicriteria network is a graph where each edge has a vector of traversal costs, e.g., travel time, distance, toll fees, ascension in road networks. Therefore, there might exist multiple optimal paths between two nodes of the network. To find a set of path alternatives, path skyline queries compute all paths optimal under an arbitrary monotone combination function. As an alternative, linear path skyline queries restrict the result set to paths optimal under a linear combination function. In general, algorithms for computing alternative paths in large multicriteria networks employ lower bound estimations of the cost for reaching the target from a given node for each cost criterion. To generate these bounds, the established method [4] runs a all-source single-target Dijkstra search for each of the d cost criteria. For networks with more than two cost criteria, this requires d separate all-source searches, each visiting the whole network. In this paper, we present

ParetoPrep as an alternative method for lower bound computation when searching for optimal path alternatives. ParetoPrep computes optimal lower bounds for all criteria in a single graph traversal. Furthermore, ParetoPrep only visits a very limited part of the graph, which further accelerates the computation. Thus, ParetoPrep can be used in arbitrarily large networks where visiting all nodes would cause an infeasible overhead. To show that ParetoPrep still visits enough nodes to support optimal path queries, we prove that any node on a pareto optimal path is visited and bounded by ParetoPrep. Hence, ParetoPrep is an efficient preprocessing step for all algorithms computing multiple paths between a pair of nodes w.r.t. varying cost functions. In our experiments, we show that ParetoPrep considerably reduces query times and the visited part of the network for the state-of-the-art path computation algorithms ARSC [13] and LSCH [16].

Acknowledgements. This research has received funding from the Shared E-Fleet project (in the IKTII program), by the German Federal Ministry of Economics and Technology (grant no. 01ME12107).

References

1. Andersen, K.A., Skriver, A.J.: A label correcting approach for solving bicriterion shortest-path problems. Comput. Oper. Res. **27**, 507–524 (2000)
2. Balteanu, A., Jossé, G., Schubert, M.: Mining driving preferences in multi-cost networks. In: Nascimento, M.A., Sellis, T., Cheng, R., Sander, J., Zheng, Y., Kriegel, H.-P., Renz, M., Sengstock, C. (eds.) SSTD 2013. LNCS, vol. 8098, pp. 74–91. Springer, Heidelberg (2013)
3. Borzsonyi, S., Kossmann, D., Stocker, K.: The skyline operator. In: Proceedings of the 17th International Conference on Data Engineering (ICDE), Heidelberg, Germany (2001)
4. Chew, K.L., Tung, C.T.: A multicriteria pareto-optimal path algorithm. Eur. J. Oper. Res. **62**, 203–209 (1992)
5. Ehrgott, M., Gandibleux, X.: A survey and annotated bibliography of multiobjective combinatorial optimization. OR-Spektrum **22**(4), 425–460 (2000)
6. Ehrgott, M., Raith, A.: A comparison of solution strategies for biobjective shortest path problems. Comput. Oper. Res. **36**, 1299–1331 (2009)
7. Goldberg, A.V., Harrelson, C.: Computing the shortest path: A* search meets graph theory. Technical report MSR-TR-2004-24, Microsoft Research (2004)
8. Graf, F., Kriegel, H.-P., Renz, M., Schubert, M.: Mario: Multi attribute routing in open street map (2011)
9. Guo, C., Jensen, C.S., Kaul, M., Yang, B., Shang, S.: Stochastic skyline route planning under time-varying uncertainty. In: ICDE 2014, pp. 136–147 (2014)
10. Guo, C., Jensen, C.S., Yang, B., Guo, C., Ma, Y.: Toward personalized, context-aware routing. VLDB J. **24**(2), 297–318 (2015)
11. Hansen, P.: Bicriterion path problems. In: Fandel, G., Gal, T. (eds.) Multiple Criteria Decision Making Theory and Application. Lecture Notes in Economics and Mathematical Systems, vol. 177, pp. 109–127. Springer, Heidelberg (1980)
12. Ishwar, M., Mote, J., Olson, D.L.: A parametric approach to solving bicriterion shortest path problems. Eur. J. Oper. Res. **53**, 81–92 (1991)

13. Kriegel, H.P., Renz, M., Schubert, M.: Route skyline queries: a multi-preference path planning approach. In: ICDE 2010, pp. 261–272 (2010)
14. Machuca, E., Mandow, L.: Multiobjective heuristic search in road maps. Expert Syst. Appl. **39**, 6435–6445 (2012)
15. Müller-Hannemann, M., Weihe, K.: Pareto shortest paths is often feasible in practice. In: Brodal, G.S., Frigioni, D., Marchetti-Spaccamela, A. (eds.) WAE 2001. LNCS, vol. 2141, pp. 185–197. Springer, Heidelberg (2001)
16. Shekelyan, M., Jossé, G., Schubert, M.: Linear path skylines in multicriteria networks. In: ICDE15, pp. 459–470 (2015)
17. Shekelyan, M., Jossé, G., Schubert, M., Kriegel, H.-P.: Linear path skyline computation in bicriteria networks. In: Bhowmick, S.S., Dyreson, C.E., Jensen, C.S., Lee, M.L., Muliantara, A., Thalheim, B. (eds.) DASFAA 2014, Part I. LNCS, vol. 8421, pp. 173–187. Springer, Heidelberg (2014)
18. Skriver, A.J.: A classification of bicriterion shortest path (BSP) algorithms. Asia Pac. J. Ope. Res. **17**, 199–212 (2000)
19. Stewart, B., Chelsea, I.: White. Multiobjective a*. J. ACM **38**, 775–814 (1991)
20. Tarapata, Z.: Selected multicriteria shortest path problems: an analysis of complexity, models and adaptation of standard algorithms. Int. J. Appl. Math. Comput. Sci. **17**, 269–287 (2007)

Reverse Query and Indexing

Relaxed Reverse Nearest Neighbors Queries

Arif Hidayat$^{(\boxtimes)}$, Muhammad Aamir Cheema, and David Taniar

Faculty of Information Technology, Monash University, Clayton, Australia
{arif.hidayat,aamir.cheema,david.taniar}@monash.edu

Abstract. Given a set of users U, a set of facilities F, and a query facility q, a reverse nearest neighbors (RNN) query retrieves every user u for which q is its closest facility. Since q is the closest facility of u, the user u is said to be influenced by q. In this paper, we propose a *relaxed* definition of influence where a user u is said to be influenced by not only its closest facility but also every other facility that is *almost* as close to u as its closest facility is. Based on this definition of influence, we propose relaxed reverse nearest neighbors (RRNN) queries. Formally, given a value of $x > 1$, an RRNN query q returns every user u for which $dist(u, q) \leq x \times NNDist(u)$ where $NNDist(u)$ denotes the distance between a user u and its nearest facility. Based on effective pruning techniques and several non-trivial observations, we propose an efficient RRNN query processing algorithm. Our extensive experimental study conducted on several real and synthetic data sets demonstrates that our algorithm is several orders of magnitude better than a naïve algorithm as well as a significantly improved version of the naïve algorithm.

1 Introduction

People usually prefer the facilities in their vicinity. Hence, they are influenced by nearby facilities. A *reverse nearest neighbors* (RNN) query [1–4] aims at finding every user that is influenced by a query facility q. Formally, given a set of users U, a set of facilities F and a query facility q, an RNN query returns every user $u \in U$ for which the query facility q is its closest facility. The set containing RNNs, denoted as $RNN(q)$, is also called the influence set of q.

Consider the example of Fig. 1 that shows four McDonald's restaurants (f_1 to f_4) and three users (u_1 to u_3). In the context of RNN queries, the users u_2 and u_3 are both influenced by f_1 because f_1 is their closest McDonald's. Therefore, u_2 and u_3 are the RNNs of f_1, i.e., $RNN(f_1) = \{u_2, u_3\}$. Similarly, it can be confirmed that $RNN(f_2) = \emptyset$, $RNN(f_3) = \emptyset$, $RNN(f_4) = \{u_1\}$.

A *reverse k nearest neighbors* (RkNN) query [5–10] is a natural extension of the RNN query and uses a *relaxed* notion of influence. Specifically, in the context of an RkNN query, a user u is considered to be influenced by its k closest facilities. Hence, an RkNN query q returns every user $u \in U$ for which q is among its k closest facilities. In the example of Fig. 1, assuming $k = 2$, $R2NN(f_2) = \{u_1, u_2, u_3\}$ because f_2 is one of the two closest facilities for all of the three users. Similarly, $R2NN(f_1) = \{u_2, u_3\}$, $R2NN(f_3) = \emptyset$ and $R2NN(f_4) = \{u_1\}$.

© Springer International Publishing Switzerland 2015
C. Claramunt et al. (Eds.): SSTD 2015, LNCS 9239, pp. 61–79, 2015.
DOI: 10.1007/978-3-319-22363-6_4

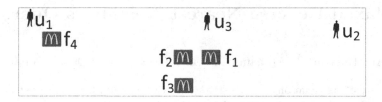

Fig. 1. Illustration of the reverse nearest neighbor query and its variants

RkNN queries have numerous applications [1] in location based services, resource allocation, profile-based management, decision support etc. Consider the example of a supermarket. The people for which this supermarket is one of the k closest supermarkets are its potential customers and may be influenced by targeted marketing or special deals. Due to its significance, RNN queries and its variants have received significant research attention in the past decade (see [6] for a survey).

In this paper, we propose an alternative definition of influence and propose a variant of RNN queries called *relaxed reverse nearest neighbors* (RRNN) query. This definition is motivated by our observation that an RkNN query may not properly capture the notion of influence as explained below.

1.1 Motivation

Consider the example of a person living in a suburban area (e.g., u_2 in Fig. 1) who does not have any McDonald's nearby. Her nearest McDonald's is f_1 which is say 30 Km from her location. In the context of R2NN query, u_2 is influenced by f_1 and f_2 – her two nearest facilities. However, we argue that it is also influenced by f_3 because a user who needs to travel a minimum of 30 Km to visit a McDonald's may also be willing to travel to a McDonald's store 31 Km far from her location.

Similarly, consider the example of another person living in a suburb (e.g., u_1 in Fig. 1) who has only one McDonald's nearby (f_4) assuming that all other McDonald's (e.g., f_1 to f_3) are in downtown area and are quite far. In the context of R2NN queries, the user u_1 is considered to be influenced by both f_4 and f_2 because these are her two closest facilities. However, we argue that the user u_1 is only influenced by f_4 because the other facilities are significantly farther than $dist(u_1, f_4)$, e.g., a user who has a McDonald's within 1 Km is not very likely to visit a McDonald's that is say 30 Km from her location.

As shown above, the definition of influence used in RkNN queries considers only the relative ordering of the facilities based on their distances from u and ignores the actual distances of the facilities from u. Motivated by this, in this paper, we propose a relaxed reverse nearest neighbors (RRNN) query that relaxes the definition of influence using a parameter x (called the x factor in this paper) and considers the relative distances between the users and the facilities.

Definition 1. *Let $NNdist(u)$ denote the distance between u and its nearest facility. Given a value of $x > 1$, a user u is said to be influenced by a facility f, if $dist(u, f) \leq x \times NNdist(u)$.*

Relaxed Reverse Nearest Neighbors (RRNN) Query. Given a value of $x > 1$, an RRNN query q returns every user u for which $dist(u, q) \leq x \times NNdist(u)$, i.e., return every user u that is influenced by q according to Definition 1. The set of RRNNs of a query q is denoted as $RRNN_x(q)$. Note that an RRNN query is the same as an RNN query if $x = 1$.

In the example of Fig. 1, assuming $x = 1.2$, RRNN of f_2 are the users u_2 and u_3, i.e., $RRNN_{1.2}(f_2) = \{u_2, u_3\}$. Similarly, $RRNN_{1.2}(f_1) = \{u_2, u_3\}$, $RRNN_{1.2}(f_3) = \{u_2\}$ and $RRNN_{1.2}(f_4) = \{u_1\}$.

Remark. RkNN queries and RRNN queries assume that the distance is the main factor influencing a user. This assumption holds in many real world scenarios. For instance, the users looking for nearby fuel stations are usually not concerned about price (or even rating) because all fuel stations have similar price (or even the same price because, in some countries, the fuel prices are regulated by the government). Similarly, users interested in McDonald's restaurants are mainly influenced by the distance because other attributes such as price, menu, and ratings are the same for all stores. Nevertheless, in the case where the users are influenced by other attributes, reverse top-k queries [11,12] can be used to compute the influence using a scoring function involving multiple attributes such as distance, price, and rating. This is a different line of research and is not within the scope of this paper.

1.2 Contributions

We make the following contributions in this paper.

1. We complement the RkNN queries by proposing a new definition of influence that uses the x factor to provide more meaningful results by considering the relative distances between the users and the facilities.
2. As we show in Sect. 3, the pruning techniques used to solve RkNN queries cannot be applied or extended for RRNN queries. This is mainly because, in our problem settings, a facility f may not be able to prune the users that are quite far from f (see Sect. 3 for details). Based on several non-trivial observations, we propose efficient pruning techniques that are proven to be *tight*, i.e., given a facility f used for pruning, the pruning techniques guarantee to prune every point that can be pruned by f. We then propose an efficient algorithm that utilizes these pruning techniques to efficiently compute the RRNNs.
3. We conduct an extensive experimental study on three real data sets and several synthetic data sets to show the effectiveness of our proposed techniques. Since existing techniques cannot be extended to answer RRNN queries, we compare our algorithm with a naïve algorithm (called RQ) as well as a significantly improved version of RQ (called IRQ). The experimental results show

that our algorithm is several orders of magnitude better than both of the competitors. Furthermore, we note that the results of an RRNN query are the same as the RkNN ($k = 1$) query when x is quite close to 1. Therefore, we also compare our algorithm (by setting $x = 1 + 10^{-0.6}$) with the most notable RNN algorithms. Although our algorithm solves a more challenging version of the problem, our experiments show that it performs reasonably well compared to RNN algorithms.

The rest of the paper is organized as follows. We present the problem definition and an overview of the related work in Sect. 2. The pruning techniques are discussed in Sect. 3. Section 4 describes our algorithm to solve RRNN queries. An extensive experimental study is provided in Sect. 5 followed by conclusions and directions for future work in Sect. 6.

2   Preliminaries

2.1   Problem Definition

Similar to RkNN queries, RRNN queries can also be classified into *bichromatic* RRNN queries and *monochromatic* RRNN queries.

Bichromatic RRNN Query. Given a set of users U, a set of facilities F, a query facility q (which may or may not be in F), and a value of $x > 1$, a bichromatic RRNN query returns every user $u \in U$ for which $dist(u, q) \leq x \times NNdist(u)$ where $NNDist(u)$ denotes the distance between u and its nearest facility in F.

Monochromatic RRNN Query. Given a set of facilities F, a query facility q (which may or may not be in F), and a value of $x > 1$, a monochromatic RRNN query returns every facility $f \in F$ for which $dist(f, q) \leq x \times NNdist(f)$ where $NNDist(f)$ denotes the distance between f and its nearest facility in $\{F - f\}$.

In Fig. 1, the monochromatic RRNNs of f_2 (assuming $x = 1.5$) are f_1 and f_3. Monochromatic queries aim at finding the facilities that are influenced by the query facility. Consider a set of police stations. For a given police station q, a monochromatic query returns the police stations for which q is a nearby police station. Such police stations may seek assistance (e.g., extra policemen) from q in case of an emergency event.

Although our techniques can be easily applied to monochromatic RRNN queries, in this paper, we focus on bichromatic RRNN queries because the bichromatic version has more applications in real world scenarios. Similar to the existing work on RNN queries, we assume that both the facility and user data sets are indexed by R*-tree [13]. The R*-tree that indexes the set of facilities (resp. users) is called facility (resp. user) R*-tree. Since most of the applications of the RNN query and its variants are in location-based services, similar to the existing RNN algorithms [6], the focus of this paper is on two dimensional location data.

2.2   Related Work

The RkNN query has been extensively studied [2–5,7–10,14–19] ever since it was introduced in [1]. Below, we briefly describe two widely used pruning strategies.

Half-Space Based Pruning [5]. A perpendicular bisector between a facility f and a query q divides the whole space into two halves. Let $H_{f:q}$ denote the half-space that contains f and $H_{q:f}$ be the half-space that contains q. A user u that lies in $H_{f:q}$ cannot be the RNN of q because $dist(u, f) < dist(u, q)$. Consider the example of Fig. 2, where the half-space $H_{a:q}$ is the shaded area. The users u_1 and u_2 cannot be the RNN of q because they lie in $H_{a:q}$. This observation can be extended for RkNN queries. Specifically, a user u cannot be the RkNN of q if it lies in at least k such half-spaces. In Fig. 2, assuming $k = 2$, the user u_2 cannot be R2NN of q because it lies in $H_{a:q}$ and $H_{b:q}$. In other words, the area $H_{a:q} \cap H_{b:q}$ (the dark shaded area) can be pruned.

Six-Regions Based Pruning [2]. In six-regions based pruning approach, the space around q is divided into six equal regions of $60°$ each (see P_1 to P_6 in Fig. 3). Let d_i^k be the distance between q and its k-th nearest facility in a partition P_i. It can be proved that a user u lying in a partition P_i cannot be the RkNN of q if $dist(u, q) > d_i^k$. Based on this observation, the k-th nearest facility in each partition P_i is found and the distance d_i^k is used to prune the search space. For instance, in Fig. 3, the shaded area can be pruned if $k = 1$, i.e., the users u_1 and u_2 are pruned.

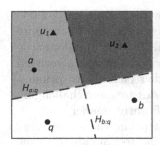 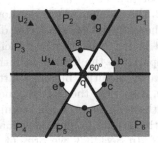 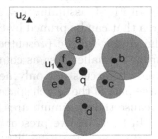

Fig. 2. Half-space pruning   **Fig. 3.** Six-regions prun-   **Fig. 4.** Challenges
ing

It has been shown [5] that the half-space based approach prunes more area than the six-regions based pruning. However, the advantage of the six-regions based pruning is that it is computationally less expensive. Six-region [2] and SLICE [10] are the most notable algorithms that use six-regions based pruning whereas TPL [5], FINCH [20], InfZone [8,21], and TPL++ [6] are some of the remarkable algorithms that employ half-space based pruning. The details of these algorithms can be found in a recent survey paper [6].

To the best of our knowledge, none of the existing algorithms can be applied or trivially extended to answer RRNN queries studied in this paper. The idea of relative distances has been discussed in [22] in the context of k nearest neighbors queries. However, this is a survey study and a solution was not proposed.

3 Pruning Techniques

Given a facility f, a user u cannot be the RRNN of q if $dist(u, q) > x \times dist(u, f)$. In such case, we say that the facility f prunes the user u. In this section, we will present the pruning techniques that use a facility f or an MBR of the facility R*-tree to prune the users. First, we highlight the challenges.

3.1 Challenges

Existing pruning techniques cannot be applied or extended for the RRNN queries due to the unique challenges involved. For instance, the algorithms to solve RkNN queries can prune most of the search space by considering only the nearby facilities surrounding q. Consider the example of Fig. 4 where the six-regions approach finds the nearest facility to the query q in each of the six partitions and the shaded area can be pruned.

However, in the case of RRNN queries, the nearby facilities surrounding the query q are not sufficient to prune a large part of the search space. Assuming $x = 2$, in partition P_3 (see Fig. 4), while the user u_1 can be pruned by f the user u_2 cannot be pruned by f. In other words, the users that are further from a facility f are less likely to be pruned by it.

In Fig. 4, assuming $x = 2$, the six shaded circles show the maximum possible area that can be pruned by the six facilities a to f (the details on how to compute the circles will be presented later). Note that the facilities that are close to q prune a smaller area as compared to the farther facilities. Hence, the algorithm needs to access not only nearby facilities but also farther facilities to prune a large part of the search space. Also, note that RRNN queries are more challenging because the maximum area that can be pruned is significantly smaller.

In Sect. 3.2, we present the pruning techniques that prune the space using a data point, i.e. a facility f. In Sect. 3.3, we present the techniques to prune the space using an MBR of the facility R*-tree. Efficient implementation of the pruning techniques is discussed in Sect. 3.4.

3.2 Pruning Using a Facility Point

Before we present our non-trivial pruning technique, we present the definition of a pruning circle.

Definition 2 (Pruning circle). *Given a query q, a multiplication factor $x > 1$ and a point p, the pruning circle of p (denoted as C_p) is a circle centered at c with radius r where $r = \frac{x \cdot dist(q,p)}{x^2 - 1}$ and c is on the line passing through q and p such that $dist(q, c) > dist(p, c)$ and $dist(q, c) = \frac{x^2 \cdot dist(q,p)}{x^2 - 1}$.*

Consider the example of Fig. 5 that shows the pruning circle C_f of a facility f assuming $x = 2$. The centre of c is located on the line passing through q and f such that $dist(q,c) = \frac{4 \cdot dist(q,f)}{3}$, $dist(q,c) > dist(f,c)$ and radius $r = \frac{2 \cdot dist(q,f)}{3}$. The condition $dist(q,c) > dist(f,c)$ ensures that c lies towards f on the line passing through q and f, i.e., f lies between the points c and q as shown in Fig. 5. Next, we introduce our first pruning rule in Lemma 1.

Lemma 1. *Every user u that lies in the pruning circle C_f of a facility f cannot be the RRNN of q, i.e., $dist(u,q) > x \times dist(u,f)$.*

Proof. Given two points v and w, we use $\overline{vw}$ to denote $dist(v,w)$. Consider the example of Fig. 5. Since u is inside the circle C_f, $\overline{uc} < r$. Assume that $\overline{uc} = n \cdot r$ where $0 \leq n < 1$. Since $r = \frac{x \cdot \overline{qf}}{x^2 - 1}$, we have $\overline{uc} = n \cdot r = n \cdot \frac{x \cdot \overline{qf}}{x^2 - 1}$.

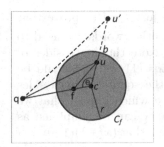
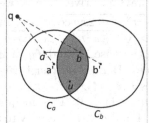
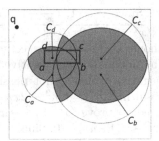

Fig. 5. Lemma 1   **Fig. 6.** Lemma 3   **Fig. 7.** Pruning using MBR

Considering the triangle $\triangle quc$, $\overline{qu} = \sqrt{(\overline{qc})^2 + (\overline{uc})^2 - 2 \cdot \overline{uc} \cdot \overline{qc} \cdot \cos \theta}$. Since $\overline{uc} = n \cdot \frac{x \cdot \overline{qf}}{x^2 - 1}$ and $\overline{qc} = \frac{x^2 \cdot \overline{qf}}{x^2 - 1}$, we have

$$\overline{qu} = \sqrt{(\frac{x^2 \cdot \overline{qf}}{x^2 - 1})^2 + n^2(\frac{x \cdot \overline{qf}}{x^2 - 1})^2 - 2n(\frac{x \cdot \overline{qf}}{x^2 - 1})(\frac{x^2 \cdot \overline{qf}}{x^2 - 1}) \cdot \cos \theta}$$

$$= \sqrt{(\frac{x \cdot \overline{qf}}{x^2 - 1})^2 (x^2 + n^2 - 2 \cdot x \cdot n \cdot \cos \theta)} \tag{1}$$

$$= (\frac{x \cdot \overline{qf}}{x^2 - 1})\sqrt{x^2 + n^2 - 2xn \cos \theta}$$

Similarly considering $\triangle fcu$, $\overline{fu} = \sqrt{(\overline{fc})^2 + (\overline{uc})^2 - 2 \cdot \overline{uc} \cdot \overline{fc} \cdot \cos \theta}$. Since $\overline{fc} = \overline{qc} - \overline{qf}$ and $qc = \frac{x^2 \cdot \overline{qf}}{x^2 - 1}$, we get $\overline{fc} = \frac{\overline{qf}}{x^2 - 1}$. We can obtain the value of $\overline{fu}$ by replacing the values of $\overline{fc}$ and $\overline{uc}$.

$$\overline{fu} = \sqrt{(\frac{\overline{qf}}{x^2-1})^2 + n^2(\frac{x \cdot \overline{qf}}{x^2-1})^2 - 2 \cdot n(\frac{x \cdot \overline{qf}}{x^2-1}) \cdot (\frac{\overline{qf}}{x^2-1}) \cdot \cos\theta}$$

$$= (\frac{\overline{qf}}{x^2-1})\sqrt{1 + n^2x^2 - 2nx\cos\theta} \qquad (2)$$

Note that the user u can be pruned if $dist(u,q) > x \times dist(u,f)$. Therefore, we need to show $\overline{qu} - x \cdot \overline{fu} > 0$. The left side of this inequality can be obtained using the values of $\overline{qu}$ and $\overline{fu}$ from Eqs. (1) and (2), respectively.

$$\overline{qu} - x \cdot \overline{fu} = \frac{x.\overline{qf}}{x^2-1}(\sqrt{x^2 + n^2 - 2xn\cos\theta} - \sqrt{1 + x^2n^2 - 2xn\cos\theta}) \qquad (3)$$

Since $x > 1$, $(\frac{x.\overline{qf}}{x^2-1})$ is always positive. Hence, we just need to prove that $(\sqrt{x^2 + n^2 - 2xn\cos\theta} - \sqrt{1 + x^2n^2 - 2xn\cos\theta} > 0$. In other words, we need to show $(\sqrt{x^2 + n^2 - 2xn\cos\theta} > \sqrt{1 + x^2n^2 - 2xn\cos\theta}$. Note that both sides of this inequality are positive (otherwise $\overline{qu}$ and $\overline{fu}$ in Eqs. (1) and (2) would be negative which is not possible). Hence, we can take the square of both sides resulting in $x^2 + n^2 - 2xn\cos\theta > 1 + x^2n^2 - 2xn\cos\theta$ which implies that we need to prove $(x^2 + n^2 - x^2n^2 - 1) > 0$. This inequality can be simplified as $(x^2-1)(1-n^2) > 0$. Since $x > 1$ and $n < 1$, it is easy to see that $(x^2-1)(1-n^2) > 0$ which completes the proof. □

Note that although the pruning technique itself is non-trivial, applying this pruning rule is not expensive, i.e., to check whether a user u can be pruned or not, we only need to compute its distance from the centre c and compare it with the radius r. Next, we show that this pruning rule is *tight* in the sense that any user u' that lies outside C_f is guaranteed not to be pruned by the facility f.

Lemma 2. *Given a facility f and a user u' that lies on or outside its pruning circle C_f, then $dist(u',q) \le x \times dist(u',f)$, i.e. u' cannot be pruned by f.*

Proof. Consider the user u' in Fig. 5. Since u' is on or outside the pruning circle, it satisfies $\overline{u'c} = n \cdot r$, where $n \ge 1$. The proof is similar to the proof of Lemma 1 except that we need to show that $\overline{u'q} - x.\overline{fu'} \le 0$, i.e., we need to show $(x^2 - 1)(1 - n^2) \le 0$ which is obvious given that $x > 1$ and $n \ge 1$. □

Note that the pruning circle C_f is larger if $dist(q,f)$ is larger which implies that the facilities that are farther from the query prune larger area. For instance, in Fig. 6, the pruning circle C_b is bigger than the pruning circle C_a.

3.3   Pruning Using the Nodes of Facility R*-tree

In this section, we present our techniques to prune the search space using the intermediate or leaf nodes of the facility R*-tree. These pruning techniques reduce the I/O cost of the algorithm because the algorithm may prune the search space using a node of the R*-tree instead of accessing the facilities in its sub-tree.

A node of the facility R*-tree is represented by a minimum bounding rectangle (MBR) that encloses all the facilities in its sub-tree. Without accessing the contents of the node, we cannot know the locations of the facilities inside it except that each side of the MBR contains at least one facility. We utilize this information to devise our pruning techniques. Specifically, we use all four sides of the MBR and use each side (i.e., line segment) to prune the search space. Lemma 3 presents the pruning rule and Fig. 6 provides an illustration.

Lemma 3. *Given a query q, a multiplication factor $x > 1$, and a line $\overline{ab}$ representing a side of an MBR, a user u cannot be the RRNN of q if it lies inside both of the pruning circles C_a and C_b, i.e., u can be pruned if u lies in $C_a \cap C_b$.*

Proof. Let $maxdist(p, \overline{ab})$ denote the maximum distance between a point p and a line $\overline{ab}$. Note that $maxdist(u, \overline{ab}) = max(dist(u,a), dist(u,b))$. Since u lies in both C_a and C_b, $dist(u,q) > x \times dist(u,a)$ and $dist(u,q) > x \times dist(u,b)$ (according to Lemma 1). In other words, $dist(u,q) > x \times maxdist(u, \overline{ab})$. Since there is at least one facility f on the line $\overline{ab}$, $dist(u,f) \leq maxdist(u, \overline{ab})$. Hence, $dist(u,q) > x \times dist(u,f)$ which implies that the user u can be pruned.   □

In Fig. 6, the shaded area can be pruned by using the line $\overline{ab}$. he next lemma shows that this pruning rule is also tight.

Lemma 4. *Given a line $\overline{ab}$ such that the only information we have is that there is at least one facility f on $\overline{ab}$, a user u cannot be pruned if it lies outside either C_a or C_b.*

Proof. Without the loss of generality, assume that u lies outside C_a. Now assume that there is exactly one facility f on the line $\overline{ab}$ and it lies at the end point a. Since f lies on a, $C_a = C_f$ which implies that u is outside C_f. Hence, u cannot be pruned by f (Lemma 2).   □

To prune the search space using an MBR, we apply Lemma 3 on each of side s_i of the MBR. Specifically, a user u can be pruned if, for *any* side s_i of the MBR, u lies in *both* of the pruning circles of the end points of s_i. Consider the example of Fig. 7 where an MBR $abcd$ is shown along with the pruning circles of the corners of the MBR (see C_a to C_d). Let A_i denote the area pruned by a side s_i of the MBR. In Fig. 7, the shaded area can be pruned which corresponds to $\cup_{i=1}^{4} A_i$ where $A_1 = C_a \cap C_b$, $A_2 = C_b \cap C_c$, $A_3 = C_c \cap C_d$, and $A_4 = C_d \cap C_a$.

3.4 Implementation of the Pruning Techniques

In the previous sections, we discussed how to prune the search space using a facility point or an MBR of the facility R*-tree. In this section, we discuss how to efficiently and effectively implement the pruning techniques.

Assume that we have a set of facilities and MBRs to be used for pruning the search space. Let A_i denote the area pruned by a facility point or a side of an MBR. Let $\mathcal{A} = \{A_i, \cdots, A_n\}$ be the total area that can be pruned by using the set of facilities and MBRs. In this section, we present Algorithm 1 that efficiently checks whether an entry e of user R*-tree (i.e., a point or an MBR) can be pruned by $\mathcal{A}$ or not, i.e., whether e lies inside $\mathcal{A}$ or not. Before we discuss the details of Algorithm 1, we describe how to prune a user MBR e using a single pruning area $A_i \in \mathcal{A}$. Since e is an MBR, it is possible that e only partially lies in A_i. Ideally, we should be able to prune the part of the MBR that lies inside A_i. In our algorithm, we process the MBR e such that the area that lies inside A_i is trimmed. Below are the details on how to do this.

Case 1: A_i corresponds to the area pruned by a facility. Consider the example of Fig. 8 where A_i corresponds to the circle C_a. Note that only a part of the rectangle R lies in the circle. In such case, we conservatively approximate the area that can be pruned. Specifically, we use a function $\mathtt{TrimEntry}(C_a, R)$ that trims the MBR R using a circle C_a and returns R_a that corresponds to the minimum bounding rectangle of the part of R that lies outside C_a, i.e., R_a cannot be pruned by C_a. In Fig. 8, R_a is the shaded area. In Fig. 9, R_b (the light shaded area) is returned by $\mathtt{TrimEntry}(C_b, R)$. The function $\mathtt{TrimEntry}(C_a, R)$ can be implemented as follows. Let I be the set of intersection points between a circle C_a and a rectangle R. Let $\mathcal{C}$ be the corners of R that lie outside C_a. The trimmed entry R_a is the minimum bounding rectangle enclosing the points in $I \cup \mathcal{C}$.

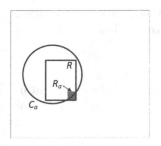 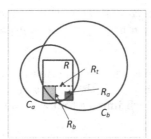 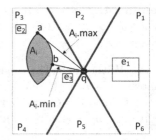

Fig. 8. Trimming an MBR   **Fig. 9.** Pruning an entry   **Fig. 10.** Observations 1 & 2

Case 2: A_i corresponds to the area pruned by a side of an MBR. Consider the example of Fig. 9 where A_i corresponds to the area pruned by a line $\overline{ab}$, i.e., $A_i = C_a \cap C_b$. In this case, we find the part of the MBR R that cannot be pruned by A_i as follows. Let $R_a = \mathtt{TrimEntry}(C_a, R)$ (see the dark shaded area) and $R_b = \mathtt{TrimEntry}(C_b, R)$ (see the light dotted area) in Fig. 9. The unpruned part of R is the minimum bounding rectangle enclosing both R_a and R_b, e.g., R_t shown in thick broken lines in Fig. 9 cannot be pruned by $C_a \cap C_b$.

Algorithm 1 shows the details of how to prune an entry e using a set of pruned areas $\mathcal{A}$. The output of the algorithm is the part of e that cannot be

---

Algorithm 1. PruneEntry($e, \mathcal{A}$)

Input: e: the entry to be pruned, $\mathcal{A}$: the set of pruned areas
Output: Return the part of e that cannot be pruned by $\mathcal{A}$

1: **for each** $A_i \in \mathcal{A}$ **do**
2:   **if** A_i is related to a facility f **then**
3:     $R \leftarrow$ TrimEntry(C_f, e)
4:   **else if** A_i is related to a line $\overline{ab}$ **then**
5:     $R_a \leftarrow$ TrimEntry(C_a, e)
6:     $R_b \leftarrow$ TrimEntry(C_b, e)
7:     $R \leftarrow$ minimum bounding rectangle enclosing both R_a and R_b
8:   $e \leftarrow R$
9:   **if** e is empty **then**
10:     return ϕ
11: return e

---

pruned by $\mathcal{A}$. Each entry A_i is iteratively accessed from $\mathcal{A}$ and the entry e is trimmed using the details described earlier (lines 2 to line 7). The trimmed part R is assigned to e which is to be further trimmed in the next iteration (line 8). At any stage, if e is empty, the algorithm terminates by returning ϕ (line 10) which indicates that the whole entry e can be pruned by $\mathcal{A}$. When all entries A_i in $\mathcal{A}$ have been accessed, the algorithm returns e.

We remark that although the trimming significantly improves the I/O cost (2 to 3 times) of the algorithm, the overall CPU time is also increased due to the overhead of trimming. This must be taken into consideration when making the decision on whether to use trimming or not, e.g., the trimming should not be used if the main focus is to optimize CPU cost.

Improving Algorithm 1. Note that Algorithm 1 accesses every entry $A_i \in \mathcal{A}$ regardless of whether A_i can prune a part of e or not. Now, we discuss how to improve the efficiency of Algorithm 1 by ignoring the entries A_i that cannot prune e. Similar to six-regions approach [2] and SLICE [10], we divide the whole space around q in t equally sized partitions, e.g., see the partitions P_1 to P_6 in Fig. 10. Our technique is based on the following two simple observations.

Observation 1. Let $\mathcal{P}$ be the set of partitions overlapped by a pruned area A_i. An entry e can be pruned by A_i only if e overlaps with at least one partition in $\mathcal{P}$. Consider the example of Fig. 10 where the area A_i is shown shaded and overlaps with partitions P_3 and P_4. Since the entry e_1 does not overlap with P_3 or P_4, it cannot be pruned by A_i.

Observation 2. Let $A_i.max$ and $A_i.min$ denote the maximum and minimum distances between q and the pruned area A_i, respectively. Figure 10 shows $A_i.max = dist(q, a)$ and $A_i.min = dist(q, b)$. We remark that $A_i.max$ and $A_i.min$ can be computed following the ideas presented in [23,24]. Note that an entry e cannot be pruned by A_i if $mindist(q, e) > A_i.max$ or $maxdist(q, e) < A_i.min$. For instance, the entry e_2 cannot be pruned by A_i because $mindist(q, e_2) > A_i.max$. Similarly, the entry e_3 cannot be pruned because $maxdist(q, e_3) < A_i.min$.

Let $A_i.interval$ denote an interval from $A_i.min$ to $A_i.max$ and $e.interval$ denote an interval from $mindist(q, e)$ to $maxdist(q, e)$. Observation 2 shows that an entry e can be pruned by A_i only if $e.interval$ overlaps with $A_i.interval$. We use an interval tree [25] to efficiently retrieve every A_i for which $A_i.interval$ overlaps with $e.interval$. Specifically, for each partition P_i, we maintain an interval tree $\mathcal{T}_i$ that contains $A_j.interval$ for every $A_j \in \mathcal{A}$ that overlaps with P_i. To check whether an entry e (that overlaps with a partition P_i) can be pruned by $\mathcal{A}$, we issue an interval query on $\mathcal{T}_i$ with input interval $e.interval$. Let $\mathcal{A}_e$ denote the set containing every area A_j returned by the interval query $e.interval$. In Algorithm 1, we use $\mathcal{A}_e$ instead of $\mathcal{A}$. Note that the cost of interval query is $O(m + \log n)$ where n is the number of intervals stored in the interval tree and m is the number of intervals that overlap with the input interval.

4   Algorithm

Our algorithm consists of three phases namely *pruning*, *filtering* and *verification*. In the pruning phase, we use the facility R*-tree to prune the search space, i.e., compute $\mathcal{A}$. In the filtering phase, the users that lie in the pruned space are pruned and the remaining users are inserted in a candidate list called L_{cnd}. Finally, in the verification phase, each candidate user $u \in L_{cnd}$ is verified to check whether it is a RRNN of q or not.

Pruning Phase. Algorithm 2 presents the details of the pruning phase. The algorithm initializes a heap h with the root of the facility R*-tree. The entries are iteratively de-heaped from the heap and are processed as follows. If a de-heaped entry e is pruned (i.e., the entry e' returned by Algorithm 1 is empty), we ignore it (lines 5 and 6). Otherwise, we process it as follows.

Algorithm 2. Pruning

Input: facility R*-tree, and a query q
Output: The set of pruned areas $\mathcal{A}$

1: $\mathcal{A} \leftarrow \phi$
2: insert root of facility R-tree in a h
3: **while** h is not empty **do**
4:    de-heap an entry e
5:    $e' \leftarrow PruneEntry(e, \mathcal{A})$ ▷ *Algorithm 1*
6:    **if** $e' \neq \phi$ **then**   ▷ *e is not pruned*
7:        **if** e is an intermediate or leaf node **then**
8:            **for** each side $\overline{ab}$ of e **do**
9:                create $A_i = C_a \cap C_b$ and insert in $\mathcal{A}$
10:           **for** each child c of e **do**
11:               **if** c overlaps with e' **then** insert c in the heap
12:       **else**   ▷ *e is a facility point*
13:           create $A_i = C_e$ and insert in $\mathcal{A}$

Algorithm 3. Filtering

Input: user R*-tree, query q, and $\mathcal{A}$
Output: a list of candidates L_{cnd}

1: $L_{cnd} \leftarrow \phi$
2: insert root of user R*-tree in a stack S
3: **while** S is not empty **do**
4:    retrieve top entry e from S
5:    $e' \leftarrow PruneEntry(e, \mathcal{A})$ ▷ *Algorithm 1*
6:    **if** $e' \neq \phi$ **then**   ▷ *e is not pruned*
7:       **if** e is an intermediate or leaf node **then**
8:          **for** each child c of e **do**
9:             **if** c overlaps with e' **then** insert c in stack S
10:        **else**   ▷ *e is a user*
11:           insert e in L_{cnd}

If e is an intermediate or leaf node of the R*-tree, for each side of e, we create a pruning area A_i and insert it in $\mathcal{A}$ (line 9). We also insert its children in the heap h. Note that a child c of e that does not overlap with e' can be pruned because it lies in the pruned area. Hence, only the children that overlap with e' are inserted in the heap (line 11). If e is a facility point, we create the pruning circle C_e and add it to $\mathcal{A}$ (line 13). The algorithm terminates when the heap becomes empty.

Filtering Phase. Algorithm 3 describes the filtering phase. A stack S is initialized with the root entry of the user R*-tree. Each entry e is iteratively retrieved from S and processed as follows. If e can be pruned by $\mathcal{A}$, it is ignored (lines 5 and 6). Otherwise, if it is an intermediate or leaf node, its children that overlap with e' are inserted in the stack (line 9). If e is a user, it is inserted in L_{cnd} (line 11). The algorithm stops when the stack S becomes empty.

Verification Phase. In the verification phase, each candidate user $u \in L_{cnd}$ is verified as follows. Note that a user u is a RRNN if and only if there is no facility f for which $dist(u, f) < \frac{dist(u,q)}{x}$. A circular boolean range query is issued with centre at u and radius $r = \frac{dist(u,q)}{x}$ that returns true if and only if there exists a facility in the circle. The boolean range query is conducted on the facility R*-tree as in previous works [7] and u is reported as an answer if it returns false.

5 Experiments

5.1 Experimental Setup

To the best of our knowledge, there is no prior algorithm to solve RRNN queries. We consider a naïve algorithm (RQ) and make reasonable efforts to devise a significantly improved version of RQ, as explained below.

Range Query (RQ). For each user u, a boolean range query with range $dist(u, q)/x$ is issued on the facility R*-tree (as described in the verification phase above).

Improved Range Query (IRQ). Note that an intermediate or leaf node entry e_u of the user R*-tree cannot contain any RRNN if there exists at least one facility f such that $mindist(e_u, q) > x \times maxdist(e_u, f)$, i.e., e_u can be pruned. Based on this, to check whether e_u can be pruned or not, we use a function $\texttt{isPruned}(e_u)$ that is implemented as follows. The facility R*-tree is traversed in ascending order of $maxdist(e_u, e_f)$ where e_f denotes an entry in the facility R*-tree. The entry e_u is pruned as soon as we find an entry e_f for which $mindist(e_u, q) > x \times maxdist(e_u, e_f)$. To further improve the I/O and CPU cost of $\texttt{isPruned}(e_u)$, we do not access the sub-tree of a facility entry e_f if $mindist(e_u, q) < x \times mindist(e_u, e_f)$ because no child of e_f can prune e_u.

The IRQ algorithm is the same as Algorithm 3 except that (1) "**if** $\texttt{isPruned}(e)$ **then**" replaces lines 5 and 6 of Algorithm 3; and (2) at line 11, the user is reported as an answer instead of inserting it in L_{cnd}. Note that IRQ does not have a pruning and verification phase because it merges all these phases in one algorithm. In our experiments, we observed that the performance of IRQ can be further improved if $\texttt{isPruned}(e_u)$ is only applied to leaf entries of the user R*-tree. This is because the intermediate nodes are highly unlikely to be pruned and result in un-necessary I/O. We included this optimization in IRQ.

All algorithms were implemented in C++ and experiments were run on Intel Core $I5$ 2.3 GHz PC with 8 GB memory running on Debian Linux. Experimental settings are quite similar to the existing work [6]. Specifically, we use the same real data sets containing 175, 812 points from North America (called NA data set hereafter), 2.6 million points from Los Angeles (LA) and 25.8 million points from California (CA). We also generate several synthetic data sets containing 1, 000 to 20 million points following normal distributions. The default real data set is LA containing 2.6 million points. Unless mentioned otherwise, each data set is randomly divided into two sets of almost equal size, one corresponding to the facilities and the other to the users. The page size of each R*-Tree [13] is set to 4, 096 Bytes. We randomly select 100 points from the facility data set and treat them as query points. The cost reported in the experiments correspond to the average cost of a single RRNN query. We vary the value of x from 1.1 to 4 and the default value is 1.5.

5.2   Evaluating Performance

Effect of Buffers. All three algorithms need to traverse facility R*-tree every time a boolean range query is issued to verify a candidate user. Hence, the buffers may reduce the I/O cost. We study the effect of the number of buffers on each algorithm. Each buffer page can hold one node of the R*-tree and we use random eviction strategy. In Fig. 11, we report the I/O cost of each algorithm on LA data set for different number of buffers. As expected, the I/O cost of each algorithm decreases with the increase in number of buffers. Note that IRQ is up to two orders of magnitude better than RQ and our algorithm is up to three orders of magnitude better than IRQ. Similar to [6], we use 100 buffer pages for each algorithm in the rest of the experiments.

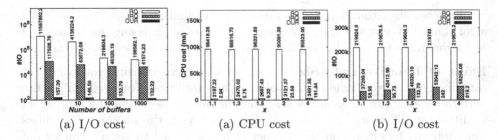

(a) I/O cost (a) CPU cost (b) I/O cost

Fig. 11. Number of buffers   **Fig. 12.** Effect of the x factor (LA data set)

Effect of the x Factor. In Fig. 12, we study the effect of the x factor on the three algorithms. Specifically, Fig. 12(a) shows the CPU cost and Fig. 12(b) shows the I/O cost of the three algorithms for varying values of x. In terms of both CPU and I/O cost, our algorithm is up to three orders of magnitude better than IRQ and up to four orders of magnitude better than RQ. The cost of our algorithm and IRQ is higher for larger x factor because the pruning area shrinks as the x factor increases which results in a larger number of candidates and RRNNs. Note that the cost of RQ is not significantly affected by the x factor mainly because it needs to verify every user regardless of the value of x.

Effect of Data Set Size. In Fig. 13(a) and (b), we study the effect of data set size on the performance of the three algorithms. Specifically, we conduct experiments on three real data sets: NA (175,000 points), LA (2.6 million points) and CA (25.8 million points). Our algorithm outperforms the other two algorithms and the gap between the three algorithms increases as the data set size increases (please note that log-scale is used in both figures). For example, Fig. 13(a) shows that our algorithm is around 25 times faster than IRQ on NA data set and 330 times faster on CA data set. Similarly, Fig. 13(b) shows that the I/O cost of our algorithm is around 12 times lower than IRQ for NA data set and almost 430 times lower for CA data set. Also, as expected the cost of each algorithm increases as the data set size increases. This is mainly because the size of each R*-tree increases and more entries are required to be processed.

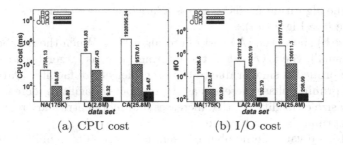

(a) CPU cost (b) I/O cost

Fig. 13. Performance comparison on different real data sets

Since our algorithm is up to several orders of magnitude better than the other algorithms, in the rest of the experiments, we focus on analysing the behavior of our algorithm and omit the cost of the other algorithms for better illustration.

Effect of Relative Data Size. In the previous experiments, each data set contained almost the same number of users and facilities. Next, we analyse the performance of our algorithm where the number of users and the number of facilities are different. Specifically, in Fig. 14 we vary the number of facilities from 1000 to 1 million and the number of users is fixed to $100K$. The sets of facilities and users are generated using normal distribution. Figure 14(a) and (b) show the CPU and I/O cost of our algorithm, respectively. Figure 14(c) shows the number of candidates, number of RRNNs and the number of entries (facility points and MBRs) used for pruning.

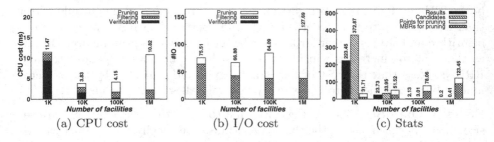

(a) CPU cost   (b) I/O cost   (c) Stats

Fig. 14. Effect of varying the number of facilities ($100\,K$ users)

Figure 14(a) shows that the CPU cost of our algorithm is larger if the number of facilities is too small or too large as compared to the number of users. The reason is as follows. When the number of facilities is too small (e.g., $1,000$), the total area that can be pruned is smaller due to the lower density of the facilities. This results in a larger number of candidates and RRNNs (as shown in Fig. 14(c)). Hence, the verification cost of the algorithm is larger as shown in Fig. 14(a). On the other hand, when the number of facilities is too large (e.g., 1 million), the pruning phase is the dominant cost of the algorithm. This is because the algorithm needs to access a larger number of entries to prune the search space (see Fig. 14(c)).

Figure 14(b) shows the I/O cost of our algorithm. When the number of facilities is too small, the I/O cost of the filtering phase is larger because the area that can be pruned is smaller due to the lower density of facilities data set. The I/O cost of pruning phase increases as the number of facilities increases. This is because the size of facility R*-tree increases and more entries are required to be accessed to prune the search space.

In Fig. 15, we vary the number of users from $1,000$ to 1 million and fix the number of facilities to $100\,K$. Figure 15(a) shows that the CPU cost of the algorithm increases as the number of users increases. This is because the filtering and verification cost of the algorithm increases for larger set of users, e.g., the

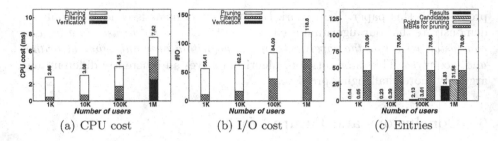

(a) CPU cost   (b) I/O cost   (c) Entries

Fig. 15. Effect of varying the number of users ($100K$ facilities)

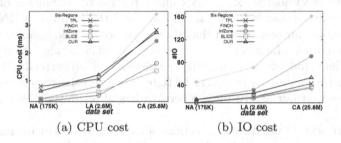

(a) CPU cost   (b) IO cost

Fig. 16. Comparison with state-of-the-art RNN algorithms

number of candidate users and RRNNs increases (as shown in Fig. 15(c)). Similarly, Fig. 15(b) shows that the I/O cost of the algorithm also increases for larger number of users. This is because the filtering requires traversing a larger user R*-tree which results in requiring to access more nodes of the users.

Figure 15(c) also shows the effectiveness of the proposed pruning techniques. Note that the number of candidates is much smaller as compared to the total number of users. Furthermore, almost 65 % of the candidates are the relaxed reverse nearest neighbors. We remark that the verification I/O cost of our algorithm is negligible mainly because most of the nodes accessed during verification are already present in the buffer (from pruning phase or the previously issued boolean range queries).

Efficiency Compared with RNN Algorithms. As stated earlier, there is no previous algorithm to solve RRNN queries and the existing algorithms to solve RNN queries cannot be trivially extended. Although we made significant efforts to devise the second competitor IRQ, our algorithm is up to three orders of magnitude better than it. In the absence of a well-known competitor, readers may find it harder to evaluate the efficiency of an algorithm. Therefore, we compare our algorithm with the most well-known RNN algorithms, namely SLICE [10], InfZone [8], TPL [5], FINCH [20] and six-regions [2]. For our algorithm, we set $x = 1 + 10^{-6}$ because we note that the results of an RRNN query is the same as those of an RNN query if x is very close to 1.

Figure 16 shows that the performance of our algorithm is comparable to the most popular RNN algorithms which shows the effectiveness of the techniques proposed in this paper. We remark that this experiment is conducted only to demonstrate that our algorithm is efficient and *it should not be used to draw any conclusion regarding the superiority of our algorithm over any other algorithm and vice versa*. This is because our algorithm solves an inherently different and arguably more challenging problem.

6   Conclusions and Future Work

In this paper, we propose a variant of RNN queries called relaxed reverse nearest neighbors (RRNN) queries. An RRNN query relaxes the definition of influence using the relative distances between the users and the facilities. RRNN queries are motivated by our observation that RkNN queries may be unable to properly capture the notion of influence. We propose an efficient algorithm based on several efficient and effective pruning techniques and non-trivial observations. The pruning techniques are proved to be tight. The extensive experimental study demonstrates that our algorithm is several orders of magnitude better than the competitors.

There are several interesting directions for future work. For example, it will be interesting to study the relaxed version of reverse top-k queries by using the idea of relative scores, i.e., return every user for whom the query product is almost as good as her most preferred product. Also, continuous RRNN queries for moving objects is another interesting research direction, e.g., continuously report the drivers for which my fuel station is an RRNN. RRNN queries for other distance metrics such as road network distances also need to be explored.

Acknowledgments. The research of Muhammad Aamir Cheema is supported by ARC DE130101002 and DP130103405.

References

1. Korn, F., Muthukrishnan, S.: Influence sets based on reverse nearest neighbor queries. In: SIGMOD, pp. 201–212 (2000)
2. Stanoi, I., Agrawal, D., Abbadi, A.E.: Reverse nearest neighbor queries for dynamic databases. In: ACM SIGMOD Workshop, pp. 44–53 (2000)
3. Cheema, M.A., Lin, X., Wang, W., Zhang, W., Pei, J.: Probabilistic reverse nearest neighbor queries on uncertain data. IEEE Trans. Knowl. Data Eng. **22**, 550–564 (2010)
4. Stanoi, I., Riedewald, M., Agrawal, D., Abbadi, A.E.: Discovery of influence sets in frequently updated databases. In: PVLDB, pp. 99–108 (2001)
5. Tao, Y., Papadias, D., Lian, X.: Reverse knn search in arbitrary dimensionality. In: PVLDB, pp. 744–755 (2004)
6. Yang, S., Cheema, M.A., Lin, X., Wang, W.: Reverse k nearest neighbors query processing: experiments and analysis. In: PVLDB, pp. 605–616 (2015)
7. Wu, W., Yang, F., Chan, C.Y., Tan, K.L.: FINCH: evaluating reverse k-nearest-neighbor queries on location data. In: PVLDB, pp. 1056–1067 (2008)

8. Cheema, M.A., Lin, X., Zhang, W., Zhang, Y.: Influence zone: efficiently processing reverse k nearest neighbors queries. In: ICDE, pp. 577–588 (2011)
9. Cheema, M.A., Zhang, W., Lin, X., Zhang, Y., Li, X.: Continuous reverse k nearest neighbors queries in euclidean space and in spatial networks. VLDB J. **21**, 69–95 (2012)
10. Yang, S., Cheema, M.A., Lin, X., Zhang, Y.: SLICE: reviving regions-based pruning for reverse k nearest neighbors queries. In: ICDE, pp. 760–771 (2014)
11. Vlachou, A., Doulkeridis, C., Kotidis, Y., Nørvåg, K.: Reverse top-k queries. In: ICDE, pp. 365–376 (2010)
12. Cheema, M.A., Shen, Z., Lin, X., Zhang, W.: A unified framework for efficiently processing ranking related queries. In: EDBT, pp. 427–438 (2014)
13. Beckmann, N., Kriegel, H., Schneider, R., Seeger, B.: The r*-tree: an efficient and robust access method for points and rectangles. In: Proceedings of the 1990 ACM SIGMOD International Conference on Management of Data, Atlantic City, NJ, 23–25 May, pp. 322–331 (1990)
14. Emrich, T., Kriegel, H.-P., Kröger, P., Renz, M., Züfle, A.: Incremental reverse nearest neighbor ranking in vector spaces. In: Mamoulis, N., Seidl, T., Pedersen, T.B., Torp, K., Assent, I. (eds.) SSTD 2009. LNCS, vol. 5644, pp. 265–282. Springer, Heidelberg (2009)
15. Singh, A., Ferhatosmanoglu, H., Tosun, A.S.: High dimensional reverse nearest neighbor queries. In: CIKM (2003)
16. Achtert, E., Kriegel, H.P., Kröger, P., Renz, M., Züfle, A.: Reverse k-nearest neighbor search in dynamic and general metric databases. In: EDBT, pp. 886–897 (2009)
17. Sharifzadeh, M., Shahabi, C.: Vor-tree: R-trees with voronoi diagrams for efficient processing of spatial nearest neighbor queries. PVLDB **3**(1), 1231–1242 (2010)
18. Cheema, M.A., Lin, X., Zhang, Y., Wang, W., Zhang, W.: Lazy updates: an efficient technique to continuously monitoring reverse knn. In: PVLDB, pp. 1138–1149 (2009)
19. Bernecker, T., Emrich, T., Kriegel, H.P., Renz, M., Züfle, S.Z.A.: Efficient probabilistic reverse nearest neighbor query processing on uncertain data. In: PVLDB, pp. 669–680 (2011)
20. Wu, W., Yang, F., Chan, C.Y., Tan, K.L.: Continuous reverse k-nearest-neighbor monitoring. In: MDM, pp. 132–139 (2008)
21. Cheema, M.A., Zhang, W., Lin, X., Zhang, Y.: Efficiently processing snapshot and continuous reverse k nearest neighbors queries. VLDB J. **21**(5), 703–728 (2012)
22. Taniar, D., Rahayu, W.: A taxonomy for nearest neighbour queries in spatial databases. J. Comput. Syst. Sci. **79**(7), 1017–1039 (2013)
23. Cheema, M.A., Brankovic, L., Lin, X., Zhang, W., Wang, W.: Multi-guarded safe zone: an effective technique to monitor moving circular range queries. In: ICDE, pp. 189–200 (2010)
24. Cheema, M.A., Brankovic, L., Lin, X., Zhang, W., Wang, W.: Continuous monitoring of distance-based range queries. IEEE Trans. Knowl. Data Eng. **23**(8), 1182–1199 (2011)
25. Cormen, T.H., Leiserson, C.E., Rivest, R.L., Stein, C., et al.: Introduction to Algorithms, vol. 2. MIT Press, Cambridge (2001)

Influence-Aware Predictive Density Queries Under Road-Network Constraints

Lasanthi Heendaliya$^{(\boxtimes)}$, Michael Wisely, Dan Lin, Sahra Sedigh Sarvestani, and Ali Hurson

Department of Computer Science, Missouri University of Science and Technology, Rolla, MO, USA
{heendaliyal,mwwcp2,lindan,sedighs,hurson}@mst.edu

Abstract. Density query is a very useful query type that informs users about highly concentrated/dense regions, such as a traffic jam, so as to reschedule their travel plans to save time. However, existing products and research work on density queries still have several limitations which, if can be resolved, will bring more significant benefits to our society. For example, we identify an important problem that has never been studied before. That is none of the existing works on traffic prediction consider the influence of the predicted dense regions on the subsequent traffic flow. Specifically, if road A is estimated to be congested at timestamp t_1, the prediction of the condition on other roads after t_1 should consider the traffic blocked by road A. In this paper, we formally model such influence between multiple density queries and propose an efficient query algorithm. We conducted extensive experiments and the results demonstrate both the effectiveness and efficiency of our approach.

1 Introduction

Sitting in road traffic congestion is obviously not a pleasant experience for a traveler. The impacts of traffic congestions indeed expand beyond the inconvenience and include environmental, economical, and safety issues [2,18]. This work aims to find solutions to traffic congestion problems by leveraging mobile devices and their popularity among the community.

Some common strategies related to relieving the traffic congestion problem include providing current traffic information (like Google Maps) [4,9,23] or modeling future traffic conditions with past data [17,20,21]. However, these existing approaches still have several limitations which, if can be resolved, will bring more significant benefits to our society. Specifically, such limitations include the following. First, existing approaches that provide only current traffic information do not offer many options for a driver. Because, based on current information, it may already be too late for some vehicles very close to the traffic congestion to divert to a new route. Therefore, several research works [13] have been proposed to predict traffic conditions with mobile object database queries. Unfortunately, most of them simplify the problem by considering objects moving on Euclidean space rather than under road-network constraints, making them hard to

© Springer International Publishing Switzerland 2015
C. Claramunt et al. (Eds.): SSTD 2015, LNCS 9239, pp. 80–97, 2015.
DOI: 10.1007/978-3-319-22363-6_5

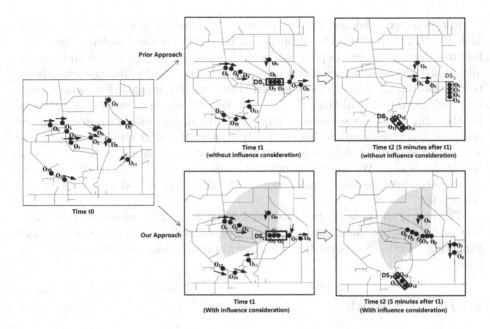

Fig. 1. An example of traffic influence effect

be directly adopted in real scenarios. Very few works predict traffic conditions under the road-network constraints. The most recent one is the Predictive Line Query (PLQ) [11]. However, PLQ only returns predicted traffic information of a user specified road. It would not be able to identify all possible dense regions (i.e., traffic congestions) automatically.

Besides the aforementioned limitations, there is another important issue that has been neglected by all existing works on predictive density queries, which is the influence of a predicted dense region on the subsequent traffic flow. Specifically, if road A is forecasted to be congested at timestamp t, the prediction of the condition on other roads after t should consider the traffic being blocked by road A during the period of congestion. To have a better understanding, let us consider a more concrete example shown in Fig. 1. In Fig. 1, the leftmost diagram shows the positions and moving directions of vehicles (denoted as black points) at timestamp t_0. Without considering traffic influence, a density query will predict a dense road segment DS_1 (highlighted by the rectangle) at timestamp t_1 and then another two (DS_2 and DS_3) at timestamp t_2 (say 5 min later). However, if observed carefully, vehicles stopped by the congestion at DS_1 are unlikely to travel to DS_2 since traffic would not be clear within 5 min. As a result, DS_2 may not have any congestion at all at timestamp t_2. This scenario explains that predicting dense areas on a given timestamp could be inaccurate unless the influence of former possible dense areas are taken into consideration. Our goal is to model such influence (as shown by the gray area) and provide more realistic traffic prediction.

In this paper, we define a new type of query, called *Influence-aware Predictive Density (IPD) Queries*. Our proposed IPD query has the following three key features that are unseen in prior works: (i) it automatically identifies and reports all possible dense areas, in terms of road segments, considering the underlying road network; (ii) it provides predicted traffic density information, which users will find more practical than the current density information; (iii) it accounts for the influence of dense regions on other nearby dense regions to produce more accurate traffic estimation. To efficiently answer the IPD queries, we propose a three-phase query algorithm along with several heuristics to further prune the search space. We have conducted experiments using real road maps and the experimental results demonstrate both effectiveness and efficiency of our approach.

The rest of the paper is organized as follows. Section 2 reviews related works. Section 3 formally defines the density query problem. Section 4 presents the query algorithm. Section 5 reports the experimental results and Sect. 6 concludes the paper.

2   Related Work

Generally speaking, a density query aims to retrieve all regions with a density (i.e., the number of moving objects per square unit) that exceeds a given threshold. It is worth noting that the density query is different from the range query in that the input to the range query is the location of a query region, whereas the input to the density query is the density threshold and the size (but not the location) of the dense region. The remainder of this section discusses the past work on the density query problem and its solutions.

Existing works on density queries can be roughly classified into two categories: (i) density queries in Euclidean space; (ii) density queries in road networks. Most of existing works on density queries mainly consider objects moving freely in Euclidean space. The first work is by Hadjieleftheriou et al. [16] who proposed two versions of density queries: Snapshot Density Queries (SDQ) and Period Density Queries (PDQ). The SDQ identifies dense regions for a specific time instance in the future, while the PDQ identifies dense regions in a time interval. In their approach, the entire space is divided into a grid of equal-sized cells, and density regions are reported in terms of cells. Such approach ignores possible dense regions located in the middle of multiple cells and causes a so-called answer loss problem as pointed out by Jensen et al. in [13]. To resolve the answer loss problem, Jensen et al. [13] redefined the density query and propose a two-phase query algorithm to predict dense regions that can be located anywhere and are not constrained to partitioning cells. Unlike previous works where the dense regions are square shaped, Ni and Ravishankar [19] define a pointwise dense regions (PDR) that allows the dense region to be of any shape and any size. They also partition the space into grid while their search algorithm ensures that a 4-cell block is searched each time and hence also avoids the answer loss problem. All the aforementioned query algorithms consider the snapshot version of the density query. The algorithms proposed in [10,22] support the continuous density queries. Similar to the snapshot queries, the continuous density queries also take the density threshold and size of the dense region as input. In order to continuously identify dense regions, the algorithm repeatedly divides the entire space into quadrants until the quadrant is no larger than the given size of the dense region. Then, each quadrant is labeled as either dense or not. This is the main limitation of such an approach.

In addition to the aforementioned density query algorithms, to which movements are considered in the Euclidean space, very few works [15] have been conducted on density queries restricting the movements to road networks. One such work is by Lai et al. [15] who propose the Effective Road-Network Density Query (e-RNDQ). The definition of density under road network constraint is now the number of moving objects per road segment rather than per square unit. Furthermore, the distance between any two neighboring objects in a dense road segment should not exceed the given distance threshold. This condition prevents skewed object distribution in a query result. Lai et al. propose a clustering-based algorithm to obtain the query results. The main limitation is that they only identify dense road segment at current timestamp but do not support any predictive queries.

In summary, our work is different from existing works in the following aspects: (i) it is the first time that the traffic influence is considered during multiple dense region exploration; (ii) it is the first work, to the best of our knowledge, that supports predictive density queries on road networks.

3   Problem Statement

Without loss of generality, we consider uni-directional or bi-directional roads with separate lanes for each direction. In other words, it is assumed that the high traffic density of one direction does not affect the traffic on the other direction. Under this assumption, in what follows we formally define our proposed *Influence-aware Predictive Density Query*.

Definition 1. *[Density] The density of a road segment r is the number of objects per unit length (e.g., meter) per lane on the road, represented as $DS(r) = \frac{N}{m \cdot len(r)}$, where N is the number of objects on road r, $len(r)$ is the length of road r, and m is the number of lanes.*

Definition 2. *[Dense Road Segment] Given a density threshold ρ, the road segment r is dense if the $density(r)$ at time t is greater than the threshold ρ.*

We now proceed to define a new concept, namely *mutually independent dense road segments*, which is the base of the influence-aware predictive density query.

Definition 3. *[Mutually Independent Dense Road Segments] Given any two dense road segments R_a and R_b with occurrence time t_a and t_b ($t_a < t_b$) respectively, R_a and R_b are mutually independent of each other if one of the following conditions is satisfied:*

1. $t_b - t_a > \sigma$, where σ is the threshold that describes the typical time taken to clear a traffic jam;
2. For any $o \in O_a$, $Dist(o, R_b) > v_{max} \cdot (t_b - t_a)$, where O_a is the set of objects on road R_a at time t_a, $Dist(o, R_b)$ computes the shortest road-network distance between object o and the closer end of road R_b, and v_{max} is the maximum moving speed of objects.

The definition of mutually independent dense road segments aims to ensure that objects that contribute to the density of one road segment will not affect the density computation of another road. Specifically, the first condition checks if the occurrence of dense regions are far enough apart in terms of time that objects stuck on road R_a may already be freed when computing the density for road R_b. The second condition avoids considering any subsequent dense road segment caused by the similar set of objects that recently contributed to a dense road segment. For example, consider an object o passes by road R_a at t_a and then R_b at t_b. If R_a is predicted as a dense road segment at t_a, we should not consider o when computing the density of road R_b at t_b since o has stuck or been slowed down by the traffic congestion.

Definition 4. *[Influence-aware Predictive Density (IPD) Query] Given a road map G, a density threshold ρ and a time window t_{max}, an IPD query computes a list of predicted mutually independent dense road segments $\{DS_1, DS_2, DS_3, ..., DS_n\}$, where the occurrence times $t_1 \leq t_2 \leq t_3 \leq \cdots \leq t_n(t_n \leq t_{max})$; t_i is the occurrence time of DS_i.*

It is worth noting that dense road segments caused by the moving objects that have already been accounted for in antecedent dense road segment are excluded from further consideration in IPD queries. In other words, only the earliest occurring dense road segment of a chain of dense segments is considered each time.

4   Influence-Aware Predictive Density Query Algorithm

In this section, we first introduce the data structure that is utilized to support the influence-aware predictive density (IPD) queries and then elaborate the query algorithm.

4.1   Data Structure

To answer IPD queries, we need indexing structures to manage the information of objects moving on road networks. There have been several such kind of indexes such as IMORS [14], ANR-tree [6], R-TPR$^{\pm}$-tree [8], and TPR^{uv} [7]. Here, we employ the most recent one, the R^D-tree [12], which is also our prior work. Although our main contribution of this work lies in the query algorithm (in Sect. 4.2) and not the data structure, we briefly introduce the data structure here to facilitate a better understanding of the whole algorithm.

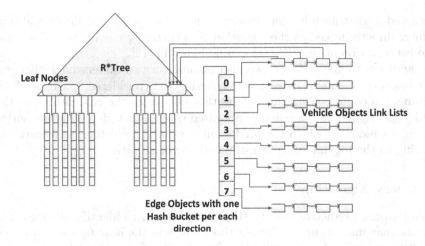

Fig. 2. The R^D-tree

The R^D-tree indexes two types of data: road-network information and object location information. The road network is represented as a graph $\mathcal{G} = (\mathcal{V}, \mathcal{E})$, where $\mathcal{V}$ is the set of vertices and $\mathcal{E}$ is the set of edges. Each edge $e = \{v_1, v_2\} \in \mathcal{E}$ represents a road segment in the network where $v_1, v_2 \in V$; v_1 and v_2 are starting and ending nodes of the road segment, respectively. Furthermore, each road segment is associated with two parameters: l and s, where l is the length of the edge and s is the maximum possible speed on that edge.

A moving object O is represented by the tuple $\{o_{id}, x_t, y_t, o_e^t, o_e^{t+1}, o_v^t, o_{gd}, t\}$, where o_{id} is the unique object ID, x_t and y_t are the coordinates of the moving object at the latest update timestamp t, o_e^t is the current road segment that the object is on, o_e^{t+1} is the next road segment that the object is headed to, o_v^t is the object's velocity (or speed), and o_{gd} is the object's travel destination. It is assumed that most moving objects are willing to disclose their tentative traveling destinations to the service provider (server) in order to obtain high-quality services, albeit their destinations may change during the trip.

Figure 2 illustrates the overall structure of the R^D-tree. The R^D-tree is composed of an R*-tree [3] and a set of hash tables. The road-network information is indexed by the R*-tree. Each entry in the non-leaf node is in the form of $(node\_MBR, child\_ptr)$, where $node\_MBR$ is the MBR (Minimum Bounding Rectangle) covering the MBRs of all entries in its children pointed to by the $child\_ptr$. Leaf nodes in R*-tree pointing to hash tables represent moving objects on each road segment. Each entry in the leaf node is in the form of $(edge\_MBR, obj\_ptr)$, where $edge\_MBR$ is the MBR of a road segment and obj_ptr links to a hash table storing objects moving on this edge. Each hash table has an N_d hash bucket, where N_d is the number of traveling directions. Each bucket has two linked lists that provide a finer grouping for objects based on their traveling directions. Moving objects with similar traveling directions are hashed to the same hash bucket and stored in one of the sorted linked lists maintained in that hash bucket. Moreover, for easy update, each object also has a pointer directly linked to the edge that it is currently moving on. The details of the construction of the R^D-tree can be found in [12].

In addition to the R^D-tree, we also maintain a two-dimensional histogram that comprises of square-shaped cells covering the considered space. Each cell maintains the counts of moving objects that may cross the cell within the time period $[t_{now}; t_{now} + H]$ for equally calibrated timestamps. Here H is the horizon – the time window in which the prediction is valid. The histogram is initialized according to the moving object's estimated traveling path.

4.2 Query Algorithm

Influence-aware predictive density (IPD) queries aim to identify all dense road segments that may occur at different timestamps in the near future and also do not influence each other as defined in Sect. 3. The query algorithm consists of three phases: filtering, refinement, and refreshing.

The Filtering Phase. The filtering phase utilizes the histogram to quickly identify potential grid cells that may contain dense road segments. Recall that the histogram stores the estimated number of objects in the corresponding cell at each timestamp starting from the current timestamp. When considering whether a cell may contain a dense road segment, we do not simply use the original count of objects in the cell, as in previous works. Instead, we consider the *adjusted cell density* (Definition 5) in order to take into account the road topology. Specifically, the adjusted cell density estimates the average number of objects per unit length of road segments. For example, if a cell contains very few roads but a large number of objects, it is very likely that the cell contains a dense road segment. For time efficiency, the adjusted cell density can be generated along with the computation of the count of objects when building the histogram. For storage efficiency, the histogram can be compressed similarly as that in [13].

Definition 5. *[Adjusted Cell Density] Let N_c^t be the number of objects in cell c at timestamp t, and l_c^t be the total length of road segments in cell c. The adjusted cell density ACD_c^t is computed as follows:*

$$ACD_c^t = \frac{N_c^t}{l_c^t}$$

The filtering phase starts checking the adjusted cell density of each cell at the earliest timestamp stored in the histogram. If a cell's density is above a threshold ρ_c, the cell will be inserted to a priority queue to be sent to the refinement phase. The challenging issue here is to determine the proper value of the threshold ρ_c. If we simply set ρ_c to the same value as the density threshold ρ given by the query issuer, we may miss the dense road segment that spans multiple non-dense cells and have many false negatives. If we set ρ_c to a very low value, we will not miss the dense road segment but the filtering phase will lose the pruning power and keep too many non-dense cells for further examination, which in turn will increase the overall query cost. Therefore, we model this effect using the following linear regression function. The goal is to identify the best value of ρ_c that balances both the query cost and the number of false negatives. Specifically, in Eq. 1, $Cost_q$ and FN are the estimated query cost and false negatives for a given ρ_c, α is a weight value, and $Cost_{max}$ is the query cost to retrieve all objects in the entire space. Here, $Cost_q$ is the estimated cost of the second refinement phase which is determined by the number of queries on cells in the queue. The lower the ρ_c, the fewer the cells to be further examined and lower $Cost_q$. $Cost_{max}$ is used to normalize the value of the first part of the equation to the range of 0 and 1, so that it is comparable to the false negatives. FN is estimated using the number of road segments spanning cells. The goal is to minimize the "Penalty" to determine ρ_c.

$$Penalty(\rho_c) = \alpha \cdot \frac{Cost_q}{Cost_{max}} + (1 - \alpha) \cdot FN \qquad (1)$$

After all cells at the same timestamp are considered, we move to the next timestamp and the same filtering process continues until all timestamps in the histogram have been examined. The final priority queue will have a list of cells ordered in an ascending order of the timestamps. Cells at the same timestamp are ordered by descending density.

More importantly, each cell c in the queue also maintains a list of influenced cells whose priority is lower than its own, along with the number of vehicles coming from the cell c to the influenced cell. This list will later be used to quickly prune non-dense cells as discussed in the refreshing phase.

The Refinement Phase. The refinement phase takes a further look at the candidate dense cells obtained from the filtering phase to see whether these cells actually contain dense road segments. The refinement phase starts from the highest prioritized candidate cell from the priority queue and moves to the next highest prioritized cell of the same timestamp, and so on. After all candidate cells at the same timestamp are examined, the *refreshing phase* (Sect. 4.2) will be activated in which the entries in the queue and their priority will be updated. Then the highest prioritized cell from the updated queue is selected and sent back to the refinement phase. The iteration between the refinement and refreshing continues until the priority queue is empty. In what follows, we elaborate the three main steps taken during the refinement.

The first step is to find the road segments containing the objects that may pass by the candidate cell at the timestamp t_d (as stored in the priority queue) when this cell may be dense. Note that these road segments may not be the road segments located in the candidate cell. This is because we are predicting future dense road segments, and objects on road segments outside the candidate cell at current timestamp may enter this cell at a future timestamp. In order to identify these related road segments, we perform a square-shaped ring query on the R^D-tree as shown in Fig. 3, where the square shaped ring is represented by the shaded area between solid-line squares. The dimension of the square-shaped ring is determined according to the road network information. Specifically, the lengths *innerL* and *outerL* are the distances to the closest and the farthest objects that may be able to enter the candidate cell according to the road speed limits.

$$innerL = \frac{cell\ width}{2} + Dist(speed_{min} \cdot (t_d - t_c)) \tag{2}$$

$$outerL = \frac{cell\ width}{2} + Dist(speed_{max} \cdot (t_d - t_c)). \tag{3}$$

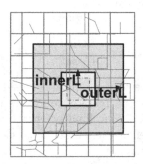

Fig. 3. Squared shaped ring query

The second step is to retrieve objects in the relevant hash bucket of each road segment found by the previous step. Recall that each road segment is associate with multiple buckets containing objects traveling towards different destinations. Intuitively, if an object is not heading toward the candidate cell, we do not need to retrieve it. To determine which bucket needs to be checked, we consider the relevancy of the object's traveling angle and the candidate cell as shown in Fig. 4. The Fig. 4 illustrates two cases where the number of hash buckets is 8 and different distances to the candidate cell (due to the difference in times when the density is computed for: t_a and $t_b > t_a$). As shown in the figure, the number of hash buckets selected to examine candidate cell A's density is 3 (hash bucket 1, 2, and 3) where that for candidate cell B is only two (bucket 0 and 1). The use of bucket selection helps improve the overall query performance by pruning irrelevant objects for further consideration.

Since objects returned from the second step may still contain objects that do not contribute to any dense road segment, the final refinement step is to compute the exact traveling routes of these candidate objects and then identify the truly dense road segments. Specifically, each road segment in the candidate cell is associated with a counter. For each candidate object's path, we increase the counter of the road segment passed by the object by one. This ensures that each

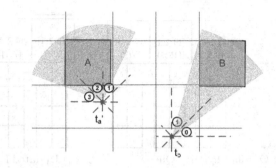

Fig. 4. Two examples of modified hash bucket selection

path is examined only once. After analyzing all the candidate objects' routes, the road segments with count above the density threshold will be reported.

The Refreshing Phase. The refreshing phase aims to compute quarantine areas of identified dense road segments. Since objects occurring on one dense road segment are impossible to occur on another at the same time, it is not necessary to run refreshing phase after each identified road segment. Instead, the refreshing runs after all the dense road segments have been identified for each timestamp considered to improve efficiency. It consists two main steps: (i) compute quarantine areas; (ii) rejuvenate the priority queue.

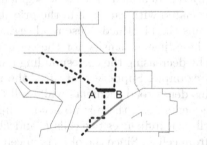

Fig. 5. Influenced road segments in the quarantine area of road $\overline{AB}$

A *quarantine area* is defined for each identified dense road segment within the same timestamp. The area contains the dense road segments and the segments that the congestion would propagated to. Figure 5 shows an example of the quarantine area regarding a dense road segment $\overline{AB}$ at time t_a. The dashed-lined road segments are the road segments that will be affected by $\overline{AB}$. More specifically, the road segments in the quarantine area contain objects stuck in $\overline{AB}$ at time t_a for the computation of their density at a near future timestamp t_b when the traffic may not be cleared. Therefore, the computation of density of road segments in the quarantine area should ignore the objects stuck in $\overline{AB}$.

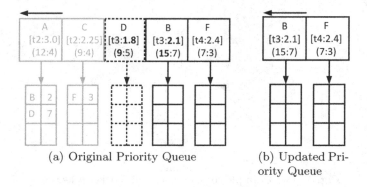

(a) Original Priority Queue   (b) Updated Priority Queue

Fig. 6. Updated queue

The objects on the road segments in the quarantine area are disregard from subsequent dense area identification. The formal definition of quarantine area is given by Definition 6, where the value of n is determined based on the typical time that a traffic congestion can be cleared.

Definition 6. *[Quarantine Area] Given a road network $\mathcal{G}(\mathcal{V}, \mathcal{E})$ and a set of dense road segments $\mathcal{S}$; where $\mathcal{S} \subset \mathcal{E}$. The quarantine area of $\mathcal{S}$ is a set of road segments, $\mathcal{Q} = \bigcup_{i \in |\mathcal{S}|} (\mathcal{S}_i \cup \mathcal{S}_i^n)$. Here $\mathcal{S}_i^n$ is the n -hop adjacent edges of ith edge in $\mathcal{S}$.*

The second step of the refreshing is to rejuvenate the priority queue by discarding cells influenced by the identified dense road segment. This step leverages the influence cell list associated with each cell in the priority queue. Specifically, for each cell c that contains the identified dense road segment at the timestamp considered, we update the adjusted density for the cells that overlap with the quarantine area of cell c by decreasing the corresponding number of objects stuck in c. Figure 6 shows an example. Suppose that after the refinement phase, we know that cell A contains dense road segments while cell C does not. Since cell C has no dense road segments in it, there is no need to update its entry in the priority queue. As for cell A, it influences two cells B and D. Cell B contains two objects that will travel from cell A. Since the objects in cell A are stopped due to the high density of cell A, the total number of expected objects in cell B would be decreased, and the new adjusted density of cell B becomes $15/7 = 2.1$. Similarly, cell D's new density is 1.8. Assuming that the cell D's density is now below the density threshold ρ_c, it would be removed from the priority queue for subsequent computations. In this way, the refreshing phase helps avoid unnecessary computations.

5   Performance Study

In this section, we evaluate the performance of our proposed influence-aware predictive density (IPD) query algorithm by varying a number of parameters on different datasets. Since the IPD query is the only approach that predicts dense road segments under road network constraints, we compare it with a baseline approach which simply examines each object's shortest path at different timestamps to directly compute the dense road segments. The baseline approach also implements the concept of quarantine area by discarding the objects on identified dense road segments from subsequent computation. Both algorithms were implemented and tested on a 2.40 GHz Intel®Xeon®E5620 CPU desktop with 11 Gigabytes of memory. The page size is 4 K bytes. The R^D-tree implementation of R^D-tree in our approach is the same as that in [12][1]. The internal nodes of a tree are pinned in a LRU buffer of 50 pages.

Table 1. Statistics of the road maps

State	Land area	Number of road segments	Average road segment length
IA	55,857	3392	356
AZ	66,455	4935	383
WA	113,594	1442	628
CA	155,779	8062	225

The datasets used for testing are generated by the commonly adopted Brinkhoff generator [5]. The generator was fed with four different US state maps: IA, WA, AZ, and CA. The states differ in total land area, number of road segments, and average road segment length, which results in different mobile object distributions. The statistics of the chosen states are given in Table 1. The number of moving objects in each dataset ranges from 10 K to 100 K. Average traveling time of each data set is 60 min. The chosen input parameters and their values are presented in Table 2 with the default value in bold. The efficiency and effectiveness are measured in terms of the average I/O cost (i.e., the number of page accesses) and the number of identified dense road segments from current time to the query life (i.e., the predictive time window), respectively.

5.1   Effect of Number of Moving Objects

In the first round of experiments, we evaluate the query performance by varying the number of moving objects from 10 K to 100 K while keeping other parameters as default (in Table 2). As shown in Fig. 7(a), it is expected that the query cost increases with the number of moving objects since more data need to be retrieved from disk and examined. Our IPD query algorithm significantly outperforms the baseline approach and the performance gap increases with the number of moving objects. This can be attributed to our proposed filtering algorithm and bucket selection algorithm that help reduce the number of objects to be examined and

---

[1] R^D-tree adopts its R*-tree simulator from [1].

Table 2. Parameters and their values

Parameters	Values
Number of mobile objects	10 K, 20 K, . . . , **50 K**, . . . , 100 K
Road network topology	IA, AZ, **WA**, CA
Predictive time window (minutes)	10, 20, **30**, 50
Cell density threshold (ρ_c)	**0.05**, 0.1, 0.15, . . . , 1
Road density threshold (ρ)	0.2, 0.4, 0.6, 0.8, **1**
Grid size (d)	10, 20, **30**, 40, 50
Vehicles equipped with the system	25 %, 50 %, 75 %, **100 %**
Timestamp interval (minutes)	**5**

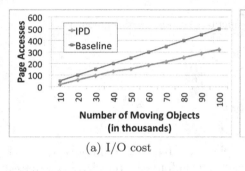
(a) I/O cost

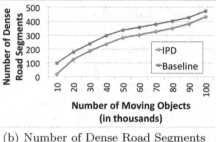
(b) Number of Dense Road Segments

Fig. 7. Effect of the number of moving objects

the refreshing phase that further prunes the candidate cells which have been influenced by the identified road segments.

Figure 7(b) compares the number of dense road segments found by the two approaches. From the figure, we can observe that our IPD query reports fewer number of dense road segments than the baseline approach. This is expected as discussed in Sect. 4.2. Specifically, we adopt a cell density threshold in the filtering phase. Higher ρ_c will prune more cells and yield better query performance but may introduce false negatives. Fortunately, we also observe that the percentage of such difference becomes smaller for bigger datasets. In a small dataset, such effect is more severe since missing one object may make a road segment to be non-dense easily. In the next experiment, we will take a closer look at how the cell density threshold affect the performance.

5.2   Effect of Cell Density Threshold

We now study the effect of cell density threshold ρ_c. Recall that ρ_c is used to prune those cells that are highly unlikely to contain dense road segments. As aforementioned, higher ρ_c may yield fewer number of candidate cells to be further examined and hence reduce query cost. However, it is possible that when

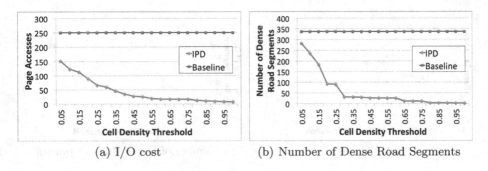

(a) I/O cost (b) Number of Dense Road Segments

Fig. 8. Effect of the cell density threshold

ρ_c is high, the cells with low density and contain part of a dense road segment (i.e., a dense road segment that spans multiple cells) may be left out, resulting in false negatives. Figure 8 reports the experimental results when varying the cell density threshold from 0.05 to 1. As expected, the query cost of our approach decreases quickly when the cell density threshold increases. Meanwhile the number of missing dense road segments increases. In order to minimize false negatives, we adopt the cell density threshold as 0.05 in our experiments.

5.3 Effect of Road Density Threshold

In this set of experiments, we study the effect of the road density threshold given by the query. The results are shown in Fig. 9. We can observe that our IPD algorithm again outperforms the baseline approach in terms of query efficiency and identified similar number of dense road segments in all cases. Moreover, we also observe that the query cost of our IPD algorithm increases with the road density threshold while the baseline approach has constant performance. This is because the baseline approach always checks all objects' travel paths when computing the density of road segments. In our approach, the higher the road density threshold, the fewer the number of dense road segments. Correspondingly, there will be fewer number of quarantine areas and fewer number of objects that can be pruned.

5.4 Effect of Cell Size

We proceed to study the effect of the cell size. As shown in Fig. 10, the cell size does not affect the baseline approach since the baseline approach does not utilize cells for pruning. As for our IPD algorithm, it achieves better performance when the cell size is small. This is probably because the adjusted density of smaller cells is more useful for determining whether there is a potential dense road segment. When the area of a cell is large, the road topology in the single cell becomes more complex and the pruning becomes less effective. On the other hand, when the cell is large, there is fewer chances to have dense road segments across multiple

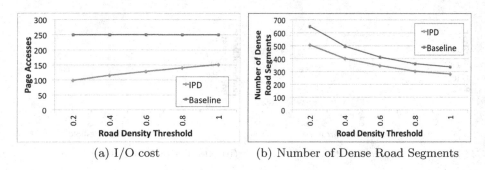

(a) I/O cost   (b) Number of Dense Road Segments

Fig. 9. Effect of the road density threshold

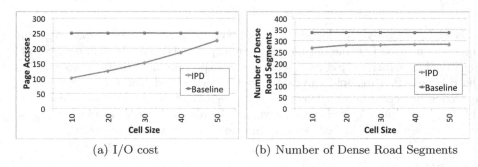

(a) I/O cost   (b) Number of Dense Road Segments

Fig. 10. Effect of the cell size

cells and hence it helps reduce false negatives. In our experiments, we choose the cell size to be 30 as it balances both performance and accuracy.

5.5   Effect of Predictive Time Window

The effect of predictive time window was also studied. The results are shown in Fig. 11. It also shows the efficiency of the IPD algorithm. In fact, the IPD has

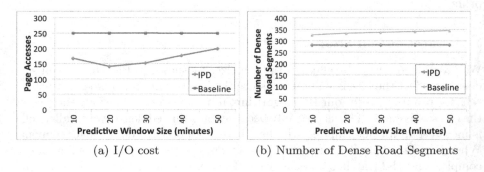

(a) I/O cost   (b) Number of Dense Road Segments

Fig. 11. Effect of the predictive time window

lower page accesses compared to that of the baseline algorithm. The IPD shows an increase in the page accesses for longer predictive window lengths. This can be expected as the area covered by the square-ring is also increased. The IPD exhibits fewer unidentified road segments. In fact, the number of unidentified dense road segments increases when the predictive time window is lengthier. It is again due to relaxation of the cell threshold.

5.6   Effect of Road Network Topology

This round of experiments evaluate the effect of the road topology on the query performance. As shown in Fig. 12, our approach always achieves better query efficiency than the baseline approach when different road maps are considered. In addition, the results also indicate that the map topology does affect the performance. In general, maps with fewer roads across multiple cells tend to yield better performance. Moreover, we also observe that the numbers of dense road segments identified by the two approaches demonstrate the same trend, which proves the effectiveness of our approach.

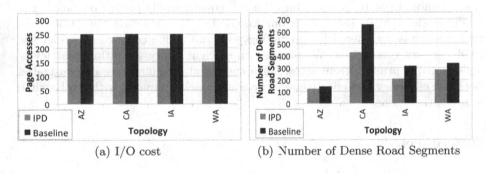

(a) I/O cost   (b) Number of Dense Road Segments

Fig. 12. Effect of the road network topology

5.7   Effect of Percentage of Vehicles Equipped with the System

In the last set of experiments, we aim to examine an interesting and realistic scenario when not all vehicles subscribe to traffic prediction services. That means, the system will estimate the traffic based on a subset of vehicles. It is expected the fewer the number of vehicles equipped with the system, the less accurate the traffic prediction will be. To compensate for the missed information, we adjust the system parameters by lowering both the cell density threshold and road segment density threshold. For example, given a density threshold $\rho = 1$, we set $\rho_c = 0.0375$ and $\rho' = 0.75$. Our findings are reported in Fig. 13, where "Adjusted IPD" refers to the approach with adjusted new threshold. We can observe that the adjusted IPD has similar query cost as the original one but much better accuracy in terms of the number of dense road segments being identified.

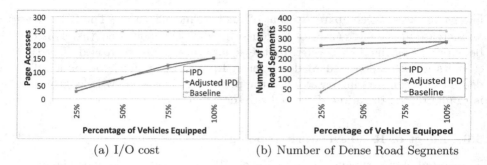

(a) I/O cost   (b) Number of Dense Road Segments

Fig. 13. Effect of the percentage of vehicles equipped with the system

6   Conclusion

In this paper, we define a new type of density query, namely Influence-aware Predictive Density (IPD) queries, with the goal to take into account the impact of a traffic congestion on the traffic flow, i.e., objects stuck in the traffic congestion should not be counted into the subsequent traffic prediction before the traffic is cleared. To the best of our knowledge, it is the first time that traffic influence is considered for predicting potential traffic congestions under the road network constraints. We propose an efficient query algorithm that leverages multiple pruning techniques. Our experimental results have demonstrated both the efficiency and effectiveness of our approach compared with the baseline approach.

References

1. Achtert, E., Kriegel, H.-P., Schubert, E., Zimek, A.: Interactive data mining with 3d-parallel-coordinate-trees. In: Proceedings of the ACM SIGMOD International Conference on Management of Data (2013)
2. Barth, M., Boriboonsomsin, K.: Real-world carbon dioxide impacts of traffic congestion. Transport. Res. Rec. J. Transport. Res. Board **2058**, 163–171 (2008)
3. Beckmann, N., Kriegel, H.-P., Schneider, R., Seeger, B.: The R*-tree: an efficient and robust access method for points and rectangles (1990)
4. Bok, K.S., Yoon, H.W., Seo, D.M., Kim, M.H., Yoo, J.S.: Indexing of continuously moving objects on road networks. IEICE Trans. Inf. Syst. **E91–D**, 2061–2061 (2008)
5. Brinkhoff, T.: A framework for generating network-based moving objects. GeoInformatica **6**, 153–180 (2004)
6. Chen, J.-D., Meng, X.-F.: Indexing future trajectories of moving objects in a constrained network. J. Comput. Sci. Technol. **22**(2), 245–251 (2007)
7. Fan, P., Li, G., Yuan, L., Li, Y.: Vague continuous K-nearest neighbor queries over moving objects with uncertain velocity in road networks. Inf. Syst. **37**(1), 13–32 (2012)

8. Feng, J., Lu, J., Zhu, Y., Mukai, N., Watanabe, T.: Indexing of moving objects on road network using composite structure. In: Apolloni, B., Howlett, R.J., Jain, L. (eds.) KES 2007, Part II. LNCS (LNAI), vol. 4693, pp. 1097–1104. Springer, Heidelberg (2007)

9. Feng, J., Lu, J., Zhu, Y., Watanabe, T.: Index method for tracking network-constrained moving objects. In: Lovrek, I., Howlett, R.J., Jain, L.C. (eds.) KES 2008, Part II. LNCS (LNAI), vol. 5178, pp. 551–558. Springer, Heidelberg (2008)

10. Hao, X., Meng, X., Xu, J.: Continuous density queries for moving objects. In: Proceedings of the Seventh ACM International Workshop on Data Engineering for Wireless and Mobile Access, MobiDE 2008 (2008)

11. Heendaliya, L., Lin, D., Hurson, A.: Continuous predictive line queries for on-the-go traffic estimation. In: Hameurlain, A., Küng, J., Wagner, R., Decker, H., Lhotska, L., Link, S. (eds.) TLDKS XVIII. LNCS, vol. 8980, pp. 80–114. Springer, Heidelberg (2015)

12. Heendaliya, L., Lin, D., Hurson, A.R.: Predictive line queries for traffic forecasting. In: Database and Expert Systems Applications (2012)

13. Jensen, C.S., Lin, D., Beng, C.O., Zhang, R.: Effective density queries on continuously moving objects. In: Proceedings of the 22nd International Conference on Data Engineering (2006)

14. Kyoung-Sook, K., Si-Wan, K., Tae-Wan, K., Ki-Joune, L.: Fast indexing and updating method for moving objects on road networks. In: Proceedings of the 4th International Conference on Web Information Systems Engineering Workshops (2003)

15. Lai, C., Wang, L., Chen, J., Meng, X., Zeitouni, K.: Effective Density queries for moving objects in road networks. In: Dong, G., Lin, X., Wang, W., Yang, Y., Yu, J.X. (eds.) APWeb/WAIM 2007. LNCS, vol. 4505, pp. 200–211. Springer, Heidelberg (2007)

16. Gunopulos, D., Hadjieleftheriou, M., Kollios, G., Tsotras, V.J.: On-line discovery of dense areas in spatio-temporal databases. In: International Symposium on Advances in Spatial and Temporal Databases, SSTDn (2003)

17. Min, W., Wynter, L.: Real-time road traffic prediction with spatio-temporal correlations. Transp. Res. Part C $19(4)$, 606–616 (2011)

18. Morgan, L.: The effects of traffic congestion (2014)

19. Ni, J., Ravishankar, C.V.: Pointwise-dense region queries in spatio-temporal databases. In: IEEE 23rd International Conference on Data Engineering (2007)

20. Quek, C., Pasquier, M., Lim, B.B.S.: Pop-traffic: a novel fuzzy neural approach to road traffic analysis and prediction. IEEE Trans. Intell. Transp. Syst. $7(2)$, 133–146 (2006)

21. Smith, B.L., Williams, B.M., Oswald, R.K.: Comparison of parametric and non-parametric models for traffic flow forecasting. Transp. Res. Part C $10(4)$, 303–321 (2002)

22. Wen, J., Meng, X., Hao, X., Xu, J.: An efficient approach for continuous density queries. Front. Comput. Sci. $6(5)$, 581–595 (2012)

23. Yiu, M.L., Tao, Y., Mamoulis, N.: The Bdual-tree: indexing moving objects by space filling curves in the dual space. VLDB J. $17(3)$, 379–400 (2008)

Uncertain Voronoi Cell Computation Based on Space Decomposition

Tobias Emrich[1], Klaus Arthur Schmid[1]([✉]), Andreas Züfle[1],
Matthias Renz[1], and Reynold Cheng[2]

[1] Institute for Informatics, Ludwig-Maximilians-Universität München,
München, Germany
{emrich,schmid,zuefle,renz}@dbs.ifi.lmu.de
[2] Department of Computer Science, University of Hong Kong, Hong Kong, China
ckcheng@cs.hku.hk

Abstract. The problem of computing Voronoi cells for spatial objects whose locations are not certain has been recently studied. In this work, we propose a new approach to compute Voronoi cells for the case of objects having rectangular uncertainty regions. Since exact computation of Voronoi cells is hard, we propose an approximate solution. The main idea of this solution is to apply hierarchical access methods for both data and object space. Our space index is used to efficiently find spatial regions which must (not) be inside a Voronoi cell. Our object index is used to efficiently identify Delauny relations, i.e., data objects which affect the shape of a Voronoi cell. We develop three algorithms to explore index structures and show that the approach that descends both index structures in parallel yields fast query processing times. Our experiments show that we are able to approximate uncertain Voronoi cells much more effectively than the state-of-the-art, and at the same time, improve runtime performance.

1 Introduction

The extensive use of social media, s.a. smartphones, and social networks produce a huge flood of geo-spatial and geo-spatio-temporal data. This data allows to assess information about the current positions of mobile entities, such as friends in social networks, unoccupied cabs in a taxi application, or the current position of users in augmented reality games. However, our ability to unearth valuable knowledge from large sets of spatial and spatio-temporal data is often impaired by the quality of the data.

- Data may be imprecise, due to measurement errors, for instance in applications using sensor measurements such as location-based services.
- Data records can be obsolete. For example, ties of friendship bind and break over time, without necessarily reflecting such changes in a social network; in location-based services, users may update their location infrequently, due to bad connectivity or to preserve battery.

© Springer International Publishing Switzerland 2015
C. Claramunt et al. (Eds.): SSTD 2015, LNCS 9239, pp. 98–116, 2015.
DOI: 10.1007/978-3-319-22363-6_6

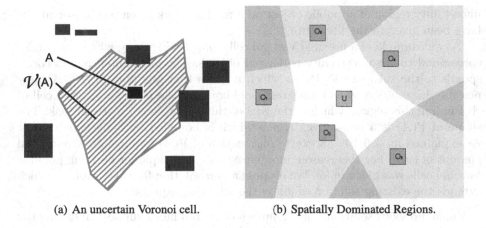

(a) An uncertain Voronoi cell.   (b) Spatially Dominated Regions.

Fig. 1. Uncertain Voronoi cells.

- Data can be obtained from unreliable sources, such as crowd-sourcing applications, where data is obtained from individual users, which may incur inaccurate or plain wrong data, deliberately or due to human error.
- To prevent privacy threats and to protect user anonymity, users often consent
  to relay just a cloaked indication of their whereabouts [1] abstracted as an
  uncertainty region enclosing (but apparently not centered at) their current
  position.

Simply ignoring these notions of imprecise, obsolete, unreliable and cloaked data,
thus pretending that the data is accurate, current, reliable and correct is a common source of false decision making. The research challenge in handling uncertainty in spatial and spatio-temporal data is to obtain reliable results despite the
presence of uncertainty. In this work, we revisit the problem of reliably answering
nearest-neighbor queries in uncertain data. The problem of finding the closest
uncertain object, which has applications such as taxi-customer matching, has
gained much attention in recent years [2–5]. Following a common approach in
uncertain data management, these approaches assume that uncertain objects are
represented by rectangular or circular uncertainty regions, which are guaranteed
to enclose the true (but unknown) position of the respective spatial objects. Following the approach of [6], we carry the concept of Voronoi cells to uncertain
data. The idea of [6] is to approximate the *possible Voronoi cell* $\mathcal{V}(O)$ of an
object O, which is defined as the space where a query point q can possibly have
O as its nearest neighbor. Applications for possible Voronoi cells include geo-
location-based services, such as taxi assignments: The possible Voronoi cell of
an individual taxi cab c covers the space of a city where customers may possibly
have c as their nearest taxi. In such an application, as we see in taxi-GPS data
sets such as the T-drive dataset [7,8], the GPS position $c(t)$ of a cab c at a time
t may be highly obsolete, due to infrequent GPS updates. Models to infer the

uncertainty region of a mobile object on a road network given past observations have been given in the literature [9].

As an example of a possible Voronoi cell, consider Fig. 1(a), where rectangles correspond to the uncertainty regions of objects. The highlighted region corresponds to the subspace $\mathcal{V}(A)$, for which it holds that any point $q \in \mathcal{V}(A)$ may possibly have object A as its nearest neighbor, i.e., the possible Voronoi cell of A. Finding this region, which is the goal of this paper, is not a trivial task: The shape of $\mathcal{V}(A)$ is a non-convex region which is bounded by hyperbolic curves. As explained in [3,6,10], an exact construction of $\mathcal{V}(A)$ requires an exponential amount of time. For this reason, an approximate technique for deriving possible Voronoi cells was given in [6]. We propose a new solution for this problem, which extends the existing solution of [6] by the following aspects:

- Unlike previous solutions, our approach offers full index support, indexing the object space using an R^*-tree [11] and indexing the data space using a kd-trie [12].
- Rather than approximating the Voronoi cell $\mathcal{V}(o)$ by a single rectangle ([6]), we use a set of kd-trie partitions, which allows much higher approximation quality. This gain in approximation quality not only improves query times, as our experiments show, but can also be used to gain a detailed visual exploration of possible Voronoi cells.
- Our experiments further show that our provided index support for both data and space enables the scaling of uncertain Voronoi cell computation to large databases.

2   Related Work

The problem of answering nearest neighbor queries on uncertain data generally involves two steps: A filter approach and a refinement step. In the filter step, a (possibly small) set of objects is returned that contains all objects having a non-zero probability of being the result object. In the refinement step, the exact probability of each candidate object is computed. The refinement step is the main research topic of [13–15], showing how to compute exact probabilities of an object to be the nearest neighbor of a query object, given the probability density functions of objects. In contrast, other existing work focuses on the filter step, applying spatial filter steps in order to identify object that are guaranteed to have a zero probability to be the result object [3,5,6]. In this work, we focus on the filter step, i.e., the step of finding objects having a non-zero probability to be the nearest neighbor of an object using Voronoi-cells.

The idea of using Voronoi diagrams to answer nearest neighbor (NN) queries over points has been widely studied [16] . In this context, Voronoi diagrams have been used to support nearest neighbor queries in geo-spatial applications [17], location-based services [18,19], in spatial data streams [20] and in distributed spatial environments [21] as well as in spatial network environments [22]. To support nearest neighbor queries on uncertain data, initial approaches have been presented in [2,13]. However, in these work, only the database objects are assumed to be uncertain, whereas the query object is assumed to be a point. In [3] a solution to compute possible Voronoi-cells for the case of circular uncertainty regions has been presented. This exact approach has exponential construction and storage cost. Due to this computational drawback, an approximate solution was presented in [6]. The aim of this approach is to approximate the true (but unknown) possible Voronoi-cell $\mathcal{V}(O)$ of an uncertain object O using two rectangle: A single conservative rectangle $h(O)$ which is guaranteed to completely contain $\mathcal{V}(O)$, and a single progressive rectangle $l(O)$ which is guaranteed to be completely contained by $\mathcal{V}(O)$. These two approximation rectangles are obtained by iteratively expanding the progressive rectangle $l(O)$, and iteratively shrinking the conservative rectangle $h(O)$. However, considering examples such as shown in Fig. 1, it is evident that such approximations may be rather inaccurate. Thus, $h(O)$ may cover a large body of space not belonging to $\mathcal{V}(O)$, while $l(O)$ may miss a large body of $\mathcal{V}(O)$, even in the case where $h(O)$ is the smallest conservative bounding rectangle and $l(O)$ is the largest progressive bounded rectangle.[1] Furthermore, an approach for nearest neighbor search on moving uncertain objects has been presented in [4]. A problem common to [3] and [4] is that their solutions are customized for 2D data, making extensive use of intersection and rotation operations of 2D hyperbolic curves. Our approach, as well as the approach of [6] is applicable to arbitrary dimensionality. In comparison to [6], the main contribution of this work is that we can accurately approximate an arbitrarily shaped possible Voronoi-cell, rather than using a single rectangular approximation only. This allows to answer nearest-neighbor queries more efficiently, since less candidates have to be checked, and it allows to more precisely illustrate the Voronoi-region of an uncertain object.

3   Problem Definition

Figure 1(b) shows how the possible Voronoi cell $\mathcal{V}(U)$ of an uncertain object U is defined. Each shaded region in Fig. 1(b) corresponds to a pruning region $S_A(U)$, i.e., the smallest region such that for any $q \in S_A(U)$, object A must be closer to q than U. Formally,

Definition 1 (Nearest Neighbor Pruning Region). *Let $\mathcal{D} = \{O_1, ..., O_N\}$ be an uncertain database where each object $O_i \in \mathcal{D}$ is represented by a rectangular uncertainty region in $\mathcal{R}^d$. Let $dist(.,.)$ denote any L_p norm.[2] For any $A, B \in \mathcal{D}$, we define the nearest neighbor pruning region where any point must be closer to A than to B as follows:*

$$S_A(B) := \{q \in \mathcal{R}^d : maxDist(q, A) < minDist(q, B)\},$$

where $maxDist(q, A)$ and $minDist(q, B)$ denote the maximum and minimum distance between a point q and a rectangle A or B, respectively, as defined in [23].

---

[1] The later case can not be guaranteed by the approach of [6] due to the numeric nature of their approach.

[2] We use Euclidean distance in all examples and illustrations, but any L_p norm can be applied.

Table 1. Table of notations.

Notation	Meaning	Notation	Meaning
$\mathcal{D}$	The database	$S = \mathcal{R}^d$	d-dimensional data space
$U \in \mathcal{D}$	an uncertain object	$\mathcal{V}(U)$	possible Voronoi cell of U
$\mathcal{I_D}$	Hierarchical data index	$\mathcal{I_S}$	Hierarchical space index
$\mathcal{G}$	d-dimensional grid	$g_i \in \mathcal{G}$	Rectangular grid cell
$S_A(B) \subseteq \mathcal{R}^d$	The region where object A dominates object B		
$Dom(A, B, R)$	Predicate that is true iff rectangle R is fully contained $S_A(B)$. Can be evaluated efficiently [24].		
$PDom(A, B, R)$	Predicate that is true iff rectangle R intersects $S_A(B)$. Can be evaluated efficiently [24].		
$h \subseteq \mathcal{R}^d$	Rectangular Space Index Entry obtained from $\mathcal{I_S}$: Partition of Space for which we want to decide if it belongs to $\mathcal{V}(U)$		
$e \subseteq \mathcal{R}^d$	Rectangular Data Index Entry obtained from $\mathcal{I_D}$: Spatial approximation of a set of data objects if e is non-leaf entry, Uncertainty region of a single data object if e is a leaf entry.		

Fig. 1(b) shows five nearest neighbor pruning regions $S_{O_1}(U), ..., S_{O_5}(U)$ as shaded regions. Using Definition 1, we can now define the possible Voronoi cell $\mathcal{V}(U)$ of an object U as the space that does not intersect any nearest neighbor pruning region associated with U, formally:

Definition 2 (Possible Voronoi Cell). *Let $U \in \mathcal{D}$ be an uncertain object. Then the possible Voronoi cell $\mathcal{V}(U)$ is defined as*

$$\mathcal{V}(U) = \mathcal{R}^d \setminus \bigcup_{O \in \mathcal{D} \setminus \{U\}} S_O(U).$$

In Fig. 1(b), the white (i.e., non-shaded) region corresponds to the Voronoi cell $\mathcal{V}(U)$. The problem tackled in this paper is to compute $\mathcal{V}(U)$ for a given object $U \in \mathcal{D}$ efficiently.

4   Spatial Domination Revisited

The concept of spatial domination and efficient techniques to verify it were introduced in [24]. Spatial domination describes the spatial relation of three rectangles to each other. Since the spatial domination can also be utilized for the computation of uncertain voronoi cells, we briefly want to review the concept. Notations used throughout this paper are explained in Table 1.

Definition 3 (Spatial Domination). *Let $A, B, R \subseteq \mathcal{R}^d$ be rectangles in a d-dimensional space and dist() be a distance function defined on that space.*

The rectangle A dominates B w.r.t. R iff for all points $r \in R$ it holds that every point $a \in A$ is closer to r than any point $b \in B$, i.e.

$$Dom(A, B, R) \Leftrightarrow \forall r \in R, \forall a \in A, \forall b \in B : dist(a, r) < dist(b, r)$$

Informally speaking, $Dom(A, B, R)$ is thus true if A is "certainly" closer to R than B. In addition the concept of partial spatial domination was introduced.

Definition 4 (Partial Spatial Domination). *Let $A, B, R \subseteq \mathcal{R}^d$ be rectangles in a d-dimensional space and dist() be a distance function defined on that space. The rectangle A dominates B partially w.r.t. R , denoted by PDom(A, B, R) if A dominates B for some, but not all $r \in R$, i.e.*

$$PDom(A, B, R) \Leftrightarrow (\exists r \in R : \forall a \in A, \forall b \in B : dist(a, r) < dist(b, r)) \wedge$$
$$(\exists r \in R : (\exists a \in A, \exists b \in B : dist(a, r) \leq dist(b, r)) \wedge$$
$$(\exists a \in A, \exists b \in B : dist(a, r) \geq dist(b, r))).$$

In [5] it was shown that spatial domination can be utilized when the rectangles conservatively approximate uncertain objects. In this case Dom(A, B, R) means P("R is closer to A than to B") = 1 and PDom(A, B, R) means $0 \leq$ P("R is closer to A than to B") ≤ 1. Using the Dom()- and the PDom()-function it is thus possible to decide the location of a rectangle w.r.t. the uncertain bisector of two uncertain objects. The uncertain bisector between two uncertain objects A and B (conservatively approximated by rectangles) defines three spaces: In $S_A(B) = \{s \in \mathcal{S} : Dom(A, B, \{s\})\}$ all objects are certainly closer to A than to B, in $S_B(A) = \{s \in \mathcal{S} : Dom(B, A, \{s\})\}$ object are certainly closer to B than to A and in the space in between no certain decision can be made. This relation is shown in Fig. 2. We are thus able to decide where the rectangle R is located w.r.t. the bisector $S_B(A)$ and $S_A(B)$ of A and B respectively by performing the Dom() and the PDom() function [24]. The following six cases are defined using a function DomCase(A, B, R) as follows.

Definition 5 (Domination Cases). *Let A and B be rectangles. For any rectangle R, one of the following cases holds:*

Case 1: *R is fully contained in $S_A(B)$ iff $Dom(A, B, R)$;*
Case 2: *R intersects $S_A(B)$ but not $S_B(A)$ iff $PDom(A, B, R) \wedge \neg PDom(B, A, R)$;*
Case 3: *R intersects neither $S_A(B)$ nor $S_B(A)$ iff*
    $\neg Dom(A, B, R) \wedge \neg PDom(A, B, R) \wedge \neg PDom(B, A, R) \neg Dom(B, A, R)$;
Case 4: *R intersects $S(B)$ but not $S(A)$ iff $\neg PDom(A, B, R) \wedge PDom(B, A, R)$;*
Case 5: *R is fully contained in $S(B)$ iff $Dom(B, A, R)$;*
Case 6: *R intersects both $S(A)$ and $S(B)$ iff $PDom(A, B, R) \wedge PDom(B, A, R)$;*

Figure 2 depicts all possible cases. Here, each rectangle R_i corresponds to Case i in Definition 5. Note that the materialization of the pruning regions $S_A(B)$ and $S_B(A)$ is a hard problem [6]. Nevertheless, the function DomCase(A, B, R) allows to efficiently decide between the six possible domination cases defined above. In the next section we will show how to use these relations in order to compute uncertain Voronoi cells.

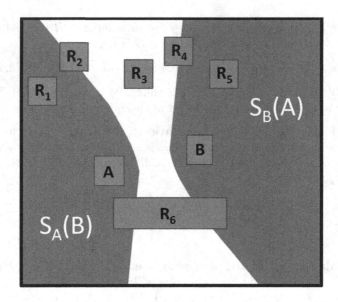

Fig. 2. Domination relation

5   Possible-Voronoi Cell Approximation

Computing the possible-Voronoi cell is a daunting task for two reasons: First, it is challenging to find the objects in the database that have an effect on its shape. Second, the representation of the cell is hard since it consists of many linear and parabolic parts that grow exponentially with the dimensionality. This section will present four algorithms that apply the concept of spatial domination to efficiently approximate the possible-Voronoi cell $\mathcal{V}(U)$ of an object U as tight as possible. The first, naive, algorithm divides the space into equi-distant grid cells and labels the cells according to their membership to the possible-Voronoi cell. The second algorithm, additionally uses an R*-tree to index the data objects to avoid exploration of irrelevant objects. The third algorithm uses a kd-trie to index the grid cells, in order to identify large regions of space which can not be part of $\mathcal{V}(U)$ or which must be part of $\mathcal{V}(U)$. The fourth algorithm uses both a kd-trie to index the space and an R-tree to index the data. For the later algorithm, the main challenge is to smartly descend both hierarchical index structures in parallel, to minimize the computational overhead.

5.1   Naive Solution

A straightforward way of computing $\mathcal{V}(U)$ is to apply an equi-distant d-dimensional grid to partition the data space. For each cell g_i we decide weather it belongs to $\mathcal{V}(U)$ or not.

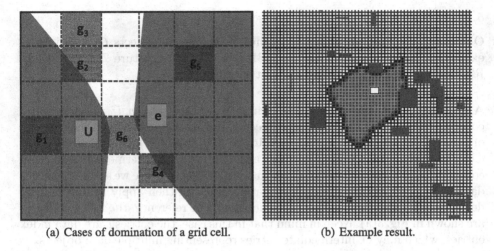

(a) Cases of domination of a grid cell.   (b) Example result.

Fig. 3. Illustration of the Naive approach.

Algorithm. The algorithm takes as input the target object U, $\mathcal{D}$ and a grid $\mathcal{G}$ covering the space of $\mathcal{D}$. We iterate over all grid cells $g \in \mathcal{G}$ and in order to decide whether g_i is part of the UV cell of U, domination against all objects $O \in \mathcal{D} \setminus U$ has to be checked. All possible cases of domination of a grid-cell g are depicted in Fig. 3(a). To determine if a grid-cell is (i) completely outside of $\mathcal{V}(U)$ or (ii) completely inside $\mathcal{V}(U)$ or (iii) a boarder cell, we can apply the six cases of Definition 5 as follows:

(i) If $\exists O \in \mathcal{D} \setminus U : Dom(O, U, g_i)$ then g_i is not part of $\mathcal{V}(U)$. This corresponds to **Case 5** of Definition 5 and cell g_5 in Fig. 3(a).

(ii) Otherwise, if $\exists O \in \mathcal{D} : PDom(O, U, g_i)$ then at least a small part of g_i can be part of $\mathcal{V}(U)$. This case corresponds to the cases of cells g_4 and g_6 in Fig. 3(a), i.e., **Case 4** or **Case 6** of Definition 5.

(iii) Otherwise we can conclude that g_i can be completely contained in $\mathcal{V}(U)$, since for database object, U, it holds that g corresponds to one of the remaining cases **Case 1**, **Case 2** and **Case 3** of cells g_1, g_2 or g_3, respectively, as shown in Fig. 3(a)

The set of all grid cells satisfying (iii) define a lower bound of $\mathcal{V}(U)$, and all grids cells satisfying (ii) or (iii) define an upper bound of $\mathcal{V}(U)$. An exemplary result of this approach for a small database of uncertain objects is depicted in Fig. 3(b). Here, the space grid is shown, where (i) unfilled cells are guaranteed to be outside of $\mathcal{V}(U)$, (ii) black cells are guaranteed to be on the border of $\mathcal{V}(U)$ and (iii) blue cells are guaranteed to be inside $\mathcal{V}(U)$. In the next subsection, we show how we can obtain this result in a more efficient way. Thus note that the algorithms presented in the following subsections compute the same result approximation, but in a more efficient way.

5.2 Indexing $\mathcal{D}$

Obviously, checking an object U against all uncertain objects $O \in \mathcal{D}$ is very expensive. Instead, we can use an MBR based index structure $\mathcal{I}_{\mathcal{D}}$ (such as an R*-Tree) to organize the objects hierarchically.

Algorithm. The algorithm takes as input the target object U, $\mathcal{I}_{\mathcal{D}}$ and a grid covering the space of $\mathcal{I}_{\mathcal{D}}$. For each grid cell g_i the algorithm traverses the entries e of $\mathcal{I}_{\mathcal{D}}$ in a best first manner [25] according to $MinDist(e, U)$. Note that the entry e can be a single uncertain object (i.e., a leaf-entry) or an intermediate node that conservatively approximates multiple uncertain objects. Since we assume that our data index uses rectangular approximations, we can then apply Definition 5 to decide which index entries have to be accessed. For reference, the following cases are shown in Fig. 3(a). Keep in mind that in this case, the entries e are data index entries, which may be intermediate entries representing multiple data objects.

Case 1: $Dom(U, e, g_1)$: e and none of its children can exclude g_1 from the UV-cell $\mathcal{V}(U)$. Thus, e don't has to be resolved and g_1 can be part of $\mathcal{V}(U)$.

Case 2: $PDom(U, e, g_2)$: same as case 1.

Case 3: $\neg PDom(U, e, g_3) \wedge \neg PDom(e, U, g_3)$: As long as e is not a leaf entry (an object), there might exist a child of e which excludes g_3 from the UV-cell, thus e has to be resolved. If e is a leaf entry g_3 is labeled as candidate for being part of $\mathcal{V}(U)$

Case 4: $PDom(e, U, g_4)$: same as case 3.

Case 5: $Dom(e, U, g_5)$: g_5 (and all child nodes of g_5) cannot be part of $\mathcal{V}(U)$.

Case 6: $PDom(U, e, g_6) \wedge PDom(e, U, g_6)$: same as case 1.

5.3 Indexing $\mathcal{S}$

Instead of indexing the data objects one could also think of indexing the space containing the grid cells. We propose to use a tree based index structure (denoted as $\mathcal{I}_{\mathcal{S}}$ to organize the data space (e.g. Quadtree, kd-trie). For each entry $h \in \mathcal{I}_{\mathcal{S}}$ it can be checked if it is part of the UV cell of U.

Algorithm. The algorithm takes as input the target object U, $\mathcal{I}_{\mathcal{S}}$, $maxdepth$ and a list of all data objects $O \in \mathcal{D}$. The entries $h \in \mathcal{I}_{\mathcal{S}}$ are traversed in a depth-first manner. For each entry h we check all $O \in \mathcal{D}$ to decide if the traversal has to go deeper (to the children of h) or its subtree can be discarded for further processing. The parameter $maxdepth$ defines the maximum depth that $\mathcal{I}_{\mathcal{S}}$ is traversed. Thus the larger $maxdepth$, the finer the granularity of the UV-cell approximation.

We can again distinguish the same cases as in Sect. 5.1:

1. If $\exists O \in \mathcal{D} : Dom(O, U, h)$ (**Case 5**) then h is not part of the UV cell of U and it does not have to be resolved further.

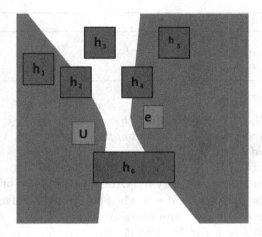

Fig. 4. Cases of domination for a data index entry e.

2. Otherwise if $\exists O \in \mathcal{D} : PDom(O, U, h)$ (**Case 4** or **Case 6**) then at least a small part of h can be part of the UV cell of U. Thus we have to resolve h further. If h is on the *maxdepth*-level we label it as candidate to be part of $\mathcal{V}(U)$.
3. Otherwise (**Cases 1–3**) we can conclude that h can be completely contained in the UV cell of U. In this case we label h as candidate to be part of $\mathcal{V}(U)$ and don't have to resolve it, even if h is not on the *maxdepth*-level.

5.4   Indexing $\mathcal{D}$ and $\mathcal{S}$

It seems apparent to combine the ideas of Sects. 5.2 and 5.3 and utilize both index structures ($\mathcal{I}_{\mathcal{D}}$ and $\mathcal{I}_{\mathcal{S}}$) to boost the performance. The non trivial task is the definition of a traversal order to minimize necessary operations.

Prelude. Our approach is basically a depth-first traversal of $\mathcal{I}_{\mathcal{S}}$. Additionally we define $AS_{\mathcal{D}}$ to be the active set of entries of the index $\mathcal{D}$. Each entry $h \in \mathcal{I}_{\mathcal{S}}$ has its own active set and passes it on to its children (always removing irrelevant entries $e \in AS_{\mathcal{D}}$). $AS_{\mathcal{D}}$ contains all entries of $\mathcal{D}$ which have already been seen and not yet resolved during the traversal of the algorithm. For each entry $h \in \mathcal{I}_{\mathcal{S}}$ we first try to identify one of the two following properties (cf Fig. 4):

Case 5: $\exists e \in AS_{\mathcal{D}} : Dom(e, U, h) \Rightarrow h$ is not part of the UV cell of U.
Case 1: $\forall e \in AS_{\mathcal{D}} : Dom(U, e, h) \Rightarrow h$ can be part of $\mathcal{V}(U)$.

If neither of the two conditions hold, either the current entry h or an entry $e \in AS_{\mathcal{D}}$ has to be resolved. Here we propose the following heuristics:

Case 2: $PDom(U, e, h) \Rightarrow$ resolve e or h depending on which one covers more space.
  Intuition: uncertain area becomes small if both constructing objects are small

---

Algorithm 1. UV-Cell computation

Require: $U, \mathcal{I_D}, \mathcal{I_S}$
 1: $AS_{\mathcal{D}} = \text{windowQuery}^*(U, \mathcal{I_D})$
 2: $\text{UVCellCheck}(U, \mathcal{I_S}.root, AS_{\mathcal{D}})$

Case 3: $\neg PDom(U, e, h) \wedge \neg PDom(e, U, h) \Rightarrow$ resolve e.
 Intuition: Resolving h can not yield any new information, since any child of h must also yield Case 3.
Case 4: $PDom(e, U, h) \Rightarrow$ resolve h if we find another data entry for which Case 4 holds (for this space entry h). Otherwise resolve e or h depending on which one covers more space. If e is a leaf entry only resolve h.
 Intuition: If more than one data entry constructs Case 4, chances are good that large portions of h can be decided.
Case 6: $PDom(U, e, h) \wedge PDom(e, U, h) \Rightarrow$ resolve h. (cf Fig. 4, case 6)
 Intuition: Resolving e can not yield any new information

Clearly, at one point there may be multiple data entries in the activate set of a space node h, which may yield different cases. It may be smart to prioritize the refinement of some data entries. In a nutshell, a data entry should be chosen which maximizes the chance that we can guarantee that h is not part of $\mathcal{V}(U)$. We propose to choose an entry e according to the following priority schema:

1. directory entries are prioritized over leaf entries.
2. prioritize cases in order 5, 4, 6, 3, 2, 1.
3. prioritize entries according to mindist to query

For ease of presentation of our algorithm, we define the function $maxprio(U \in \mathcal{D}, h \in \mathcal{I_S}, E \subseteq \mathcal{I_D})$ which maps an uncertain object U, a space region h and a set of data index entries E to the object which has the highest priority corresponding to the heuristics above.

Algorithm 1: Takes as parameters the object U for which the UV-cell is to be computed; the database $\mathcal{D}$ indexed by an R^*-tree $\mathcal{I_D}$; and the Quadtree/KD-trie $\mathcal{I_S}$ indexing the space. The idea of Algorithm 1 is to build an initial active set $AS_{\mathcal{D}}$ that is reasonable for all space partitions $h_i \in \mathcal{I_S}$ to come during query processing. For this we perform a window-query-like operation. $\text{windowQuery}^*(U, \mathcal{I_D})$ basically performs a window query on $\mathcal{I_D}$, but discards entries $e \in \mathcal{D}$ that fall in the window (since these entries cannot help to decide the borders of $\mathcal{V}(U)$). The result are now all entries $e \in \mathcal{I_D}$ that have been seen during the window-query but have not been resolved. This set is then used as an initial *active set*(denoted as $AS_{\mathcal{D}}$) in the recursive Algorithm 2 which is initiated by Algorithm 1.

Algorithm 2: This algorithm requires the uncertain object U for which the UV-cell is being computed, *one* region of the result space h(initially the root of

Algorithm 2. UVCellCheck

Require: $U, h, AS_\mathcal{D}$
1: e_{max} //*entry with maximum priority*
2: **for all** $e \in AS_\mathcal{D}$ **do**
3:     **if** $Dom(e, U, h)$ **then**
4:         h is not part of UVCell
5:         return
6:     **else if** $Dom(U, e, h)$ **then**
7:         $AS_\mathcal{D} = AS_\mathcal{D} \setminus e$
8:     **else**
9:         $e_{max} = maxprio(e_{max}, e)$
10:     **end if**
11: **end for**
12: **if** $AS_\mathcal{D}$ is empty **then**
13:     h is part of UVCell
14:     return
15: **end if**
16: **if** $case(e_{max}, U, h) \mathrel{!}= 6$ **then**
17:     $AS_\mathcal{D} = AS_\mathcal{D} \setminus e_{max} \cup e_{max}.children$
18: **end if**
19: //*redundant calculations can be reduced in the following*
20: **if** $case(e_{max}, U, h) = 4$ or 6 &&$\neg$ maxdepth **then**
21:     **for all** $h_c \in h.children$ **do**
22:         UVCellCheck(U, h_c, $AS_\mathcal{D}$.clone())
23:     **end for**
24: **else**
25:     UVCellCheck(U, h, $AS_\mathcal{D}$)
26: **end if**

---

the kd-tree), and the active set $AS_\mathcal{D}$ containing a set of $\mathcal{I}_\mathcal{D}$-entries. The algorithm works as follows:

- In a loop (lines 2–11) the algorithm first searches for the entry e defining the most prioritized case (8–10). Of course we can stop further consideration of h if we find an entry e which defines case 5 (lines 3–5). On the other hand side if an entry e defines case 1, it can never disqualify the current h thus can be excluded from $AS_\mathcal{D}$ (lines 6–7).
- In lines 12–14 we check if all entries in the active set $AS_\mathcal{D}$ have been pruned. If that is the case, no object may possible prune h and thus h must be a true hit, i.e. fully contained in the Voronoi cell.
- Now we decide whether we want to refine e_{max} or h, depending on the case (c.f. Fig. 4 and Definition 5).

Case 4: there is a chance that refining h may allow child entries of h to be pruned, and refining e_{max} may allow child entries of e_{max} to prune all of h. Therefore, we refine both entries in this case.

Case 6: refining e cannot possibly allow us to prune h. However, refining h may allow us to either prune children of h or to return children of h as true hits. Thus we refine h.

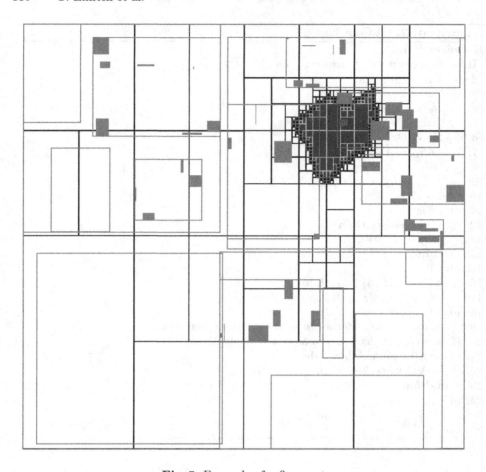

Fig. 5. Example of refinement

Case 3: no children of h can possibly be pruned.[3] Thus we split e_{max}, which may allow h to be pruned.

Case 2: we refine h.

– Finally, space index entries h which must be completely contained in $\mathcal{V}(U)$ are identified as entries having only **Cases 1–3** in their active set. Computation breaks if this is the case. After splitting the objects according to the rules above. We recursively restart the algorithm with the new objects.

Figure 5 illustrates in which manner the algorithm resolves entries of $\mathcal{I}_\mathcal{D}$ and $\mathcal{I}_\mathcal{S}$. The figures shows all pages and objects of $\mathcal{I}_\mathcal{D}$ which have been seen during the computation of the possible Voronoi-cell $\mathcal{V}(U)$ of the green objects U. Refined data objects are represented by filled red rectangles and refined directory nodes are

---

[3] recall that if $e_{max}^\mathcal{D}$ corresponds to case 3, then there exists no R^*-entry such that case 4 holds.

represented by unfilled red rectangles. Furthermore, refined entries of $\mathcal{I}_S$ are shown as (i) unfilled black rectangles if they are guaranteed to be fully outside of $\mathcal{V}(U)$, (ii) as black rectangles if on the border of $\mathcal{V}(U)$, and (iii) as blue rectangles if completely inside $\mathcal{V}(U)$. We can observe that in areas far away from the UV cell, $\mathcal{I}_S$ is resolved coarse whereas at the border of the cell it is resolved at very fine granularity. The entries of $\mathcal{I}_D$ are also only resolved around the UV cell. Note that although the number of resolved objects seems large, most of the objects are only needed for a small fraction of the computations, especially on coarser levels of $\mathcal{I}_S$. Finally, note that a nice side effect of this computation is that we obtain a tight superset of the (uncertain-) delaunay neighbors of U. This can be achieved by memorizing the objects O for which Case 4 or Cast 6 (see Definition 5) holds.

6 Experiments

Our experimental evaluation investigates algorithm behaviour w.r.t. maximum kd-trie depth, database size, object extent and dimension. Extent is a parameter to control the size of the uncertain objects (object MBR) and corresponds to the maximum extent of an object in one dimension. Experiments use synthetically generated datasets as well as an excerpt from the T-Drive trajectory dataset [7,8] which we modified to fit the scope. We implemented all approaches in the ELKI framework [26], which also provided an R-tree implementation.

Dataspace is always normalized to [0,1] per dimension. In synthetic data, objects are uniformly distributed over space with a randomly assigned side length between 0 and maximum extent. Data points from the real world dataset were sampled as a single snapshot of the world, on the afternoon of February 2nd, 2008. Therefore, one data point corresponds to the position of one taxicab within the city of Beijing, China. After removing some outliers, this dataset contains 890 separate entities. To suit our application of location obfuscation, sample locations were randomized using a Gaussian distribution based on this object's location. A single sample from this distribution is then set as center of the object's new MBR, with its extent set to 6σ of this object's Gaussian (3 to each direction). On said city scale, an extent of 0.01 would equal an area of 100 m side length.

Table 2. Default settings.

Parameter	default value	Notation	Algorithm
Dimension	2	DI	Data index traversal (Sect. 5.2)
db size	1000	SI	Space index traversal (Sect. 5.3)
extent	0.01	DSI	Data & Space Index traversal (Sect. 5.4)
tree depth	14	SR	Single rectangle (Implementation of [6])

Table 2 denotes input parameters and their default settings, as well as an explanation of our algorithm notation. If not otherwise specified, the following experiments use these input values. Those setups focusing on approximation quality use DSI exemplarily for all algorithms from Sects. 5.2–5.4, since result quality is the same. Naturally, our real world dataset T-Drive has inherent values that override parameters, namely dimension and size of database. The standard depth of 14 refers to a maximum of 14 splits in our index structure, corresponding to $16384 (= 2^{14})$ individual grid cells. Applied to a city scale of 10 by 10 kilometers, each grid cell side would measure some 78 m. As the proposed approach is later scaled up to a depth of 22, grid cells correspond to an area of only 4.8 by 4.8 meters, which on a city scale is extremely precise.

6.1   Approximation Quality

Our first evaluation explores how well the generated bounds approximate a cell. For this, we set the tree depth for our implementation to various levels between 5 and 22, corresponding to the number of splits. Evidently, smaller grid cells can more closely follow the outline of a UV-cell.

Figure 6 visualizes how upper and lower bounds converge with higher tree granularities. The dark blue line refers to the upper bound of DSI, the orange line to its lower bound, each represented by the total volume of their cells. The hatched space in between the two lines refers to the range in which the true cell volume must be located. As a point of reference, upper and lower bounds from the *Single Rectangle* (SR) approach have also been denoted in the same graphic, with the area shaded in grey corresponding to the approximation error. Since SR does not use an index, its results remain unchanged for all tree granularities.

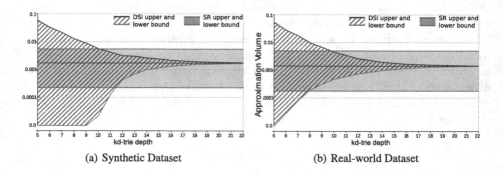

(a) Synthetic Dataset   (b) Real-world Dataset

Fig. 6. Approximation quality for DSI and SR

Performance was tested on different datasets. Figure 6(a) represents average results for runs on synthetic data, while Fig. 6(b) contains the results for our real world dataset. While overall performance is fairly comparable, DSI provides a usable lower bound remarkably early, with as little as 8 tree splits necessary to outperform SR. SR itself shows fairly similar behaviour on both datasets, with results looking even more similar than they are due to logarithmic scale.

6.2   Algorithmic Runtime

Runtime experiments were conducted while modifying database population and dimensionality, between our three different traversal approaches compared to *SR* as well as for *DSI* alone to cover larger ranges of database size (others have been excluded due to their worse performance). Although the taxi dataset is not applicable here since we modify parameters that are inherent to specific datasets, the semantics still stand: inserting more objects into a database of the same geometric expansion could represent offering more taxis for hire in a city, hence changing the nearest neighbor situation in most of the places. Therefore, the maximum object extent remained unchanged for all database sizes, since obfuscation of one's location is independent of the world's object density.

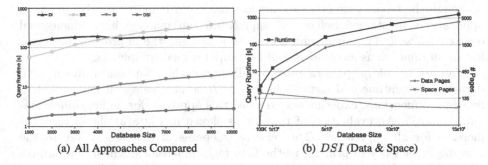

(a) All Approaches Compared   (b) *DSI* (Data & Space)

Fig. 7. A runtime comparison for all algorithms over different sizes of *DB*

In Fig. 7, run times to calculate one *UV*-cell are denoted over different database sizes. Figure 7(a) contains results for the approaches *Dataindex Traversal (DI), Spaceindex traversal (SI), Data and Space Index Traversal (DSI)* and *SR*. Note how *DI* shows a relatively constant, high runtime since for each query, every grid cell g_i is explored, independently of database population. *SR* starts off better, but since it features pairwise comparisons without the use of an index, it does not scale well for higher numbers ob database objects. *SI* clearly shows how such an index improves performance drastically, but also scales up rather fast. *DSI* also increases in runtime for higher dimensional datasets, but at generally much lower absolute values than the other approaches. Also, *DSI* increases at a lower rate. This is because the combined approach of data and space index allows for early pruning of large portions of the database.

As query performance generally deteriorates for larger datasets (or remains at high values in the case of *DI*), further scaling experiments were conducted using *DSI* only. Figure 7(b) shows the results of database populations from 10 K to 15 Million objects. To avoid gross overlapping of objects, object extent has been lowered to 0.001 for these runs. The left axis again refers to the average time to perform one UV-cell calculation, which corresponds to the blue data line. We observe a slightly superlinear scaling, confirming our theoretical observations that (i) adding more objects leads to linearly more intersections with

Voronoi cells, which are at least as big as U, and (ii) a linear increase in object count causes logarithmic tree index growth. This results in a combined log-linear growth in runtime.

The right scale denotes average page views during cell calculation, with the orange line referring to pages of the data index, and the green line for pages of the space index. Note that data index exploration roughly follows runtime development, while the space index is used less for larger databases. This is easily explained by a constant tree depth, resulting in a constant resolution of space. With a higher database population, the likelyhood of all relevant objects being enclosed in a small space increases.

6.3   Effect of Data Dimensions

Although the trivial case of a two-dimensional world is most intuitive for most applications mentioned before, all approaches can manage high-dimensional datasets as well. The main limitation here is keeping the approximation error low in all dimensions at once, as well as computational complexity.

Figure 8 displays performance of all approaches for multi-dimensional datasets. As runtime and memory usage of SR do not scale well for more than five data dimensions, experiments excluded this approach for higher dimensionalities than 5. An evaluation of runtime as shown in Fig. 8(a) shows constant increase for all approaches. The relative steepness of increase is due to the growing inefficiency of pruning methods in high dimensions, which deteriorates searches toward a linear scan, which itself has quadratic complexity.

Approximation quality for higer dimensions is shown in Fig. 8(b). As mentioned before, fitting a bound to a more and more complex object leaves much room for approximation error. Therefore, volumes of upper and lower bounds diverge more for higher dimensions. Displayed here are bounds for SR up to dimension 5 (grey) and two different settings of our DSI approach, once with a depth of 14 (blue) and a depth of 20 (orange). As expected, a higher depth allows for more tree splits per dimension and thus a better approximation.

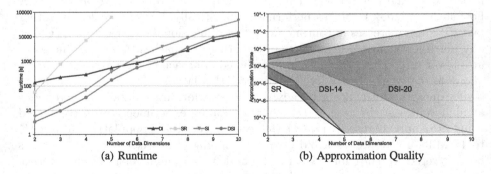

(a) Runtime   (b) Approximation Quality

Fig. 8. A comparison for increasing data dimensions.

6.4   Conclusions

In this work, we proposed an index-supported approach to approximate the shape of a possible Voronoi-cell to support nearest neighbor queries on uncertain data. Our approache uses an R^*-tree as a hierarchical access method to efficiently find the set of uncertain objects that influence the possible Voronoi-cell of an uncertain object U, i.e., the set of Delauny-neighbors of U. In addition, we propose to use a kd-trie as a hierarchical access method to identify regions of space which must (not) be part of a Voronoi-cell. Compared to the state-of-the-art of computing uncertain Voronoi-cells, our approach allows for much higher approximation quality, since our result approximation consists of a set of rectangular kd-trie nodes, rather than a single bounding rectangle. As future work, we want to extend our ideas to find *certain Voronoi-cells*, that is regions, where a query object has a probability of one of having some object U as its nearest neighbor. Furthermore, we want to extend our solution to the case of k'th-order Voronoi-cells to support k-nearest neighbor queries. Even in the case of certain data, k'th-order Voronoi-cells become complexly shaped, having a representation complexity exponential in k. However, since we are using space approximation techniques, rather than computing exact bounds, we can avoid this computational drawback.

Acknowledgements. Part of the research leading to these results has received funding from the Deutsche Forschungsgemeinschaft (DFG) under grant number RE 266/5-1 and from the DAAD supported by BMBF under grant number 57055388. Reynold Cheng was supported by the Research Grants Council of Hong Kong (RGC Project (HKU 711110)).

References

1. Chow, C.Y., Mokbel, M.F., Aref, W.G.: Casper*: query processing for location services without compromising privacy. ACM TODS **34**(4), 24 (2009)
2. Beskales, G., Soliman, M.A., Ilyas, I.F.: Efficient search for the top-k probable nearest neighbors in uncertain databases. VLDB Endow. **1**(1), 326–339 (2008)
3. Cheng, R., Xie, X., Yiu, M.L., Chen, J., Sun, L.: Uv-diagram: A voronoi diagram for uncertain data. In: ICDE, pp. 796–807. IEEE (2010)
4. Ali, M.E., Tanin, E., Zhang, R., Kotagiri, R.: Probabilistic voronoi diagrams for probabilistic moving nearest neighbor queries. DKE **75**, 1–33 (2012)
5. Bernecker, T., Emrich, T., Kriegel, H.P., Mamoulis, N., Renz, M., Züfle, A.: A novel probabilistic pruning approach to speed up similarity queries in uncertain databases. In: Proceedings of the ICDE, pp. 339–350 (2011)
6. Zhang, P., Cheng, R., Mamoulis, N., Renz, M., Zufle, A., Tang, Y., Emrich, T.: Voronoi-based nearest neighbor search for multi-dimensional uncertain databases. In: ICDE, pp. 158–169. IEEE (2013)
7. Yuan, J., Zheng, Y., Zhang, C., Xie, W., Xie, X., Sun, G., Huang, Y.: T-drive: driving directions based on taxi trajectories. In: SIGSPATIAL, pp. 99–108 (2010)
8. Yuan, J., Zheng, Y., Xie, X., Sun, G.: Driving with knowledge from the physical world. In: SIGKDD, pp. 316–324 (2011)
9. Emrich, T., Kriegel, H.P., Mamoulis, N., Renz, M., Züfle, A.: Querying uncertain spatio-temporal data. In: ICDE, pp. 354–365. IEEE (2012)

10. Emrich, T., Kriegel, H.-P., Kröger, P., Renz, M., Züfle, A.: Incremental reverse nearest neighbor ranking in vector spaces. In: Mamoulis, N., Seidl, T., Pedersen, T.B., Torp, K., Assent, I. (eds.) SSTD 2009. LNCS, vol. 5644, pp. 265–282. Springer, Heidelberg (2009)
11. Beckmann, N., Kriegel, H.P., Schneider, R., Seeger, B.: The R*-tree: an efficient and robust access method for points and rectangles, vol. 19. ACM (1990)
12. Orenstein, J.A., Merrett, T.H.: A class of data structures for associative searching. In: ACM SIGACT-SIGMOD, pp. 181–190. ACM (1984)
13. Cheng, R., Kalashnikov, D.V., Prabhakar, S.: Querying imprecise data in moving object environments. In: IEEE TKDE (2004)
14. Li, J., Saha, B., Deshpande, A.: A unified approach to ranking in probabilistic databases. VLDB Endow. $2(1)$, 502–513 (2009)
15. Bernecker, T., Kriegel, H.P., Mamoulis, N., Renz, M., Zuefle, A.: Scalable probabilistic similarity ranking in uncertain databases. TKDE $22(9)$, 1234–1246 (2010)
16. Aurenhammer, F.: Voronoi diagrams-a survey of a fundamental geometric data structure. ACM CSUR $23(3)$, 345–405 (1991)
17. Sharifzadeh, M., Shahabi, C.: Vor-tree: R-trees with Voronoi diagrams for efficient processing of spatial nearest neighbor queries. VLDB Endow. $3(1–2)$, 1231–1242 (2010)
18. Zheng, B., Xu, J., Lee, W.C., Lee, L.: Grid-partition index: a hybrid method for nearest-neighbor queries in wireless location-based services. VLDB J. $15(1)$, 21–39 (2006)
19. Nutanong, S., Zhang, R., Tanin, E., Kulik, L.: The V*-Diagram: a query-dependent approach to moving kNN queries. VLDB Endow. $1(1)$, 1095–1106 (2008)
20. Sharifzadeh, M., Shahabi, C.: Approximate Voronoi cell computation on spatial data streams. VLDB J. $18(1)$, 57–75 (2009)
21. Akdogan, A., Demiryurek, U., Banaei-Kashani, F., Shahabi, C.: Voronoi-based geospatial query processing with mapreduce. In: IEEE CloudCom, pp. 9–16 IEEE (2010)
22. Kolahdouzan, M., Shahabi, C.: Voronoi-based K nearest neighbor search for spatial network databases. In: VLDB Endowment, pp. 840–851 (2004)
23. Roussopoulos, N., Kelley, S., Vincent, F.: Nearest neighbor queries. In: ACM SIGMOD, vol, 24, pp. 71–79 ACM (1995)
24. Emrich, T., Kriegel, H.P., Kröger, P., Renz, M., Züfle, A.: Boosting spatial pruning: On optimal pruning of MBRs. In: Proceedings of the SIGMOD, pp. 39–50 (2010)
25. Hjaltason, G.R., Samet, H.: Ranking in spatial databases. In: Proceedings of the SSD, pp. 83–95 (1995)
26. Achtert, E., Kriegel, H.P., Schubert, E., Zimek, A.: Interactive data mining with 3D-parallel-coordinate-trees. In: Proceedings of the SIGMOD, pp. 1009–1012 (2013)

Navigation and Routing

Towards Fast and Accurate Solutions to Vehicle Routing in a Large-Scale and Dynamic Environment

Yaguang Li[1]([✉]), Dingxiong Deng[1], Ugur Demiryurek[1], Cyrus Shahabi[1], and Siva Ravada[2]

[1] Department of Computer Science, University of Southern California, Los Angeles, California
{yaguang,dingxiod,demiryur,shahabi}@usc.edu
[2] Oracle, USA
siva.ravada@oracle.com

Abstract. The delivery and courier services are entering a period of rapid change due to the recent technological advancements, E-commerce competition and crowdsourcing business models. These revolutions impose new challenges to the well studied vehicle routing problem by demanding (a) more ad-hoc and near real time computation - as opposed to nightly batch jobs - of delivery routes for large number of delivery locations, and (b) the ability to deal with the dynamism due to the changing traffic conditions on road networks. In this paper, we study the Time-Dependent Vehicle Routing Problem (TDVRP) that enables both efficient and accurate solutions for large number of delivery locations on real world road network. Previous Operation Research (OR) approaches are not suitable to address the aforementioned new challenges in delivery business because they all rely on a time-consuming a priori data-preparation phase (i.e., the computation of a cost matrix between every pair of delivery locations at each time interval). Instead, we propose a *spatial-search-based* framework that utilizes an on-the-fly shortest path computation eliminating the OR data-preparation phase. To further improve the efficiency, we adaptively choose the more promising delivery locations and operators to reduce unnecessary search of the solution space. Our experiments with real road networks and real traffic data and delivery locations show that our algorithm can solve a TDVRP instance with 1000 delivery locations within 20 min, which is 8 times faster than the state-of-the-art approach, while achieving similar accuracy.

1 Introduction

The vehicle routing problem (VRP) aims to find a set of routes at a minimal cost (e.g., total distance or travel time) for a set of geographically dispersed delivery locations which are assigned to a fleet of delivery vehicles. Each location is visited only once, by only one vehicle, and each vehicle has a limited capacity. VRP is an NP-hard combinatorial optimization problem. Exact algorithms based on

© Springer International Publishing Switzerland 2015
C. Claramunt et al. (Eds.): SSTD 2015, LNCS 9239, pp. 119–136, 2015.
DOI: 10.1007/978-3-319-22363-6_7

branch-and-bound or dynamic programming are slow and only capable of solving relatively small instances (e.g., less than 30 delivery locations), thus heuristics are mainly used in practice.

While VRP and its variations (e.g., VRP with time windows and VRP with multiple depots) have been extensively studied in the literature [1], in recent years we are witnessing a renewed interest to this problem due to two very important transformations. First, the traffic data at a very high resolution have become available that can significantly enhance the accuracy of the routes assigned to delivery vehicles and can consequently result in considerable benefit. For example, according to UPS [2] the company can save $50 million a year if the average daily travel distance of its drivers can be reduced by one mile, which is typically less than 1 % of the daily travel distance of a delivery vehicle. Second, due to the increasing popularity of on-line shopping (i.e., E-commerce), there is a growing need for fast delivery to very large number of customers; to the point that some E-commerce companies (e.g., Google Express [3]) are developing their own proprietary delivery solutions to stay ahead of the competition. This is because the existing solutions provided by the major delivery companies (e.g., FedEx and UPS) assume that the delivery orders (and their locations) are known (at least a day) in advance. It is also not hard to envision an Uber-type application for deliveries in near-future, democratizing the delivery business.

These two transformations have challenged the traditional approaches to VRP. The basis of all the traditional approaches are to utilize some sort of Operation Research (OR) technique (e.g., integer programming) to solve VRP. Consequently, as an input to all these approaches, a pairwise distance matrix is required, which contains the distances between every two delivery locations. Without the dynamism resulting from traffic data and in the world where delivery plans were prepared the night before for a small number of delivery locations, creating such a matrix a priori was acceptable. However, considering travel-time as the "distance", different time intervals in the day require different distance matrices (due to traffic congestions), which increases the complexity of preparing the input for the OR approaches. The increase in the complexity along with more delivery locations and less time to prepare the delivery plans render the OR data-preparation phase impractical.

Therefore, in this paper, we take a completely different approach to solve VRP by utilizing the lessons learned from the field of spatial-databases. First, considering the vehicle routing problem in time-dependent road networks, we compute network distance (i.e., travel-time) on-the-fly utilizing a time-dependent shortest path technique from the spatial-database literature [4]. Note that although some OR-based approaches are developed for the Time-Dependent version of VRP [5,6] (called TDVRP hereafter), they still rely on the time-consuming data preparation phase. In fact, the complexity of that phase becomes even worse because now it requires the computation of the pairwise distance matrix for each and every time interval in a day. We, however, completely eliminate the data preparation phase by on-demand calculation of shortest path between two delivery locations and at the same time caching the partial results from the expansion of shortest path computation for future use. As a byproduct, our proposed approach could start finding the solutions as soon as the delivery requests are received.

Second, we improve another phase in the OR approach known as *local search*, by exploiting the spatial information of the locations in the search. In particular, the local search starts from an initial solution and iteratively moves to new solution by selecting from a neighborhood of the current solution through "the move operators". The main bottleneck of the local search is that it needs large number of iterations to find neighborhood solutions and this number grows exponentially with the number of delivery locations. We observe that local search relies on blind evaluation of delivery locations and move operators towards finding neighborhood solutions in which they treat each delivery locations and operators equally. However, we argue that not all delivery locations are equally important: some delivery locations are more promising to generate the effective neighborhood solutions and hence we assign weights to each delivery location. The delivery locations with higher weights are more likely to be chosen and their weights are adjusted adaptively based on their previous performance. A similar idea applies to the operators: operators with higher weights are more likely to be applied. Consequently, our algorithm leverages a spatially guided search by selecting promising delivery locations and operators first, which significantly reduces the running time while generating high-quality solution.

We conducted extensive experiments on real world road network of Los Angeles with real traffic data. Experimental results show that (1) by leveraging the real time-dependent traffic pattern, we can reduce the travel cost of routes by 7 % on average with respect to its static counterparts, and (2) our algorithm can solve TDVRP with 1000 delivery locations within 20 min, which is 8 times faster than the state-of-the-art approach, while achieving similar accuracy.

The remainder of this paper is organized as follows. In Sect. 2, we formally define our Vehicle Routing Problem in Time-dependent road network. In Sect. 3, we present our spatial-search-based framework to solve this problem. Experiment results are reported in Sect. 4. In Sect. 5, we review the related work and Sect. 6 concludes the paper.

2 Problem Definition

In this section, we formally define the vehicle routing problem in time-dependent road networks. We model the road network as a time-dependent weighted graph where the non-negative weights are time-dependent travel times (i.e., positive piece-wise linear functions of time) between the nodes.

Definition 1 (Time-dependent Graph). *A Time-dependent Graph (G_T) is defined as $G_T = (V, E)$, where V and E represent set of nodes and edges, respectively. For every edge $e(v_i, v_j)$, there is a cost function $c(v_i, v_j, t)$ which specifies the travel cost from v_i to v_j at time t.*

Definition 2 (Time-dependent Travel Cost). *Let $\{s = v_1, v_2, \cdots, v_k = d\}$ represents a path which contains a sequence of nodes where $e(v_i, v_{i+1}) \in E$ and $i = 1, \cdots, k - 1$. Given a G_T, a path (s, d) from source s to destination d, and a departure-time at the source t_s, the time-dependent travel cost $TT(s \rightsquigarrow d, t_s)$ is* *the time it takes to travel along the path. Since the travel time of an edge varies depends on the arrival time to that edge (i.e., arrival dependency), the travel time is computed as follows:*

$$TT(s \rightsquigarrow d, t_s) = \sum_{i=1}^{k-1} c(v_i, v_{i+1}, t_i)$$

where $t_1 = t_s$, $t_{i+1} = t_i + c(v_i, v_{i+1}, t_i), i = 1, \cdots k$.

Definition 3 (Time-dependent Shortest Path). *Given a G_T, s, d and t_s, the time-dependent shortest path $TDSP(s, d, t_s)$ is a path with the minimum travel-time among all paths from s to d starting at time t_s.*

Definition 4 (Vehicle Routing Problem in Time-dependent Road Network). *Given a time-dependent graph G_T, a depot $v_d \in V$, a start time t_s, k delivery vehicles with capacity C, and a set of delivery locations $V_c \subset V$, each delivery location $v_i \in V_c$ has a demand of d_i, TDVRP aims to find k routes with the minimum total time-dependent travel cost subject to the following constraints:*

- *all routes start and end at the depot;*
- *each delivery location in V_c is visited exactly once by exactly one vehicle;*
- *the total demands of delivery locations in a route must not exceed C;*

Note that the input TD road network G_t could have more than 100 thousand nodes and edges, the delivery locations are a small subset of nodes in the given road network and the travel time between each delivery location is not known in advance. This is different with the typical input of VRP, which requires a pairwise cost matrix between each pair of delivery locations per instance.

3   Proposed Algorithm

In this section, we first investigate one state-of-the-art local search based algorithm termed RTR [7]. RTR has shown [8] to be able to generate high quality solutions for large-scale delivery locations for static network. However, we discover that RTR has two major drawbacks. First, like other Operation Research (OR) approaches, RTR relies on the time consuming data-preparation step. In addition, we further identify that the blind evaluation of neighborhood solution dominates the search process of RTR.

To address the above two issues, we propose a Spatial-Search-Based Local Search (SSBLS) framework (Sect. 3.3) which incorporates on-the-fly shortest path computation (Sect. 3.5) into the local search framework. To avoid the unnecessary search of the solution space, we adaptively choose more promising delivery locations and move operators (Sect. 3.4). Note that our proposed improvements can be easily adopted into other local search based algorithms which iteratively generates and improves the neighborhood solutions.

3.1 Local Search Based Approach

Local search is a popular framework which is proven to provide a high-quality solution for VRP. Local search starts from an initial solution and iteratively moves to one of the neighborhood solutions based on heuristics. Typically, a local search algorithm is developed by the following general framework:

- Step 1: choose an initial solution. (*Initial solution*)
- Step 2: generate one or more neighborhood solutions by applying operators to the current solution. (*Neighborhood generation*)
- Step 3: select one solution to continue using heuristic, e.g., the first, the best or arbitrary one. (*Acceptance criteria*)
- Step 4: if the stop condition is not satisfied, e.g., the solution is not considered as optimal, then goto Step2, else stop. (*Stop condition*)

In general, the initial solution is generated based on some heuristic methods (e.g., Clarke-wright heuristic [9]), consequently a local move procedure is used to generate neighborhood solution. The definition of local move is as follows:

Definition 5 (Local Move). *Local move is the process of generating a new solution by removing k edges in the current solution and replacing them with other k edges.*

Local move is performed by applying one operator at a time to the existing solution. In this paper, we use three basic operators studied in [7], i.e., One-Point (OP), Two-Point (TP) and Two-Opt (TO), because the combination of these operators is enough to generate high-quality solution for large scale delivery locations. Specifically, each operator uses three parameters to complete the local move. Given a current solution S, a selected location v_x, and one of its neighboring location v_y, $OP(S, v_x, v_y)$ moves v_x to the new position after v_y(i.e., v_x is visited after v_y), $TP(S, v_x, v_y)$ swaps the positions of v_x, v_y in S; and finally $TO(S, v_x, v_y)$ removes the two edges $e(v_x, v'_x), e(v_y, v'_y)$ in S by replacing them with $e(v_x, v_y)$ and $e(v'_x, v'_y)$.

Figure 1 shows the process of applying One-Point operator to a solution S which contains two routes r_1 and r_2. After applying $OP(S, v_j, v_a)$, point v_j in r_2 is relocated to the position after v_a in r_1, thus generating a new neighborhood solution which contains r'_1 and r'_2. In this way, the three edges $e(v_i, v_j), e(v_j, v_k)$ and $e(v_a, v_b)$ in S are replaced by $e(v_i, v_k), e(v_a, v_j)$ and $e(v_j, v_b)$. Once the neighborhood solution is generated, the algorithm checks its feasibility by evaluating its cost and decides whether to choose this solution to continue.

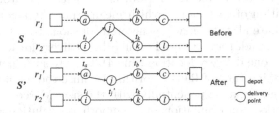

Fig. 1. Neighborhood solution generation through One-Point Move

3.2   Baseline Approach: RTR

RTR follows local search framework: (1) An initial feasible solution S is generated using the classic Clarke-Wright heuristic [9], (2) The neighborhood generation and improvement over S is shown in Algorithm 1. The algorithm interleaves with two search procedures: record-to-record and downhill search. The major difference between the two search procedures is that: for record-to-record search, when we apply one operator op to one location v_i, a non-improving solution with a small range of deviation to the current best solution is allowed to jump out of the local optimal (lines 13–14); on the other hand, only the improved solution is accepted for the downhill search. In terms of acceptance criteria, RTR uses the first-accept standard, i.e., whenever a better solution is found, the local search for the current location v_i is stopped and we move to the next location (lines 9–11). (3) RTR stops when no better solution can be found after K continuous executions of Algorithm 1.

Algorithm 1. GenerateNeighborhood(G_T, S)

Input: time-dependent graph G_T, the current solution S, and the *deviation*
 1: **for each** search procedure (i.e., record-to-record and downhill) **do**
 2:      **for each** operator op in operators **do**
 3:          **for each** location v_i in S **do**
 4:              **for each** location v_j in neighbors of v_i **do**
 5:                  $S' \leftarrow \mathbf{op}(S, v_i, v_j)$
 6:                  $cost(S') \leftarrow \mathbf{eval}(S')$
 7:                  **if** $cost(S') < cost(S_{best})$ **then**
 8:                      $S_{best} \leftarrow S'$ // store S' with the smallest value
 9:                  **if** $cost(S') < cost(S)$ **then**
10:                      $S \leftarrow S'$
11:                      **break**
12:                  // record-to-record search continues, downhill search stops here
13:              **if** record-to-record search **And** $S_{best} \leq cost(S) + deviation$ **then**
14:                  $S \leftarrow S_{best}$

Although RTR is one of the best choices for large-scale VRP, it is inefficient for Time-dependent road networks where each edge has different (time-varying) costs for each time instance throughout the day. In the following, we analyze the bottleneck in this search process. As shown in Algorithm 1, during each search procedure (record-to-record and downhill), each operator is applied to each delivery location of each route, which means that the algorithm needs to generate and evaluate all these newly generated solutions to determine whether to accept it or not. In addition, with RTR, all other delivery locations are treated as the neighbors of one delivery location (line 4). Clearly such exhaustive search dominates the running time of RTR algorithm because the number of generate-and-evaluate process grows exponentially with the size. Moreover, compared to static road networks, the evaluation of neighborhood solution with TD road network is more time consuming (line 6). This is because changing the edges in a route due to operators could lead to different arrival times of the corresponding delivery locations, which in turn changes the weight of the following edges.

To illustrate, consider Fig. 1, in static case, the total travel cost of the new solution S' can be calculated using the following equation:

$$c(S') = c(r_1') + c(r_2') = c(r_1) + \Delta c(r_1) + c(r_2) + \Delta c(r_2)$$

$\Delta c(r_1)$ and $\Delta c(r_2)$ are calculated via the following equation:

$$\Delta c(r_1) = -c(v_a, v_b) + c(v_a, v_j) + c(v_j, v_b), \Delta c(r_2) = c(v_i, v_k) - c(v_i, v_j) - c(v_j, v_k)$$

where $c(v_i, v_j)$ represents the static travel cost between delivery locations v_i and v_j, and $c(r)$ represent the static cost of route r.

In the static case, the evaluation involves five delivery locations (i.e., v_a, v_b, v_i, v_j, v_k) and can be calculated in constant time when the cost matrix is given. However, in TD case, because the arrival times of delivery location v_j and v_k are changing, the arrival times for the following delivery locations (e.g., v_c and v_l) are also changing which result in re-calculating the cost of the whole path.

3.3 Spatial-Search-Based Local Search Framework

By analyzing RTR, we find that two processes dominant its total running time and make RTR less practical for the new delivery application: (1) the neighborhood generation and evaluation process. (2) the data-preparation process which computes the travel cost between each pair of nodes at each time interval.

The above observations lead us to the road map of how to improve the efficiency of RTR while maintain the high accuracy. As shown in Fig. 2, we aim to eliminate the data preparation process as well as reduce the evaluation time[1].

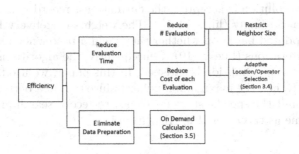

Fig. 2. Methods to improve efficiency

---

[1] Reducing the computation cost of each evaluation usually depends more on the problem setting, for example a method that utilizes time window constraints and dynamic programming to reduce evaluation cost was proposed in [6] for the TDVRPTW problem. In this paper, we work on a general TDVRP setting and thus focus on reducing the number of evaluations.

Algorithm 2. Framework of SSBLS algorithm

1: $iter \leftarrow 0, c_{min} \leftarrow maxval$
2: $S \leftarrow$ **initialSolution**()
3: **while** stop condition is not satisfied **do**
4:     **for each** search procedure (i.e., record-to-record and downhill) **do**
5:         $op \leftarrow$ **selectOperator**()
6:         $v_i \leftarrow$ **selectLocation**()
7:         **for each** location v_j in neighbors of v_i **do**
8:             $iter \leftarrow iter + 1$
9:             **if** $iter$ **mod** $I = 0$ **then**
10:                 **updateWeight**()
11:             $S' \leftarrow$ **op**(S, v_i, v_j)
12:             $c \leftarrow$ **onDemandEval**(S')
13:             **if** $c < c_{min}$ **then**
14:                 $S \leftarrow S'$
15:                 $c_{min} \leftarrow$ **min**(c_{min}, c)
16:                 **break**
17: **return** S

Towards this end, we propose a Spatial-Search-Based Local Search (SSBLS) framework, which is shown in Algorithm 2.

SSBLS starts with an initial solution which is generated using the Clarke-Wright algorithm (line 2). During each search procedure (record-to-record and downhill search), in order to generate the neighborhood solution, it adaptively selects an operator op and a delivery location v_i (lines 5–6) based on their weights (See Sect. 3.4). Subsequently, the algorithm applies the selected operator to the delivery location v_i and iterates through the neighborhood solution (lines 7–16). The newly generated solution S' is evaluated via our on-demand shortest path calculation procedure (See Sect. 3.5), and thus the algorithm determines whether to accept S'. If a solution is accepted, the current best record r and the solution S are updated accordingly (lines 14–15). The weights of delivery locations and operators are updated based on whether they yield a better neighborhood solution in every I iterations (lines 9–10). Consequently, the algorithms repeats this process until the stop condition is satisfied. In this paper, we use the same stop condition as RTR, i.e., the maximum number failure to find a better solution. To save space, we omit the specific steps for record-to-record search in Algorithm 2, which is the same as those listed in Algorithm 1.

3.4 Adaptive Point and Operator Selection

A straightforward method to improve the efficiency is to restrict the number of candidate neighbors of a delivery location. This is because when applying operators to one specific location (line 4 in Algorithm 1), RTR treats all other delivery locations as the neighbor of that location. Therefore, to reduce the number of evaluations, we can only apply operators to a fixed number of its nearest neighbors (rather than all other delivery locations) for a delivery location [7].

Besides restricting the neighborhood size, we further propose a novel approach to reduce the number of evaluations. The intuition of our algorithm is that not all delivery locations and operators are equally important in order to generate the new neighborhood solutions. Towards this end, we first define "Effective Local Move" and explain some observations based on the empirical study on real-world dataset with 100 delivery locations on Los Angeles road network. Similar observations exists for larger datasets (see more details in Sect. 4).

Definition 6 (Effective Local Move). *Effective local move is defined as the local move whose corresponding solution is accepted by the algorithm based on variable criterion, e.g., down-hill, simulate-annealing.*

Observation 1. *Only very small portion of local moves will become effective local move, most of the new solutions generated by local moves are not effective and hence not be accepted.*

In the process of local search, we record the number of evaluated local moves and the number of resulted effective moves. Table 1 shows the statistic of three types of local moves. We observe that only 0.21 % of the local moves become effective, i.e., the corresponding solutions are accepted by the algorithm. Because a large portion of the evaluation is useless, our idea to reduce the number of evaluations is via selectively evaluating the promising moves. Intuitively, some operators (e.g., One-Point) are more suited for one type of instance while others are best suited for another type of instance. In addition, instead of using the same sequence to iterate operators and customers in Algorithm 1, we believe that alternating between different customers (operators) makes the heuristic more robust. In the following, we discuss how to decide on the promising delivery locations and operators.

Table 1. Local move statistics

Local move	# Evaluated moves	# Effective moves	Percentage
One-Point Move	30300856	83071	0.27 %
Two-Point Move	27774218	25504	0.09 %
Two-Opt Move	30275671	78683	0.26 %
Total	88350745	187258	0.21 %

Promising Delivery Locations and Weights Assignment. Figure 3 shows the distribution of the effectiveness of each delivery location. The effectiveness is measured by the percentage of local move that leads to the effective local move via applying operators to one specific delivery location. From Fig. 3, we observe that some delivery locations are much more effective than the remaining locations (e.g., the effectiveness of the top 1 delivery location is at least 20 times higher than the least effective point), which leads us to the following observation and define the promising delivery locations.

Fig. 3. Distribution of the effectiveness of delivery locations (x-axis is ordered by the effectiveness)

Observation 2. *Delivery locations are not equally effective in generating better result. In other words, applying operators on some delivery locations are more likely to generate a solution with lower travel cost than others.*

To capture the effectiveness of the delivery locations, we assign weight w_i to each delivery location v_i (initially they have the same weights), and a delivery location is selected with probability that is proportional to its weight. Moreover, we adjust the weight of a delivery location in the search process based on its performance in the previous iterations, i.e., the better it performs, the more likely its weight will be increased.

Specifically, we divide the whole local search process into separate parts, each part contains I iterations. Suppose we are in the kth part of the search process, the weight of one delivery location in the $(k + 1)$th part is updated once we completed the I iteration in the kth part by the following equation (line 10 in Algorithm 2):

$$w_i^{k+1} = (1 - \eta)w_i^k + \eta\frac{\lambda_i^k}{\Lambda_i^k} \tag{1}$$

where Λ_i^k is the total number of local moves that involves delivery location v_i during the kth part, λ_i^k is the corresponding total number of effective local moves that involves v_i during the kth part, and η is learning rate which captures the trade-off about how much we should rely on the performance of the kth part or the previous $(k - 1)$ parts. If $\eta = 0$, the weight only relies on the previous $(k - 1)$ parts of the search process, otherwise if $\eta = 1$, the weight only relies on the performance on the kth part of search. Usually, $0 < \eta < 1$, and parameter tunning is required to get the best performance.

Once we have assign weights to each delivery location, the probability that one delivery location is chosen to generate the neighborhood solution is calculated in the following equation:

$$Prob(v_i) = \frac{w_i}{\sum_{j=1}^{N} w_j} \tag{2}$$

where N is the number of delivery locations, w_i the weight of v_i.

Similar to the delivery locations, operators are also not equally effective. Therefore, we apply a similar idea to the operators. Because of the space limitation, we omit the details here.

3.5   Eliminating Data-Preparation

When dealing with vehicle routing problem, existing OR based methods assume that the travel costs between each pair of delivery locations are calculated in advance, or can be calculated in $O(1)$ time (e.g., in Euclidean space). Usually, the calculated travel cost is stored in a matrix-like structure (referred as *cost matrix*).

Although the cost matrix can be calculated in static cases, it is far more time-consuming to do this precomputation in time-dependent road network. This is because it has to calculate the time-dependent travel costs between every point pair in each time interval. This process involves $O(TN^2)$ time-dependent shortest path calculation, where T is the number of time intervals, N is the number of delivery locations. Note that the shortest path calculation in time-dependent road network is costly [4].

Table 2. Percentage of accessed cells in cost matrix

# Locations	100	200	500	1000
Accessed cell	25.73 %	12.70 %	6.48 %	4.12 %

Observation 3. *Most of the precomputed travel cost is not required.*

To analyze the effectiveness of the data preparation step, we record the number of cells in cost matrix that is accessed by the algorithm for at least once. If the algorithm used the travel cost from location v_i to location v_j at the t-th time interval. then the corresponding cell, e.g., $cost[i][j][t]$, will be accessed. Table 2 shows the percentage of accessed cells in the cost matrix (neighbor size $\sigma = 40$) with different number of delivery locations. As shown in Table 2, most of the cells in the cost matrix have never been accessed. For example, less than 5 % of the cells in the matrix are accessed when the number of delivery locations is 1000. This could be explained via the following reasons:

- Most of the effective local moves are those applied to a delivery location and its nearby locations, which means that the shortest path is only required between a location and some of its close neighbors.
- For a certain pair of delivery locations, only a small portion of the time intervals is accessed. For example, suppose the vehicle departs from depot at t_0, and the minimum travel time between the depot v_d and a delivery location v_i is $c(v_d, v_i, t_0)$, then it is possible that the time-dependent travel cost between v_d and v_i with start time earlier than $t_0 + c(v_d, v_i, t_0)$ are not accessed. This is because the path from depot to v_i which bypasses other delivery location may yield a later arrival time than the path which directly passes the location v_d.

Algorithm 3. onDemandEval(S)

Input: Solution $S = (r_i)_{i=1}^{k}, r_i = (v_{i_j})_{j=1}^{|r_i|}$
1: $c \leftarrow 0$
2: **for each** route r_i in S **do**
3:      **if** **isChanged**(r_i) **then**
4:          $t \leftarrow t_0$
5:          **for each** v_{i_j} in r_i **do**
6:              **if** **isCached**($v_{i_j}, v_{i_{j+1}}, t$) **then**
7:                  $c_e \leftarrow c(v_{i_j}, v_{i_{j+1}}, t)$
8:              **else**
9:                  $c_e \leftarrow$ **TDSPAndCache**($v_{i_j}, v_{i_{j+1}}, t, \sigma$)
10:             $c \leftarrow c + c_e, \ t \leftarrow t + c_e$
11:     **else**
12:         $c \leftarrow c + $ **getCost**(r_i)
13: **return** c

Based on the above observation, we argue that data preparation process can be eliminated and replaced by an on-demand calculation strategy. The main idea is to compute shortest path only when it is needed, and at the same time the partial results are cached from the expansion of shortest path computation for future usage. Algorithm 3 shows our proposed on-demand-calculation approach to evaluate the routes of a candidate solution. When one route of the solution is changing, we recalculate the travel cost from the beginning of this route (lines 4–10). In this process, if the travel cost between two delivery locations v_i and v_j during a time interval is not cached previously, function $TDSPAndCache$ is called on-the-fly to calculate the travel cost (line 9). Meanwhile, the travel cost from v_i to its σ nearest neighbors is also calculated and cached through the expansion of v_i because these neighbors are usually close to the current delivery location and caching them could facilitate the future search process. We use the time-dependent incremental network expansion algorithm described in [10].

4   Experimental Evaluation

4.1   Experimental Settings

Datasets. The experiments are conducted in Los Angeles(LA) road network dataset which contains 111,532 vertices and 183,945 edges. The time-varying edge patterns(i.e., time-dependent edge weights) of LA road network are generated from the sensor dataset we have been collecting in the past three years: we split the day time into 60 intervals from 6am to 9pm, and for each interval we assign the aggregated sensor data travel times to the corresponding network segment. In terms of the delivery locations, they are obtained from a delivery company in Los Angeles, each delivery locations corresponds to a node in the LA road network. We generate 40 test cases from these real delivery locations, which contains 100, 200, 500 and 1000 delivery locations separately, each has 10 test cases[2].

---

[2] Note that in operation research literatures, even an instance with 500 delivery locations is considered large. Algorithms are usually tested on much smaller instances.

Baseline Approaches. We compare our proposed algorithm with the following algorithms:

- *Sweep*: a cluster-first and route-second approach [11].
- *Clarke-Wright*: a saving heuristic based approach [9] that is widely used in most industries.
- *TDRTR*: we extend *RTR* to support time-dependent road network. Specifically, we pre-compute the cost matrix and evaluate the travel cost using time-dependent edge weights.

Accuracy Evaluation Method. For accuracy comparison, TDRTR is treated as the benchmark as it usually generates the best result.Thus, we compute the *gap* between the travel time from the current solution and the solution returned by TDRTR. Formally, for a problem instance, suppose the travel time of a solution generated by another algorithm (e.g., Sweep) is c', and the travel time of the solution generated by TDRTR is c, then the *gap* $\epsilon = \frac{c'-c}{c}$.

Smaller gap means smaller travel cost and thus yields higher accuracy. Note that the value of gap can be negative, because sometimes an algorithm could achieve better solution than TDRTR.

Configuration. We first compare our proposed algorithm(SSBLS) with Sweep, Clarke-Wright and TDRTR. For SSBLS, we use the parameters σ and η which produces the best solution tuned in the experiment. To show the effectiveness of using time-dependent road network, we also apply RTR in the static road network and evaluate the retrieved solution on the time-dependent road network. The steps are listed as follows: (1) generate a static road network by averaging the travel cost of each edge during the day. (2) run RTR in the generated averaged static road network, and generate a routing plan. (3) evaluate the travel cost of the routing plan in the time-dependent road network.

We then vary the neighbor size σ, and the learning rate η in the adaptive delivery location and operator selection process. All algorithms were implemented using Java, and all the experiments were performed on a Linux machine with 3.5 GHz CPU and 16 GB RAM.

4.2 Comparisons of Different Algorithms

Figure 4 shows the accuracy comparison between algorithms with respect to number of delivery locations. As illustrated, SSBLS achieves a similar accuracy with TDRTR and performs better than RTR. In general, local search based algorithms (e.g., SSBLS, TDRTR and RTR) are much more accurate than Sweep and Clarke-Wright. For example, SSBLS is 15 % to 27 % more accurate than Sweep or Clarke-Wright algorithm. In addition, with increasing number of delivery locations, the benefit tends to grow larger.

At the same time, with comparing RTR with TDRTR, we observe that time-dependent road network based apoproach can save 6 % to 8 % travel cost. This is

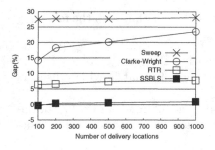
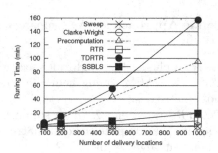

Fig. 4. Accuracy comparison   **Fig. 5.** Efficiency comparison

because in real world road network, the travel cost between two delivery locations may be quite different during rush hours and non-rush hours. Failing to consider traffic information leads to congestion, and thus larger travel cost.

Figure 5 shows the comparison of efficiency between different algorithms. We also show the precomputation (i.e., data preparation) time for TDRTR. As classic heuristics are much faster, and they can generate a feasible schedule within 1 min even for more than 1000 delivery locations. However, these heuristic algorithms suffer from low accuracy, which makes them less promising in practice. Although SSBLS is less efficient than Sweep and Clarke-Wright, SSBLS is much more faster than TDRTR. For example, SSBLS solves a problem instance with 1000 delivery locations in 20 min, which is 8 times faster than TDRTR. This is because (1) data preparation step is eliminated in SSBLS, which takes more than half of the total running time for TDRTR, as shown in Fig. 5, (2) SSBLS prefers to choose promising delivery locations and operators to generate neighborhood solution, thus reduce the unnecessary search of the solution space.

Note that although RTR is less accurate than SSBLS, RTR is a little faster than SSBLS. This is mainly because RTR is performed in the static road network. Compared to the time-dependent road network, (1) data preparation process in static road network is much faster, i.e., the number of interval is only 1 rather than 60, (2) travel cost evaluation in static network is also much faster.

In summary, considering the balance between accuracy and efficiency, SSBLS is best suitable for fast delivery with large scale delivery locations.

4.3 Effect of Neighbor Size

Figures 6 and 7 show the effect of neighbor size σ in terms of efficiency and accuracy when the number of delivery locations N are 100 and 500. For $N = 100$, $\sigma \in [10, 100]$, and for $N = 500$, $\sigma \in [10, 500]$. We use the number of evaluations to describe the relative speed. Generally, with decreasing σ, the algorithm tends to be more efficient but less accurate. For example, by setting $\sigma = 10$, the number of evaluated moves is reduced to 10 % compared with the case where $\sigma = 100$, however it suffers from a considerable drop of accuracy. In the following experiment, we set $\sigma = 40$ to balance accuracy and efficiency.

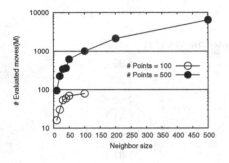

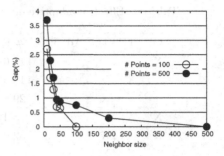

Fig. 6. Neighbor size v.s. efficiency

Fig. 7. Neighbor size v.s. accuracy

Table 3. Effect of adaptive delivery location selection ($\sigma = 40, N = 100$)

η	–	0	0.01	0.02	0.04	0.1	0.2	0.4	1
#Evaluated moves (M)	79.6	72.3	43.7	42.1	37.5	65.6	73.4	54.6	49.9
Gap (%)	0.91	0.92	0.87	0.78	0.98	1.36	1.4	1.45	1.8

4.4 Effect of Adaptive Delivery Location Selection

Table 3 shows the effect of adaptive delivery location selection. "–" means adaptive delivery location selection is not used. With increasing η, the accuracy first increases and then decreases. Thus, parameter tuning is required to achieve a good result. In addition, using adaptive delivery location selection also increases the efficiency. We use the number of evaluated moves to compare the relative efficiency of using different value of η. With adaptive delivery location selection, the algorithm prefers to conduct local moves on promising delivery locations, thus it is more likely to generate a better neighbor solution and reach the (local) optimal solution with less number of iterations. From Table 3, we find that by setting the learning rate $\eta = 0.02$, the algorithm reduces the number of evaluated moves to half while achieves a better accuracy compared with treating all the delivery locations with equal importance.

4.5 Effect of Adaptive Operator Selection

Table 4 shows the performance of different operator selection strategy. The number of evaluations describe the relative speed, and the gap represents the accuracy. In this set of experiment three operators are used: i.e., One-Point(OP), Two-Point(TP) and Two-Opt(TO). We compare the performance of using different operators, "X" in the cell means one corresponding operator is used.

As shown in Table 4, different operator strategies result in different efficiency and accuracy. Generally, the more operators we use, the more accurate results and the more number of evaluations will be conducted. However, by adaptively selecting operators in the local search process, the algorithm manages to be

Table 4. Effect of adaptive operator selection ($\sigma = 40, N = 100$)

OP	TP	TO	Adaptive	# Evaluated moves (M)	Gap(%)
X				17.6	2.99
	X			15.9	3.06
		X		29.7	4.62
X	X			53.4	1.91
X		X		32.9	1.27
	X	X		54.8	1.61
X	X	X		79.6	0.91
X	X	X	X	26.3	0.91

3 times faster than using all three operators, while not compromising the accuracy. This is because by applying promising operators, i.e. those who have good performance in the previous iterations, the algorithm is more likely to generate a better neighborhood solution, thus quickly reaching a (local) optimal, i.e., meet the stop condition with less number of evaluations.

5   Related Work

Vehicle Routing Problem (VRP) [1] is a well studied combinatorial optimization problem with different variants such as vehicle routing problem with time windows (VRPTW) [6,12], the capacitated vehicle routing problem (CVRP) [13]. Recently, the delivery and courier services are entering a period of rapid change enabled by recent technologies. On-time and fast delivery is becoming a significant differentiator for both delivery and E-commerce companies (e.g., Amazon), and hence same-day delivery or even 2-hour delivery become increasingly popular. To enable this new type of delivery, the efficiency and the accuracy are two of the most important factors. Previous methods focus either on efficiency or quality, but not both. Among the heuristic algorithms on the VRP problem, there are two main categories: classic heuristics and meta-heuristics. Classic heuristics (e.g., Clarke-Wright [9], Sweep algorithms [14]) emphasize more on quickly obtaining a feasible solution: for example, Clarke-Wright algorithm starts with an initial solution where each route only contains one delivery location, it continuously merges two routes into one route which generates the largest savings whenever it is feasible. The drawback with classic heuristics is that the solution could exists more than 20 % deviation with the best-known solution. This means that even in a small delivery instance with around 30 delivery locations, the solution calculated by classic heuristics could take 1 hour more delivery time than the best solution. In contrast, meta-heuristics perform a more thorough search of the solution space and hence gain more in solution quality but at the expense of speed. For example, even for a small delivery instance with around 100 delivery locations, some meta-heuristics [1] take more than one hour to calculate a routing plan with gap of less than 1 % with the best solution. In addition, this time usually increases exponentially when the number of delivery locations grows larger (e.g., more than 500 delivery locations), which is quite common in the real applications [7]. Compared with previous methods, our SSBLS algorithm strikes a balance between the accuracy and efficiency through restricting the neighbor size and adaptively selecting the delivery locations and local search operators.

Previous approaches mainly assume that the problem instance is defined on a complete directed graph. However, in practice, most of the VRPs take place on real road networks. Although, it is possible to transform a VRP instance on a road network into an instance on a complete directed graph, it involves large amount of shortest path computation [15, 16]. The problem becomes even more challenging when dealing with real world time-dependent road network. Previous works on time dependent VRP [17, 18] mainly use the synthetic time-varying edge weights in which they assume for each delivery pair, there exists a few fixed number (e.g., 3 or 4) of time intervals and the travel time between the delivery locations at each time interval is a constant, thus they could pre-compute a distance matrix for each delivery pair at every time interval. However, the real world time-dependent road network has a much larger number of time intervals (e.g., 60), and the travel cost between a pair of delivery locations at certain time interval is not known in advance, which usually requires a costly shortest path computation on road network. In [4, 10], a time-dependent bidirectional A* shortest path algorithm is proposed, which builds indexes and calculates tight bound in the A* search process based on the lower/upper bound graph. Even with these optimization, shortest path computation on time-dependent road network takes several hundred milliseconds on Los Angeles road network. Moreover, in the on-line delivery business, the requests may not be known in advance, which also prohibits the pre-computation of the distance matrix for each delivery problem. In this paper, we eliminate the data-preparation step via an on-demand calculation procedure, which significantly reduces the total running time of the algorithm.

6 Conclusion

In this paper, we studied the problem of Vehicle Routing in time-dependent road network. To enable fast and accurate results, we proposed a new local search framework, which eliminates the time-consuming data-preparation step required by the Operation Research methods via an on-demand-calculation strategy from the field of spatial databases. We further improved the efficiency by leveraging a guided search process to reduce the unnecessary exploration of the solution space. Our experiments with real-world dataset verified that our algorithm strikes a good balance between efficiency and accuracy, which makes it practical for the future delivery business in large-scale and dynamic environment.

Acknowledgements. This research has been funded in part by NSF grants IIS-1115153 and IIS-1320149, the USC Integrated Media Systems Center (IMSC), METRANS Transportation Center under grants from Caltrans, and unrestricted cash gifts from Oracle. Any opinions, findings, and conclusions or recommendations expressed in this material are those of the author(s) and do not necessarily reflect the views of any of the sponsors such as the National Science Foundation.

References

1. Cordeau, J.F., Gendreau, M., Laporte, G., Potvin, J.Y., Semet, F.: A guide to vehicle routing heuristics. J. Oper. Res. Soc. **53**(5), 512–522 (2002)
2. The Wall Street Journal: At UPS, the algorithm is the driver (2015). http://www.wsj.com/articles/at-ups-the-algorithm-is-the-driver-1424136536
3. Google Express. https://www.google.com/shopping/express/
4. Demiryurek, U., Banaei-Kashani, F., Shahabi, C., Ranganathan, A.: Online computation of fastest path in time-dependent spatial networks. In: Pfoser, D., Tao, Y., Mouratidis, K., Nascimento, M.A., Mokbel, M., Shekhar, S., Huang, Y. (eds.) SSTD 2011. LNCS, vol. 6849, pp. 92–111. Springer, Heidelberg (2011)
5. Malandraki, C., Daskin, M.S.: Time dependent vehicle routing problems: formulations, properties and heuristic algorithms. Transp. Sci. **26**(3), 185–200 (1992)
6. Hashimoto, H., Yagiura, M., Ibaraki, T.: An iterated local search algorithm for the time-dependent vehicle routing problem with time windows. Discrete Optim. **5**(2), 434–456 (2008)
7. Li, F., Golden, B., Wasil, E.: Very large-scale vehicle routing: new test problems, algorithms, and results. Comput. Oper. Res. **32**(5), 1165–1179 (2005)
8. Groer, C.: Parallel and serial algorithms for vehicle routing problems. ProQuest (2008)
9. Clarke, G., Wright, J.W.: Scheduling of vehicles from a central depot to a number of delivery points. Oper. Res. **12**, 568–581 (1964)
10. Demiryurek, U., Banaei-Kashani, F., Shahabi, C.: Efficient K-nearest neighbor search in time-dependent spatial networks. In: Bringas, P.G., Hameurlain, A., Quirchmayr, G. (eds.) DEXA 2010, Part I. LNCS, vol. 6261, pp. 432–449. Springer, Heidelberg (2010)
11. Laporte, G., Gendreau, M., Potvin, J.Y., Semet, F.: Classical and modern heuristics for the vehicle routing problem. Int. Trans. Oper. Res. **7**(4–5), 285–300 (2000)
12. Potvin, J.Y., Kervahut, T., Garcia, B.L., Rousseau, J.M.: The vehicle routing problem with time windows part I: tabu search. INFORMS J. Comput. **8**(2), 158–164 (1996)
13. Baldacci, R., Hadjiconstantinou, E., Mingozzi, A.: An exact algorithm for the capacitated vehicle routing problem based on a two-commodity network flow formulation. Oper. Res. **52**(5), 723–738 (2004)
14. Gillett, B.E., Miller, L.R.: A heuristic algorithm for the vehicle-dispatch problem. Oper. Res. **22**(2), 340–349 (1974)
15. Longo, H., de Aragão, M.P., Uchoa, E.: Solving capacitated arc routing problems using a transformation to the CVRP. Comput. Oper. Res. **33**(6), 1823–1837 (2006)
16. Letchford, A.N., Nasiri, S.D., Oukil, A.: Pricing routines for vehicle routing with time windows on road networks. Comput. Oper. Res. **51**, 331–337 (2014)
17. Ichoua, S., Gendreau, M., Potvin, J.Y.: Vehicle dispatching with time-dependent travel times. Eur. J. Oper. Res. **144**, 379–396 (2003)
18. Kok, A., Hans, E., Schutten, J.: Vehicle routing under time-dependent travel times: the impact of congestion avoidance. Comput. Oper. Res. **39**(5), 910–918 (2012)

Oriented Online Route Recommendation for Spatial Crowdsourcing Task Workers

Yu Li$^{(\boxtimes)}$, Man Lung Yiu, and Wenjian Xu

Department of Computing, Hong Kong Polytechnic University, Hong Kong, China
{csyuli,csmlyiu,cswxu}@comp.polyu.edu.hk

Abstract. Emerging spatial crowdsourcing platforms enable the workers (i.e., crowd) to complete spatial crowdsourcing tasks (like taking photos, conducting citizen journalism) that are associated with rewards and tagged with both time and location features. In this paper, we study the problem of online recommending an optimal route for a crowdsourcing worker, such that he can (i) reach his destination on time and (ii) receive the maximum reward from tasks along the route. We show that no optimal online algorithm exists in this problem. Therefore, we propose several heuristics, and powerful pruning rules to speed up our methods. Experimental results on real datasets show that our proposed heuristics are very efficient, and return routes that contain 82–91 % of the optimal reward.

1 Introduction

Spatial crowdsourcing platforms[1,2] publish crowdsourcing tasks that are associated with rewards and tagged with spatial / temporal attributes (e.g., location, release time and deadline). To complete a task, a worker must reach the task's location before its deadline. Popular tasks include taking photos, reporting activities / accidents, and verifying data on-site, etc.

Regarding the matching between tasks and workers, existing approaches on spatial crowdsourcing can be divided into: (i) the *server-centric* mode [15,16], where the server assigns tasks to workers based on their reported locations / regions, or (ii) the *worker-centric* mode [3,7,10], where the server publishes its tasks and let workers to choose any task freely. In this paper, we adopt the worker-centric mode as it protects the location privacy of the worker [10] and enables the worker to choose tasks autonomously from different crowdsourcing platforms which he has registered in.

The closest work to ours is the *maximum task scheduling* (MTS) problem [10]. It returns a route that covers the maximum number of tasks (in a worker's specified region, e.g., his city). Since [10] considers the MTS problem at a snapshot, it would not update the worker's route when new tasks arrive. We illustrate it in

---

The research is partly supported by grant GRF 152201/14E from Hong Kong RGC.

[1] www.clickworker.com/en/mobile-crowdsourcing.

[2] features.en.softonic.com/mobile-crowdsourcing-does-it-work.

© Springer International Publishing Switzerland 2015
C. Claramunt et al. (Eds.): SSTD 2015, LNCS 9239, pp. 137–156, 2015.
DOI: 10.1007/978-3-319-22363-6_8

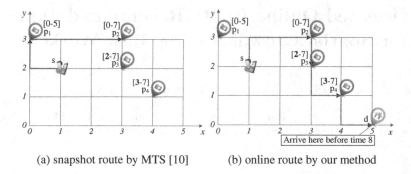

(a) snapshot route by MTS [10]   (b) online route by our method

Fig. 1. Route recommendation for the worker: each task p_i with [release time - deadline]

Fig. 1a. Assume that we use the Manhattan distance and each grid takes a time unit to travel. Each task p_i is tagged with its release time and deadline. Suppose that the worker starts from s at time 0. The MTS route is $s \to p_1 \to p_2$. The solution in [10] would not update the route when new tasks are released (e.g., p_3, p_4).

In this paper, we wish to support two extra requirements compared to [10]: ($\mathcal{R}1$) update the worker's route online with respect to newly released tasks and ($\mathcal{R}2$) align with the worker's trip, i.e., reaching a destination before expected time. It is important to support $\mathcal{R}1$ in order to assign a worker as many tasks as possible. New spatial crowdsourcing tasks are indeed being released continuously in real systems[3]. We also consider the requirement $\mathcal{R}2$ as the worker may have planned his own activities, e.g., reaching a specified destination by an expected time [17]. Such worker is willing to take crowdsourcing tasks along his trip provided that he can arrive at his destination on time.

To this end, we study the online route recommendation problem for spatial crowdsourcing workers, by taking requirements $\mathcal{R}1$ and $\mathcal{R}2$ into consideration. Figure 1b illustrates the route recommended by our method. Suppose that the worker starts from s at time 0 and plans to arrive at home $(5,0)$ at time 8. At time 0, the worker is recommended to take the task p_2. When new tasks are released (e.g., p_3, p_4), the worker is recommended to take them. In summary, our recommended route is $s \to p_2 \to p_3 \to p_4 \to d$, which covers 3 tasks and reaches the destination d on time.

To the best of our knowledge, this paper is the first on tackling the online route recommendation problem for spatial crowdsourcing workers with destination and arrival time constraints. We contribute the followings:

- We show that no algorithm can achieve a non-zero competitive ratio [2] in our online problem, meaning that the number of tasks found by any online algorithm may be arbitrarily small compared to the optimal offline solution.
- We propose two categories of heuristics (GetNextTask and Re-Route) that offer trade-offs between the response time and the number of tasks. GetNextTask greedily selects the next task to complete so it incurs a short response time.

---

[3] www.clickworker.com/en/clickworkerjob, www.lionbridge.com.

On the other hand, Re-Route produces a route with more tasks as it conducts a complete search to update the optimal route with respect to newly released tasks.

- We further propose pruning rules to reduce the response time of Re-Route.

Experiments on real datasets show that our methods take less than 1 s to update the route, and return routes that contain 82–91 % of the optimal number of tasks.

The remainder of this paper is organized as follows. We formally define our problem in Sect. 2. Then, we illustrate our proposed heuristics in Sect. 3 and present optimization techniques in Sect. 4. In Sect. 5, we test the performance of our proposed techniques on both real and synthetic datasets. Section 6 highlights the related work. Finally, we conclude our paper in Sect. 7.

2   Problem Statement

We first introduce some terminology and then define our problem formally.

Definition 1 (Task p). *We denote a task by $p_{sid,kid} = (loc, [t_p^-, t_p^+])$, where loc is the task's location, t_p^-, t_p^+ are the release time and deadline of the task, respectively. The subscripts sid and kid denote the task's server ID and task ID, respectively. A worker may complete p and collect the reward[4] if he can reach p.loc before t_p^+.*

Definition 2 (Query q). *We denote a query q by $q = (s, d, [t_q^-, t_q^+])$. s and d are the worker's start and destination locations, respectively. t_q^- and t_q^+ are the start time from s and expected arrival time at d, respectively.*

Definition 3 (Travel Time τ). *We denote the travel time as $\tau(v, u) = \frac{dist(v,u)}{speed_q}$, where $dist(v, u)$ is the distance[5] between v and u, and $speed_q$ is the (constant) travel speed of the worker for q. $\tau(R)$ denotes the travel time along a route R (via vertices on R).*

With the above terminology, we are ready to define our problem formally below.

Problem 1 (Oriented Online Route Recommendation (OnlineRR)). Let a worker's query be $q = (s, d, [t_q^-, t_q^+])$. OnlineRR aims to find a route such that it covers *the maximum number of tasks* and the worker can arrive at d by t_q^+. It may update the route according to the worker's live location and the new tasks released by crowdsourcing servers.

---

[4] The reward of a task can be collected by the same worker for only once. Similar to [10], we assume that each task has a unit reward and can be completed immediately.

[5] Our method can be applied to any distance function provided that it satisfies the triangle inequality, such as Euclidean distance, Manhattan distance, and road network distance.

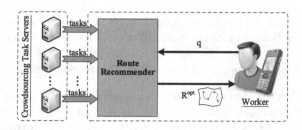

Fig. 2. System architecture

We adopt the system architecture as depicted in Fig. 2. Spatial crowdsourcing servers publish new spatial crowdsourcing tasks. A worker may install our *route recommender* on his mobile device (smartphone). The route recommender is responsible for: (i) collecting task information from different servers continuously, (ii) recommending / updating a route based on the worker's current location and available tasks.

3   Online Route Recommendation

First, we prove in Sect. 3.1 that no online algorithm can achieve a non-zero competitive ratio in OnlineRR. Then, we propose two categories of heuristic approaches for OnlineRR in Sects. 3.2 and 3.3.

3.1   Competitive Analysis

We use the competitive ratio [2] to measure the performance of online algorithms. Since OnlineRR is a maximization problem, the competitive ratio $\mathcal{CR}$ is defined as:

$$\mathcal{CR} = \min_{e \in E} \frac{count(R_{alg}(e))}{count(R_{opt}(e))} \qquad (1)$$

where E denotes the set of all problem instances, $R_{alg}(e)$ is the route recommended by an online algorithm alg for instance e, $R_{opt}(e)$ is the optimal route R_{opt} for instance e (cf. Definition 4), and $count(R_*(e))$ means the number of tasks on $R_*(e)$.

Definition 4 (Optimal route $R_{opt}(e)$ for OnlineRR). *Given a problem instance e, we denote its optimal route by $R_{opt}(e)$, which is obtained under assumption that the information of all tasks are known in advance (even before their release times).*

We show our competitive analysis below. It applies to any online algorithm, including both deterministic algorithms and randomized algorithms.

Theorem 1. *No online algorithm has a non-zero competitive ratio for OnlineRR.*

Proof. Since $\mathcal{CR} = \min_{e \in E} \frac{count(R_{alg}(e))}{count(R_{opt}(e))}$, it suffices to find a specific instance (i.e., the adversary) that makes $\mathcal{CR}$ as low as possible. Without loss of generality, in the following proof, we consider only locations on the positive half line $[0, +\infty)$.

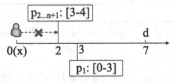
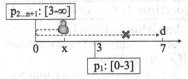

(a) Case 1: the worker cannot reach the location of $p_{2\leq i\leq n+1}$ before their deadline (i.e., time 4)

(b) Case 2: the worker cannot proceed to $p_{2\leq i\leq n+1}$ and arrive at d on time

Fig. 3. At time $m = 3$, adversaries release tasks $p_{2\leq i\leq n+1}$ with [release time - deadline]

For the query, we set $t_q^- = 0$, $s = 0$, $t_q^+ = 10$, $d = 7$. Assume that $speed_q = 1$, that is $\tau_e(v, u) = |v - u|$. We simply denote a task p by $(p.loc, [t_p^-, t_p^+])$.

At time 0, the adversary releases a task $p_1 = (3, [0, 3])$. At time $m = 3$, the adversary will check the worker's current location (say x), and then decides to further release n tasks accordingly. There are two cases: (1) $x = 0$, or (2) $x > 0$. We show that the adversary can release those n tasks to make $\mathcal{CR}$ arbitrarily small.

Case 1: $x = 0$. In this case, the adversary will release tasks $p_{2\leq i\leq n+1} = (2, [3, 4])$ (see Fig. 3a). The worker cannot complete these tasks, since he cannot reach them before their deadlines, and thus $count(R_{alg}) = 0$. But if all tasks are known in advance, the worker can wait at position 2 until all tasks are released and finish them on time $m = 3$. In this case, the competitive ratio is: $\mathcal{CR} = 0/n = 0$.

Case 2: $x > 0$. In this case, the adversary would release n tasks $p_{2\leq i\leq n+1} = (0, [m, \infty])$ (see Fig. 3b). As $m + x + d > m + d = 10 = t_q^+$, the worker cannot proceed to position 0 at time m; otherwise, he cannot reach d before t_q^+. So, the worker can finish at most the task p_1 only if he moves directly to m at time 0. However, if all tasks are known in advance, the worker could stay at 0 until time $m = 3$ to finish tasks $p_{2\leq i\leq n+1}$, and thus $count(R_{opt}) = n$. Therefore, $\mathcal{CR} \leq 1/n \to 0$ because n can be an arbitrary large value.

3.2 Greedy Task Approach

In this section, we present a greedy approach that incurs low response times.

The greedy approach works as follows. Initially, it calls GetNextTask (cf. Algorithm 1) to find the first task for the worker. Given the set of available[6] tasks P and the worker's location s_{now} at current time t_{now}, GetNextTask greedily selects the task with the highest score ψ_p. Upon reaching the chosen task, GetNextTask is involved to get the next task repeatedly until reaching d.

Due to the tasks' deadlines and the worker's expected arrival time (cf. Definitions 1, 2), the worker may complete a task p if: (i) he can reach $p.loc$ before t_p^+, and (ii) he can reach d no later than t_q^-. Therefore, we call a task to be *feasible* if it satisfies:

$$\tau(s_{now}, p) + \tau(p, d) \leq t_q^+ - t_{now} \qquad \text{and} \qquad t_{now} + \tau(s_{now}, p) \leq t_p^+ \qquad (2)$$

---

[6] Available tasks are tasks released before the current time t_{now}.

Algorithm 1. Get next best task

---

    algorithm GetNextTask (Query $q = (s_{now}, d, [t_{now}, t_q^+])$, Set of available tasks P)
1: $Cand \leftarrow$ compute the set of feasible tasks from P ▷ apply Equation 2
2: **if** $Cand \neq \emptyset$ **then**
3:     $p_{next} \leftarrow$ choose $p \in Cand$ with best score ψ_p ▷ ψ_p is a heuristic function
4:     Return p_{next}
5: **else**
6:     Apply policy $\mathcal{P}_{stay}$ or $\mathcal{P}_{go}$ until $Cand \neq \emptyset$ or $t_{now} + \tau(s_{now}, d) = t_q^+$

If there is no feasible task for q, the worker may stay or move based on a predefined policy (cf. Line 6 in Algorithm 1). In the policy $\mathcal{P}_{go}$, the worker simply moves towards the destination d. In the policy $\mathcal{P}_{stay}$, the worker waits at s_{now} until $t_{now} + \tau(s_{now}, d) = t_q^+$. When new feasible tasks are released, we resume the search and invoke GetNextTask to obtain the next task.

We illustrate several heuristics for computing the score ψ_p. Figure 4a shows the map of tasks which are labeled with release times and deadlines, and Fig. 4b shows the result route of each heuristic. In this example, we use the query $q = (s, d, [0, 10])$, the policy $\mathcal{P}_{stay}$, and the Manhattan distance.

Nearest Neighbor Heuristic (G-NN). It chooses the nearest feasible task to the worker's current location s_{now}, and thus setting $\psi_p = \tau(s_{now}, p)$. In Fig. 4, G-NN produces the route $\langle s, p_7, p_5, d \rangle$.

Earliest Deadline Heuristic (G-ED). It chooses the task with the earliest deadline, and thus setting $\psi_p = t_p^+$. In Fig. 4, G-ED recommends the route $\langle s, p_7, p_5, d \rangle$.

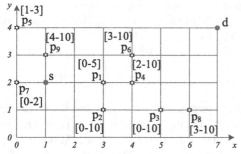

Heuristic	Route
G-NN	$\langle s, p_7, p_5, d \rangle$
G-ED	$\langle s, p_7, p_5, d \rangle$
G-MCS	$\langle s, p_1, p_4, p_6, d \rangle$
Re-Route	$\langle s, p_1, p_2, p_3, p_8, d \rangle$

(a) map of tasks with [release time - deadline]   (b) result routes (with $\mathcal{P}_{stay}$)

Route	# of tasks	Route	# of tasks	Route	# of tasks
$\langle s, p_1, d \rangle$	1	$\langle s, p_2, d \rangle$	1	$\langle s, p_3, d \rangle$	1
$\langle s, p_7, d \rangle$	1	$\langle s, p_1, p_2, d \rangle$	2	$\langle s, p_1, p_3, d \rangle$	2
$\langle s, p_2, p_3, d \rangle$	2	$\langle s, p_7, p_1, d \rangle$	2	$\langle s, p_1, p_2, p_3, d \rangle$	3
$\langle s, p_3, p_1, d \rangle$	not feasible	$\cdots$	not feasible	$\cdots$	not feasible

(c) all possible routes known at $t_{now} = 0$

Fig. 4. Example of query $q = (s, d, [0, 10])$ in OnlineRR (using Manhattan Distance)

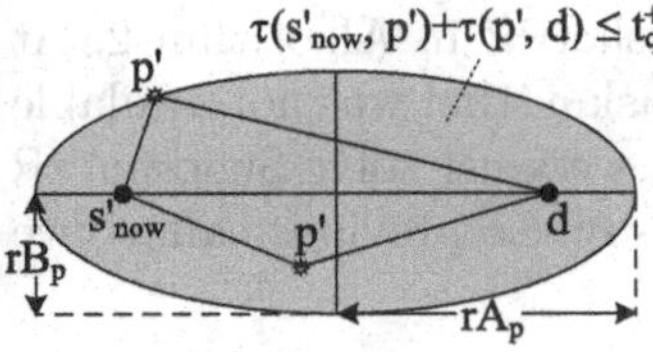

$$\tau(s'_{now}, p') + \tau(p', d) \le t_q^+ - t'_{now}$$

$$\psi_p = \pi \cdot rA_p \cdot rB_p$$
$$rA_p = (t_q^+ - t'_{now})/2$$
$$rB_p = \sqrt{rA_p^2 - (\tau(s'_{now}, d)/2)^2}$$
$$\text{where } t'_{now} = t_{now} + \tau(s_{now}, p) \text{ and } s'_{now} = p.loc$$

(a) search space (in shade)   (b) search space size calculation

Fig. 5. Feasible candidates search space for Euclidean distance metric

Maximum Candidate Space Heuristic (G-MCS). It chooses the task p that can maximize the search space of feasible tasks (Eq. 2) in future. The search space in future is obtained under the assumption that p is just completed. The space shape differs for different distance metrics, but we can use a general approach Monte Carlo [20] to compare it. If a specific distance metric is used, then the exact candidate space size can be calculated. Take Euclidean distance for example, the space size is the area of the ellipse shown in Fig. 5a, and thus we can calculate the score ψ_p using equations in Fig. 5b for Euclidean distance metric.

We illustrate how G-MCS works in Fig. 4. At time 0, the feasible tasks are p_1, p_2, p_3, p_7. Since p_1 has the highest score (ψ_{p_1}), p_1 is chosen to be visited. When the worker reaches p_1, a new task p_4 is released while p_7 expires, so the set of feasible tasks becomes $\{p_2, p_3, p_4\}$. Then p_4 is chosen as it has the highest score (ψ_{p_4}). Upon reaching p_4, the algorithm selects p_6 as it has the best score among $\{p_3, p_6, p_8\}$. After completing task p_6, there are no more feasible tasks. After waiting for two more time units, the worker moves toward d. In summary, G-MCS obtains the route $\langle s, p_1, p_4, p_6, d \rangle$.

3.3   Complete Search for Route Approach

In this section, we present a complete search approach that tends to find more tasks than the heuristics in Sect. 3.2.

Specifically, we formulate the following SnapshotRR problem, which takes the current query and the set of available tasks as input. Then, we solve SnapshotRR by enumerating all possible routes and obtain the one with the maximum number of tasks.

*Problem 2 (Snapshot Route Recommendation (*SnapshotRR*)).* Given a query $q = (s_{now}, d, [t_{now}, t_q^+])$ at the current snapshot t_{now}, SnapshotRR aims to find a route such that it covers *the maximum number of tasks* and the worker can arrive at d by t_q^+.

We illustrate this approach for the query $q = (s, d, [0, 10])$ in Fig. 4. At time 0, we apply Eq. 2 and obtain the set of feasible tasks: $P = \{p_1, p_2, p_3, p_7\}$. Figure 4c shows all possible routes (known at time 0). The optimal route at time 0 is $\langle s, p_1, p_2, p_3, d \rangle$.

We propose a simple optimization to solve SnapshotRR in Algorithm 2. At Line 3, we check whether there exists a new feasible task p (that was not available in the previous call of Algorithm 2). If such p exists, we must solve SnapshotRR again. Otherwise, the best route remains the same as in the previous call, so we need not solve SnapshotRR again.

Algorithm 2. Complete search the result route

---

    algorithm Re-Route (Query $q = (s_{now}, d, [t_{now}, t_q^+])$, Set of available tasks P)
1: Let P_{prev} be the set of available tasks in the previous call
2: **if** $P \neq \emptyset$ **then**
3:     **if** $\exists p \in P - P_{prev}$ such that p is feasible **then**   ▷ Equation 2
4:         $R \leftarrow$ Solve SnapshotRR(q, P) ▷ conduct complete search
5: **else**
6:     Apply policy $\mathcal{P}_{stay}$ or $\mathcal{P}_{go}$ until $P \neq \emptyset$ or $t_{now} + \tau(s_{now}, d) = t_q^+$

We proceed to illustrate how Re-Route works in the example in Fig. 4. At time 0, Re-Route computes the route $R_0 = \langle s, p_1, p_2, p_3, d \rangle$, and then the worker moves along R_0 to p_1. Upon reaching p_1, a new feasible task p_4 is found, so Re-Route re-calculates the route as $R_1 = \langle p_1, p_2, p_3, d \rangle$. When the worker reaches p_2, a new feasible task p_8 is found, so Re-Route updates the route to $R_2 = \langle p_2, p_3, p_8, d \rangle$. After reaching p_8, a new task p_9 is found but it is not feasible. Thus, Re-Route would not computes the route again (cf. Line 3 in Algorithm 2). Eventually, the worker moves to d. In summary, the actual route traveled by the worker is: $\langle s, p_1, p_2, p_3, p_8, d \rangle$. It covers more tasks than other heuristics (cf. Figure 4b).

Since it is expensive to solve SnapshotRR by enumerating all possible routes, we will present optimizations to solve SnapshotRR efficiently in Sect. 4.

4   Optimization for SnapshotRR

We adapt the *bi-directional search* algorithm for the Orienteering Problem with Time Windows (OPTW) problem [19] to solve our problem. For brevity in discussion, we use $q = (s, d, [t_q^-, t_q^+])$ instead of $q = (s_{now}, d, [t_{now}, t_q^+])$. We will conduct bi-directional search for SnapshotRR in three steps:

Step 1: Search sub-routes in the forward direction (from s) and store them in $\overrightarrow{\mathbb{R}}$
Step 2: Search sub-routes in the backward direction (from d) and store them in $\overleftarrow{\mathbb{R}}$
Step 3: Join sub-routes between $\overrightarrow{\mathbb{R}}$ and $\overleftarrow{\mathbb{R}}$

According to Pruning Rule 1, the bi-directional search can reduce the search space. However, the method in [19] does not exploit spatial properties in our problem. In this section, we develop more effective pruning rules to accelerate bi-directional search on SnapshotRR.

Pruning Rule 1 (Half travel time bound property proved in [19]). *In the forward (or backward) route searching from vertex s (or d), only routes R with $\tau(R) \leq \tau_{max}/2$ are maintained and extended, where $\tau_{max} = t_q^+ - t_q^-$.*

4.1 Forward Search and Backward Search

In this section, we elaborate the forward search (Step 1) and discuss adaptations for the backward search (Step 2) at the end. In the following discussion, we use R instead of $\overrightarrow{R}$ to represent a sub-route found in forward search (which will be stored in $\overrightarrow{\mathbb{R}}$) for simplicity.

We first introduce the sub-route concept and its extension operation. Then, we propose a pruning rule and a search strategy to speedup the computation. In the following, we denote the set of vertices as $V = P \cup \{s, d\}$, where P is the set of available tasks.

Sub-route Extension.

We denote a path from s to $v \in V$ as a sub-route R_v, which contains four attributes $R_v = (\tau(R_v), B_{R_v}, C_{R_v}, v)$.

- $\tau(R_v)$ represents the travel time along R_v (i.e., from s to v).
- B_{R_v} stores a sequence of tasks visited <u>before</u> on the sub-route R_v. We denote the profit of R_v as $|B_{R_v}|$ because all tasks have the same reward.
- C_{R_v} is a set of <u>candidate</u> vertices (that are feasible for visiting in future), and its calculation is discussed in Eq. 5.

During route search, for each vertex v, we store all sub-routes of the form R_v into a set $\mathbb{R}_v$. In addition, we only consider *feasible routes*. Recall that $\tau(R_v)$ represents the travel time (along R_v) from s to v. According to Eq. 2, a sub-route R_v is said to be *feasible* if:

$$\tau(R_v) \le t_v^+ - t_q^- \qquad \text{and} \qquad \tau(R_v) \le t_q^+ - t_q^- \tag{3}$$

where t_v^+ is the deadline for vertex v when v is a task, or ∞ when $v \in \{s, d\}$.

For each vertex $u \in C_{R_v}$, we can extend R_v with an arc (v, u) to form a new sub-route R_u. The component of $R_u = (\tau(R_u), B_{R_u}, C_{R_u}, u)$ is calculated as follows:

$$B_{R_u} \leftarrow \langle B_{R_v}, v \rangle \quad \text{and} \quad \tau(R_u) \leftarrow \tau(R_v) + \tau_e(v, u) \tag{4}$$

The set C_{R_u} contains each candidate vertex p that satisfies:

$$p \in C_{R_v}(\heartsuit) \quad \text{and} \quad p \notin B_{R_u}(\diamondsuit)$$
$$\tau(u, p) \le t_p^+ - t_q^- - \tau(R_u) \quad \text{and} \quad \tau(u, p) \le (t_q^+ - t_q^-)/2 - \tau(R_u)(\clubsuit, \spadesuit, \blacklozenge)$$
$$\tau(u, p) + \tau(p, d) \le t_q^+ - t_q^- - \tau(R_u)(\clubsuit, \blacktriangledown) \tag{5}$$

which involve the constraints in Eq. 4 ($\clubsuit$), Eq. 3 ($\spadesuit$), triangle inequality ($\heartsuit$), the constraint that each task can be visited only once ($\diamondsuit$), the worker's arrival time t_q^+ ($\blacktriangledown$) and Pruning Rule 1 ($\blacklozenge$).

We illustrate sub-route extension in Fig. 6. Assume that $q = (s, d, [0, 10])$ and $P = \{p_1, p_2, \cdots, p_7\}$. We consider Manhattan distance in this example. First, we compute the candidate set of s. By Pruning Rule 1, we only consider tasks within $10/2 = 5$ units from s (i.e., tasks in the dotted diamond in Fig. 6). Thus, tasks p_3, p_7 are not feasible. The tasks p_4 and p_5 are not feasible as they violate constraints on the task's deadline and the worker's arrival time, respectively. Thus, we obtain the candidate set of s as $C_s = \{p_1, p_2, p_6\}$, and compute the sub-route for s as $R_s = (0, \emptyset, C_s, s)$. Next, we append arcs (s, p_1), (s, p_2), (s, p_6) into R_s to generate three new sub-routes: $R_1 = (1, \langle p_1 \rangle, \{p_2, p_6\}, p_1)$, $R_2 = (3, \langle p_2 \rangle, \{p_6\}, p_2)$, $R_6 = (5, \langle p_6 \rangle, \emptyset, p_6)$.

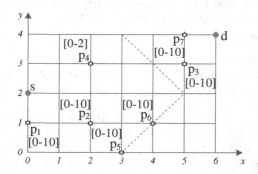

Fig. 6. Example query $q = (s, d, [0, 10])$ for SnapshotRR problem (using Manhattan distance)

Dominate Test Pruning.

We develop the following pruning rule to further reduce the search space.

Pruning Rule 2 (Dominating Pruning). *Let $R_v = (\tau(R), B_{R_v}, C_{R_v}, v)$ and $R'_v = (\tau(R'_v), B_{R'_v}, C_{R'_v}, v)$ be two feasible routes associated with v. We can prune R'_v if:*

$$\tau(R_v) \leq \tau(R'_v) \quad and \quad |C_{R'_v} \cap B_{R_v}| \leq |B_{R_v}| - |B_{R'_v}|$$

Proof. Among all full routes with R'_v as the prefix, let $R'_{opt} = \langle s, B_{R'_v}, R'_{tail}, d \rangle$ be the maximum reward route. With the given condition $\tau(R_v) \leq \tau(R'_v)$, after traveling along R_v, we can still follow all tasks in R'_{tail} and arrive at d by t_q^+. There exists a route $R_{exist} = \langle s, B_{R_v}, R_{tail}, d \rangle$ where $R_{tail} = R'_{tail} - B_{R_v}$. R_{exist} ensures that the reward of each task is gained at most once as B_{R_v} and R_{tail} have no common tasks.

Since $R'_{tail} \subseteq C_{R'_v}$, we have $|R'_{tail}| = |R_{tail}| + |R'_{tail} \cap B_{R_v}| \leq |R_{tail}| + |C_{R'_v} \cap B_{R_v}|$.

By combining the above with the given condition $|C_{R'_v} \cap B_{R_v}| \leq |B_{R_v}| - |B_{R'_v}|$, we derive: $|B_{R_v}| + |R_{tail}| \geq |B_{R'_v}| + |C_{R'_v} \cap B_{R_v}| + |R_{tail}| \geq |B_{R'_v}| + |R'_{tail}|$. As the reward of R_{exist} (extended from R_v) is greater than or equal to that of R'_{opt} (extended from R'_v), we can prune the subroute R'_v.

Search Strategy.

Our strategy is to identify sub-routes with better reward values in order to utilize pruning rule 2. To do so, we introduce the concept of upper bound reward:

Definition 5 (Vertex upper bound reward $\$_v^+$). *Given a sub-route $R_v = (\tau(R_v), B_{R_v}, C_{R_v}, v)$, we define its upper bound reward as $\$_{R_v}^+ = |B_{R_v}| + |C_{R_v}|$. The upper bound reward of vertex $v \in V$ is defined as: $\$_v^+ = \max\{\$_{R_v}^+ \mid R_v \in \overrightarrow{\mathbb{R}_v}\}$.*

Initially, we begin the search from a sub-route at s. We iteratively extend sub-routes found so far and apply pruning rule 2 to discard unpromising sub-routes. During the search, we employ a heap H to process vertices in descending order of $\$_v^+$.

Table 1. Forward space search

Iteration	Selected Vertex	Extended Route R	Modified $\mathbb{R}$	Heap H
1	s	$(0, \emptyset, \{p_1, p_2, p_6\}, s)$	$\overrightarrow{\mathbb{R}_{p_1}} = \{(1, \langle p_1 \rangle, \{p_2, p_6\}, p_1)\}$	$(p_1, 3)$
			$\overrightarrow{\mathbb{R}_{p_2}} = \{(3, \langle p_2 \rangle, \{p_6\}, p_2)\}$	$(p_2, 2)$
			$\overrightarrow{\mathbb{R}_{p_6}} = \{(5, \langle p_6 \rangle, \emptyset, p_6)\}$	$(p_6, 1)$
2	p_1	$(1, \langle p_1 \rangle, \{p_2, p_6\}, p_1)$	$\overrightarrow{\mathbb{R}_{p_2}} = \{(3, \langle p_1, p_2 \rangle, \{p_6\}, p_2)\}$	$(p_2, 3)$
			$\overrightarrow{\mathbb{R}_{p_6}} = \{(5, \langle p_1, p_6 \rangle, \emptyset, p_6)\}$	$(p_6, 2)$
3	p_2	$(3, \langle p_1, p_2 \rangle, \{p_6\}, p_2)$	$\overrightarrow{\mathbb{R}_{p_6}} = \{(5, \langle p_1, p_2, p_6 \rangle, \emptyset, p_6)\}$	$(p_6, 3)$
4	p_6	$\emptyset$	$\emptyset$	$\emptyset$
$\overrightarrow{\mathbb{R}}$		$(5, \langle p_1, p_2, p_6 \rangle, \emptyset, p_6)$, $(3, \langle p_1, p_2 \rangle, \{p_6\}, p_2)$, $(1, \langle p_1 \rangle, \{p_2, p_6\}, p_1)$, $(0, \emptyset, \{p_1, p_2, p_6\}, s)$		

We illustrate this method on the example in Fig. 5 and show the running steps in Table 1. Iteration 1 corresponds to the extension of the sub-route R_s at s, which we have discussed before. We obtain three new subroutes R_1, R_2, R_6, insert them in their corresponding route sets $\overrightarrow{\mathbb{R}_p}$, and also enheap p_1, p_2, p_6 into H. In each subsequent iteration, we deheap the vertex $v \in H$ with the largest $\$_v^+$, and extend its sub-routes R_v in the descending order of $|B_{R_v}|$.

In iteration 2, we generate a new sub-route $(3, \langle p_1, p_2 \rangle, \{p_6\}, p_2)$ and apply Pruning Rule 2 to discard the previous subroute at p_2, i.e., $(3, \langle p_2 \rangle, \{p_6\}, p_2)$. Similarly, the previous sub-routes for p_6: $(5, \langle p_6 \rangle, \emptyset, p_6)$ and $(5, \langle p_1, p_6 \rangle, \emptyset, p_6)$ are pruned in iterations 2 and 3, respectively.

The forward search terminates when H becomes empty, i.e., no sub-routes can be extended. It returns the set $\overrightarrow{\mathbb{R}}$ of all surviving sub-routes.

Algorithm 3 illustrates the pseudo code of route search in forward direction. It is self-explanatory and summarizes what we have discussed above.

Backward Search. Route space search in backward direction is similar to that in forward direction. The pruning rules, searching strategies, and dominating testing discussed for forward search can be modified for backward search directly.

4.2 Route Join

In this section, we elaborate on how to join sub-routes obtained in the forward search and the backward search. Let $\overrightarrow{R_v} = (\tau(\overrightarrow{R_v}), B_{\overrightarrow{R_v}}, C_{\overrightarrow{R_v}}, v)$ and $\overleftarrow{R_u} = (\tau(\overleftarrow{R_u}), B_{\overleftarrow{R_u}}, C_{\overleftarrow{R_u}}, u)$ be two sub-routes in the forward and the backward directions, respectively. They are *feasible* to be joined if:

Algorithm 3. Forward search

  function RouteSearchFW(Query $q = (s, d, [t_q^-, t_q^+])$, Vertex set $V = P \cup \{s, d\}$)
                                                                    ▷

  Initialization
1: Create an empty set $\overrightarrow{\mathbb{R}_v}$ for each vertex $v \in V$ to store sub-routes associated with
    v
2: Calculate the candidate vertex set C_s of s ▷ Equation 5
3: $\overrightarrow{\mathbb{R}_s} \leftarrow \{(0, \emptyset, C_s, s)\}$
4: Create a max-heap $H \leftarrow \{(s, |C_s|)\}$ to store vertices whose routes will be extended

                                    ▷ Repeatedly generate feasible sub-routes

5: **while** $H \neq \emptyset$ **do**
6:     $(v, v.ub) \leftarrow$ Extract-Max(H) ▷ Searching strategy
7:     Sort routes $R \in \overrightarrow{\mathbb{R}_v}$ in the descending order of $|B_R|$ ▷ Searching strategy
8:     **for all** $R_v \in \overrightarrow{\mathbb{R}_v}$ **do**
9:         **for all** $u \in C_{R_v}$ **do**
10:             $R_u \leftarrow$ Extend(R_v, q, u) ▷ Equation 4, 5 Pruning Rule 1
11:             RemoveDominate$(\overrightarrow{\mathbb{R}_u}, R_u)$ ▷ Pruning Rule 2
12:             **if** $R_u \in \overrightarrow{\mathbb{R}_u}$ **then**   ▷ R_u not pruned
13:                 **if** $(u, u.ub) \notin H$ **then**
14:                     Insert $(u, \$_{R_u}^+)$ into H
15:                 **else**
16:                     $u.ub \leftarrow \max\{u.ub, \$_{R_u}^+\}$
17: Return $\overrightarrow{\mathbb{R}} \leftarrow$ all routes in each nonempty $\overrightarrow{\mathbb{R}_v}$

$$\tau(\overrightarrow{R_v}) + \tau(v, u) \leq u.t_p^+ \quad \text{and} \quad \tau(\overrightarrow{R_v}) + \tau(\overleftarrow{R_u}) + \tau(v, u) \leq t_q^+ - t_q^-$$
$$B_{\overrightarrow{R_v}} \cap B_{\overleftarrow{R_u}} = \emptyset \tag{6}$$

We denote the joined route as $R^{join} = \langle s, B_{\overrightarrow{R_v}}, rev(B_{\overleftarrow{R_u}}), d \rangle$, where $rev(B_{\overleftarrow{R_u}})$ refers to a list of vertices in $B_{\overleftarrow{R_u}}$ but in the reversed order. Its reward is: $|B_{\overrightarrow{R_v}}| + |B_{\overleftarrow{R_u}}|$.

We develop two optimization techniques to accelerate the join procedure. First, we apply pruning rule 3 to skip the *feasible* checking (cf. Eq. 6) for pairs of sub-routes. Second, we sort sub-routes in the descending order of their $|B_R|$. This helps us find a tigher $\$^{best}$ earlier, and in turn boosts the power of Pruning Rule 3.

Pruning Rule 3 (Reward bound pruning). *Let* $\$^{best}$ *be the maximum reward on all joined routes found so far. If* $|B_{\overrightarrow{R}}| + |B_{\overleftarrow{R}}| \leq \$^{best}$, *then we need not join* $\overrightarrow{R}$ *and* $\overleftarrow{R}$.

Continuing with the example in Fig. 6, we illustrate the join procedure in Table 2. First, we sort forward sub-routes $\overrightarrow{R} \in \overrightarrow{\mathbb{R}}$ and backward sub-routes

$\overleftarrow{R} \in \overleftarrow{\mathbb{R}}$ in descending order of $|B_R|$. For each pair of $\overrightarrow{R}$ and $\overleftarrow{R}$, if it survives Pruning Rule 3, then we conduct feasible checking and then join the pair. After joining the forward sub-route $\overrightarrow{R} = (5, \langle p_1, p_2, p_6 \rangle, \emptyset, p_6)$ with the backward sub-route $\overleftarrow{R} = (2, \langle p_7, p_3 \rangle, \{p_6\}, p_3)$, we update $\$^{best}$ to 5. All remaining pairs are pruned according to Pruning Rule 3. The best route (known at this snapshot) is $\langle s, p_1, p_2, p_6, p_3, p_7, d \rangle$.

Table 2. Route join

	sub-routes sorted in the descending order of $|B_R|$			
$\overrightarrow{\mathbb{R}}$	$(5, \langle p_1, p_2, p_6 \rangle, \emptyset, p_6)$, $(3, \langle p_1, p_2 \rangle, \{p_6\}, p_2)$, $(1, \langle p_1 \rangle, \{p_2, p_6\}, p_1)$, $(0, \emptyset, \{p_1, p_2, p_6\}, s)$			
$\overleftarrow{\mathbb{R}}$	$(5, \langle p_7, p_3, p_6 \rangle, \emptyset, p_6)$, $(2, \langle p_7, p_3 \rangle, \{p_6\}, p_3)$, $(1, \langle p_7 \rangle, \{p_3, p_6\}, p_7)$, $(0, \emptyset, \{p_3, p_6, p_7\}, d)$			
route join iterations				
iteration	candidate join pairs		join result	$\$^{best}$
	$\overrightarrow{R}$	$\overleftarrow{R}$	R^{join}	
1	$(5, \langle p_1, p_2, p_6 \rangle, \emptyset, p_6)$	$(5, \langle p_7, p_3, p_6 \rangle, \emptyset, p_6)$	not feasible (Eq. 6)	0
2	$(5, \langle p_1, p_2, p_6 \rangle, \emptyset, p_6)$	$(2, \langle p_7, p_3 \rangle, \{p_6\}, p_3)$	$R = \langle s, p_1, p_2, p_6, p_3, p_7, d \rangle$	5
...	...	...	skipped (Pruning Rule 3)	5
optimal route for this snapshot			$R = \langle s, p_1, p_2, p_6, p_3, p_7, d \rangle$	

5   Experiment

This section studies the effectiveness and efficiency of our proposed methods on both real and synthetic datasets.

5.1   Experimental Setting

We first introduce the datasets used in experiments, and then describe the performance measures for algorithms.

Datasets.

Real Datasets. Similar to [10], we obtain real check-in data in Foursquare[7] and convert them to crowdsourcing tasks in our problem. Specifically, we collect check-in data for New York city (NYC) and Los Angeles County (LA) in a month (September 2012). For each day in that month, we use all check-in items within a 90-minute duration. We take check-in items at the same location as a single task, set its release time and deadline to the earliest and the latest check-in time respectively[8]. We measure the travel time $\tau(v, u)$ as the Euclidean distance between two locations divided by the average speed. We use a walking speed 6 km/h for NYC (whose map size 789 km^2 is small), and use a driving speed 60 km/h for LA (whose map size 10,570 km^2 is large).

Synthetic Datasets. As NYC and LA have similar result trends (see Fig. 7), we use the map domain of LA to generate synthetic datasets. For each synthetic

---

[7] https://foursquare.com/.

[8] For each location with only one check-in item (say, at time t), we choose its deadline randomly in $[t, t_q^+]$, where t_q^+ refers to the query's deadline.

Table 3. Experiment parameters

Parameter	Default	Range
total number of tasks	100	$20, 50, 100, 200, 500$
$t_q^+ - t_q^-$ [minutes]	90	$30, 60, 90, 120, 150$
Gaussian x	0.1	$0.05, 0.1, 0.25, 0.5$

task, we randomly choose its release time t_p^- randomly in $[t_q^-, t_q^+]$ and then choose its deadline t_p^+ in range $[t_p^-, t_q^+]$, as we consider queries of the form $q = (s, d, [t_q^-, t_q^+])$ in our experiments. We generate two types of datasets. In each uniform dataset (UNI), task locations are randomly chosen within the map domain. In each Gaussian dataset (GAU), task locations are generated based on four Gaussian bells, with the standard deviation of Gaussian bell as x times of the map domain length. The parameter values for the number of tasks and Gaussian standard deviation x are shown in Table 3.

Platform and Performance Measures. We implemented our methods (G-NN, G-MCS, G-ED, Re-Route) in C++, and conducted experiments on an Ubuntu 11.10 machine with a 3.4 GHz Intel Core i7-3770 processor and 16 GB RAM.

We use queries of the form $q = (s, d, [t_q^-, t_q^+])$, where $t_q^+ - t_q^- = 90$ min by default. We randomly choose s, d in the map domain such that $\tau(s, d) = 45$ min. The parameter values for $t_q^+ - t_q^-$ are given in Table 3.

In each experiment, we run a set Q of 50 queries and report (i) the quality ratio for Q, and (ii) the average response time per call of a method. Specifically, we define the *quality ratio* of a method as:

$$quality\ ratio\ = \frac{1}{|Q|} \cdot \sum_{q \in Q} \frac{count(R_{method}(q))}{count(R_{opt}(q))}$$

where q is a query in Q, $R_{method}(q)$ is the route for q found by our method, $R_{opt}(q)$ is the optimal route for q found by an offline method that knows all tasks in advance[9].

We have tested the effects of policies $\mathcal{P}_{stay}$ and $\mathcal{P}_{go}$ (cf. Section 3.2) on our methods. For the same method, the quality ratios between $\mathcal{P}_{stay}$ and $\mathcal{P}_{go}$ differ only by $0.01 - 0.02$. Thus, we take the default policy in our methods as $\mathcal{P}_{stay}$.

5.2 An Experiment on Real Datasets

We plot the performance of methods on real datasets (LA and NYC) on each day from Sep/21/2012 to Sep/30/2012 in Fig. 7. Within the query period, LA and NYC contain 60 and 40 tasks on average, respectively. The optimal routes

---

[9] As mentioned in Definition 4, $R_{opt}(q)$ is obtained with assumption that all tasks' information are known in advance at time t_q^-. With this assumption, OnlineRR becomes a special case of SnapshotRR where tasks can have release time larger than t_q^- and the approach for SnapshotRR can be used to find $R_{opt}(q)$ then.

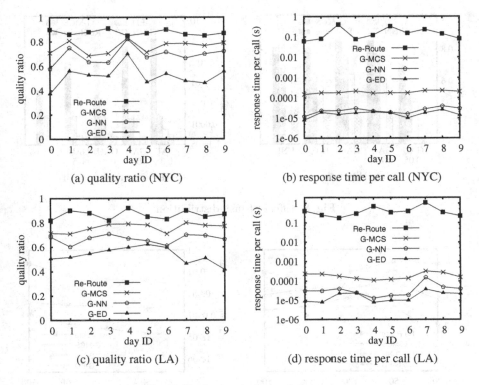

(a) quality ratio (NYC)　　　　(b) response time per call (NYC)

(c) quality ratio (LA)　　　　(d) response time per call (LA)

Fig. 7. Performance on real datasets

R_{opt} in LA and NYC cover 10 and 5 tasks on average, respectively. Figure 7a,c show the quality ratio of the methods on NYC and LA, respectively. Re-Route outperforms other methods and achieves 0.82–0.91 quality. G-MCS is the second best and obtains 0.70–0.84 quality. Although Re-Route incurs higher response time, it takes less than 1 s per call, as depicted in Fig. 7b,d. We consider such time acceptable for crowdsourcing workers. For example, for the LA dataset, Re-Route is called for 10 times (on average) during the query period (90 min). Observe that the time per call (1 s) is negligible compared to the average travel time between two tasks ($90/10 = 9$ min).

5.3 Scalability Experiments on Synthetic Datasets

Effect of Task Distribution. Figure 8 depicts the performance of methods on GAU datasets with standard deviation x and on a UNI dataset. As illustrated in Table 4a, a more skewed dataset (i.e., with smaller x) leads to an optimal route with higher reward because tasks in the same cluster are close together. Since our methods can also find routes with higher reward on a more skewed dataset, the quality ratio does not vary much (See Fig. 8a). Re-Route again outperforms other methods on the quality ratio. On the other hand, a more skewed dataset

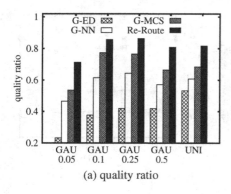
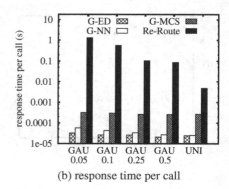

(a) quality ratio   (b) response time per call

Fig. 8. Effect of task distribution

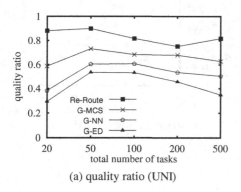
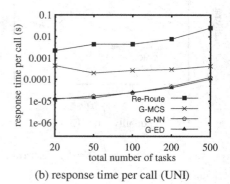

(a) quality ratio (UNI)   (b) response time per call (UNI)

Fig. 9. Effect of the Total Number of Tasks

Table 4. Reward on the optimal route

Task distribution	Gaussian	Uniform	
Parameter values	**(a)** standard deviation x	**(b)** total number of tasks	**(c)** query period $t_q^+ - t_q^-$
	0.05, 0.1, 0.25, 0.5	20, 50, 100, 200, 500	30, 60, 90, 120, 150
Reward of R_{opt}	$12.57, 9.39, 6.84, 4.72$	$1.7, 3.14, 5.26, 7.94, 13.2$	$1.62, 3.26, 5.26, 6.92, 8.92$

induces more feasible candidate tasks in Re-Route, and thus it incurs higher response time. Nevertheless, Re-Route takes at most around 1 s per call in Fig. 8b, which is acceptable for crowdsourcing workers.

Since the trend on quality is consistent across different task distributions, we only use UNI datasets in the remaining experiments.

Effect of Total Number of Tasks. When the total number of tasks increases, both the optimal route (cf. Table 4b) and our methods' routes would cover more tasks. Thus, the quality ratio is independent of the total number of tasks, as shown in Fig. 9a. The response time of Re-Route increases slightly with the total number of tasks (see Fig. 9b), but it is still within 0.1 s per call.

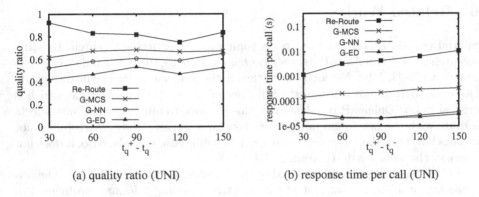

(a) quality ratio (UNI)   (b) response time per call (UNI)

Fig. 10. Effect of the query period $t_q^+ - t_q^-$

Effect of the Query Period $t_q^+ - t_q^-$. As the query period $t_q^+ - t_q^-$ widens, more tasks become feasible and thus the optimal route contains more tasks, as shown in Table 4c. We plot the performance of the methods with respect to $t_q^+ - t_q^-$ in Fig. 10. The quality ratio is independent of $t_q^- - t_q^-$ as our methods are also able to find routes with more tasks. The response time per call in Re-Route remains acceptable.

Effect of Pruning Rules on Re-Route. We proceed to test the effect of optimization techniques (cf. Sect. 4) on the response time per call of Re-Route. We consider two variations of Re-Route: (i) DISABLE applies only pruning rule 1 (in Ref. [19]), and (ii) ENABLE applies all three pruning rules in Sect. 4.

As DISABLE is very slow, we scale down the total number of tasks in this experiment, and terminate it if it takes more than 300 s per call. We show the response time per call of DISABLE and ENABLE on both UNI and GAU datasets in Fig. 11. Observe that ENABLE runs much faster than DISABLE, implying that our pruning rules are able to shrink the search space significantly.

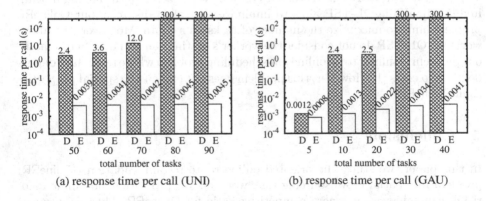

(a) response time per call (UNI)   (b) response time per call (GAU)

Fig. 11. Effect of pruning rules on Re-Route (E for ENABLE, D for DISABLE)

6 Related Work

Spatial crowdsourcing is an emerging topic in crowdsourcing research. Existing researches are divided into the *server-centric* mode [9,15,16,18,22] and the *worker-centric* mode [3,7,10]. We focus on the latter one as discussed in the introduction. However, [3,7] do not consider the influence of the worker's travel time, which is critical in our OnlineRR problem. The closest work to ours is [10], which selects a route with the maximum number of tasks for a worker. However, [10] does not discuss how to update a route with respect to online task arrivals. Also, it does not consider the worker's destination and deadline.

Our OnlineRR problem is related to the orienteering problem [13,23]. The orienteering problem is a variant of the selective traveling salesman problem [11], where (i) not all requests need to be completed, and (ii) the cost is the sum of the total travel time and the penalty of rejected requests. The orienteering problem is well studied [13,23], but only several works [5,8,12,19] consider the Orienteering Problems with each request having a Time Window (OPTW). Those works focus on the offline scenario but not the online scenario. While there exist approximation algorithms for OPTW offline [5,8,12], OnlineRR is an online problem and does not permit any online algorithm to achieve a non-zero competitive ratio.

Righini et al. [19] propose an exact bi-directional search algorithm for OPTW, which can be adapted to solve our SnapshotRR problem. Unlike our solution, this algorithm does not exploit spatial properties to prune unpromising sub-routes. In Sect. 4, we have developed two pruning rules and a search strategy that are specific for SnapshotRR.

Other related route planning problems include the trip planning problem [17] and the optimal sequenced route problem [21]. They require finding the shortest route that passes through specific types of points-of-interests. On the other hand, our problem needs to maximize the number of tasks on a route subject to the tasks' deadlines and the worker's deadline.

OnlineRR problem is also related to online traveling salesman problem (OL-TSP) [4,14]. Few works have studied OL-TSP with each request having a deadline [6,24]. While OL-TSP aims to minimize the travel distance, our OnlineRR problem aims to maximize the number of tasks on a route. Moreover, the above works on OL-TSP do not consider the worker's destination and deadline. Finally, our problem is similar to an online job-scheduling problem whose tasks have dependent setup costs [1]. However, this problem does not exploit the spatial properties as in OnlineRR.

7 Conclusion

In this paper, we study the oriented online route recommendation (OnlineRR) problem for spatial crowdsourcing task workers. We prove that no online algorithm can achieve a non-zero competitive ratio for OnlineRR. Then we propose several heuristics for OnlineRR and optimizations to speedup the computation.

According to our experimental findings, Re-Route produces routes with the highest quality (0.82–0.91) in acceptable response time per call (0.1–1 s), whereas G-MCS returns routes with the second highest quality (0.70–0.84) at real-time (below 1 ms). Workers preferring to save smartphone battery power may choose G-MCS as it has less computation cost. OnlineRR will be extended to consider the task diversity and task novelty in the future.

References

1. Allahverdi, A., Ng, C.T., Cheng, T.C.E., Kovalyov, M.Y.: A survey of scheduling problems with setup times or costs. Eur. J. Oper. Res. **187**, 985–1032 (2008)
2. Allan, B., Ran, E.-Y.: Online Computation and Competitive Analysis. Cambridge University Press, Cambridge (1998)
3. Alt, F., Shirazi, A.S., Schmidt, A., Kramer, U., Nawaz, Z.: Location-based crowdsourcing: extending crowdsourcing to the real world. In: NordiCHI, pp. 13–22 (2010)
4. Ausiello, G., Feuerstein, E., Leonardi, S., Stougie, L., Talamo, M.: Algorithms for the on-line travelling salesman. Algorithmica **29**, 560–581 (2001)
5. Bansal, N., Blum, A., Chawla, S., Meyerson, A.: Approximation algorithms for deadline-tsp and vehicle routing with time-windows. In: Symposium on Theory of, Computing, pp. 166–174 (2004)
6. Blom, M., Krumke, S.O., Paepe, W.E.D., Stougie, L.: The online tsp against fair adversaries. INFORMS J. Comput. **13**, 138–148 (2001)
7. Bulut, M.F., Yilmaz, Y.S., Demirbas, M.: Crowdsourcing location-based queries. In: PERCOM Workshops, pp. 513–518 (2011)
8. Chekuri, C., Korula, N.: Approximation algorithms for orienteering with time windows. CoRR (2007)
9. Chen, Z., Fu, R., Zhao, Z., Liu, Z., Xia, L., Chen, L., Cheng, P., Cao, C.C., Tong, Y., Zhang, C.J.: gmission: A general spatial crowdsourcing platform. In: PVLDB, pp. 1629–1632 (2014)
10. Deng, D., Shahabi, C., Demiryurek, U.: Maximizing the number of worker's self-selected tasks in spatial crowdsourcing. In: SIGSPATIAL, pp. 314–323 (2013)
11. Eiselt, H.A., Gendreau, M., Laporte, G.: Location of facilities on a network subject to a single-edge failure. Networks **22**, 231–246 (1992)
12. Frederickson, G.N., Wittman, B.: Approximation algorithms for the traveling repairman and speeding deliveryman problems. Algorithmica **62**, 1198–1221 (2012)
13. Gavalas, D., Konstantopoulos, C., Mastakas, K., Pantziou, G.E.: A survey on algorithmic approaches for solving tourist trip design problems. J. Heuristics **20**, 291–328 (2014)
14. Jaillet, P., Wagner, M.R.: Online routing problems: value of advanced information as improved competitive ratios. Transp. Sci. **40**, 200–210 (2006)
15. Kazemi, L., Shahabi, C.: Geocrowd: enabling query answering with spatial crowdsourcing. In: SIGSPATIAL, ppp. 189–198 (2012)
16. Kazemi, L., Shahabi, C., Chen, L.: Geotrucrowd: trustworthy query answering with spatial crowdsourcing. In: SIGSPATIAL, pp. 304–313 (2013)
17. Li, F., Cheng, D., Hadjieleftheriou, M., Kollios, G., Teng, S.-H.: On trip planning queries in spatial databases. In: Medeiros, C.B., Egenhofer, M., Bertino, E. (eds.) SSTD 2005. LNCS, vol. 3633, pp. 273–290. Springer, Heidelberg (2005)

18. Pournajaf, L., Xiong, L., Sunderam, V.S., Goryczka, S.: Spatial task assignment for crowd sensing with cloaked locations. In: MDM, pp. 73–82 (2014)
19. Righini, G., Salani, M.: Decremental state space relaxation strategies and initialization heuristics for solving the orienteering problem with time windows with dynamic programming. Comput. Oper. Res. **36**, 1191–1203 (2009)
20. Rubinstein, R.Y., Kroese, D.P.: Simulation and the Monte Carlo method. Wiley, Hoboken (2011)
21. Sharifzadeh, M., Kolahdouzan, M.R., Shahabi, C.: The optimal sequenced route query. VLDB J. **17**, 765–787 (2008)
22. To, H., Ghinita, G., Shahabi, C.: A framework for protecting worker location privacy in spatial crowdsourcing. In: PVLDB, pp. 919–930 (2014)
23. Vansteenwegen, P., Souffriau, W., Oudheusden, D.V.: The orienteering problem: a survey. Eur. J. Oper. Res. **209**, 1–10 (2011)
24. Wen, X., Xu, Y., Zhang, H.: Online traveling salesman problem with deadlines and service flexibility. J. Comb. Optim. 1–18 (2013)

Knowledge-Enriched Route Computation

Georgios Skoumas[1]($\boxtimes$), Klaus Arthur Schmid[2], Gregor Jossé[2],
Matthias Schubert[2], Mario A. Nascimento[3], Andreas Züfle[2],
Matthias Renz[2], and Dieter Pfoser[4]

[1] National Technical University of Athens, Athens, Greece
gskoumas@dblab.ece.ntua.gr
[2] Ludwig-Maximilians-Universität München, Munich, Germany
{schmid,josse,schubert,zuefle,renz}@dbs.ifi.lmu.de
[3] University of Alberta, Edmonton, Canada
nascimento@ualberta.ca
[4] George Mason University, Fairfax, USA
dpfoser@gmu.edu

Abstract. Directions and paths, as commonly provided by navigation
systems, are usually derived considering absolute metrics, e.g., finding
the shortest or the fastest path within an underlying road network. With
the aid of Volunteered Geographic Information (VGI), i.e., geo-spatial
information contained in user generated content, we aim at obtaining
paths that do not only minimize distance but also lead through more
popular areas. Based on the importance of landmarks in Geographic
Information Science and in human cognition, we extract a certain kind
of VGI, namely spatial relations that define *closeness* (*nearby, next to*)
between pairs of *points of interest* (POIs), and quantify them following a probabilistic framework. Subsequently, using Bayesian inference we
obtain a crowd-based *closeness* confidence score between pairs of POIs.
We apply this measure to the corresponding road network based on an
altered cost function which does not exclusively rely on distance but also
takes crowdsourced geo-spatial information into account. Finally, we propose two routing algorithms on the enriched road network. To evaluate
our approach, we use Flickr photo data as a ground truth for popularity. Our experimental results – based on real world datasets – show that
the paths computed w.r.t. our alternative cost function yield competitive
solutions in terms of path length while also providing more "popular"
paths, making routing easier and more informative for the user.

1 Introduction

User generated content has benefited many scientific disciplines by providing a
wealth of new data. Technological progress, especially smartphones and GPS
receivers, has facilitated contributing to the plethora of available information.
OpenStreetMap[1] constitutes the standard example and reference in the area
of VGI. Authoring geo-spatial information typically implies coordinate-based,

---

[1] https://www.openstreetmap.org/.

© Springer International Publishing Switzerland 2015
C. Claramunt et al. (Eds.): SSTD 2015, LNCS 9239, pp. 157–176, 2015.
DOI: 10.1007/978-3-319-22363-6_9

quantitative data. Contributing quantitative data requires specialized applications (often part of social media platforms) and/or specialized knowledge, as is the case of OpenStreetMap (OSM).

The broad mass of users contributing content, however, are much more comfortable using *qualitative information*. People typically do not use geo-coordinates to describe their spatial motion, for instance when traveling or roaming. Instead, they use qualitative information in the form of toponyms (landmarks) and spatial relationships ("near", "next to", "close by", etc.). Hence, there is an abundance of geo-spatial information (freely) available on the Internet, e.g., in travel blogs, largely unused. In contrast to quantitative information, which is mathematically measurable, qualitative information is based on personal cognition. Therefore, accumulated and processed qualitative information may better represent the human way of thinking.

This is of particular interest when considering the "routing problem" (equivalent to "path computation"). Traditional routing queries use directions from systems that only take inherent cost measure of the underlying road network into account, e.g., distance or travel time. In human interaction, such information is usually enhanced with qualitative information (e.g. "the street next to the church", "the bridge North of the Eiffel tower"). Combining traditional routing algorithms with crowdsourced geo-spatial references we aim to more properly represent human perception while keeping it mathematically measurable.

In [1], the authors analyze the important role of landmarks for the representation of geographic space in human mind, i.e., people tend to describe their position in space based on landmarks and relations between them. Based on this fact, in this work, we enrich a road network with information about spatial relations between pairs of Points of Interest (POIs) extracted from user generated data (travel blog data). Using these relations, we obtain routes that are easier to interpret and follow, possibly rather resembling a route that a person would provide.

Fig. 1. Shortest (continuous) and alternative paths (dot dashed and dotted) alongside POIs in the city of Paris. This result is an output of some of the algorithms presented in this paper.

As an example, consider the routing scenario in Fig. 1 which is set in the city of Paris, France. The continuous line represents the conventional shortest path from starting point "Gare du Nord" to the target at "Quai de la Rapée" while the dot dashed and dotted lines represent alternative paths computed by the algorithms introduced in this paper. The triangles in this example denote touristic landmarks and sights. For instance, the dot dashed path on the bottom right passing through recognizable locations such as "Place de la République", "Cirque d'hiver" and "la Bastille", as proposed by our algorithms, is considerably easier to describe and follow, and might yield more interesting sights for tourists than the shortest path.

The major challenge in this contribution is the extraction of crowdsourced geo-spatial information from textual data and the enrichment of an existing road network with this information. The enriched road network is subsequently used to provide paths between a given start and target that satisfy the claim of higher popularity (which is formally introduced in Sect. 3), while only incurring a minor additional spatial distance. In addition to this main application, we note that our techniques can furthermore be used to automatically provide interesting touristic routes in any place where information about POIs is available. The transition from textual information to routing in networks is not at all straightforward, therefore we employ and develop various methods from different angles of computing science. In a pre-processing step, we first mine VGI from user generated texts, by employing Natural Language Processing (NLP) methods in order to determine spatial entities (POIs) and spatial relations between them (see Sect. 2). Furthermore, due to the inherent uncertainty of crowdsourced data, we employ probability distributions to quantitatively model spatial relations mined from the text (see Sect. 2.2). Having this information available, we propose and approach for "popular" path computation. To summarize, our contributions are as follows:

- We introduce a Bayesian inference-based transition from the modeled spatial relations to spatial *closeness* confidence measurements according to the crowd (see Sect. 3.1).
- We define a new cost criterion which is used to enrich an underlying road network with the aforementioned confidence measurements (see Sect. 3.2).
- We extend our previously presented road network enrichment approach (see [2]) with a skyline-based road network enrichment approach.
- Finally, we propose two algorithms which use the enriched road network to compute actual paths (see Sect. 4).

2   Pre-processing: Spatial Relation Extraction and Modeling

This section highlights our approach on qualitative data extraction from texts and presents a probabilistic approach for representing spatial relationships based on distance and orientation features. Key ingredients of our approach are NLP methods for information extraction from texts and algorithms that train probabilistic models, which are required due to the inherent uncertainty of crowdsourced data. Our discussion below includes a short description of NLP tools we use to extract spatial relations between POIs, the features we used to model spatial relations as probability distributions, and a short analysis of the modeling approach used in [3]. These models are necessary to assess the quality of spatial relations extracted from text which will be used in Sect. 3.2 for the enrichment of the underlying road network.

2.1   Spatial Relation Extraction from Texts

In this work, we choose travel blogs as a rich source for (crowdsourced) geospatial data. This selection is based on the fact that people tend to describe their experiences in relation to their trips and places they have visited, which results in "spatial" narratives. To gather such data, we use classical Web crawling techniques and compile a database consisting of 250,000 texts, obtained from 20 travel blogs.

Obtaining qualitative spatial relations from text involves the detection of (i) POIs (or toponyms) and (ii) spatial relationships linking the POIs. The employed approach involves geoparsing, i.e., the detection of candidate phrases, and geocoding, i.e., linking the phrases to actual coordinate information.

For the relation extraction task we follow the approach used in [4] where a Natural Language Processing Toolkit (NLTK) (cf. [5]) based spatial relation extraction approach is presented. NLTK is a leading platform for analyzing raw natural language data. The search for spatial relations in texts results into triplets of the form (P_i, R^k, P_j), where p_i and p_j are named entities (landmarks) and R^k is the spatial relation that intervenes between P_i and P_j. Following this path, we managed to extract 500,000 POIs from the aforementioned travel blog text corpus. For the geocoding of the POIs, we rely on the GeoNames[2] geographical gazetteer data, which contains over 10 million POI names worldwide and their coordinates. This procedure associates (whenever possible) POIs extracted from travel blogs with geographical coordinates. Using the GeoNames gazetteer we were able to geocode about 480,000 out of the 500,000 extracted POIs and to end up with about 600,000 triplets of the form (P_i, R^k, P_j) worldwide.

For our experiments we want to focus on regions with high triplet density in order to get meaningful results. Therefore, we focus on the cities of Paris and New York. The triplets we extracted for each of these two cities define what we call *Spatial Relationship Graph*, i.e., a spatial graph in which nodes represent POIs and edges are spatial relationships between them. Let us point out that for the scope of this work, i.e., a combination of short and enriched routes, we only consider distance and topological relations that denote closeness (near, close, next to, at, in etc.). The use of relations that denote direction, e.g., North, South etc., or remoteness, e.g., away from, far from etc., is an open direction for future work.

2.2   Modeling Spatial Relations

Feature Extraction In order to train probabilistic models, we need informative features. We model each spatial relation in terms of *distance* and *orientation* as

---

[2] http://www.geonames.org/.

presented in [3]. Therefore, we extract occurrences of a spatial relation (such as "near") from travel blogs. For each occurrence, we create a two-dimensional spatial feature vector $D = (D_d, D_o)^\mathsf{T}$ where D_d denotes the distance and D_o denotes the orientation between P_i and P_j. Specifically, assuming a projected (Cartesian) coordinate system, the distance between two POIs P_i and P_j is computed as the Euclidean metric between the two respective coordinates. The orientation is established as the counterclockwise rotation of the x-axis, centered at point P_j, to point P_i. This way, we end up with a set of two-dimensional feature vectors $\mathcal{D}_{\text{rel}} = \{D_1, D_2, \ldots, D_n\}$ for each spatial relation. We will use the set of two-dimensional feature vectors in order to train a probabilistic model for each spatial relation.

Probabilistic Modeling. As described in [3], by using a set of two-dimensional feature vectors for each spatial relation such as "near" or "into", we can train Gaussian Mixture Models (GMMs), which have been extensively used in many classification and general machine learning problems [6].

In general, a GMM is a weighted sum of M-component Gaussian densities as $p(d|\lambda) = \sum_{i=1}^{M} w_i g(d; \mu_i, \Sigma_i)$ where d is a l-dimensional data vector (in our case $l = 2$), w_i are the mixture weights, and $g(d; \mu_i, \Sigma_i)$ is a Gaussian density function with mean vector $\mu_i \in \mathbb{R}^l$ and covariance matrix $\Sigma_i \in \mathbb{R}^{l \times l}$. To fully characterize the probability density function $p(d|\lambda)$, one requires the mean vectors, the covariance matrices and the mixture weights. These parameters are collectively represented as $\lambda = \{w_i, \mu_i, \Sigma_i\}$ for $i = 1, \ldots, M$.

Let $\mathcal{R} = \{R^1, \ldots, R^n\}$ denote the set of all spatial relations that we take into account. In our setting, each relation R^k is modeled under a probabilistic framework by a 2-dimensional GMM, trained on each relation's set of two-dimensional feature vectors $\mathcal{D}_{\text{rel}}$. For the parameter estimation of each GMM, we use Expectation Maximization (EM) [7]. EM enables us to update the parameters of a given M-component mixture with respect to a feature vector set $\mathcal{D}_{\text{rel}} = \{D_1, \ldots, D_m\}$ with $1 \leq j \leq m$ and $D_j \in \mathbb{R}^l$, such that the log-likelihood $\mathcal{L} = \sum_{j=1}^{m} \log(p(D_j|\lambda))$ increases with each re-estimation step, i.e., EM re-estimates model parameters λ until convergence. Further details on modeling spatial relations under a probabilistic framework are given in [3].

This procedure results in a trained GMM of the form $p_k(D|\lambda)$, for each spatial relation $R^k, 1 \leq k \leq n$. Given a distance and orientation vector, we can use this model to estimate the probability that a particular relation exists. Based on this information, by bayesian inference we derive a closeness score for pairs of POIs. This procedure is described in the next section.

3   Road Network Enrichment

In this section, we describe our approach to enrich an actual road network with crowdsourced geo-spatial information. Our discussion below includes a description of how we transform a *Spatial Relationship Graph*, as presented in Sect. 2.1, into a weighted graph, and how we use the edge weights of the weighted graph in order to modify the edge costs of a real road network.

3.1   From Relationship to Weighted Graphs

As presented in Sect. 2, the spatial relation extraction procedure results in a relationship graph between POIs. A simple example of such a graph is shown in Fig. 2. In general, let $\mathcal{P} = \{P_1, \ldots, P_m\}$ denote the set of nodes representing the POIs, and let $\mathcal{R} = \{R^1, \ldots, R^n\}$ denote the pre-defined set of spatial closeness relations, represented by spatial NLP expressions like "next to" or "close by".

Furthermore, let $R_{i,j} \subseteq \mathcal{R}$ denote the set of relations extracted from the text between two distinct nodes P_i and P_j. Note that R^k denotes an abstract relation, while $R_{i,j}$ denotes a set of occurrences of relations between a pair of nodes. Let $D_{i,j}$ denote the spatial feature vector (distance and orientation), between two distinct POIs P_i and P_j (as presented in Sect. 2.2). Finally, let $\mathcal{D} := \bigcup_{i \neq j \wedge R_{i,j} \neq \emptyset} D_{i,j}$ denote the set of all spatial feature vectors between all pairs of POIs which have non-empty sets of relations.

We want to estimate the posterior probability of a class $R^k \in R_{i,j}$ based on the spatial feature data $D_{i,j}$ between two POIs P_i and P_j. This is given by Eq. 1. Here, $p(D_{i,j}|R^k)$ denotes the likelihood of $D_{i,j}$ given relation R^k based on the trained GMM (presented as $p(D|\lambda)$ Sect. 2.2), while $P(R^k)$ denotes the prior probability of relation R^k given only the observed relations $R_{i,j}$.

$$P(R^k|D_{i,j}) = \frac{p(D_{i,j}|R^k)P(R^k)}{\sum\limits_{l=1}^{n} p(D_{i,j}|R^l)P(R^l)} \tag{1}$$

In a traditional classification problem the spatial relation R^k between a pair of POIs would be classified to the spatial relation model with the highest posterior. In contrast to this approach, we consider each posterior probability $P(R^k|D_{i,j})$ as a measure of confidence of the existence of relation R^k between P_i and P_j. Remember that all the relations we consider reflect terms of spatial closeness.

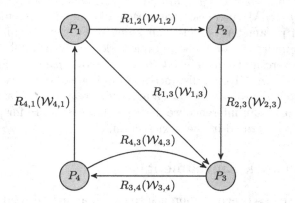

Fig. 2. Simple relationship graph. Nodes represent POIs and each edge represents the set of relations $R_{i,j}$ through which its adjacent nodes P_i and P_j are connected. Each of these sets is mapped onto the closeness score $\mathcal{W}_{i,j}$, turning the relationship into a weighted graph.

We combine all these posteriors into one measure which we refer to as *closeness score* $\mathcal{W}_{i,j}$ of the pair of POIs P_i and P_j, defined in Eq. 2.

$$\mathcal{W}_{i,j} = \frac{1}{|\mathcal{R}|} \cdot \sum_{i=1}^{|R_{i,j}|} \frac{P(R^k|D_{i,j})}{\max_k\{P(R^k|\mathcal{D})\}} \tag{2}$$

Here, we sum all the posteriors $P(R^k|D_{i,j})$ normalized by the maximum posterior of each relation in the relationship graph and we normalize the summation by the total number of spatial relations in the relationship graph. This is done for all pairs P_i, P_j where $R_{ij} \neq \emptyset$. We refer to these pairs as *close* since at least one of our relations, reflecting closeness, exists. As is illustrated in Fig. 2, assigning the respective weights $\mathcal{W}_{i,j}$ to the edges of the relationship graph, we obtain a weighted graph. Note that $\mathcal{W}_{i,j} \in [0, 1]$ but typically $0 < \mathcal{W}_{i,j} \ll 1$. In Sect. 5 the influence of $\mathcal{W}_{i,j}$ on the results is examined, in particular, different scalings are tested. In this weighted relationship graph, denoted by H^*, there exists a vertex for each POI and an edge (P_i, P_j) (equipped with weights $\mathcal{W}_{i,j}$ and Euclidean distances d_{ij}) for each pair of POIs P_i, P_j that are close in the above sense ($R_{i,j} \neq \emptyset$).

3.2   From Weighted Graphs to Road Network Enrichment

Now that we have extracted and statistically condensed the crowdsourced data into a closeness score, we need to apply the obtained closeness scores to the underlying network. We have investigated several strategies and have decided upon a compromise between simplicity and effectiveness. We will present two road network enrichment approaches and we propose two algorithms on routing with enriched graphs. The first enrichment approach, also analyzed in our previous work in [2], is based on Djikstra shortest path computation while the second is based on Skyline path computation.

Initially, let $G = (V, E, d)$ denote the graph representing the underlying road network, i.e., the vertices $v \in V$ correspond to crossroads, dead ends, etc., the edges $e \in E = V \times V$ represent roads connecting vertices. Furthermore, let $d : E \rightarrow \mathbb{R}_0^+$ denote the function which maps every edge onto its distance. We assume that $\mathcal{P} \subseteq V$, i.e., each POI is also a vertex in the graph. This is only a minor constraint since we can easily map each POI to each nearest node on the graph or introduce pseudo-nodes. Our two enrichment methods are described below.

Djikstra Shortest Path Approach. For each pair of spatially connected POIs, P_i, P_j, we compute the shortest path connecting P_i and P_j in G, which we denote by $r(i, j)$. We then define a new cost function $c : E \rightarrow \mathbb{R}_0^+$ which modifies the previous cost $d(e)$ of an edge as follows:

$$c(e) = d(e) \cdot \prod_{e \in r(i,j)} (1 - \alpha \mathcal{W}_{i,j}) \tag{3}$$

where $e \in r(i,j)$ iff e is an edge within the shortest path from P_i to P_j and where $\alpha \in [0,1]$ is a weight scaling factor to control the balance between the spatial distance $d(e)$ and the modification caused by the closeness score $\mathcal{W}_{i,j}$. In the case of $\alpha = 0$, we obtain the unadapted edge weight $c(e) = d(e)$. Summarizing, the more shortest paths between POI pairs run through e, the lower its adjusted cost $c(e)$. The reason for enriching the shortest paths is that they represent the most intuitive connections between any two points in a road network.

We now define the *enriched graph* $G^* = (V, E, c)$. It consists of the original vertices and edges and is equipped with the new cost function which implies the re-weighting of edges. Any path computation algorithm in G^* (e.g. a Dijkstra search) therefore favors edges which are part of shortest paths between POIs which are close according to the crowd. When computing the cost of a path on G^*, as before, we sum the respective edge weights which now differ from the original edge weights (due to the altered cost function). We refer to this procedure of incorporating the crowdsourced information as *D-enrich*.

Path Skyline Approach. One shortcoming of *D-enrich* is the assumption that the crowd unanimously favors exactly one path to connect a pair of POIs P_i and P_j, namely the shortest path. Especially in multicriteria networks which comprise of a set of cost criteria, e.g., travel time, energy consumption, road tolls, optimality is usually defined as a personal trade-off between the given criteria. For example: How much additional time has to be spent to avoid a toll road? However, defining this trade-off numerically as a vector of preferences is not reasonable, and even if it would be, finding the personally preferred trade-offs for all users is in general not possible. Therefore, the best practice is to present a set of alternative paths to the user. The most established and very comprehensive set of alternative paths is the so-called path skyline [8]. This set contains all paths which are non-dominated in the following sense: The cost vector u dominates a cost vector v, denoted $u \prec_{\mathrm{dom}} v$, if u has a smaller cost value than v in at least one dimension i and v does not have a smaller cost value than u in any dimension j. Hence, the path skyline comprises all paths which are optimal under some monotone combination function of the cost criteria. Hence, the path skyline contains all optimal paths for all possible trade-offs between the cost criteria.

To enrich our road network, we compute the path skyline (w.r.t. distance and travel time) as proposed in [9] between each pair of spatially connected POIs P_i and P_j in G, denoted by $s(i,j)$. Although the paths contained in $s(i,j)$ differ from one another, they often share some edges. Simply following each path for enrichment might unnecessarily favor edges contained in many skyline paths. Therefore, we adjust the weights of edges independent of the number of skyline paths in which they occur. Let $S_{i,j} \subset E$ denote the set of all distinct edges which are part of at least one skyline path from P_i to P_j. Analogously to *D-enrich*, we define the cost function $c : E \to \mathbb{R}_0^+$ to modify the original cost $d(e)$ of an edge, as before. While the adjusted cost function is the same as before (see Eq. 3), the set of edges with adjusted costs is a superset, i.e., $S_{i,j} \supseteq r(i,j)$.

We now define the *enriched graph* $G^{**} = (V, E, c)$. It consists of the original vertices and edges equipped with the altered cost function reflecting a re-weighting of edges contained in skyline paths. Any path computation algorithm in G^{**} (e.g. a Dijkstra search) therefore favors edges which are part of the Sky-line paths between POIs which are close according to the crowd. We refer to this procedure of incorporating the crowdsourced information as *S-enrich*.

3.3 Influence of Adjusted Costs

In order to measure the influence of the adjusted cost values along a computed path $p = (e_1, \ldots, e_r)$ on an enriched graph (G^* or G^{**}), we introduce the *enrich-ment ratio* (ER) function er.

$$er(p) = \frac{1}{d(p)} \sum_{i=1}^{r} c(e_i) \qquad (4)$$

Here, $d(\cdot)$ and $c(\cdot)$ are as in the previous two sections. By normalizing with the total length of the path, we are able to compare the spatial connectivity of paths independent of length as well as start and target nodes. Here, a lower ratio implies higher closeness score values along the edges of the path. If none of the edges of a path is part of any shortest or skyline path between POIs, its enrichment ratio is 1, while the (highly unlikely) optimal enrichment ratio is 0. On the enriched graphs G^* and G^{**} we may now define our path computation algorithms.

4 Path Computation on Enriched Graphs

Now that we have a measure quantifying the enrichment of a path, we investigate the effect of *D-enrich* and *S-enrich* on the actual path computation. For this purpose, we present two approaches which make use of the enriched network and the weighted relationship graph H^* (Sect. 3.1). In Sect. 5 they are compared to the conventional shortest paths within the original graph, as obtained with Dijkstra's algorithm, which we denote by Dij-G.

Note that for the evaluation procedure, all paths in this paper are computed by Dijkstra's algorithm because our main focus is not the routing itself but the incorporation of textual information into existing road networks. If desired, speed-up techniques, such as preprocessing steps and/or other search algorithms, could easily be employed.

Our first approach, given start and target nodes, executes a Dijkstra search in the enriched road network graph G^* or G^{**} w.r.t. the adjusted cost function. Depending on the enrichment used, *D-enrich* or *S-enrich*, we refer to the first algorithm as Dij-G* or Dij-G**, respectively.

Our second approach, uses the enriched road network graphs G^* or G^{**} as well as the weighted relationship graph H^*. Given start and target nodes within the enriched graph (G^* or G^{**}), entry and exit nodes within H^* are determined.

Subsequently, we route within H^*, i.e., from POI to POI, again using Dijkstra's algorithm. Depending on the enrichment used, *D-enrich* or *S-enrich*, we refer to the second approach we want to present as Dij-H* or Dij-H**, respectively. Note that in both cases we use the same graph H^*, but we refer to the *S-enrich* case as Dij-H** in order to differentiate the two methods.

All approaches, return paths connecting start and target. But while Dij-G computes the shortest path in the original graph G, all the approaches compute the shortest paths in the enriched graphs w.r.t. the adjusted cost function c. By construction of c, it favors edges which are part of the Dijkstra shortest paths or the skyline paths, between close POIs. Dij-H* and Dij-H** in contrast, do not only favor these edges, but are restricted to them. Having found entry and exit nodes within H^*, Dij-H* and Dij-H** hop from POI to POI in direction of the target. Hence, Dij-G, Dij-G*, Dij-G**, Dij-H*, Dij-H** in that order, represent an increasing binding to the extracted relations. Dij-G is not bound to the relations at all, while Dij-G* and Dij-G** (by the adjusted cost function) favors "relation-edges", and Dij-H* and Dij-H** are strictly bound to the relations and the graph formed by them.

Let us formalize Dij-H* (Dij-H** can be formalized in the same way). Given start and target node in G^* (or G^{**} for the Dij-H** case), it first determines the so-called entry and exit nodes to and from H^*. However, to exclude POIs which would imply a significant detour, we restrict the set of valid POIs, i.e., we restrict the search to a subgraph of H^*, denoted as h^*. Figure 3 illustrates our computationally inexpensive implementation of a query ellipse that allows for some deviation in the middle of the path as well as for minor initial and final detours.

The pseudo-code for the second approach is given in Algorithm 1. Here, we present only the Dij-H* case, since Dij-H** works in the same way by utilizing the G^{**} graph. After selecting the valid set of POIs (Step 2), entry and exit nodes to and from H^* are determined, i.e., the closest POIs to start and target node, respectively (Steps 4 and 5). Entry and exist nodes connect the road network G^* to the relationship graph H^*. Subsequently, the shortest path in h^* from entry to exit node is computed using Dijkstra's algorithm w.r.t. the Euclidean distance (Step 5). Note that a shortest path within H^* is a sequence of POIs. We there-

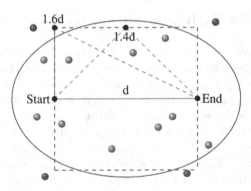

Fig. 3. Restriction of relationship graph H^* to a subgraph h^*, in order to avoid implausible detours. The green dots represent POIs, i.e., nodes of H^* which are also in h^*, the blue ones are left out (color figure online).

fore map this sequence onto G^* by computing the shortest paths between the consecutive pairs of POIs in G^* w.r.t. the adjusted cost function (Step 8). Also, we

---

Algorithm 1. Dij-H*

Input: Enriched Graph G^*, Spatial Relationship Graph H^*, start s, target t
Output: Path p between s and t

1 **begin**
2      $h^* \leftarrow$ subgraph of H^* in bounding ellipse
3      $p \leftarrow$ empty path
4      $P_{\text{entry}} \leftarrow$ select POI $P \in h^*$ closest to s
5      $P_{\text{exit}} \leftarrow$ select POI $P \in h^*$ closest to t
6      $p_h \leftarrow$ Dijkstra$(h^*, P_{\text{entry}}, P_{\text{exit}})$
7   predecessor $\leftarrow s$
8   **foreach** *POI P on path p_h* **do**
9         $v \leftarrow$ select node $v \in G^*$ representing P
10        $p.$APPEND$($Dijkstra$(G^*, \text{predecessor}, v))$
11   predecessor $\leftarrow v$
12   **end**
13     $p.$APPEND$($Dijkstra$(G^*, last, t))$
14   **return** p
15 **end**

---

compute the shortest paths in G^* from start to entry node and exit to target node. Concatenating these paths (start to entry, POI to POI, exit to target), we return a full path.

5 Experimental Evaluation

In this section, we want to investigate the effect and impact of the network enrichment. We compare the results of the conventional Dijkstra search, Dij-G, to the results of Dij-G* and Dij-H*, which use the Djikstra shortest path enriched (*D-enrich*) graph G^*, and the results of Dij-G** and Dij-H**, which use the skyline path enriched (*S-enrich*) graph G^{**}. All approaches are evaluated on real world datasets. Besides comparing the computed path w.r.t. their enrichment ratio (ER) and length (as presented in Sect. 3.2), we introduce a measure of popularity based on Flickr data, which is explained in the following section. All the text processing parts were implemented in Python while modeling parts were implemented in Matlab. Network enrichment and path computation tasks were conducted using the Java-based MARiO Framework [10] on an Intel(R) Core(TM) i7-3770 CPU at 3.40 GHz and 32 GB RAM running Linux (64 bit).

5.1 Enrichment Ratio, Distance and Popularity Evaluation

Our experiments are set in two cities, Paris and New York. These regions have comparatively high density of spatial relations, Flickr photo data, and OSM data, which accounts for an exact representation of the road networks. As mentioned before, we compare the output of Dij-G, Dij-G*, Dij-H*, Dij-G** and

Table 1. Statistics for the weighted relationship graphs, Flickr datasets and road networks of Paris and New York respectively.

	Relationship Graph (H^*)		Flickr		Road Network (G)	
Dataset	# POI Pairs	# Relations	# Photos	# Max Photos per Vertex	# Vertices	# Edges
Paris	400	$2K$	$400K$	100	$550K$	$300K$
New York	300	$1.5K$	$90K$	200	$220K$	$120K$

Dij-H** w.r.t. to the paths they return, more precisely, w.r.t. ER and length of these paths. Since ER is a measure introduced in this paper, we use Flickr data as an independent ground truth. We are aware that to cognitive aspects (like the importance of sights or the value of landmarks) there is no absolute truth. However, in order to be able to draw comparisons, we presume that if the dataset is large enough, the bias can be neglected. We use a geotagged Flickr photo dataset, provided by the authors in [11], to assign a number of photos to each vertex of the underlying road network. The number of Flickr photos assigned to each vertex is referred to *popularity*. In our settings, every photo which is within the 20-meter radius of a vertex, contributes to the popularity of that vertex. The popularity of a path is computed by the summation of all popularity values along this path.

The sizes of the weighted relationship graphs H^*, road network and Flickr photo data for both cities are shown in Table 1. Regarding the weighted relationship graphs, we provide the number of unique POI pairs extracted from the travel blog corpus and the number of spatial (closeness) relations extracted between them, as was presented in Sect. 2. Regarding Flickr data, we provide the total number of geotagged photos in each city and the maximum number of photos assigned to one vertex of the road network. Finally, regarding the road network, we provide the total number of edges and vertices. Note that although the datasets differ in terms of density (w.r.t. to relations and Flickr photos), our algorithms provide similar results.

We present two experimental settings: In Setting (i) we examine the influence of different scalings of the closeness score $\mathcal{W}_{i,j}$ in terms of enrichment ratio, path length increase (distance) and popularity. Setting (ii) investigates the influence of the path length, i.e., the distance between start and target is varied, again in terms of enrichment ratio, path length increase (distance) and popularity. In both settings we present the ER performance of the algorithms separately from their performance in terms of distance and popularity as ER is a measure that mainly proves that our network enrichment approach works properly, i.e., ER should increase with the increase of the influence of $\mathcal{W}_{i,j}$ on the network and the increase of the path length. Hence, based on our own measure (ER) we validate that the proposed approach works properly.

In Setting (i), for 100 randomly chosen pairs of start and target nodes the respective shortest paths within the actual road network are computed using

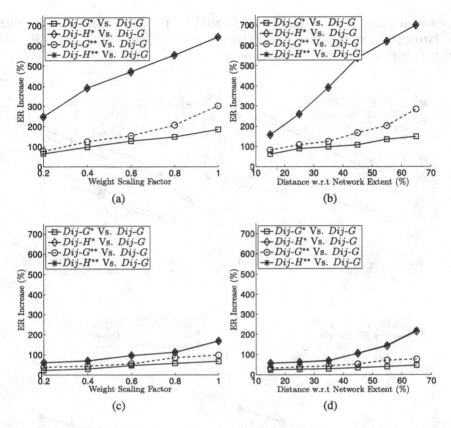

Fig. 4. (a), (b) show ER increase for algorithms Dij-G* and Dij-H*for Paris dataset for Settings *i* and *ii* respectively. (c), (d) show ER increase for algorithms Dij-G* and Dij-H*for New York dataset for Settings *i* and *ii* respectively.

Dijkstra's algorithm, Dij-G. Continuing, for the same start and target pairs, we run Dij-G*, Dij-H*, Dij-G** and Dij-H**. Subsequently, for each pair the difference w.r.t. ER, distance and popularity is computed, and finally averaged over all pairs. We require the distance between start and target nodes to be at least 30 % and at most 50 % of the Euclidean extent of the network (approximately 6 km to 10 km), in order to exclude paths which start and end in the outskirts of the city (where there are few to no POIs). Figure 4 ((a), (c)) show the influence of the closeness score $\mathcal{W}_{i,j}$ on ER for the datasets of Paris and New York respectively. As we increase the impact of $\mathcal{W}_{i,j}$, we observe an increase of ER for all four cases in comparison to Dij-G in both datasets. For the Paris dataset, the increase in ER is in the range of 80 % to 250 % for the Dij-G* and Dij-G**, with the latter performing better, and in the range of 250 % to 620 % for Dij-H* and Dij-H** with the latter performing better. For the New York dataset, the

increase in ER is in the range of 20 % to 80 % for the Dij-G* and Dij-G**, with
the latter performing better, and in the range of 80 % to 150 % for Dij-H* and
Dij-H**, with the latter performing better.

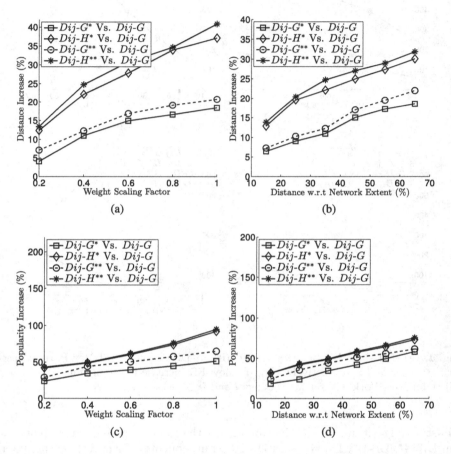

Fig. 5. (a), (c) show Distance and Flickr popularity increase for algorithms Dij-G*and
Dij-H*for Paris dataset for experimental Setting i. (b), (d) show Distance and Flickr
popularity increase for algorithms Dij-G*and Dij-H*for Paris dataset for experimental
Setting ii.

Moreover, the first column of Figs. 5 and 6 ((a), (c)) shows the influence of
weight scaling factor $\mathcal{W}_{i,j}$ on distance and popularity. As we increase $\mathcal{W}_{i,j}$ from
0.2 to 1.0, we observe an increase of distance and popularity for both cases in
comparison to Dij-G in both datasets. The increase among all datasets, in terms
of path length is in the range of 3 % to 16 % for Dij-G* and Dij-G**, and in
the range of 7 % to 38 % for Dij-H* and Dij-H**. Additionally, the increase in
popularity is in the range of 30 % to 120 % for Dij-G* and Dij-G**, and in the
range of 40 % to 160 % for Dij-H* and Dij-H**.

It is clear that Dij-G* and Dij-G** always perform better than Dij-H* and Dij-H** in terms of path length increase, but Dij-H*and Dij-H**perform always better in terms of ER and popularity. This is because Dij-H*and Dij-H** route directly through the POIs, causing greater detours, but passing along highly weighted parts of the enriched graphs (G^*or G^{**}), which mostly coincide with dense Flickr photo regions. Moreover, it is clear that *S-enrich* always performs better than *D-enrich*, in terms of ER and popularity with a very short increase, of about 2–3 % in path length. This validates that skyline enrichment provides competitive paths in terms of distance (minorincrease) and popularity (significant increase).

Continuing, in Setting (*ii*) we vary the distance between and target, relative to the extent of the whole network. We consider five different distance brackets of shortest paths in the original graph G, the first one ranging from 10 % to 20 %, the last one ranging from 50 % to 60 % of the extent of the whole network. For 100 randomly chosen pairs of start and target nodes (within the respective distance bracket) paths with Dij-G, Dij-G*, Dij-G**, Dij-H* and Dij-H** are computed. As before, for each pair the difference w.r.t. ER, distance and popularity is computed and averaged over all pairs. Figure 4 ((b), (d)) show the increase of ER as we proceed through the distance brackets for both datasets. The second column of Figs. 5 and 6 ((b), (d)) show the results in terms of distance and popularity increase. As we proceed through the distance brackets, we observe an increase of the distance and popularity for all cases in comparison to Dij-G in both datasets. The increase among all datasets, in terms of path length, is in the range of 3 % to 18 % for Dij-G* and Dij-G**, and in the range of 5 % to 30 % for Dij-H* and Dij-H**. Finally, the increase in terms of popularity is in the range of 10 % to 70 % for Dij-G* and Dij-G**, and in the range of 30 % to 140 % for Dij-H* and Dij-H**. As in our previous experimental setting, it is clear that Dij-G* and Dij-H* always perform slightly better (only 2–3 %) in terms of path length increase, while Dij-G** and Dij-H* always outperform Dij-G* and Dij-H* in terms of enrichment ratio and popularity. This underlines the validaty of *S-enrich*, as it provides significantly more popular paths while only incurring minor detours (2–3 % in terms of path length).

Here, we may conclude that both *D-enrich* and *S-enrich* approaches show convincing results. Both cases yield significant increase in terms of ER as well as in terms of the independent Flickr-based measure popularity, while increasing path length only slightly. In the best case, ER increase amounts to almost 700 % while popularity increase amounts to almost 160 % (in comparison to the conventional shortest paths, as computed by Dij-G), while the worst case increase in path length is about 38 % with most cases being less than 10 %. Overall, *D-enrich* works slightly (2–3 %) better in terms of path length while the *S-enrich* is always significantly better (more than 10 % in most of the cases) in terms of popularity scores. Consequently, we can claim that spatial relations, extracted from crowdsourced information, can indeed be used to enrich actual road networks and define an alternative kind of routing which reflects what people perceive as "close".

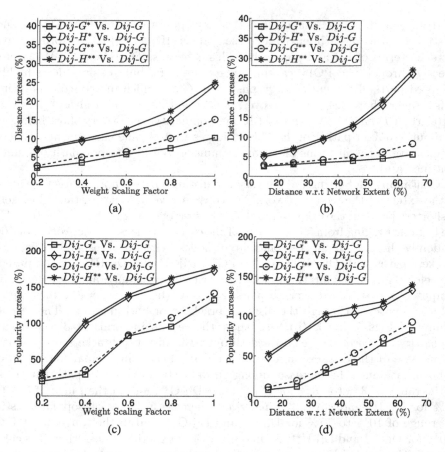

Fig. 6. (a), (c) show Distance and Flickr popularity increase for algorithms Dij-G* and Dij-H* for New York dataset for experimental Setting i. (b), (d) show Distance and Flickr popularity increase for algorithms Dij-G* and Dij-H* for New York dataset for experimental Setting ii.

Finally, Fig. 7 illustrates the trade-off (mean distance and popularity increase overall experiments) that we take by deviating from the shortest path in order to obtain more interesting paths. This figure shows the relative increase in distance and popularity of the paths returned by our proposed approaches, compared to the baseline approach Dij-G. Here, we use letter D to refer to the distance increase while we use letter P to refer to popularity increase. For both datasets, we can observe that by road network enrichment we can obtain a significant increase in popularity of up to 120 % for the meager price of no more than 25 % additional distance incurred in both experimental settings. With the proposed *S-enrich* approach we achieve to significantly increase popularity while keeping the distance increase almost in the same levels with the *D-enrich* approach.

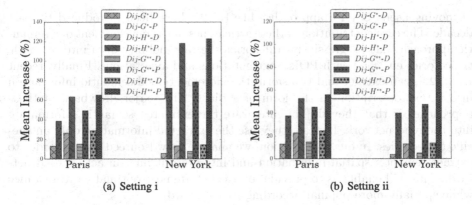

Fig. 7. Trade-off between distance and popularity increase of paths

6 Related Work

Research areas relevant to this work include: (i) qualitative routing and (ii) mining of semantic information from moving object trajectories and trajectory enrichment with extracted semantic information. In what follows, we discuss previous work in both of these areas.

While finding shortest paths in road networks is a thoroughly explored research area, qualitative routing has hardly been explored. Nevertheless, providing meaningful routing directions in road networks is a research topic of great importance. In various real world scenarios, the shortest path may not be the ideal choice for providing directions in written or spoken form, for instance when in an unfamiliar neighborhood, or in cases of emergency. Rather, it is often more preferable to offer "simple" directions that are easy to memorize, explain, understand and follow. However, there exist cases where the simplest route is considerably longer than the shortest. The authors in [12] and [13] try to tackle the problem of efficient routing by using cost functions that trade off between minimizing the length of a provided path while also minimizing the number of turns on the provided path. The major shortcoming of these approaches is that they focus almost exclusively on road network data without taking into account any kind of qualitative information, i.e., information coming from the user. Opposed to that, we try to approach the problem of efficient routing by integrating spatial knowledge coming from the crowd thus enriching an actual road network.

The discovery of semantic places through the analysis of raw trajectory data has been investigated thoroughly over the course of the last years. The authors in [14–16] provide solutions for the semantic place recognition problem and categorize the extracted POIs into pre-defined types. Moreover, the concept of "semantic behavior" has recently been introduced. This refers to the use of semantic abstractions of the raw mobility data, including not only geometric patterns but also knowledge extracted jointly from the mobility data as well as the underlying geographic and application domains in order to understand the actual behaviour of moving users. Several approaches like [17,21] have been introduced the last decade. The core contribution of these articles lies in the development of a semantic approach that progressively transforms the raw mobility data into semantic trajectories enriched with POIs, segmentations and annotations. Finally, recent work [22], can extract and transform the aforementioned semantic information into a text description in the form of a diary. The major drawback of these approaches is that they do not intergrate the extracted semantic information into the road network. Instead, they use the extracted information only on specific trajectories. In our contribution, we analyze crowdsourced data in order to extract semantic spatial information and intergrate it into an actual road network. This will enable us to provide routes that are near-optimal w.r.t. distance while spatially more popular according to the crowd.

7   Conclusions and Outlook

In this work we presented new approaches to computing knowledge-enriched paths within road networks. We incorporated novel methods to extract spatial relations between pairs of POIs, such as "near" or "close by", from crowdsourced textual data, namely travel blogs. We quantified the extracted relations using probabilistic models to handle the inherent uncertainty of user-generated content. Based on these models, we proposed a new cost function to enrich real world road networks, based on Djikstra and skyline path computation. The new cost function reflects the closeness aspect according to the crowd. In contrast to existing approaches, we did not enrich previously computed paths with semantical information, but the entire network. Continuingly, two routing algorithms were presented taking this closeness aspect into account. Finally, we evaluated our ideas on two real world road network datasets, i.e., Paris, France, and New York City, USA. We used metadata from geotagged Flickr photos as a ground truth to support our initial goal of providing more popular paths. All our approaches performed very well by providing slightly longer paths but with significantly higher values of popularity. For future work, we are researching alternative methods for aggregating all categories of spatial relations. Furthermore, we would like to investigate ways to suggest the popular path descriptions to the user based on the POIs they will encounter underway.

Acknowledgements. The research leading to these results has received funding from the EU FP7 project GEOSTREAM (grant No. FP7-SME-2012-315631) as well as the Shared-E-Fleet project by the German Federal Ministry of Economics and Technology (grant No. 01ME12107), the Deutsche Forschungsgemeinschaft (DFG) under grant number RE 266/5-1 and from the DAAD supported by the BMBF under grant number 57052426. Mario A. Nascimento has been partially supported by NSERC Canada. Dieter Pfoser has been partially supported by NGA NURI (grant No. HM02101410004).

References

1. Richter, K.F., Winter, S.: Cognitive aspects: how people perceive, memorize, think and talk about landmarks. In: Landmarks, pp. 41–108. Springer International Publishing, Cham (2014)
2. Skoumas, G., Schmid, K.A., Jossé, G., Züfle, A., Nascimento, M.A., Renz, M., Pfoser, D.: Towards knowledge-enriched path computation. In: Proceedings of the 22nd ACM International Conference on Advances in Geographic Information Systems, 485–488 (2014)
3. Skoumas, G., Pfoser, D., Kyrillidis, A.: On quantifying qualitative geospatial data: a probabilistic approach. In: Proceedings of the Second ACM International Workshop on Crowdsourced and Volunteered Geographic Information, pp. 71–78 (2013)
4. Skoumas, G., Pfoser, D., Kyrillidis, A.T.: Location estimation using crowdsourced geospatial narratives. In: CoRR abs/1408.5894 (2014)
5. Loper, E., Bird, S.: NLTK: The natural language toolkit. In: Proceedings of the ACL 2002 Workshop on Effective Tools and Methodologies for Teaching Natural Language Processing and Computational Linguistics , vol. 1, pp. 63–70 (2002)
6. Bishop, C.M.: Pattern Recognition and Machine Learning (Information Science and Statistics). Springer-Verlag New York Inc, Secaucus (2006)
7. Dempster, A.P., Laird, N.M., Rubin, D.B.: Maximum likelihood from incomplete data via the em algorithm. J. Roy. Stat. Soc. Ser. B **39**, 1–38 (1977)
8. Borzsony, S., Kossmann, D., Stocker, K.: The skyline operator. In: Proceedings of the 17th International Conference on Data Engineering, pp. 421–430 (2001)
9. Shekelyan, M., Jossé, G., Schubert, M.: Paretoprep: fast computation of path skylines queries. In: CoRR abs/1410.0205 (2014)
10. Graf, F., Kriegel, H.-P., Renz, M., Schubert, M.: MARiO: multi-attribute routing in open street map. In: Pfoser, D., Tao, Y., Mouratidis, K., Nascimento, M.A., Mokbel, M., Shekhar, S., Huang, Y. (eds.) SSTD 2011. LNCS, vol. 6849, pp. 486–490. Springer, Heidelberg (2011)
11. Mousselly-Sergieh, H., Watzinger, D., Huber, B., Döller, M., Egyed-Zsigmond, E., Kosch, H.: World-wide scale geotagged image dataset for automatic image annotation and reverse geotagging. In: Proceedings of the 5th ACM Multimedia Systems Conference, pp. 47–52 (2014)
12. Sacharidis, D., Bouros, P.: Routing directions: keeping it fast and simple. In: Proceedings of the 21st ACM International Conference on Advances in Geographic Information Systems, pp. 164–173 (2013)
13. Westphal, M., Renz, J.: Evaluating and minimizing ambiguities in qualitative route instructions. In: Proceedings of the 19th ACM International Conference on Advances in Geographic Information Systems, pp. 171–180 (2011)
14. Lv, M., Chen, L., Chen, G.: Discovering personally semantic places from GPS trajectories. In: Proceedings of the 21st ACM International Conference on Information and Knowledge Management, pp. 1552–1556 (2012)
15. Yan, Z., Chakraborty, D., Parent, C., Spaccapietra, S., Aberer, K.: Semitri: a framework for semantic annotation of heterogeneous trajectories. In: Proceedings of the 14th International Conference on Extending Database Technology, pp. 259–270 (2011)
16. Palma, A.T., Bogorny, V., Kuijpers, B., Alvares, L.O.: A clustering-based approach for discovering interesting places in trajectories. In: Proceedings of the ACM Symposium on Applied Computing, pp. 863–868 (2008)

17. Alvares, L.O., Bogorny, V., Kuijpers, B., de Macedo, J.A.F., Moelans, B., Vaisman, A.: A model for enriching trajectories with semantic geographical information. In: Proceedings of the 15th Annual ACM International Symposium on Advances in Geographic Information Systems, pp. 22:1–22:8 (2007)
18. Parent, C., Spaccapietra, S., Renso, C., Andrienko, G., Andrienko, N., Bogorny, V., Damiani, M.L., Gkoulalas-Divanis, A., Macedo, J., Pelekis, N., Theodoridis, Y., Yan, Z.: Semantic trajectories modeling and analysis. ACM Comput. Surv. **45**, 42:1–42:32 (2013)
19. Spaccapietra, S., Parent, C.: Adding meaning to your steps. In: Proceedings of the 30th International Conference on Conceptual Modeling, pp. 13–31 (2011)
20. Yan, Z., Chakraborty, D., Parent, C., Spaccapietra, S., Aberer, K.: Semantic trajectories: mobility data computation and annotation. ACM Trans. Intell. Syst. Technol. **4**, 49:1–49:38 (2013)
21. Yan, Z., Spremic, L., Chakraborty, D., Parent, C., Spaccapietra, S., Aberer, K.: Automatic construction and multi-level visualization of semantic trajectories. In: Proceedings of the 18th International Conference on Advances in Geographic Information Systems, pp. 524–525 (2010)
22. Feldman, D., Sugaya, A., Sung, C., Rus, D.: iDiary: from GPS signals to a text-searchable diary. In: Proceedings of the 11th ACM Conference on Embedded Networked Sensor Systems, pp. 6:1–6:12 (2013)

Trajectory Analysis

Efficient Point-Based Trajectory Search

Shuyao Qi[1(✉)], Panagiotis Bouros[2], Dimitris Sacharidis[3], and Nikos Mamoulis[1]

[1] Department of Computer Science, The University of Hong Kong, Hong Kong, China
{syqi2,nikos}@cs.hku.hk
[2] Department of Computer Science, Humboldt-Universität zu Berlin,
Berlin, Germany
bourospa@informatik.hu-berlin.de
[3] Faculty of Informatics, Technische Universität Wien, Wien, Austria
dimitris@ec.tuwien.ac.at

Abstract. Trajectory data capture the traveling history of moving objects such as people or vehicles. With the proliferation of GPS and tracking technology, huge volumes of trajectories are rapidly generated and collected. Under this, applications such as route recommendation and traveling behavior mining call for efficient trajectory retrieval. In this paper, we first focus on distance-based trajectory search; given a collection of trajectories and a set query points, the goal is to retrieve the top-k trajectories that pass as close as possible to all query points. We advance the state-of-the-art by combining existing approaches to a hybrid method and also proposing an alternative, more efficient range-based approach. Second, we propose and study the practical variant of bounded distance-based search, which takes into account the temporal characteristics of the searched trajectories. Through an extensive experimental analysis with real trajectory data, we show that our range-based approach outperforms previous methods by at least one order of magnitude.

1 Introduction

The proliferation of GPS and tracking technology has brought to availability huge volumes of trajectories from real moving objects such as mobile phone users, vehicles and animals. Searching such a collection of trajectories finds several applications, including route recommendation, behavior mining, and in transportation systems [1,2]. Different from conventional retrieval tasks which identify similar trajectories to a given one or those crossing a specific spatial region, in this paper we focus on *point-based search*, which retrieves trajectories based on given points. In particular, taking as input a set of query points Q (e.g., a particular set of POIs), the *distance-based trajectory search* studied in [3,4] retrieves the trajectories that pass as close as possible to all query points. Specifically, the distance of a trajectory t to Q is computed by summing up, for each query point $q \in Q$, its distance to the nearest point in t.

Work supported by grant HKU 715413E from Hong Kong RGC, and by the European Social Fund and Greek National Funds through the NSRF Research Program Thales.

C. Claramunt et al. (Eds.): SSTD 2015, LNCS 9239, pp. 179–196, 2015.
DOI: 10.1007/978-3-319-22363-6_10

Consider for instance a collection of touristic trajectories; a travel agency issues a distance-based query to survey or recommend popular routes that pass close to specific sightseeing attractions. As another example, query set Q could contain traffic congestion points; in this case, the traffic department seeks to discover the causes of the congestion by analyzing the trajectories that pass near the points in Q. In the context of surveillance and security applications, Q may contain locations of crime scenes, and hence the police department issues a distance-based query to investigate the correlation of these crime locations by identifying suspects who moved close to all of them.

Contributions. This paper tackles two problems under the point-based trajectory search. First, we thoroughly study the efficient evaluation of *distance-based trajectory search*. We review in detail existing algorithms IKNN [3] and GH/QE [4]. These methods follow a *candidate generation* and *refinement* paradigm, and invoke a nearest neighbor (NN) search centered at each query point to examine the trajectories in ascending order of their distance to Q. By analyzing the pros and cons of these methods, we design a hybrid NN-based algorithm which consistently outperforms IKNN and GH/QE by over an order of magnitude. Going one step further, we tackle the inherent shortcomings of the NN-based approach itself, namely (a) the increased I/O cost due to independently running multiple NN searches and (b) the increased CPU cost for continuously maintaining a priority queue for each NN search. We propose a novel *spatial range-based* approach, which is up to 2 times faster than our hybrid algorithm.

Second, we observe that the distance-based search ranks trajectories solely on how close they pass to the query points in Q, ignoring however other qualitative characteristics of the retrieved results. To fill this gap, we introduce a practical variant of distance-based trajectory search, which also takes into account the temporal aspect of the trajectories. Specifically, this *bounded distance-based search* filters out non-interesting trajectories, whose points closest to Q span a time interval greater than a user-defined threshold.

Outline. The rest of the paper is organized as follows. Section 2 formally defines the distance-based and bounded distance-based trajectory search while Sects. 3 and 4 address their efficient evaluation. Then, Sect. 5 discusses problem variants where (a) the trajectories are ranked both on their distance to the query points and the time interval they span, and (b) Q is a sequence of query points, instead of a set. Section 6 presents our experimental analysis. Finally, Sect. 7 outlines related work, while Sect. 8 concludes the paper.

2   Problem Definition

Let T be a collection of trajectories. A trajectory in T is defined as a sequence of spatio-temporal points $\{p_1, \ldots, p_n\}$, each represented by a $\langle latitude, longitude, timestamp \rangle$ triple. The input of *point-based trajectory search* over collection T is a set of m spatial query points $Q = \{q_1, \ldots, q_m\}$. Given a query point $q_j \in Q$ and a trajectory $t_i \in T$, we define the $\langle p_{ij}^*, q_j \rangle$ *matching pair* based on the

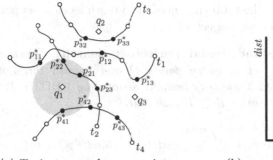

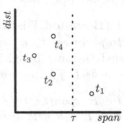

(a) Trajectory and query points   (b) *span-dist* plot of trajectories

Fig. 1. Distance-based trajectory search with 4 trajectories, $T = \{t_1, \ldots, t_4\}$, and 3 query points, $Q = \{q_1, \ldots, q_3\}$; t_1, t_2 is the result to 2-DTS(T, Q), while t_2, t_3 the result to 2-BDTS(T, Q, τ)

nearest to q_j point p_{ij}^* of trajectory t_i, i.e., $p_{ij}^* = \arg\min_{p \in t_i} dist(p, q_j)$, where $dist(\cdot, \cdot)$ denotes the distance (e.g., Euclidean) between two points in space. We then define the *distance* of a trajectory to Q based on the matching pairs for every query point q_j as:

$$dist(t_i, Q) = \sum_{q_j \in Q} dist(p_{ij}^*, q_j) \tag{1}$$

Consider the example in Fig. 1(a), where query points are represented as diamonds, and trajectory points as circles; filled circles indicate matched points of the trajectory to query points. For trajectory t_1, point p_{11}^* is its closest point to query point q_1, and hence $\langle p_{11}^*, q_1 \rangle$ represents a matching pair. The other matched trajectory points of t_1 are p_{12}^* and p_{13}^*. Note that it is possible for a trajectory point to be matched with multiple query points. This is the case with trajectory t_3, where p_{32}^* is the closest point to both q_1 and q_2, i.e., $p_{31}^* \equiv p_{32}^*$. We now formally define the *distance-based trajectory search* problem [3,4].

Problem 1 (Distance-based Trajectory Search). *Given a collection of trajectories T and a set of query points Q, the k-Distance-based Trajectory Search, denoted by k-DTS(T, Q), retrieves a subset of k trajectories $R \subseteq T$ such that for each $t \in R$ and $t' \in T \setminus R$, $dist(t, Q) \leq dist(t', Q)$ holds.*

Returning to the example of Fig. 1(a), trajectory t_1 has the lowest distance to Q, followed by t_2, t_3 and t_4; hence, the result to 2-DTS(T, Q) is t_1, t_2.

Next, we introduce a novel point-based trajectory search problem by also taking into account the temporal aspect of the trajectories. Let P_i^* be the set of all matching pairs for a trajectory t_i, sorted ascending on the timestamp of the involved trajectory points. We define the *span* of trajectory t_i with respect to Q, denoted by $span(t_i, Q)$, as the length of the time interval between the first and the last pair in P_i^*, or equivalently:

$$span(t_i, Q) = \max_{q_x, q_y \in Q} (timestamp(p_{ix}^*) - timestamp(p_{iy}^*)) \tag{2}$$

Intuitively, $span(t_i, Q)$ equals the total time needed to reach as close as possible to all query points in Q, following trajectory t_i.

Problem 2 (Bounded Distance-based Trajectory Search). *Given a collection of trajectories T, a set of query points Q and a span threshold τ, the k-Bounded Distance-based Trajectory Search, denoted by k-BDTS(T,Q,τ), retrieves the subset of k trajectories $R \subseteq T$ such that:*

- *for each $t \in R$, $span(t, Q) \leq \tau$ holds, and*
- *for each $t' \in T \smallsetminus R$ with $span(t', Q) \leq \tau$, $dist(t, Q) \leq dist(t', Q)$ holds.*

Returning to Fig. 1(a), assume for simplicity that trajectory points are reported in fixed time intervals. As a result, the span of a trajectory is proportional to the number of its points from the first to the last matched point (excluding the first). For example, $span(t_1, Q) = 4$ as there are 4 points from p_{11}^* and up to p_{13}^*. Similarly, we obtain the spans of t_2, t_3, t_4 as 2, 1, 2, respectively. Figure 1(b) plots the trajectories in the *span-dist* plane. DTS ignores the span values and simply returns the trajectories with the lowest *dist* coordinate. In contrast, BDTS introduces a threshold, e.g., $\tau = 3$, on the span of the trajectories, depicted as the dashed vertical line. Trajectories to the right of this line, i.e., t_1, do not qualify as BDTS results. Therefore, the result of 2-BDTS is t_2, t_3, i.e., the trajectories with the 2 lowest distances among those left of the line. Notice that BDTS may not return the trajectory with the lowest distance to Q if its span exceeds the threshold; e.g., t_1 in Fig. 1.

Depending on the application, one may consider alternative definitions for point-based trajectory search that take into account both the distance and the span metrics. We briefly overview one of them in Sect. 5, where we also discuss trajectory search given a sequence of query points, instead of a set.

3   Distance-Based Trajectory Search

We first discuss trajectory search based on the distance to a set of query points. Section 3.1 revisits existing work, while Sects. 3.2 and 3.3 present our NN-based and spatial range-based methods, respectively.

3.1   Existing Methods

Methods IKNN [3] and GH/QE [4] have previously tackled distance-based trajectory search. Note that in [3] the problem was defined with respect to the similarity of a trajectory t_i to the set of query points Q, defined as $sim(t_i, Q) = \sum_{q_j \in Q} e^{-dist(p_{ij}^*, q_j)}$. In what follows, we describe the straightforward adaptation of the IKNN algorithm for the distance metric of Eq. (1) (which was also used in [4]). The adaptation of GH/QE and our methods (Sects. 3.2 and 3.3) to the similarity metric of [3] is also straightforward and therefore, omitted. Moreover, the relative performance of all methods is identical independent of the metric used.

Algorithm 1. IKNN

Input	: collection of trajectories T, set of query points Q, number of results k
Output	: result set R
Variables	: candidate set C, k-th distance upper bound UB_k, distance lower bound LB

1 initialize $C \leftarrow \emptyset$, $UB_k \leftarrow \infty$ and $LB \leftarrow 0$;
2 **while** $UB_k > LB$ **do**
3   **for each** $q_j \in Q$ **do**
4   δ_j-NN$(q_j) \leftarrow$ the next δ_j nearest trajectory points to q_j;
5   update C with δ_j-NN(q_j);
6   update UB_k and LB ▷ Equations (3) and (4)
7 $R \leftarrow$ Refine$_{\text{DTS}}(k, T, Q, C)$;
8 **return** R;

All existing methods adopt a *candidate generation* and *refinement* evaluation paradigm. During the first phase, a set of candidate trajectories is determined by incrementally retrieving the nearest trajectory points to the query points in Q. For this purpose, the methods utilize a single R-tree to index all trajectory points. A candidate trajectory t is called a *full match* if the matching pairs of t to all query points in Q have been identified; otherwise, t is a *partial* match. As soon as the candidate set is guaranteed to include the final results (even as partial matches), candidate generation is terminated, and the refinement phase is then employed to identify and output the results.

The IKNN Algorithm. Note that the IKNN algorithm comes in two flavors; in the following, we consider the one based on best-first search, as it was shown in [3] to be both faster and require fewer I/O operations. Algorithm 1 shows the pseudocode of IKNN. During candidate generation (Lines 2–6), the algorithm iterates over the points of Q in a round robin manner. For each query point q_j, the (next) batch of nearest to q_j trajectory points is retrieved using the R-tree index, in Line 4. The nearest neighbor search retrieves a different number of trajectory points δ_j per query point q_j, in order to expedite the termination of this first phase (details in [3]). Based on the newly identified matching pairs that involve q_j, the set of candidates C is then updated in Line 5 by either adding new partial matches or filling an empty slot for existing. For each partial match t_i in C, IKNN computes an *upper bound* of its distance to Q by setting the distance of t_i to every unmatched query point equal to the diameter of the space (maximum possible distance between two points):[1]

$$\overline{dist}(t_i, Q) = \sum_{q_j \in Q_i} dist(p_{ij}^*, q_j) + |Q \setminus Q_i| \cdot DIAM, \tag{3}$$

where set $Q_i \subseteq Q$ contains all the query points already matched to a point in trajectory t_i. We denote by UB_k the k-th smallest among the distance bounds for the trajectories in C. In addition, IKNN computes a *lower bound* LB of the distance to Q for all unseen trajectories (i.e., those not contained in C), by aggregating the distance of the farthest (retrieved so far) trajectory point to each query point in Q. Formally:

---

[1] Under the similarity-based definition of DTS in [3], IKNN sets empty "slots" to 0.

Algorithm 2. GH

    Input   : collection of trajectories T, set of query points Q, number of results k
    Output   : result set R
    Variables : candidate set C, global heap H
1 initialize $C \leftarrow \emptyset$ and $H \leftarrow \emptyset$;
2 **while** C contains less than k full matches **do**
3   pop $\langle p_{ij}, q_i \rangle$ from H ▷ Get the globally nearest trajectory point to some query point
4   update C with $\langle p_{ij}, q_i \rangle$;
5   push to H the next nearest trajectory point to q_i;
6 $R \leftarrow \text{Refine}_{\text{DTS}}(k, T, Q, C)$;
7 **return** R;

Algorithm 3. QE

    Input   : collection of trajectories T, set of query points Q, number of results k
    Output   : result set R
    Variables : candidate set C, global heap H, distance lower bound LB
1 initialize $C \leftarrow \emptyset$, $H \leftarrow \emptyset$ and $LB \leftarrow 0$;
2 **while** C contains less than k full matches with $dist(\cdot, Q) \geq LB$ **do**
3   pop $\langle p_{ij}, q_i \rangle$ from H ▷ Get the globally nearest trajectory point to some query point
4   update C with $\langle p_{ij}, q_i \rangle$;
5   push to H the next nearest trajectory point to q_i;
6   complete the most promising partial matches in C ▷ Equation (5)
7   update LB ▷ Equation (6)
8 $R \leftarrow \text{Refine}_{\text{DTS}}(k, T, Q, C)$;
9 **return** R;

$$LB = \sum_{q_j \in Q} dist(p_j^\delta, q_j) \tag{4}$$

where p_j^δ is the last trajectory point returned by the NN search centered at q_j.

The candidate generation phase of IKNN terminates when $UB_k \leq LB$; in this case, none of the unseen trajectories can have smaller distance to Q compared to the candidates in C. Last, IKNN invokes $\text{Refine}_{\text{DTS}}$ to produce the results. Briefly, the function examines candidates in ascending order of a lower bound on their distance, retrieving them from disk to compute $dist(\cdot, Q)$ (details in [3]).

The GH/QE Algorithms. Different from IKNN, the methods in [4] retrieve trajectory points in ascending order of the distance to their closest query point. Specifically, a *global heap* H is used to retrieve at each iteration the *globally* nearest trajectory point p_{ij} to some query point q_j, and then, to update candidate set C, accordingly. Algorithm 2 shows the pseudocode of GH. The candidate generation phase of GH is terminated as soon as set C contains k full matches (proof of correctness in [4]). Note that these full matches are not necessary among the final results identified in Line 6 during the refinement phase.

In practice, the order imposed by global heap H cannot guarantee a good performance unless both trajectory and query points are uniformly distributed in space. For instance, if a particular query point is very close to many trajectories, GH will generate a large number of partial matches with only that slot filled. Consequently, it will take longer to produce the k full matches needed to terminate the generation phase, and at the same time a large number of candidates would

have to be refined. A similar problem occurs when a query point is located away
from the trajectories.

To address these issues, Tang et al. [4] proposed an extension to GH termed QE,
which periodically fills the empty slots for the partially matched trajectories with
the highest potential of becoming results. These are then retrieved from disk,
and their actual distance is computed. A trajectory has high potential if it has
(i) few empty slots and (ii) small distance in each filled slot with respect to the
next point to be retrieved for that slot. These factors are captured respectively
by the denominator and enumerator of the following equation:

$$potential(t_i) = \frac{\sum_{q_j \in Q_i} \left(dist(p_j^H, q_j) - dist(p_{ij}^*, q_i)\right)}{|Q \setminus Q_i|} \tag{5}$$

where set $Q_i \subseteq Q$ contains all the query points already matched to a point in t_i,
p_j^H is the next nearest trajectory point to q_j contained in heap H and p_{ij}^* is the
nearest to q_j point in trajectory t_i.

Algorithm 3 shows the pseudocode of QE. The candidate generation phase
of QE terminates when candidate set C contains k full matches (similar to GH),
provided however that their distance to Q is smaller than the distance of all
unseen trajectories (Line 2) (proof of correctness in [4]). To determine this, QE
computes in Line 7, a *lower bound LB* of the distance for the unseen trajectories
(similar to IKNN) by aggregating the distance of the next nearest trajectory point
to every query point, i.e., the contents of heap H:

$$LB = \sum_{q_j \in Q} dist(p_j^H, q_j) \tag{6}$$

3.2   A Hybrid NN-based Approach

The DTS problem can be viewed as a top-k query [5,6]. For each query point q_j,
consider a sorted trajectory list T_j, where each trajectory is ranked according
to its distance to the query point. Then, the objective is to determine the top-k
trajectories that have the highest aggregate score, i.e., distance, among the lists.
However, as these lists are not given in advance and constructing them is costly,
the goal is to progressively materialize them, until the result is guaranteed to be
among the already seen trajectories.

Following the top-k query processing terminology, a *sorted access* on list T_j
corresponds to the retrieval of the next nearest trajectory to query point q_j,
which in turn may involve multiple trajectory point NN retrievals. In contrast, a
random access for trajectory t_i on list T_j corresponds to the retrieval of t_i from
disk and the computation of its distance to q_j; in practice, once t_i is retrieved,
its distance to all query points can be computed at negligible additional cost.

Methods IKNN, GH and QE employ various ideas from top-k query processing
(an overview of this field is presented in Sect. 7). Particularly, IKNN performs
only sorted accesses and prioritizes them in a manner similar to Stream–Combine
[7]. Similarly, GH performs only sorted accessses but follows an unconventional

strategy for prioritizing them, which explains its poor performance on our tests in Sect. 6. On the other hand, QE additionally performs random accesses following a strategy similar to the CA algorithm [5] to select which trajectory to retrieve.

In the following, we present the NNA algorithm, which combines the strengths of IKNN and QE. In short, it builds upon the Quick–Combine top-k algorithm [8] performing both sorted and random accesses to generate the candidate set. NNA has the following features. First, similar to IKNN, the algorithm retrieves in a round robin manner, batches of nearest trajectory points to each query point in Q. This addresses the weaknesses of GH when dealing with non-uniformly distributed data. Second, after performing the nearest neighbor search centered at each query point, NNA fills the slots of the trajectories with the highest potential according to Eq. (5), similar to QE. Finally, NNA employs the termination condition of IKNN for the candidate generation phase. In practice, NNA extends Algorithm 1 by completing the most promising partial matches in C (similar to QE), between Lines 5 and 6. Hence, it is able to compute tighter bounds compared to IKNN and thus terminate the generation phase earlier. In addition, it produces fewer candidates than IKNN, reducing the cost of the refinement phase.

3.3   A Spatial Range-Based Approach

We identify two shortcomings of all the NN-based methods previously described. First, each NN search is implemented independently, which means that R-tree nodes and trajectory points may be accessed multiple (up to $|Q|$) times, which increases the total I/O cost. Second, each NN search is associated with a priority queue, whose continuous maintenance increases the total CPU cost.

Our novel *Spatial Range-based* algorithm, denoted by SRA, addresses both these shortcomings. Similar to the NN-based approaches, it follows a generation and refinement paradigm. However, to generate the candidate set, it issues a spatial range search of expanding radius centered at each query point in Q. All searches operate on a common set N of R-tree nodes, which avoids accessing nodes more than once and hence saves I/O operations. Moreover, set N needs not be sorted according to any distance, eliminating costly priority queue maintenance tasks. The range-based search for each query point q_j is associated with *current radius* r_j, and is also assigned a *maximum radius* θ_j. As the algorithm progresses, current radius r_j increases while maximum radius θ_j decreases. Candidate generation terminates as soon as $r_j > \theta_j$ for some query point q_j.

Algorithm 4 shows the pseudocode of SRA. In Lines 2–4, SRA initializes the current and maximum radius for each query point. For the latter, an upper bound UB_k to the k-th smallest distance to Q is computed. In particular, SRA invokes a sum-aggregate nearest neighbor (sum-ANN) procedure [9] retrieving trajectory points in ascending order of $\sum_{q_j \in Q} dist(\cdot, q_j)$. Assuming that this procedure retrieves point p_i of trajectory t_i, the sum-aggregate value is an upper bound to the distance of t_i, i.e., $dist(t_i) \le \sum_{q_j \in Q} dist(p_i, q_j)$. Hence, once points from k distinct trajectories have been retrieved, SRA can determine a value for UB_k.

During the candidate generation phase in Lines 5–13, SRA first selects the query point $q_c \in Q$ with the fewest retrieved points so far, and increases its

Algorithm 4. SRA

Input   : collection of trajectories T, set of query points Q, number of results k
Output   : top-k list of trajectories R
Variables : candidate set C, k-th distance upper bound UB_k, current r_i and maximum θ_i
                 search radius for each $q_i \in Q$, set of R-tree nodes N

```
1  initialize C ← ∅ and N ← R-tree root node;
2  compute UB_k invoking a sum-ANN(T, Q);
3  for each q_j ∈ Q do
4  |  initialize r_j ← 0 and θ_j ← UB_k;
5  while r_j ≤ θ_j for all q_j ∈ Q do
6  |  select current q_c;
7  |  r_c ← r_c + ξ                                          ▷ Increase r_c to expand search around q_c
8  |  expand from N all nodes that intersect with the disc of radius r_c centered at q_c;
9  |  S ← trajectory points within spatial range r_c found during expansion;
10 |  update C with S;
11 |  update UB_k                                            ▷ Equation (7)
12 |  for each q_j ∈ Q do
13 |  |  update θ_j ← UB_k − Σ_{q_ℓ ∈ Q∖{q_j}} r_ℓ           ▷ Reduce maximum radius
14 R ← Refine_DTS(k, T, Q, C);
15 return R;
```

radius by a fixed ξ^2, so that each location retrieves more or less the same number of points. Then, it extends the range search centered at q_c to new radius r_c. In particular, all nodes in N that intersect with the search frontier are expanded, i.e., replaced by their children (Line 8). During the expansion, all trajectory points within the frontier are collected in set S (Line 9). Upon completion of the expansion, set N contains no R-tree node or point within r_c distance to q_c, or with distance to q_c greater than θ_c, and N will be re-used in further iterations.

After the expansion, SRA uses the newly seen trajectory points in S to properly update candidate set C. Note that for each trajectory t_i in C, SRA keeps $|Q|$ slots storing the closest trajectory points $t_i.p_j$ seen so far to each query point q_j. A slot is marked *matched* if the corresponding matching pair has been determined, i.e., when $t_i.p_j \equiv p_{ij}^*$. SRA in Line 10 performs the following tasks for each point p_x in S; let t_i be the trajectory p_x belongs to. For each slot q_j that is not *matched*, SRA checks whether p_x is closer to q_j than $t_i.p_j$, and updates the slot with p_x if true. If the slot for the current query point q_c was among those examined, it is marked as *matched*. The benefits of this update strategy are twofold. First, it guarantees that no matching trajectory point will be missed, even though SRA does not access p_x again (removed from N) for $q_j \neq q_c$. At the same time, it also helps to derive a tighter upper bound for the distance of t_i:

$$\overline{dist}(t_i, Q) = \sum_{q_j \in Q_i} dist(p_{ij}^*, q_j) + \sum_{q_j \in Q \setminus Q_i} dist(t_i.p_j, q_j). \tag{7}$$

Compared to Eq. (3) utilized by IKNN and NNA, Eq. (7) computes a tighter bound on unmatched slots. Based on these bounds, a tighter value for UB_k can be established (Line 11).

---

[2] In the future, we plan to investigate variable ξ_j values based on current radius r_j and the trajectory point density around q_j, inspired by determining δ_j value in [3].

Algorithm 5. INCREMENTAL

Input	: collection of trajectories T, set of query points Q, span threshold τ, number of results k
Output	: result set R
Variables	: candidate set C, number of intermediate results λ

1 initialize $C \leftarrow \emptyset$, $R \leftarrow \emptyset$ and $\lambda \leftarrow 0$;
2 **while** $|R| < k$ **do**
3   increase λ by $k - |R|$;
4      $C \leftarrow$ next candidate set of λ-DTS(T, Q);
5      $R \leftarrow R \cup$ Refine$_{\text{BDTS}}(k, T, Q, C, \tau)$;
6 **return** R;

To better explain the procedure in Line 10, we use the example of Fig. 1(a) for $k = 2$. SRA has just started and thus C is empty. Assume that the current query point is $q_c = q_1$, and let $r_1 = 0 + \xi$ be the radius of the shaded disk depicted in the figure. As a result, set S in Line 9 contains trajectory points $\{p_{21}^*, p_{22}^*, p_{41}^*\}$. Moreover, candidate set C contains t_2 and t_4. For trajectory t_2, p_{21}^* is settled as the matching point to q_1 because $dist(p_{21}^*, q_1) < dist(p_{22}^*, q_1)$ and no unseen point of t_2 can be closer. On the other hand, the matching points to q_2, q_3 cannot be yet determined, but we can use p_{21}^* and p_{22}^* to bound t_2's distances to q_2 and q_3. Therefore, the slots for t_2 become $\langle \mathbf{p_{21}^*}, p_{22}^*, p_{21}^* \rangle$, where bold indicates a *matched* slot. Moreover, an upper bound to the distance of t_2 is determined as $\overline{dist}(t_2, Q) = dist(p_{21}^*, q_1) + dist(p_{22}^*, q_2) + dist(p_{21}^*, q_3)$. Similarly, we obtain the slots for t_4 as $\langle \mathbf{p_{41}^*}, p_{41}^*, p_{41}^* \rangle$.

As a last step, SRA updates the maximum radius for all query points with respect to the new UB_k in Lines 12–13. Observe that SRA's termination condition for candidate generation is essentially identical to that of IKNN. Any trajectory not in the candidate set C must have distance to each q_j at least θ_j, and thus distance at least equal to $LB = \sum_{q_j \in Q} \theta_j$. The termination condition of Line 5, $r_j > \theta_j$ for some q_j, and the update of θ_j, imply that, when candidate generation concludes, $UB_k \leq LB$.

Finally, the performance of SRA can be enhanced following the key idea of QE to further improve the $\overline{dist}(t_j, Q)$ bound and therefore, UB_k. We denote this extension to the SRA algorithm by SRA+. Specifically, in between Lines 10 and 11 in Algorithm 4, SRA+ fills the empty slots of the trajectories in C with the highest potential as computed using Eq. (5).

4   Bounded Distance-Based Trajectory Search

We next address the bounded distance-based trajectory search. Recall from Sect. 2 that k-BDTS(T, Q, τ) is equivalent to a k-DTS(T', Q) distance-based query over the subset $T' \subseteq T$ containing only trajectories with $span(\cdot, Q) \leq \tau$. However, as $span(t, Q)$ can be computed only after all the matching pairs of a trajectory t to Q are identified, the major challenge is to limit the number of invalid partial matches generated, i.e., those with the $span(\cdot, Q) > \tau$. In the following, we address this issue in two alternative ways.

The idea behind the incremental approach, denoted as INCREMENTAL, is to progressively construct the result set R by utilizing the generation phase of a DTS method as a "black" box. Algorithm 5 illustrates INCREMENTAL; note that any of the algorithms in Sect. 3 can be used as the underlying DTS method. At each round, INCREMENTAL asks for the missing $k - |R|$ trajectories to complete the result set R in Lines 3–4. For this purpose, a λ-DTS(T, Q) search is processed, with the λ value been increased at each round by $k - |R|$; during the first round $\lambda = k$. Each time λ is updated in Line 3, the DTS method in Line 4 does not run from scratch. It continues the candidate generation using a new termination condition with respect to the updated λ in order to expand candidate set C. Last, in Line 5, Refine$_{\text{BDTS}}$ examines the new candidates to update result set R by computing their $dist(\cdot, Q)$ and eliminating trajectories with $span(\cdot, Q) > \tau$.

Intuitively, INCREMENTAL takes a conservative approach to bounded distance-based trajectory search. As it is unable to predict which partial matches could provide a valid trajectory (full match) with $span(\cdot, Q) \leq \tau$, a refinement phase is needed to "clean" the candidate set. Hence, INCREMENTAL may involve several rounds of generation and refinement phases. To address these issues, we propose the ONE–PASS approach which involves a single generation and refinement round. The idea is again to build upon a DTS method but by extending its candidate generation phase in two ways. First, for each partial match t_i in candidate set C, ONE–PASS computes a lower bound of $span(t_i, Q)$ based on the points of t_i matching the current subset of query points $Q_i \subset Q$, as follows:

$$\underline{span}(t_i, Q) = \begin{cases} 0, & if\ |Q_i| = 1 \\ span(t_i, Q_i), & otherwise \end{cases} \tag{8}$$

Every partial match with $\underline{span}(\cdot, Q) > \tau$ can be safely pruned. Second, the original termination is triggered only after candidate set C contains at least k valid full matches, i.e., with $span(\cdot, Q) \leq \tau$. This is because the k-th upper bound UB_k of existing candidates can be computed only through full matches. For example, candidate generation of ONE–PASS based on SRA+ terminates as soon as at least k valid full matches are identified and $r_j > \theta_j$ holds for some query point q_j.

5   Discussion

We discuss alternative definitions and variants to the point-based search problems introduced in Sect. 2.

Distance and Span-based Trajectory Search. Although taking into account their temporal span, the bounded distance-based search still ranks the trajectories solely on their distance to the query points in Q. As an alternative, we may rank the results with respect to a linear combination of the *span-dist* metrics:

$$f(t, Q) = \alpha \cdot dist(t, Q) + (1 - \alpha) \cdot span(t, Q) \tag{9}$$

where α weights the importance of each metric. With Eq. (9), we introduce the *k-Distance & Span-based Trajectory Search*, denoted by k-DSTS(T, Q) which returns the subset of k trajectories $R \subseteq T$ with the lowest $f(\cdot, Q)$ value.

All methods discussed in Sect. 3 can be extended for k-DSTS(T, Q) by replacing $dist(\cdot, Q)$ with $f(\cdot, Q)$. Note that the upper bound $\overline{f}(t, Q)$ of a partial match t can be computed by setting $\overline{span}(t, Q)$ equal to the total duration of the trajectory t. In contrast, as no matching pairs are identified for the unseen trajectories, the lower bound LB or the θ_j values are defined similar to the DTS methods, i.e., essentially setting the lower bound of span to zero. In Sect. 6.4, we experimentally investigate the efficient evaluation of DSTS.

Order-aware Trajectory Search. Similar to [3], we also consider a variation of the trajectory search when a visiting order is imposed for the query points. In this variation, the matched trajectory point p^*_{ij} to query point q_j, is not necessarily the nearest to q_j point of trajectory t_i. Consider for example trajectory t_2 in Fig. 1. The depicted $p^*_{22}, p^*_{21}, p^*_{23}$ for DTS cannot be the matched points in the $q_1 \rightarrow q_2 \rightarrow q_3$ order-aware DTS, as they violate the visiting order. Instead, the matched points that preserve the imposed visiting order are $p^*_{22}, p^*_{22}, p^*_{23}$, where p^*_{22} is matched with q_1 although $dist(p^*_{22}, q_1) > dist(p^*_{21}, q_1)$. The distance of a trajectory to sequence Q is recursively defined as follows:

$$dist_o(t, Q) = \begin{cases} \min \begin{cases} dist_o(t, T(Q)) + dist(H(t), H(Q)) - DIAM \\ dist_o(T(t), Q) \end{cases} & \text{if } t \neq \varnothing, Q \neq \varnothing \\ |Q| \cdot DIAM & \text{if } t = \varnothing \\ 0 & \text{if } Q = \varnothing \end{cases} \quad (10)$$

where $H(S)$ is the first point (head) in a sequence S, $T(S)$ indicates the tail of S after removing $H(S)$, $\varnothing$ denotes the empty sequence, and $DIAM$ represents the diameter of the space. The distance can be computed by straightforward dynamic programming [3]. To derive an upper bound on a partial matched trajectory t_i, we consider only the subsequence Q_i of Q that contains the matched query points, i.e., $\overline{dist_o}(t_i, Q) = dist_o(t_i, Q_i)$. For order-aware BDTS, distance and its upper bound are the same as in order-aware DTS. Note, however that the lower bound on span (Equation (8)) does not apply as the matching are not yet finalized. For order-aware DSTS evaluation, $f_o(t, Q)$ and its upper bound are defined in a similar manner to order-aware DTS. In Sect. 6, we experimentally investigate the order-aware variants of all three trajectory search problems.

6 Experimental Analysis

We evaluate our methods for point-based trajectory search. All algorithms were implemented in C++ and the tests run on a machine with Intel Core i7-3770 3.40 GHz and 16 GB main memory running Ubuntu Linux.

6.1 Setup

We conducted our analysis using real-world trajectories from the GeoLife Project [10–12]. The collection contains 17,166 trajectories with 19 m points in Beijing,

Table 1. POIs in Beijing

Category	Cardinality
Restaurants	51,971
Hotels	10,620
Pharmacies	6,963
Schools	6,618
Banks	6,057
Police stations	2,509
Supermarkets	2,356
Gas stations	1,916
Post offices	1,125

Table 2. Experimental parameters (default values in bold)

Description	Parameter	Values
Number of results	k	1, 5, **10**, 50, 100
Number of query points	$|Q|$	2, 4, **6**, 8, 10
Span threshold ratio	τ/τ_{min}	1, 1.5, **2**, 2.5, 3
Linear combination factor	α	0, 0.25, **0.5**, 0.75, 1

recording a broad range of outdoor movement. To generate our query sets, we considered around 90 k points of interest (POIs) of various types, located inside the same area covered by the trajectories (see Table 1 for details). A query set Q is formed by randomly selecting a combination of $|Q|$ types and a particular POI from each type. We assess the performance of all involved methods measuring their CPU and I/O cost, and the number of candidates they generate over 1,000 distinct query sets Q, while varying (i) the number of returned trajectories k and (ii) the number of query points $|Q|$. In case of BDTS queries, we additionally vary the span threshold via the τ/τ_{min} ratio, where τ_{min} is the minimum possible time required to travel among the query points in Q at a constant velocity of 50 km/h. Finally, for DSTS queries, we also vary the weight factor α of Eq. (9). Table 2 summarizes all parameters involved in our study.

6.2   Distance-Based Trajectory Search

Figure 2 reports the CPU cost, the I/O cost and the number of generated candidates for the DTS methods. As expected the processing cost of all methods goes up as the values of k and $|Q|$ increase. The tests clearly show that SRA+ is overall the most efficient evaluation method. We also make the following observations.

First, we observe that IKNN always outperforms GH/QE; note that this is the first time the methods from [3,4] are compared. Naturally, GH comes as the least efficient method; due to the examination order imposed by global heap H, the algorithm is unable to cope with the skewed distribution of the real-world data. QE manages to overcome the shortcomings of GH by completing the empty slots of the most promising candidates. Yet, compared to IKNN, QE is less efficient due to its weak termination condition for the generation phase; recall that at least k full matches are needed for this purpose which also results in generating a larger number of candidates, as shown in Fig. 2(c) and (f). The advantage of IKNN over GH/QE justifies our decision to build the hybrid NNA method upon the round robin-based candidate generation of IKNN which retrieves nearest neighbor points in batches, and its powerful threshold-based termination condition. NNA is indeed the most efficient NN-based method, in fact with an order of magnitude improvement over IKNN and GH/QE on both CPU and I/O cost. Finally, Fig. 2 clearly shows the advantage of the spatial range-based evaluation approach over

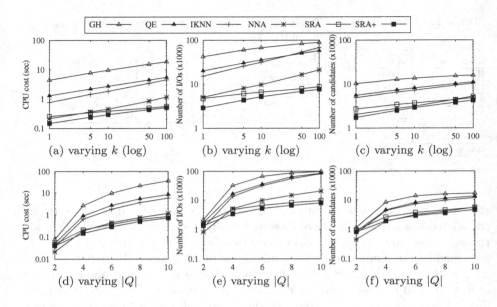

Fig. 2. Performance comparison for Distance-based Trajectory Search

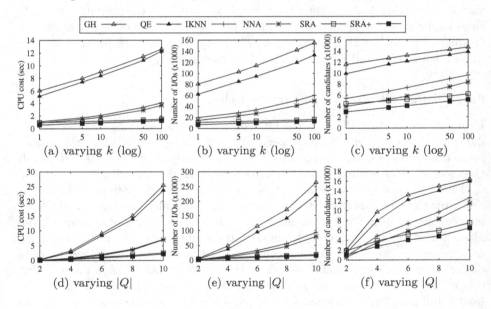

Fig. 3. Performance comparison for Distance-based Trajectory Search (order-aware)

the NN-based one. SRA is always faster while incurring fewer disk page accesses than IKNN, and in a similar manner, SRA+ outperforms NNA.

We also experimented with the order-aware variant of DTS. Figure 3 depicts similar results to Fig. 2; the spatial range-based evaluation approach is again superior to the NN-based and overall, SRA+ is the most efficient method.

Nevertheless, it is important to notice that the advantage of completing the most proposing candidates is smaller compared to Fig. 2, in terms of the CPU cost. Specifically, observe how close is the running time of GH to QE, of IKNN to NNA and of SRA to SRA+, in Fig. 3(a) and (d). This is expected as completing partial matches employs dynamic programming to compute $dist_o(\cdot, Q)$.

6.3   Bounded Distance-Based Trajectory Search

Next, we investigate the evaluation of BDTS queries while varying the k, $|Q|$ and τ/τ_{min} parameters. Based on the findings of the previous section, we use the SRA+ algorithm as the underlying DTS method. Note that due to lack of space we omit results for the order-aware variant of BDTS; the results however are similar. Figure 4 clearly shows that ONE–PASS outperforms INCREMENTAL in all cases. As expected, the conservative approach of INCREMENTAL generates a larger number of candidates by performing multiple rounds of generation and refinement which results in both higher running time and more disk page accesses. Last, notice that the evaluation of BDTS becomes less expensive for both methods while increasing τ/τ_{min}, as the number of invalid candidates progressively drops.

6.4   Distance and Span-Based Trajectory Search

Finally, we study the evaluation of DSTS queries. For this experiment, we extended the most dominant method from [3, 4], i.e., IKNN, and our methods NNA, SRA and SRA+ following the discussion in Sect. 5. The results in Fig. 5 demonstrate, similar to the DTS case, the advantage of both the spatial range-based approach and the SRA+ algorithm which is overall the most efficient evaluation method. Due to lack space, we again omit the figure for the order-aware variant of DSTS as the results are identical to Fig. 5.

7   Related Work

Apart from the studies [3, 4] for distance-based search on trajectories detailed in Sect. 3.1, our work is also related to *top-k* and *nearest neighbor* queries.

Top-k Queries. Consider a collection of objects, each having a number of scoring attributes, e.g., rankings. Given an aggregate function γ (e.g., SUM) on these scoring attributes, a top-k query returns the k objects with the highest aggregated score. To evaluate such a query, a *sorted* list for each attribute a_i organizes the objects in decreasing order of their value to a_i; requests for *random accesses* of an attribute value based on object identifiers may be also possible. Ilyas et al. overviews top-k queries in [6] providing a categorization of the proposed methods. Specifically, when both sorted and random accesses are possible, the TA/CA [5] and Quick–Combine [8] algorithms can be applied. TA retrieves objects from the sorted lists in a round-robin fashion while a priority queue to organizes the best k objects so far. Based on the last seen attribute values, the algorithm defines an upper score bound for the unseen objects, and terminates if current k-th highest aggregate score is higher than this threshold. TA assumes that the costs of the

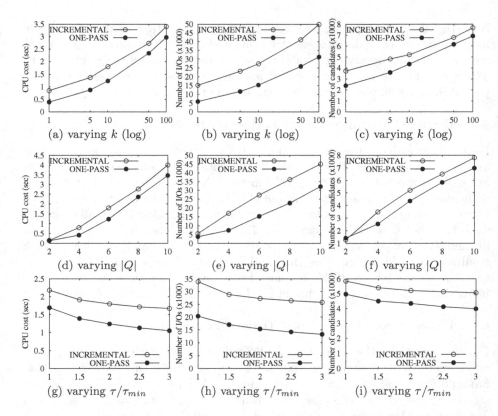

Fig. 4. Performance comparison for Bounded Distance-based Trajectory Search

two different access methods are the same. As an alternative, CA defines a ratio between these costs to control the number of random accesses, which in practice are usually more expensive than sorted accesses. Hence, the algorithm periodically performs random accesses to collect unknown values for the most "promising" objects. Last, the idea behind Quick–Combine is to favor accesses from the sorted lists of attributes which significantly influence the overall scores and the termination threshold. In contrast, when only sorted accesses are possible, the NRA [5] and Stream–Combine [7] algorithms can be applied. Intuitively, Stream–Combine operates similar to Quick–Combine without performing any random accesses. In Sect. 3.1, we discuss how the methods in [3,4] build upon previous work on top-k queries to address distance-based search on trajectories.

Nearest Neighbor Queries. There is an enormous amount of work on the *nearest neighbor* (NN) query (also known as similarity search), which returns the object that has the smallest distance to a given query point; k-NN queries output the k nearest objects in ascending distance. Roussopoulos et al. proposed a depth-first approach to k-NN query in [13] while Hjaltason et al. enhanced the evaluation with a best-first search strategy in [14]. An overview of index-based approaches can be found in [15]; efficient methods for metric spaces, e.g., [16], and high-dimensional data, e.g., [17], have also been proposed.

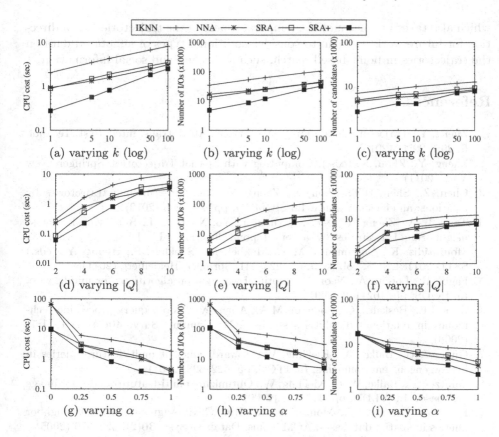

Fig. 5. Performance comparison for Distance and Span-based Trajectory Search

For a set of query points, the *aggregate nearest neighbor* (ANN) query [9] retrieves the object that minimizes an aggregate distance to the query points. As an example, for the MAX aggregate function and assuming that the set of query points are users, and distances represent travel times, ANN outputs the location that minimizes the time necessary for all users to meet. In case of the SUM function and Euclidean distances, the optimal location is also known as the Fermat-Weber point, for which no formula for the coordinates exists.

8   Conclusions

In this paper, we studied the efficient evaluation of point-based trajectory search. After revisiting the existing methods (IKNN and GH/QE), which examine the trajectories in ascending order of their distance to the queries points, we devised a hybrid algorithm which outperforms them by a wide margin. Then, we proposed a spatial range-based approach; our experiments on real-world trajectories showed that this approach outperforms any NN-based method. Besides improving the performance of distance-based search, we also introduced and investigated the evaluation of a practical variant for point-based trajectory search, which also takes into account the temporal aspect of the trajectories. As a direction for future work, we plan to consider additional types of annotated data on the trajectories in point-based search, such as textual and social information.

References

1. Zheng, Y.: Trajectory data mining: an overview. ACM Trans. Intell. Syst. Technol. **6**, 29:1–29:41 (2015)
2. Zheng, Y., Zhou, X. (eds.): Computing with Spatial Trajectories. Springer, New York (2011)
3. Chen, Z., Shen, H.T., Zhou, X., Zheng, Y., Xie, X.: Searching trajectories by locations: an efficiency study. In: SIGMOD, pp. 255–266 (2010)
4. Tang, L.-A., Zheng, Y., Xie, X., Yuan, J., Yu, X., Han, J.: Retrieving k-nearest neighboring trajectories by a set of point locations. In: Pfoser, D., Tao, Y., Mouratidis, K., Nascimento, M.A., Mokbel, M., Shekhar, S., Huang, Y. (eds.) SSTD 2011. LNCS, vol. 6849, pp. 223–241. Springer, Heidelberg (2011)
5. Fagin, R., Lotem, A., Naor, M.: Optimal aggregation algorithms for middleware. In: PODS, pp. 102–113 (2001)
6. Ilyas, I.F., Beskales, G., Soliman, M.A.: A survey of top-k query processing techniques in relational database systems. ACM Comput. Surv. **40**(4), 11:1–11:58 (2008)
7. Güntzer, U., Balke, W., Kießling, W.: Towards efficient multi-feature queries in heterogeneous environments. In: ITCC, pp. 622–628 (2001)
8. Güntzer, U., Balke, W.T., Kießling, W.: Optimizing multi-feature queries for image databases. In: VLDB, pp. 419–428 (2000)
9. Papadias, D., Tao, Y., Mouratidis, K., Hui, C.K.: Aggregate nearest neighbor queries in spatial databases. ACM Trans. Database Syst. **30**(2), 529–576 (2005)
10. Zheng, Y., Zhang, L., Xie, X., Ma, W.Y.: Mining interesting locations and travel sequences from GPS trajectories. In: WWW 2009, pp. 791–800 (2009)
11. Zheng, Y., Li, Q., Chen, Y., Xie, X., Ma, W.: Understanding mobility based on GPS data. In: UbiComp 2008: Proceedings of the 10th International Conference on Ubiquitous Computing, UbiComp 2008, Seoul, Korea, 21–24 September, pp. 312–321 (2008)
12. Zheng, Y., Xie, X., Ma, W.: Geolife: a collaborative social networking service among user, location and trajectory. IEEE Data Eng. Bull. **33**(2), 32–39 (2010)
13. Roussopoulos, N., Kelley, S., Vincent, F.: Nearest neighbor queries. In: Proceedings of the 1995 ACM SIGMOD International Conference on Management of Data, San Jose, California, 22–25 May, pp. 71–79 (1995)
14. Hjaltason, G.R., Samet, H.: Distance browsing in spatial databases. ACM Trans. Database Syst. **24**(2), 265–318 (1999)
15. Böhm, C., Berchtold, S., Keim, D.A.: Searching in high-dimensional spaces: index structures for improving the performance of multimedia databases. ACM Comput. Surv. **33**(3), 322–373 (2001)
16. Jagadish, H.V., Ooi, B.C., Tan, K., Yu, C., Zhang, R.: iDistance: an adaptive b$^+$-tree based indexing method for nearest neighbor search. ACM Trans. Database Syst. **30**(2), 364–397 (2005)
17. Tao, Y., Yi, K., Sheng, C., Kalnis, P.: Quality and efficiency in high dimensional nearest neighbor search. In: SIGMOD, pp. 563–576 (2009)

Visibility Color Map for a Fixed or Moving Target in Spatial Databases

Ishat E. Rabban[1]([✉]), Kaysar Abdullah[1], Mohammed Eunus Ali[1], and Muhammad Aamir Cheema[2]

[1] Department of Computer Science and Engineering,
Bangladesh University of Engineering and Technology, Dhaka, Bangladesh
{ieranik,kaysar,eunus}@cse.buet.ac.bd
[2] Faculty of Information Technology, Monash University, Clayton, Australia
aamir.cheema@monash.edu

Abstract. The widespread availability of 3D city models enables us to answer a wide range of spatial visibility queries in the presence of obstacles (e.g., buildings). Example queries include *"what is the best position for placing a billboard in a city?"* or *"which hotel gives the best view of the city skyline?"*. These queries require computing and differentiating the visibility of a target object from each viewpoint of the surrounding spe. A recent approach models the visibility of a *fixed* target object from the surrounding area with a visibility color map (*VCM*), where each point in the space is assigned a color value denoting the visibility measure of the target. In the proposed *VCM*, a viewpoint is simply discarded (i.e., considered as non-visible) if an obstacle even slightly blocks the view of the target from the viewpoint, which restricts its applicability for a wide range of applications. To alleviate this limitation, in this paper, we propose a scalable, efficient and comprehensive solution to construct a *VCM* for a *fixed* target that considers the *partial* visibility of the target from viewpoints. More importantly, our proposed data structures for the fixed target support incremental updates of the *VCM* if the target moves to near-by positions. Our experimental results show that our approach is orders of magnitude faster than the straightforward approach.

1 Introduction

3D city models are increasingly available through popular mapping services such as Google Maps, Google Earth and OpenStreetMap. We envision that these 3D datasets will provide a new platform for answering many real-life user queries, e.g., visibility queries in the presence of 3D obstacles, that form the basis of a large class of location based applications. For example, an advertisement company may want to check the visibility of its billboard from the surrounding areas before deciding on billboard's position; a tourist may want to check visibility of beautiful city skylines from available apartments; and a security company may want to find the suitable positions for surveillance cameras.

All of the above applications require computing and differentiating the visibility of (from) a target object from (of) the surrounding area. For example,

C. Claramunt et al. (Eds.): SSTD 2015, LNCS 9239, pp. 197–215, 2015.
DOI: 10.1007/978-3-319-22363-6_11

a target billboard may be more visible from one location than another due to different factors such as distance, viewing angle and obstacles. Also, a billboard may be seen from many viewpoints but is readable only from viewpoints closer to the target. Thus, our target applications require modeling the visibility as a continuous notion, i.e., one needs to compute the visibility of the target for every viewpoint in the space. In this paper, we propose efficient techniques to compute the visibility of a target object from the surrounding continuous space, which we call a visibility color map (*VCM*).

A *VCM* is a surface color map, where every viewpoint in a 3D space is assigned a color value denoting the *visibility measure* of the target from that viewpoint. In a recent work [1], Choudhury et al. proposed a technique to compute a *VCM* for a fixed target. Two major limitations of this technique are as follows: (i) They do not take the partial visibility into account, i.e., a viewpoint is simply discarded (i.e., considered as non-visible) if an obstacle even slightly blocks the view of the target from that viewpoint. For example, if only a small part of a billboard cannot be seen from a viewpoint, the viewpoint is declared as non-visible, though a major part of the billboard is readable from the viewpoint. Moreover, in a real 3D city environment, since many viewpoints are partially obstructed due to huge number of obstacles, only a small portion of the viewpoints surrounding the target constitute the *VCM*, which is not desirable for real life applications. (ii) They do not consider the case of a moving target, and thus a slight change of the target's position invalidates the entire *VCM*.

There is no straightforward way to incorporate the partial visibility and moving target into the existing technique due to the following reasons. First, since the proposed technique in [1] uses simple tangents between extreme points of an obstacle and the target, it cannot be converted to assess the partial visibility while computing the *VCM*. In this paper, we take the partial visibility into account, which is the correct form of the *VCM* for a fixed target and is a much harder problem with wider acceptability than [1]. Second, if the position of the target changes, the entire *VCM* needs to be reconstructed as the proposed data structure in [1] does not support incremental updates of the *VCM*.

To alleviate the above limitations, in this paper, we propose an efficient technique to construct the *VCM* for both fixed and moving target using real datasets comprising a large number of obstacles. One key idea of our approach is to identify the *potentially visible set* (*PVS*) of obstacles from the large obstacle set, by removing obstacles that cannot affect the construction of the *VCM*. To find the *PVS*, we adopt the concept of projection from computer graphics [2] and make it scalable and workable for a large number of obstacles indexed using an R-tree [3] in the database. After finding the *PVS*, we determine the *visibility states* of several *boundary points* on the target by considering the occlusion effect of the obstacles. The visibility state of a boundary point on the target represents which *cell*s are (not) visible. Finally, we add the effects of distance and angle between the target and each cell to compute the visibility of every cell.

To extend the above approach for a moving target, we rely on a pre-computation based idea that assumes an extended buffer area around the target and computes the *PVS* and visibility states for the extended region. Once the target moves to a near-by position, our proposed data structure is incrementally updated to generate the *VCM* for the new target position.

We have evaluated the performance of our solution with both real and synthetic 3D datasets. The experimental results show that our approach is on average 10^6 times faster than the straightforward approach.

In summary, we make the following contributions:

- We devise an effective algorithm to construct the *VCM* for a fixed target in the presence of a large set of obstacles considering the partial visibility of the target.
- We propose an efficient way to reconstruct the *VCM* for a moving target.
- We conduct experiments with real 3D datasets to demonstrate the effectiveness and efficiency of our solution.

2 Related Works

The notion of visibility is fundamental to various fields including computational geometry, computer graphics, urban planning, architecture and spatial databases. In this section we briefly discuss existing works on visibility.

2.1 Visibility in Computational Geometry and Computer Graphics

Visibility computation in computational geometry involves determining visibility graphs [4] and visibility polygons [5,6]. In computer graphics, a visibility map is a graph describing a view of the scene including its topology. Various methods for constructing a visibility map for a fixed [7–9] and moving [10] viewpoint have been developed. In all the above approaches, visibility is defined from a point source and consequently a binary notion, i.e. a point in the space is declared as either visible or non-visible from the viewpoint.

Visibility of/from an extended region, i.e. from-region visibility, was studied by Kim et al. [11]. His method determines the subset of the whole space which is completely visible from a region, but does not handle the case of partial visibility. Works done by Durand et al. [2] and Koltun et al. [12,13] focus on determining the set of obstacles visible from an extended region. However, they do not provide any measure of the visibility of the region from the space.

2.2 Visibility in Urban Planning and Architecture

Visibility related problems are actively studied in the fields of urban planning and architecture. Relevant contributions include [14,15]. These approaches treat visibility as a binary notion and are only applicable to cases where the number of obstacles is small enough to fit into the main memory. Urban planners and architects make use of software systems to visualize and render 3D data, such as Google Sketchup [16], AutoCAD [17] and Maya [18]. These softwares do not provide any functionality for quantifying visibility of a 3D object. By incorporating our techniques to construct the *VCM*, these applications can be equipped to answer many realistic visibility queries which require quantification of the visibility of an extended target object.

2.3 Visibility in Spatial Queries

Visibility problems studied in context of spatial databases include nearest neighbor queries [19–21] and maximum visibility queries [22,23]. The variants of nearest neighbor queries find the nearest object in an obstructed scene from a single query point or all points on a line segment, where results are ranked according to the distances from the query point. Maximum visibility query finds a subset of query points that provides the best view of an extended target.

Construction of the *VCM* for the entire space for a fixed position of the target object is studied by Choudhury et. al. [1]. But they do not handle the case of partial visibility and their solution is not applicable for a moving target as we outline in the introduction. In this paper, we devise a method to construct the *VCM* of the entire dataspace for a moving target, which also handles the case of partial visibility in the presence of a large set of obstacles.

3 Problem Formulation

To construct a *VCM*, we need to produce a color map of the dataspace where each point in the space is assigned a value that corresponds to the visibility measure of the target from that point. We formally define the *VCM* as follows:

Definition 1 *VCM. Given a d-dimensional dataspace R^d (d=2 or 3) and a set O of obstacles in the dataspace, the VCM is a color map, where for each point p in R^d, there exists a visibility color v_p in [0,1]. The color v_p corresponds to the visibility of a given target object T from p. Here, higher value of v_p corresponds to higher visibility of T from p and vice versa.*

In an earlier attempt [1] to construct the *VCM* for a fixed target, a point gets a nonzero color if the target is *entirely* visible from that point. However, the target can be partially visible from a point because of obstruction by the obstacles. In this paper, while assigning visibility color to a point, we consider the case of partial visibility by determining what portion of the target is visible from that point.

We also address the problem of reconstructing the *VCM* efficiently as the position of the target changes, i.e., for a moving target. Let VCM_p denote the *VCM* when the target is at position p. We formulate a method to incrementally reconstruct $VCM_{p'}$ efficiently for all $p' \in P$, where P is the set of all candidate positions.

4 Preliminaries

In this section we discuss some basic ideas on which our solution is built upon. First we describe the factors affecting the visibility and then we discuss the partitioning scheme of the dataspace that we have used in our solution. The examples illustrated in this section are in 2D, which can be easily extended for a 3D scenario.

4.1   Factors Affecting Visibility Color

For a particular position of the target, the visibility measure (or, color), v_p, of a viewpoint p is obtained by considering two different visibility measures: *orientation based visibility*, v_p^{or}, and *obstruction based visibility*, v_p^{ob}. Both of these values are in the range of *[0,1]*. After computing v_p^{or} and v_p^{ob}, we can estimate the visibility measure of a viewpoint p as $v_p = v_p^{or} * v_p^{ob}$.

Orientation based visibility captures the effect of both distance and the angle between the target T and the viewpoint, and is measured as the visual angle [24]. The visual angle, α_p, is the imposed angle by T at p as shown in Fig. 1(a).

To compute the obstruction based visibility measure, v_p^{ob}, for a viewpoint p, we consider a number of equally spaced boundary points on the surface of T. Let B_p^T be the set of boundary points which are visible from a point p in the absence of obstacles and B_p^O be the set of boundary points which are visible from p in the presence of all obstacles. Then $v_p^{ob} = |B_p^O|/|B_p^T|$. Here, the notation $|.|$ stands for cardinality of a set.

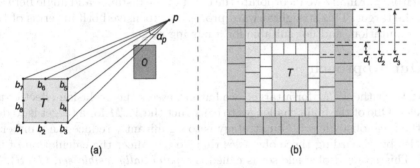

(a)   (b)

Fig. 1. (a) Factors affecting visibility color of point p around target T. (b) A sample partitioning of space. First 3 equidistant stripes are shown in different shades of gray.

In the scenario described in Fig. 1(a), there are 8 boundary points $b_1, b_2, \ldots, b_8$ on target T in a 2D space and the obstacle set O contains one obstacle, o. Here $B_p^T = \{b_3, b_4, b_5, b_6, b_7\}$, $B_p^O = \{b_5, b_6, b_7\}$ and $v_p^{ob} = 3/5 = 0.6$. Thus, v_p^{ob} measures what portion of the target T is visible from p.

4.2   Partitioning into Cells

The d-dimensional data-space R^d consists of infinitely many points and assigning a visibility color, v_p, to each viewpoint p in R^d is a prohibitively expensive task. To address this problem, we partition the whole space into a finite number of *equi-visible cells* as proposed by Choudhury et al. [1], and assign a single visibility color to each cell. The visibility color of cell c is denoted by v_c. Each cell is constructed in a way so that the deviation in visibility of T from the viewpoints inside a cell, measured as visual angle, is not visually perceivable.

To find equi-visible cells, we first partition the space based on distance and then based on the angle between the target and the viewpoints. All the viewpoints having a distance between d_{i-1} and d_i from the target, where $0 < i \leq k$ and k is a positive integer, constitute the ith *equidistant stripe*, which is denoted by S_i. Here k is the number of equidistant stripes. Then each equidistant stripe is further divided into cells based on the angle between the viewpoints and the target. The detailed process of the partitioning is described in [1]. For simplicity, we consider each cell as a rectangular region as shown in Fig. 1(b).

5   A Straightforward Approach

In this section, we present a straightforward approach to construct the *VCM* for a fixed target. Let us consider a cell c with midpoint p. To determine visibility color, v_c, of c, we need to know the number of boundary points, which are visible from p in the presence of obstacles. To determine the visibility between p and a boundary point b, we check the line segment joining b and p against all obstacles. If no obstacle intersects that line segment, b is visible from p. Otherwise b is not visible from p. Finally we incorporate the effect of the distance and angle between p and the target. This straightforward process is expensive both in terms of I/O and computation, and not suitable for a moving target.

6   Our Approach

To construct the *VCM* for a target, we have to assess the occlusion effect of the obstacles. One of the main challenges to construct the *VCM* for a target is to deal with the huge obstacle set. Our strategy is to significantly reduce the number of obstacles by discarding those obstacles that do not affect the calculation of the *VCM*. This reduced obstacle set is called the *potentially visible set, (PVS)*. To determine the *PVS*, we adopt a projection based idea of computer graphics [2] and propose additional adjustments to determine the occlusion effect of a large set of obstacles. We store the obstacles in an R-tree and perform a plane sweep algorithm to determine the combined occlusion effect of multiple obstacles. The process of determining the *PVS* is described in Sect. 6.1.

After determining the *PVS*, the next challenge is to efficiently compute the view of the target from each cell of the partitioned of dataspace. The visibility measure of each cell has two components: (i) orientation based visibility measure, which can be computed using simple equations as described in Sect. 4.1, (ii) obstruction based visibility measure, which needs to consider *what portion* of the target is visible from the cell in the presence of obstacles. As the next step of our algorithm, we determine the *visibility states*, indicating which cells are visible from a particular boundary point of the target (Sect. 6.2). Visibility states of all boundary points are then combined to measure the obstruction based visibility measure of each cell (Sect. 6.3). To compute the *VCM* for a fixed target, orientation and obstruction based visibility values are combined to obtain the visibility color value for every cell (Sect. 6.4). In Sect. 6.5, we discuss the construction of the *VCM* for a moving target.

6.1   Determining PVS

In this section, we formulate a methodology to prune out those obstacles which do not affect the calculation of the VCM. Thus we obtain a reduced set of obstacles, which we call the potentially visible set (PVS). To efficiently determine the PVS, we index all obstacles in an R-tree [3]. An R-tree consists of a hierarchy of minimum bounding rectangles ($MBRs$), where each MBR corresponds to a tree node and bounds all the $MBRs$ in its sub-tree. Data objects (obstacles, in our case) are stored in leaf nodes. Before going to the details of determining PVS, we first discuss some terminologies related to visibility and projection.

Definition 2 *Target-Obstacle Visibility. Given a target T and a set of obstacles O, T and an obstacle $o \in O$ are defined to be visible to each other if and only if there exists a pair of points p_T on T and p_o on o such that no obstacle $o' \in O$ and $o' \neq o$, intersects or touches the line segment joining p_T and p_o.*

Visibility between the target and an obstacle is bidirectional, i.e., if the target is visible from a particular obstacle o, then o is visible from the target and vice versa. An obstacle cannot affect the computation of the VCM if it is not visible from the target. So we can ignore the obstacles which are not visible from the target and obtain a reduced set of obstacles, the PVS, which we denote by O_v.

To determine the PVS, we adopt a projection based idea proposed by Durand et al. [2]. This work employs a conservative occlusion culling technique combining occlusion effects of multiple obstacles on the target visibility. A plane sweep in each principal axis direction is performed to identify obstacles that are not visible from the target. We add further modifications to this approach so that it fits our purpose of dealing with a large obstacle set indexed in an *R-tree*.

Preliminaries of Projection. In this section we describe several key concepts regarding projection. For ease of explanation, we have assumed 2D scenario with axis aligned rectangular target and obstacles. However, our approach is applicable to any convex target and obstacles in 2D and 3D spaces. We also assume that the field of view (FOV) is 90 degree centering the positive X direction. Other directions along the principal axes can be treated similarly. In subsequent sections, we treat an R-tree node (i.e., an MBR) or an obstacle as an *object*.

First, we define the *near distance* and the *far distance*. The *near distance* and the *far distance* of an object are respectively the smallest and the largest of the x ordinate values of all points of the object. We denote the near distance and the far distance of an object o, by o_n and o_f respectively. Now we discuss the idea of projection of an object [2]. The *projection* of an object is computed on a projection plane in 3D (or, projection line in 2D) with respect to the target. If the projection plane is in between the target and the object, then the projection of the object is the union of all views from any point of the target. If the projection plane is in the opposite side of the object from the target then the projection of the object is the intersection of all views from any point of the target. The projection of an object o, on the projection line $x = l$, is denoted by P_o^l.

Now we describe the concepts of *aggregated projection* and *re-projection*. We denote the aggregated projection at the projection line $x = l$, by A_l. A_l reflects the combined occluding effect of all obstacles with far distances less than or equal to l, i.e., obstacles which are entirely in front of the sweep line $x = l$. As a result, A_l consists of several disjoint projections on $x = l$.

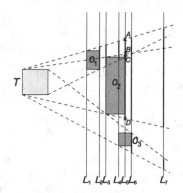

Fig. 2. Bold segments stand for aggregated projections.

The process of computing the aggregated projection is as follows. Initially, we set the aggregated projection as null, $A_{init} = \emptyset$. Then we incrementally update the aggregated projection as we encounter obstacles in the increasing order of their far distance.

Suppose we know the aggregated projection on line $x = l_p$, A_{l_p}, and we want to calculate A_{l_n}, where $l_p < l_n$. (Here subscripts p and n stand for *previous* and *next* respectively.) If no obstacle has far distance between l_p and l_n, then we obtain A_{l_n} by *re-projecting* A_{l_p} on the projection line $x = l_n$. Re-projecting an aggregated projection onto another projection plane involves projecting each disjoint projection of the aggregated projection separately onto the destination plane. Now assume that there is an obstacle o, with far distance l, where $l_p < l \leq l_n$. In such cases, first we calculate the union of P_o^l and the re-projection of A_{l_p} on $x = l$ and then re-project this combined projection on $x = l_n$. This method demonstrates that the aggregated projection needs to be recalculated only at the far distances of the obstacles.

The process of calculating aggregated projections is simulated in Fig. 2. Let, l_1, l_3 and l_4 be the near distances of o_1, o_2 and o_3 respectively, and l_2, l_5 and l_6 be the far distances of o_1, o_2 and o_3 respectively. In the figure, projection line $x = l_i$ is marked as L_i. Initially, $A_{init} = \emptyset$. We encounter the first far distance at $x = l_2$. As a result, A_{l_2} is $P_{o_1}^{l_2}$. On $x = l_5$, the re-projection of A_{l_2} on $x = l_5$ is AC and $P_{o_2}^{l_5}$ is BD. The union of AC and BD is AD. Thus, A_{l_5} is $\{AD\}$. At $x = l_6$, the re-projection of A_{l_5} on $x = l_6$ and $P_{o_3}^{l_6}$ are disjoint. Thus, A_{l_6} consists of two disjoint segments as shown in Fig. 2.

Algorithm 1. determinePVS(R,T)

```
 1  begin
 2  │   O_v, A_init, Q ⟵ ∅;
 3  │   Q.push(R.root);
 4  │   while Q ≠ ∅ do
 5  │   │   o ⟵ Q.pop();
 6  │   │   if o is entirely outside FOV then
 7  │   │   │   continue;
 8  │   │   l ⟵ o_n;
 9  │   │   determine A_l;
10  │   │   if A_l completely spans FOV then
11  │   │   │   break;
12  │   │   if P_o^l ⊆ A_l then
13  │   │   │   continue;
14  │   │   if o is an MBR then
15  │   │   │   for c ∈ o.children() do
16  │   │   │   │   Q.push(c);
17  │   │   else
18  │   │   │   O_v.push(o);
19  │   return O_v
```

The Algorithm. In this section, we formally describe the process of determining the *PVS* in the algorithm *determinePVS*. We retrieve objects from the R-tree in non-decreasing order of near distance (Fig. 3 shows an example R-tree for 8 obstacles in a 2D space). This retrieval process can be visualized as a line sweep over the 2D plane. A projection line (sweep line) perpendicular to the X axis is moved towards the positive X direction. Suppose it enters an object, o, at position $x = l$, i.e., $o_n = l$. Then the aggregated projection on the projection line $x = l$, A_l, is calculated, which reflects the combined occlusion effect of all obstacles entirely in front of the sweep line. If the projection of o on the sweep line, P_o^l, is a subset of A_l, then o is occluded by the obstacles in front of o, and consequently o is discarded (Lines 12–13). o is also discarded if o is entirely outside the field of view (Lines 6–7). If o cannot be discarded, we either declare o as a potentially visible obstacle (in case o is an obstacle, Lines 17–18) or mark o's children for later consideration (in case o is an *MBR*, Lines 14–16). The algorithm terminates when there are no more objects to process or the aggregated projection completely spans the *FOV* (Lines 10–11). Note that an obstacle, o, is discarded only if o is certainly not visible from the target, otherwise, o is considered as potentially visible.

In the algorithm *determinePVS*, movement of the sweep line is implemented by a priority queue, Q, which holds objects, i.e. *MBR*s and obstacles, in non-decreasing order of near distance. Other variables hold their usual meaning.

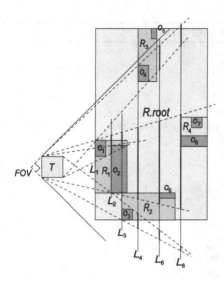

Fig. 3. The PVS computation with an R-tree. Near distance of obstacle o_i is l_i and the sweep line $x = l_i$ is represented by L_i. The bold segments on the L_is identify A_{l_i}.

In Line 9, we determine the aggregated projection at the current sweep line position, which can be calculated efficiently by considering only those obstacles that have far distances between the previous and current sweep line positions.

In Fig. 3, we present a simulation of the algorithm *determinePVS*. At projection line L_1, A_{l_1} is null. As a result, o_1 is added to O_v. On L_2, $P_{o_2}^{l_2}$ is not a subset of A_{l_2} and o_2 is declared as potentially visible. Similar is the case with o_3 and o_4. o_5 is entirely outside the field of view, and consequently discarded. o_6 is rejected because, at L_6, $P_{o_6}^{l_6}$ is a subset of A_{l_6}. The projection of the *MBR*, R_4 on L_8 is a subset of A_{d_8}. As a result, R_4, along with o_7 and o_8, is discarded. Finally, the algorithm returns the set $\{o_1, o_2, o_3, o_4\}$ as the *PVS*.

6.2   Determining Visibility State of a Point

The *visibility State* of a point, b, on the target indicates which cells in the partition are visible from b and which cells are not. The process of determining visibility state of b is equivalent to assigning a boolean value to each cell in the partition indicating whether or not the cell is visible from b. As mentioned before, to construct the *VCM*, we need to determine the visibility states of all the boundary points. In this section, we describe how to determine visibility state of a particular boundary point, b, assuming that the field of view along the positive X direction.

A cell c and a point b is *visible* to each other if and only if the line segment joining b and the midpoint of c is intersected or touched by no obstacle. We observe that the midpoints of cells in a particular equidistant stripe are situated on a straight line perpendicular to the X axis. The line joining the midpoints of the cells of S_i is called the ith *midway line*, and denoted by M_i. Recall from

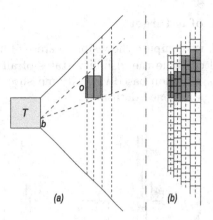

Fig. 4. (a) Aggregated near projections (bold segments) on successive mid-way lines (dotted lines). (b) Visibility state of b. Non-visible cells are darkened.

Sect. 4.2 that S_i stands for the ith equidistant stripe. Let the equation of the line M_i be $x = m_i$.

As the midway lines are vertically aligned, we can reuse the idea of the projection based strategy, constructed in Sect. 6.1, for determining the occlusion effect of obstacles around the point b. Let us first consider the point b as the target (i.e., a point target) and the PVS, returned by the algorithm $determinePVS$, as the set of obstacles. For each midway line M_i, where $i = 1, 2, \ldots, k$, we calculate the occlusion effect of all obstacles situated partially (as we explain shortly) or entirely in front of M_i. We declare a cell as not visible from b, if the cell's midpoint is occluded and vice versa.

To accurately measure the visibility state of a point, we consider the occlusion effect of an obstacle at projection lines after the near distance of the obstacle. Thus we ensure that when the projection line is between the near distance and far distance of an obstacle, the occlusion effect of the portion of the obstacle in front of the projection line is taken into account. If the occlusion effect of all obstacles situated partially or entirely in front of the projection line is taken into account, the combined projection is defined as *aggregated near projection*. Aggregated near projection at projection line $x = l$ is denoted by A_l^n.

To determine the visibility state of a boundary point, aggregated near projections are determined on successive midway lines considering O_v as the set of obstacles. In Fig. 4, a sample scenario is illustrated, where visibility state of a boundary point b is determined in the presence of an obstacle, o. Figure 4(a) shows the aggregated near projections on consecutive midway lines and Fig. 4(b) marks the cells which are declared as non-visible from b, i.e., those cells which have midpoints on the aggregated near projections.

6.3 Visibility State of a Target

In the previous section, we compute the visibility state of a point on the target. In this section, we will combine the visibility states of all boundary points on the target to find the obstruction based visibility measure of the target from all equi-visible cells of the partitioned dataspace.

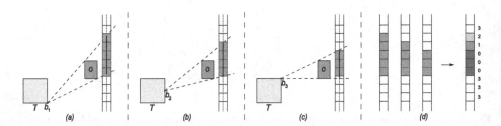

Fig. 5. Combining the visibility states of a particular equidistant stripe of 3 boundary points. Lighter shade of gray on the leftmost stripe corresponds to better view of T.

Let us consider three boundary points b_1, b_2, and b_3 on the positive X direction of the target T as shown in Fig. 5. We need to compute the visibility states of all cells on the positive X direction in the presence of an obstacle o. Figure 5(a), (b) and (c) show visibility states of a particular equidistant stripe for boundary points b_1, b_2 and b_3, respectively. We can see that 5, 4, and 3 cells of the equidistant stripe are not visible from boundary points b_1 (Fig. 5(a)), b_2 (Fig. 5(b)), and b_3 (Fig. 5(c)), respectively. If we combine these three visibility states, we obtain the obstruction based visibility measures of all cells on the equidistant stripe. Figure 5(d) shows the cells and corresponding numbers indicating how many boundary points are visible from those cells, which indicate the obstruction based visibility measure for the VCM.

6.4 VCM for a Fixed Target

Based on the above constructs, we summarize the process of VCM construction for a fixed target as follows. First, we divide the dataspace into a set of equi-visible cells. Second, for each cell c, we compute the orientation based visibility measure v_c^{or} based on the distance and angular placement of c's midpoint with respect to the target. Third, we compute the potentially visible set (PVS), O_v, by the projection based line-sweep along every axis-parallel direction. Fourth, we compute the visibility states of all the boundary points on the target by considering the occlusion effect of O_v on target visibility. Fifth, for each cell c, we compute the obstruction based visibility measure, v_c^{ob}, as $|B_c^O|/|B_c^T|$, where B_c^T is the set of boundary points that are visible from c in the absence of obstacles and B_c^O is the set of boundary points that are visible from c in the presence of obstacles. Finally, we estimate the visibility measure or the color value v_c of each cell c as $v_c^{or} * v_c^{ob}$, which constitutes the VCM for the target.

6.5 VCM for a Moving Target

In this section, we discuss our solution to construct the *VCM* for a moving target. The key concept of extending our method for a moving target is to precompute the *PVS* and visibility states for an extended buffer area around the target, and then incrementally update the *VCM* when the target moves to a near-by position. Thus we can avoid a large amount of repetitive computation and obtain the desired efficiency to construct the *VCM* for a moving target. Note that performing the pre-computation for an extended region around the target is consistent with realistic application scenarios, e.g., while placing a billboard, a user may want to check its visibility from the surrounding space. Our proposed method enables him to choose the right placement by constructing the *VCM* on-the-fly for different near-by positions of the billboard. In this paper, we consider a rectangular region that encloses the target as the buffer area. Defining the buffer region more meaningfully is in the scope of our future work.

We name the extended buffer region around the target as a *super target*, T_s and consider a number of equally spaced points in T_s as *candidate points*. We follow two pre-computation steps involving the super target and candidate points: (i) determine the *PVS* for T_s, and (ii) compute visibility states of all the candidate points by considering the *PVS* as the obstacle set.

After the pre-computation steps, we can construct the *VCM* by combining the visibility states of those candidate points that lie on the boundary of the target. When the target moves to a position inside the super target, we add the effects of visibility states of the new candidate points and remove the effects of candidate points that do not belong to the new target. We assume that the target is moved to a discrete position defined by a set of candidate points. However, if the target moves to a position where its boundary points do not superimpose with the candidate points, we approximate the given position of the target to the nearest discrete position defined by the candidate points, and construct the *VCM* accordingly. If a target moves outside the super target area, we need to calculate the new *PVS* and visibility states of the new candidate points.

The speedup in the construction of the *VCM* for a moving target comes from two sources. First, we calculate the *PVS* only once. If the target is entirely inside the super target, we can use the reduced obstacle set determined in Step (i) as the *PVS*. Second, one candidate point can be used to approximate targets at many positions. Consequently, we do not need to repeat the process of determining the *PVS* and the visibility states for any near-by positions of the target. Thus the cost of the preprocessing steps is amortized over all subsequent runs of the process for different positions of the target inside the super target.

7 Experimental Evaluation

We evaluate the performance of our proposed algorithm for constructing the *VCM* with real and synthetic datasets. At first, we compare our approach to construct the *VCM* for a fixed target with the exact straightforward approach presented in Sect. 5. Then we evaluate the efficiency of our approach for constructing the *VCM* for a moving target. The algorithms are implemented in C++ and the experiments are conducted on a core i5 2.40 GHz PC with 3GB RAM, running Microsoft Windows 7.

7.1 Experimental Setup

We conduct experiments for two real 3D datasets: (1) British[1] dataset, representing 5985 data objects obtained from British ordnance survey[2] and (2) Boston[3] dataset, representing 130,043 data objects in Boston downtown. We also conduct experiments using synthetic datasets. We vary the synthetic dataset size using both *Uniform* and *Zipf* distributions of the obstacles. In all datasets, objects are represented as 3D rectangles that are used as obstacles in our experiments. All obstacles are indexed by an R-tree, with the disk page size fixed at 1KB. We vary several parameters to evaluate our solution. The range and default value of each parameter are listed in Table 1.

Table 1. Parameters

Parameter	Range	Default
Dataset	Real, Synthetic	Real
Angular Resolution	0.5, 1, 2, 4, 8	2
Number of Boundary Points	3^2, 4^2, 5^2, 6^2, 7^2	4^2
Length of Target	15, 30, 60, 120, 240	60
Dataset Size (Synthetic)	5K, 10K, 15K, 20K, 25K	

7.2 Performance Evaluation

We compare our approach to construct the *VCM* with the exact straightforward approach described in Sect. 5. The results of our approach deviate slightly from the results of the exact method, because we have used an approximation for ease of implementation. Recall from Sect. 6.2 that while determining the visibility states, we calculated the aggregated near projections to correctly assess the occlusion effect of an obstacle, i.e. we consider the portion of the obstacles partially or completely in front of the projection lines. But in our implementation, we have calculated the projection of each obstacle at its' near distance. We have treated this projection as the occlusion effect of the obstacle and removed the obstacle from further consideration. To formulate the error incurred by this approximation, let v_c^e denote the visibility color of cell c determined by the exact

---

[1] http://www.citygml.org/index.php?id=1539.

[2] http://www.ordnancesurvey.co.uk/oswebsite/indexA.html.

[3] http://www.bostonredevelopmentauthority.org.

approach and v_c^a denote the visibility color of cell c determined by our implementation. Then the error is given by the following formula:

$$error = \frac{\sum_{c \in C} |v_c^e - v_c^a|}{|C|}$$

The performance metrics used in our experiments include: (i) total processing time, (ii) I/O cost, i.e. the number of nodes retrieved from the R-tree, and (iii) error incurred by the approximation. For each experiment, we have evaluated the results for 20 random positions of the target and reported the average.

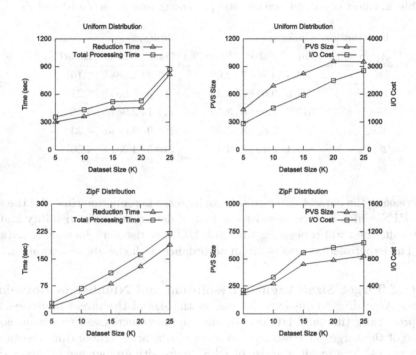

Fig. 6. Effect of dataset size on I/O performance.

7.3   Performance for Fixed Target

To evaluate the performance of our approach to construct the *VCM* for a fixed target, we vary the size of the target, dataset size, angular resolution and the number of boundary points independently and determine the total processing time, I/O cost and the incurred error. When one parameter is varied, the other parameters are kept at their default values.

Effect of Dataset Size. In this experiment we vary the dataset size from 5 K to 25 K using both Uniform and ZipF distributions and compare the performance of our solution with the straightforward exact approach. Table 2 shows the effect of dataset size on total processing time and error. Our approach runs approximately 10^5 to 10^6 times faster than the straightforward approach. The rate of error is quite low and does not change significantly with varying dataset size. The total processing time, reduction time, PVS size and I/O cost of our approach is shown graphically in Fig. 6 to analyze the I/O performance of our solution. We see that reduction time is dominating the total processing time in both datasets.

Table 2. Effect of dataset size on total processing time (in seconds) and Error

Distribution	Uniform			ZipF		
Approach	Our	Naive	Error(%)	Our	Naive	Error(%)
5K	354.3	3.4e+7	2.66	28.3	4.1e+7	2.16
10K	431.5	9.6e+7	4.98	67.9	1.6e+8	3.21
15K	519.7	1.9e+8	3.50	111.1	2.9e+8	4.46
20K	528.5	2.8e+8	4.37	161.9	3.6e+8	4.23
25K	869.4	3.4e+8	4.40	219.5	4.2e+8	4.26

Increasing the dataset size results in an increase in reduction time and the size of the PVS, which in turn increase the cost of determining the visibility states. As a result, the total processing time and I/O cost rise with increasing dataset sizes. The experimental results are in accordance with the above reasoning.

Effect of Target Size, Angular Resolution and Number of Boundary Points. According to our experiments, as the size of the target increases, the total processing time and I/O cost of our approach increase. The reason is, as the size of the target increases, the number of obstacles visible from the target also increases. As a result, the size of PVS grows with an increase in target size. Consequently, the total processing time and the I/O cost increase.

We find that the total processing time is inversely proportional to angular resolution or cell size. The reduction time does not depend on angular resolution. But the computational overhead of determining visibility states and combining the results decrease with increasing angular resolution. This is because the number of equidistant stripes decreases as the angular resolution increases.

Our experiments reveal that with an increase in number of boundary points, total processing time increases and the error decreases. Total processing time increases, because the cost of determining the visibility states is proportional to the number of boundary points. The decrease in error occurs, because with higher number of boundary points, we can obtain a more accurate measure of what portion of the target is visible from each cell.

In the above cases, our solution runs 10^5 to 10^7 times faster than the exact straightforward approach and the error rate varies from 3 % to 7 %, which is negligible. The experimental results are not shown in details for brevity.

7.4   Performance for Moving Target

To assess the efficiency of our solution for constructing the *VCM* for a moving target, we vary the ratio of the lengths of an edge of the super target and the target from 2 to 5. We determine the average total processing time for all discrete positions of the target inside the super target in two ways. First we separately run the process of constructing the *VCM* for a fixed target as described in Sect. 6.4 for each discrete position of the target inside the super target and take the average processing time, which we call *fixed target average cost*. Then we apply the method described in Sect. 6.5 to construct the *VCM* for all discrete positions inside the super target and determine the average processing time by amortizing the cost of the preprocessing steps over all discrete positions. This is called *moving target average cost*.

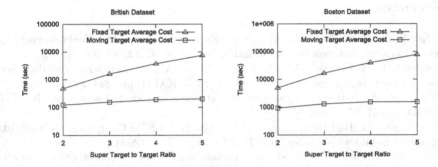

Fig. 7. Performance for moving target.

The experimental results in Fig. 7 illustrate that our pre-computation based solution for constructing the *VCM* for a moving target runs 10 to 100 times faster on average than the approach described in Sect. 6.4. As the size of the target with respect to the super target decreases, the number of discrete positions for the target inside the super target increases. As a result, the cost of the preprocessing steps becomes negligible in average and the moving target average cost becomes much smaller than the fixed target average cost.

8   Conclusion

In this paper, we have proposed an efficient and scalable technique to compute the visibility color map (*VCM*) that forms the basis of many real-life visibility queries in 2D and 3D spaces. The *VCM* quantifies the visibility of (from) a target object from (of) each viewpoint of the surrounding space and assigns colors accordingly in the presence of obstacles. Our plane-sweep based solution finds the *VCM* in three phases: finding the potentially visible obstacle set (*PVS*) from a large set of obstacles, determining the occlusion effects of obstacles in the *PVS*, and finally adding the effects of distance and angle between the target and each cell of the partitioned dataspace. Our solution works for both fixed and moving target, and handles the partial visibility of the target. When the target moves to a near-by position, our proposed data structure can be incrementally updated to generate the *VCM* on-the-fly. Experiments with real and synthetic 3D datasets demonstrate that for a fixed target, our approach outperforms the straightforward approach by 5–6 orders of magnitude in terms of total processing time. Our solution to calculate the *VCM* for a moving target runs 10 to 100 times faster than our solution for a fixed target.

Acknowledgements. This research is supported by the ICT ministry - Bangladesh innovation fund for the project "Visibility Queries in 3D Spatial Databases". Muhammad Aamir Cheema is supported by ARC DE130101002 and DP130103405.

References

1. Choudhury, F.M., Ali, M.E., Masud, S., Nath, S., Rabban, I.E.: Scalable visibility color map construction in spatial databases. Inf. Syst. **42**, 89–106 (2014)
2. Durand, F., Drettakis, G., Thollot, J., Puech, C.: Conservative visibility pre-processing using extended projections. In: SIGGRAPH, pp. 239–248 (2000)
3. Guttman, A.: R-trees: a dynamic index structure for spatial searching. In: SIG-MOD, pp. 47–57 (1984)
4. Ben-Moshe, B., Hall-Holt, O., Katz, M.J., Mitchell, J.S.B.: Computing the visibility graph of points within a polygon. In: SCG, pp. 27–35 (2004)
5. Suri, S., O'Rourke, J.: Worst-case optimal algorithms for constructing visibility polygons with holes. In: SCG, pp. 14–23 (1986)
6. Asano, T., Asano, T., Guibas, L., Hershberger, J., Imai, H.: Visibility-polygon search and euclidean shortest paths. In: SFCS, pp. 155–164 (1985)
7. James Stewart, A., Karkanis, T.: Computing the approximate visibility map, with applications to form factors and discontinuity meshing. In: Drettakis, G., Max, N. (eds.) Rendering Techniques 1998. Eurographics, pp. 57–68. Springer, Vienna (1998)
8. Grasset, J., Terraz, O., Hasenfratz, J.M., Plemenos, D.: Accurate scene display by using visibility maps. In: SCCG, pp. 180–186 (1999)
9. Bittner, J.: Efficient construction of visibility maps using approximate occlusion sweep. In: SCCG, pp. 167–175 (2002)
10. Tsai, Y.H.R., Cheng, L.T., Osher, S., Burchard, P., Sapiro, G.: Visibility and its dynamics in a PDE based implicit framework. J. Comput. Phys. **199**(1), 260–290 (2004)
11. Kim, D.S., Yoo, K.H., Chwa, K.Y., Shin, S.Y.: Efficient algorithms for computing a complete visibility region in three-dimensional space. Algorithmica **20**(2), 201–225 (1998)
12. Koltun, V., Chrysanthou, Y., Cohen-Or, D.: Hardware-accelerated from-region visibility using a dual ray space. In: Gortler, S.J., Myszkowski, K. (eds.) Rendering Techniques 2001. Eurographics, pp. 205–215. Springer, Vienna (2001)
13. Koltun, V., Chrysanthou, Y., Cohen-Or, D.: Virtual occluders: an efficient intermediate pvs representation. In: Péroche, B., Rushmeier, H. (eds.) Rendering Techniques 2000. Eurographics, pp. 59–70. Springer, Vienna (2000)

14. Hernndez, J., Garca, L., Ayuga, F.: Assessment of the visual impact made on the landscape by new buildings: a methodology for site selection. Landsc. Urb. Plan. **68**(1), 15–28 (2004)
15. Bartie, P., Reitsma, F., Kingham, S., Mills, S.: Advancing visibility modelling algorithms for urban environments. Comput. Environ. Urb. Syst. **34**(6), 518–531 (2010)
16. Google sketchup. http://www.sketchup.com
17. Autocad. http://www.autodesk.com/products/autodesk-autocad/overview
18. Maya. http://www.autodesk.com/products/autodesk-maya/overview
19. Nutanong, S., Tanin, E., Zhang, R.: Incremental evaluation of visible nearest neighbor queries. IEEE Trans. Knowl. Data Eng. **22**, 665–681 (2010)
20. Gao, Y., Zheng, B.: Continuous obstructed nearest neighbor queries in spatial databases. In: SIGMOD, pp. 577–590 (2009)
21. Gao, Y., Zheng, B., Lee, W.C., Chen, G.: Continuous visible nearest neighbor queries. In: EDBT, pp. 144–155 (2009)
22. Masud, S., Choudhury, F.M., Ali, M.E., Nutanong, S.: Maximum visibility queries in spatial databases. In: ICDE, pp. 637–648 (2013)
23. Ali, M.E., Tanin, E., Zhang, R., Kulik, L.: A motion-aware approach for efficient evaluation of continuous queries on 3d object databases. VLDB J. **19**, 603–632 (2010)
24. Kaiser, P.: The Joy of Visual Perception. York University, Toronto (1996)

Speed Partitioning for Indexing Moving Objects

Xiaofeng Xu[1]([✉]), Li Xiong[1], Vaidy Sunderam[1], Jinfei Liu[1],
and Jun Luo[2,3]

[1] Department of Mathematics/Computer Science,
Emory University, Atlanta, GA, USA
{xiaofeng.xu,lxiong,vss,jinfei.liu}@emory.edu
[2] HK Advanced Technology Center and Ecosystem Cloud Service Group,
Lenovo, Hong Kong, China
[3] Shenzhen Institutes of Advanced Technology, Chinese Academy of Sciences,
Shenzhen, China
jun.luo@siat.ac.cn

Abstract. Indexing moving objects has been extensively studied in the past decades. Moving objects, such as vehicles and mobile device users, usually exhibit some patterns on their velocities, which can be utilized for velocity-based partitioning to improve performance of the indexes. Existing velocity-based partitioning techniques rely on some kinds of heuristics rather than analytically calculate the optimal solution. In this paper, we propose a novel speed partitioning technique based on a formal analysis over speed values of the moving objects. We first formulate the optimal speed partitioning problem based on search space expansion analysis and then compute the optimal solution using dynamic programming. We then build the partitioned indexing system where queries are duplicated and processed in each index partition. Extensive experiments demonstrate that our method dramatically improves the performance of indexes for moving objects and outperforms other state-of-the-art velocity-based partitioning approaches.

1 Introduction

Over the past few decades, the rapid and continuous development of positioning techniques, such as GPS and cell tower triangulation, has enabled information to be captured about continuous moving objects, such as vehicles and mobile device users. *Location-based services* (LBSs) and *location-dependent queries* have become popular in modern human society [15]. Techniques for managing databases containing large numbers of moving objects and processing predictive queries [14,21] have been extensively studied and are becoming increasingly important in order to support many emerging applications including real-time ride sharing (e.g. Uber) and location based crowd sourcing (e.g. Waze).

By storing timestamped locations, traditional database management systems (DBMSs) can directly represent moving objects [11]. However, this approach is impractical because most applications require high update rates in order to maintain the stored locations of the moving objects up to date. Therefore, motion

C. Claramunt et al. (Eds.): SSTD 2015, LNCS 9239, pp. 216–234, 2015.
DOI: 10.1007/978-3-319-22363-6_12

functions are used instead, which significantly reduce the number of updates, for moving object databases (MODs) [10,20]. Moreover, motion functions enable MODs to perform predictive spatio-temproal queries [14,21] that retrieve near future locations of the moving objects.

Indexes are used to improve query performance of MODs. Due to high update rate in real world applications, not only query performance but also update overhead must be considered while indexing MODs. Indexes for MODs in the literature can be categorized into tree-based indexes (e.g. [4,5,8,9,14,19,21,22, 24]) and grid-based indexes (e.g. [13,16–18]). Typical tree-based indexes are balanced, i.e. the number of indexed objects within each leaf node is about the same. Therefore query performance of such structures can be estimated by the number of nodes accessed when processing a query [21]. The query performance of grid-based indexes depend on different factors, as the grid cells might contain quite different number of objects. In this work, we consider only tree-based indexes and leave grid-based ones for future work.

In most real world applications, moving objects usually exhibit particular patterns on velocities (including speed values and directions). Therefore, velocity-based partitioning can be applied to the indexes to improve performances of the indexes. Zhang et al. [23] proposed the first idea of velocity-based partitioning for indexing moving objects. In their method, they first find k velocity *seeds* which maximize the *velocity minimum bounding rectangle* (VMBR), then partition the moving objects by assigning them to the nearest seed. In this way, the moving objects are partitioned into k parts and the VMBR for each part is minimized. Nguyen et al. [12] proposed another velocity-based partitioning technique that partitions the indexes based on directions of the moving objects. This method partitions the moving objects based on their distance to the so-called *dominant velocity axes* (DVAs) in the velocity domain.

Speed values of the moving objects are always characterized by both the nature of the moving objects and the environment. For example, walking speeds for human beings range from 0 mph to 4 mph; driving speeds for vehicles in city road networks range from 0 mph to 100 mph; cruising speeds for airplanes usually range from 500 mph to 600 mph. Moreover, in most city road networks, speed values of the vehicles are also characterized by the categories of the roads. For example, most vehicles drive between 50–80 mph on highways, and 20–40 mph on street ways or even slower when the roads are busy. Such distributions of speed values of the moving objects can have significant impacts on query performances of the indexes. Query performances of typical tree-based indexes for MODs can be estimated by the average number of node accesses [21]. However, high speed moving objects will significantly enlarge the spatial areas of the index nodes containing them, which will likely incur unnecessary accesses to the low speed ones within the same nodes while processing queries. Thus partitioning the indexes by speed values of the moving objects can significantly improve query performance. Moreover, partitioning will reduce the number of objects in each index partition, which also helps accelerate update operations.

Contributions. Motivated by above observations, we propose the novel speed partitioning technique. The proposed method first computes the optimal points (ranges) for speed partitioning. Then an optional second-level partitioning, based on directions of the moving objects, is performed within each speed partition. Note that the distributions of locations and speeds might change as time elapses that leads to changes of the optimal speed partitioning. Our proposed system can handle these changes through periodical partition update routines. Moreover, the speed partitioning technique is generic and can be applied with various tree-based indexes. Contributions of this paper can be summarized as follows:

- We propose a novel method for estimating the search space expansion which can be used as a generic cost metric to estimate query performance of tree-based indexes for MODs.
- We propose the novel speed partitioning technique which minimizes search space expansion of the indexes using dynamic programming.
- Extensive experiments show that our proposed approach prominently improves update and query performance of two state-of-the-art MOD indexes (the B^x-tree and the TPR*-tree) and outperforms other state-of-the-art velocity-based partitioning techniques.

The remainder of this paper is organized as follows. In Sect. 2 we review the related works about tree-based indexes for MODs and velocity-based partitioning techniques. In Sect. 3, we introduce the concept of search space expansion and, based on which, we formulate the optimal speed partitioning problem. In Sect. 4, we present the speed partitioning technique and the partitioned indexing system. Experimental studies are presented in Sect. 5. In Sect. 6, we conclude this paper and discuss some future work.

2   Related Work

In this section, we introduce some related work about tree-based indexes for MODs that are used in this paper as well as the state-of-the-art velocity-based partitioning techniques.

2.1   The TPR/TPR*-tree

Saltenis et al. [14] proposed the TPR-tree (short for Time-Parameterized R-tree) that augments the R*-tree [1] (a variant of the R-tree [7]), with velocities to index moving objects with motion functions. Specifically, an object in the TPR-tree is indexed by its time-parametrized position with respect to its velocity vector. A node in the TPR-tree is represented by a *minimum bounding rectangle* (MBR) and the velocity on each side of the MBR which bounds all moving objects contained in the corresponding MBR at any time in the future. The TPR-tree uses time-parameterized metrics when choosing the target nodes for insertion and deletion. The time-parameterized metric is calculated as $\int_{t_l}^{t_l+H} A(t)dt$, where $A(t)$ is the metric used in the original R-trees. H is the *horizon* (the lifetime of the node) and t_l is the time of an insertion or the index creation time.

The TPR-tree uses a step-wise greedy strategy to choose the MBR where a new object is inserted. Since the objects are moving as time passes, the overlaps between MBRs become larger, which eventually makes the step-wise greedy strategy ineffective. Tao et al. proposed the TPR*-tree [21] that uses the same data structure as the TPR-tree with optimized insertion and deletion operations, which significantly reduce the overlaps between MBRs.

2.2 The B^x-tree

The B^x-tree, proposed by Jensen et al. [8], is the first indexing approach based on B^+-tree. The B^x-tree uses space-filling curves, such as Z-curves and Hilbert curves, to map the d-dimensional locations into scalars that can be indexed by B^+-trees. The time axis is partitioned into intervals of duration Δt_{mu}, which is the maximum duration in-between two updates of any object location. Each such interval is further partitioned into n equal-length *phases* and each phase is associated with a *label timestamp*. Instead of indexing the object locations at their update timestamps, the B^x-tree indexes the locations at the nearest future label timestamp. After each $\Delta t_{mu}/n$ timestamps, one phase expires and another is generated. This rotation mechanism is essential to preserve the location proximity of the objects.

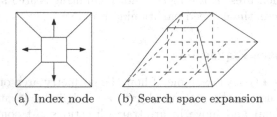

(a) Index node   (b) Search space expansion

Fig. 1. Search space expansion of an index node

2.3 Velocity-Based Partitioning

Velocity-based partitioning techniques, which utilize the velocity information from a global perspective, are used to further improve the query performance of MOD indexes. Intuitively, velocity-based partitioning can improve query performance because search space expansion (defined as the enlargement of the index nodes) [12] of the partitioned indexes considerably decreases in some scenarios.

Zhang et al. [23] firstly defined the VMBRs which represent the minimal rectangles in the velocity domain that bound the velocity vectors of all moving objects and proposed the partitioning method that minimizes the VMBRs within each partition. At the first step of this method, given the number of partitions k, the velocity vectors of exactly k moving objects that form largest VMBR are selected as seeds for the k partitions. Then each object is assigned to the partition with minimum VMBR increase. This method has some limitations.

Firstly, it is difficult to determine the number of partitions k. Secondly, the partitioning might be far from optimum since this method relies on very simple heuristics and does not perform any analysis on search space expansion.

Thi et al. [12] proposed the partitioning technique based on DVAs in the velocity domain. They applied principal component analysis and K-means clustering on the velocities of the moving objects to find k-1 DVAs. Then the velocity domain is partitioned into k partitions according to the DVAs, one partition for each DVA plus one outlier partition. Each moving object is assigned to the nearest DVA partition if the distance between its velocity vector and the DVA is smaller than a threshold, otherwise it will be assigned to the outlier partition. Through this partitioning method, the velocity domain is reduced to nearly 1-dimensional parts, which dramatically reduces the search space expansion. However, this method still requires the number of partitions k as a parameter. Moreover, the performance of this method will significantly reduce if the velocity domain has no effective DVAs.

3   The Optimization Problem

In this section, we introduce the notion of search space expansion which can be used as a generic cost metric to estimate query performance of tree-based indexes for MODs. We then present the method for computing search space expansion and formulate the optimal speed partitioning problem.

3.1   Search Space Expansion

Figure 1(a) shows a typical example of how the geometry area of an index node expands. In this figure, the moving objects are originally located in a square area (the inner one) and move in arbitrary directions. At some future time, the objects will spread in a larger square area (the outer one). We model the expansion of the node as a trapezoid prism where the top base is the original area and the bottom base is the future area of the node. Figure 1(b) illustrates such a trapezoid prism of the node in Fig. 1(a). The volume of the trapezoid prism corresponding to an index node is called the search space expansion of this node. The sum of search space expansions of all index nodes is called the search space expansion of the index. A formal definition of search space expansion is given in Definition 1.

Definition 1. *Search space expansion. Given any node in an MOD index I, its area at time t is $S(t)$. The search space expansion of the node from time 0 to any future time t_h is $\nu(t_h) = \int_0^{t_h} S(t)dt$. The search space expansion of the index is the sum of the search space expansions of all nodes: $V(t_h) = \sum_{\forall node \in I} \nu(t_h)$*

If queries are randomly generated in the predefined space domain, nodes with larger search space expansions have higher probabilities to be accessed to answer the queries [21]. Consequently, indexes with smaller search space expansion enjoy better query performance. Thus we wish to find a partitioning strategy that minimizes the search space expansion of the indexes, i.e. the volumes of all trapezoidal prisms, in order to minimize query costs.

We propose the speed partitioning technique which partitions the indexes based on speed values of the moving objects. Since the moving objects are separated based on their speed values, thus fast growing nodes for high speed objects will not affect those for low speed objects. Therefore the search space expansion of an index will be dramatically reduced if we conduct appropriate partitioning on speed values. In the next subsection, we will discuss how to achieve the optimal index partitioning based on speed values. Note that in our analysis, we only consider the search space expansions of leaf nodes, because in most scenarios the number of leaf nodes significantly exceeds that of internal nodes.

3.2   The Optimal Speed Partitioning

Our speed partitioning technique is based on solving the optimal speed partitioning problem, thus is different from and more generic than all state-of-the-art velocity-based partitioning techniques [12,23] that rely on some kinds of heuristics. We now formalize the optimal speed partitioning problem that minimizes search space expansion.

Denote $\mathcal{O} = \{o_1, o_2, \cdots, o_N\}$ as the set of moving objects and denote the speed of object o_l as v_{o_l}. Let $\Omega = \{v_1, v_2, \ldots, v_q\}$ represent the speed domain, where $v_1 < v_2 < \cdots < v_q$. Thus for all $o_l \in \mathcal{O}$, we have $v_{o_l} \in \Omega$. We note that in most applications the speed domain can be easily discretized into finite number of different speed values. Let $v_0 = v_1 - \epsilon$, where ϵ is a positive number and $\epsilon \to 0$. v_0 is a dummy speed used for simplifying notations. Let $\Omega^+ = \Omega \bigcup \{v_0\}$.

Now let $\Delta = \{\delta_0, \delta_1, \cdots, \delta_k\}$, $1 \leq \delta_i \leq q$, where $\delta_0 = 0$ and $\delta_k = q$. Therefore Δ partitions the speed domain into k (non-overlapping) parts, denoted as $\Omega_i = (v_{\delta_{i-1}}, v_{\delta_i}]$, $1 \leq i \leq k$. We say Δ is a partitioning on Ω. Meanwhile, $\mathcal{O}$ is partitioned accordingly into k parts: $P_i(1 \leq i \leq k)$, where $P_i = \{o_l : v_{o_l} \in (v_{\delta_{i-1}}, v_{\delta_i}]\}$. We denote I_i as the corresponding indexing tree, such as the B^x-tree or the TPR*-tree, for P_i. Note that k is automatically computed rather than an input of our method.

Our goal is to find the optimal partitioning, denoted as $\Delta^\star$, that minimizes the overall search space expansion of all index partitions. We can achieve this goal by solving the following minimization problem:

$$\Delta^\star = \arg\min_{\Delta} \{v_{\delta_0} < v_{\delta_1} < \cdots < v_{\delta_k} : V(t_h)\} \tag{1}$$

where $V(t_h) = \sum_{0 < i \leq k} V_i(t_h)$ represents the overall search space expansion of all index partitions and $V_i(t_h)$ the search space expansion of partition I_i. t_h is the maximum predict time for the predictive queries [14,21]. Without loss of generality, we present next how to compute $V_i(t_h)$.

According to Definition 1, in order to compute $V_i(t_h)$, we first need to compute the search space expansion of every single index node in I_i which requires

1) the initial node area, and 2) the expanding speed of each node. We present the approach to compute $V_i(t_h)$ step by step in the following paragraphs.

Generate Uniform Regions. In most real world applications, the moving objects may not be uniformly distributed. Thus before calculating the search space expansion, we first divide the space domain into subregions such that the moving objects in P_i are (close to) uniformly distributed within each subregion. Uniformity will not only significantly reduce the complexity of calculation but also help obtain more accurate estimations. We will introduce a quad tree based method to find the uniform subregions in Sect. 4. We denote the set of uniform subregions of P_i as $\mathcal{R}_i = \{R_{i1}, R_{i2}, \ldots, R_{im_i}\}$.

Compute Initial Node Area. Now we compute the initial areas of the nodes within subregion R_{ij}, where $0 < j \leq m_i$. Without loss of generality, we assume R_{ij} to be a square area with side length of D_{ij}. We also consider the index nodes as square shaped with expected side length of d_{ij} and let c represent the expected number of objects in each node. c is determined by the storage size of each node which is a parameter in our method. Since moving objects are uniformly distributed in R_{ij}, we have $\frac{d_{ij}^2}{c} \propto \frac{D_{ij}^2}{N_{ij}}$ where N_{ij} represents the number of objects in R_{ij}. Thus d_{ij} can be estimated as $d_{ij} = D_{ij}\sqrt{\frac{c}{N_{ij}}}$.

Compute Expanding Speed. Next we introduce the method for estimating expanding speeds of the index nodes in R_{ij}. Since we make no assumptions on the patterns of the moving objects' directions, we consider that the objects in each node travel at arbitrary directions. Thus every single node expands with equal speed in all directions while the expanding speed is the maximum speed value of the moving objects in the corresponding node.

Let H_{iju} represent the number of moving objects in R_{ij} whose speed values fall in the range $(v_{\delta_{i-1}}, v_u]$, where $v_u \in \Omega$ and $\delta_{i-1} < u \leq \delta_i$, formally

$$H_{iju} = |o_l \in R_{ij} : v_{\delta i-1} < v_{o_l} \leq v_u| \tag{2}$$

Since the speed values of the moving objects are independent given a certain speed distribution, expanding speed of any node in R_{ij} is v_u with the probability

$$p(i,j,u) = \frac{\binom{H_{iju}-H_{ij\delta_{i-1}}}{c} - \binom{H_{ij(u-1)}-H_{ij\delta_{i-1}}}{c}}{\binom{N_{ij}}{c}} \tag{3}$$

where $\binom{a}{b}$ is a combination number.

Compute Search Space Expansion. For each speed partition, we can apply a second-level (direction-based) partitioning into 4 quadrants as illustrated in Fig. 2 if it further improves search space expansion. Hence we compute the search space expansion of R_{ij} both with and without the second-level partitioning and select whichever achieves smaller value. When no second-level partitioning is performed, the search space expansion of a single node in R_{ij} can be calculated by

$$\nu^1(t_h) = \nu(t_h, v_u) = \int_0^{t_h} (d_{ij} + 2v_u t)^2 \, dt \tag{4}$$

When the second-level partitioning is further applied, we compute the search space expansion for each quadrant. Expected side length of the nodes in the quadrant partitions is $2d$ and the search space expansion is calculated by

$$\nu^2(t_h) = \nu(t_h, v_u) = \int_0^{t_h} (2d_{ij} + v_u t)^2 \, dt \tag{5}$$

Therefore, the expected search space expansion of all nodes in R_{ij} can be calculated by

$$V_{ij}(t_h) = \left[\frac{N_{ij}}{c}\right] \sum_{v_u \in \Omega, \delta_{i-1} < u \leq \delta_i} \nu(t_h) p(i, j, u) \tag{6}$$

where $\left[\frac{N_{ij}}{c}\right]$ computes the total number of nodes in R_{ij} and $\nu(t_h)$ represents the minimum of $\nu^1(t_h)$ and $\nu^2(t_h)$. Finally, the overall search space expansion $V(t_h)$ is calculated by

$$V(t_h) = \sum_{1 \leq j \leq k} \sum_{0 < j \leq m_i} V_{ij}(t_h) \tag{7}$$

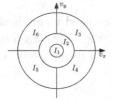

Fig. 2. Speed partitioning

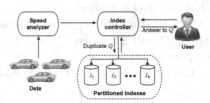

Fig. 3. System architecture of SP

4   The Partitioned Indexing System

Based on the above analysis on search space expansion, we propose the speed partitioning technique (SP) for indexing moving objects. Figure 3 illustrates the system architecture of SP. SP uses a centralized indexing system consisting of three parts: the speed analyzer, the index controller, and the partitioned indexes. The speed analyzer receives data from the moving objects and computes the optimal speed partitioning. The index controller then creates the corresponding partitioned indexes. Once receiving queries from users, the index controller duplicates the queries and push them to the index partitions. After all index partitions finish processing the queries, the index controller collects and integrates the query results and sends them back to users. We will discuss more details of SP in the remainder of this section.

Algorithm 1. $merge(\mathcal{Q})$

input : $\mathcal{Q}$: a set of quad tree nodes
output: $\mathcal{R}$: a set of uniform subregions
/* check the uniformity of the current nodes   */
1 **if** $\forall Q_j \in \mathcal{Q}$, Q_j *is uniform* **then**
2 | add the region of $\mathcal{Q}$ to $\mathcal{R}$;
3 **else**
    /* explore the child nodes   */
4 | **for** $i \leftarrow 0$ **to** 3 **do**
5 | | **foreach** $Q_j \in \mathcal{Q}$ **do**
6 | | | $C Q_j \leftarrow Q_j.child[i]$;
7 | | $merge(\mathcal{CQ})$;

4.1 The Optimal Speed Partitioning

In this subsection, we discuss how to find the optimal speed partitioning through dynamic programming. Let $\Lambda_r^\star$, $0 < r \le q$, be a sequence $(\lambda_0, \lambda_1, \cdots, \lambda_r)$ where $v_{\lambda_i} \in \Omega^\star$ and $0 = \lambda_0 < \lambda_1 \le \cdots \le \lambda_{r-1} \le \lambda_r = r$. The set of distinct values in $\Lambda_r^\star$ form the optimal partitioning of the sub speed domain of $(v_0, v_r]$, denoted as $\Delta_r^\star$. Thus our goal is to find $\Delta_q^\star$.

In order to compute $\Delta_q^\star$ using dynamic programming, we need to maintain two arrays $\mathbb{V}^\star$ and $\mathbb{T}^\star$, where $V_r^\star$ and $T_r^\star$ (the rth values of $\mathbb{V}^\star$ and $\mathbb{T}^\star$) store the search space expansion of $\Delta_r^\star$ and the rth value ($\lambda_{r-1}^\star$) in $\Lambda_r^\star$, respectively. $V_r^\star$ and $T_r^\star$ can be computed by Eqs. (8) and (9), respectively.

$$V_r^\star = \begin{cases} 0 & r = 0 \\ \min_{0 \le s < r} \{V_s^\star + V_{(v_s, v_r]}\} & 0 < r \le q \end{cases} \qquad (8)$$

$$T_r^\star = \arg\min_{0 \le s < r}\{V_s^\star + V_{(v_s, v_r]}\}, 0 < r \le q \qquad (9)$$

where $V_{(v_s, v_r]}$ is the search space expansion of partition $P_{(v_s, v_r]}$ and $P_{(v_s, v_r]} = \{o_l : v_{o_l} \in (v_s, v_r]\}$. Note that we define $V_0^\star = 0$ in order to simplify denotations. Next we discuss how to compute $V_{(v_s, v_r]}$, for all $(v_s, v_r] \subset \Omega$.

In order to compute $V_{(v_s, v_r]}$ using Eq. (7), we first need to generate the uniform subregions mentioned in Sect. 3. We propose a quad tree [6] based method

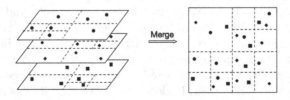

Fig. 4. An example of merge

to generate the uniform subregions for every $P_{(v_s, v_r]}$. We first divide the objects into q layers, where moving objects within the same layer have same speed values (represented by the average speed value in each layer). Each layer is divided into square subregions using a quad tree such that the objects in each subregion are uniformly distributed. We use χ^2-test (significance level 5%) to test the uniformity of each subregion. We also fix 5 as the maximum depth of the quad trees. In order to generate the uniform subregions for $P_{(v_s, v_r]}$, we need to combine the corresponding layers, layer $s + 1$ through r. We choose the most fine grained division when the divisions of different layers conflict, thus objects in the subregions of the combined layer always contain uniformly distributed objects. Figure 4(left) shows an example of such layers, where there are 3 different speed values v_1, v_2, v_3 and the objects in the 3 layers are represented as squares, diamonds, and dots, respectively. Figure 4(right) shows the result of the merge operation.

Algorithm 1 shows the pseudo code for the merge operation. This is a recursive algorithm which takes a set of $r - s$ quad tree nodes (one node for each layer) as input. If objects within all the current nodes are uniformly distributed, we add the (square) spatial region represented by the quad tree nodes into the result set (lines 1–2). Otherwise, we recursively explore the 4 child nodes (each 2-dimensional quad tree node has 4 child nodes) at the next level of the quad trees (lines 3–7). Note that the input nodes will always locate at the same positions in the corresponding quad trees for all recursive calls, since we set the root nodes of the quad trees as input of the initial call.

In order to find the optimal partitioning $\Delta_q^\star$, we need to compute $V_r^\star$ for each r $(0 < r \le q)$. As shown in Eqs. (8) and (9), we iteratively find the best s which leads to the optimal partitioning on $(v_0, v_r]$ and stores it as $T_r^\star$. During the computation for $V_r^\star$, we can use previously computed optimal results on $(v_0, v_s]$, i.e. the values of $V_s^\star$ for each s $(0 \le s < r)$. Finally, we can obtain the optimal partitioning on $(v_0, v_q]$ by tracking backwards the values in $\mathbb{T}^\star$, i.e. each $\lambda_i \in \Lambda_q^\star$ $(0 \le i \le q)$ can be computed by

$$\lambda_i = \begin{cases} 0 & i = 0 \\ T_{\lambda_{i+1}}^\star & 0 < i < q \\ q & i = q \end{cases} \tag{10}$$

Algorithm 2 shows the pseudo code of our dynamic programming based algorithm to solve the optimal speed partitioning problem. Algorithm 2 first creates the quad trees for uniform subregion generation (line 1). Then the search space expansions of partition $P_{(v_s, v_r]}$, for all $(v_s, v_r]$, are calculated (lines 2–3). Then dynamic programming is used to compute the values of $V_r^\star$ and $T_r^\star$ based on Eqs. (8) and (9) (lines 4–11). Finally, λ_0 through λ_q are computed from $\mathbb{T}^\star$ using Eq. (10) (lines 12–15). Note that we compute the search space expansion (line 3) both with and without the second-level partitioning as described in Sect. 3 and store the smaller value as $V_{(v_s, v_r]}$. The corresponding speed partition in the final result is further partitioned into four sub-partitions (one for each quadrant in the velocity domain) if it achieves smaller search space expansion.

Algorithm 2. Find the optimal speed partitioning

---

1 Create quad trees;

   /* Pre-compute search space expansion for partition $P_{(v_s, v_r]}$ using
      Equation (7)   */
2 **foreach** $(v_s, v_r] \in \Omega$ **do**
3 $\quad \lfloor \; V_{(v_s, v_r]} \leftarrow$ the search space expansion of $P_{(v_s, v_r]}$;

   /* Iteratively compute V_r^* and T_r^* using Equation (8) and (9)   */
4 $V_0^* \leftarrow 0$;
5 **for** $r \leftarrow 1$ **to** q **do**
6 $\quad min \leftarrow inf$;
7 $\quad$ **for** $s \leftarrow 0$ **to** $r - 1$ **do**
8 $\quad\quad$ **if** $V_s^* + V_{(v_s, v_r]} < min$ **then**
9 $\quad\quad\quad min \leftarrow V_s^* + V_{(v_s, v_r]}$;
10 $\quad\quad\quad T_r^* \leftarrow s$;
11 $\quad \lfloor \; V_r^* \leftarrow min$;

   /* Compute the final results using Equation (10)   */
12 $\lambda_q \leftarrow q$;
13 **for** $i \leftarrow q - 1$ **to** 1 **do**
14 $\quad \lfloor \; \lambda_i \leftarrow T_{\lambda_{i+1}}^*$;
15 $\lambda_0 \leftarrow 0$;

Figure 2 shows an example of the output of our algorithm. Actually, high speed partitions are more likely to be further partitioned into quadrants since direction has more impact on high speed partitions. The time complexity of Algorithm 2 is analyzed as follows.

Complexity Analysis. Execution time of Algorithm 2 consists of three parts: (1) creating the quad trees takes $O(N)$ time; (2) pre-computing the search space expansions for each sub speed domain takes $O(q^2)$ time; and (3) the dynamic programming part also takes $O(q^2)$ time. Thus the total time complexity of Algorithm 2 is $O(N + q^2)$. Note that the analysis relies on the condition that maximum depth of the quad trees is fixed, as mentioned earlier in this section.

4.2 Index Update

Index update of our system consists of two parts: object update and partition update. Object update corresponds to status (e.g. location and velocity) updates of the moving objects, which is essential to keep the objects' locations up-to-date. When a moving object updates its status, the index controller will determine whether it should be inserted into a different partition based on its current velocity. Then the object will be either deleted from its previous partition and inserted into the new one or simply updated in the previous partition. Note that each index partition contains only a portion of the moving objects, thus object update in the partitioned indexes takes less CPU time than that in the original index without partitioning.

Partition update corresponds to changes of the optimal speed partitioning. Since the objects are continuously moving, both their location and speed distributions might change over time. Thus we need to re-compute the uniform subregions as well as the optimal speed partitioning when necessary. We simply conduct partition updates periodically with cycle time customized according to the dataset. For example, in city road networks, location and speed distributions of the vehicles might be different between rush hours and regular hours, for which we can use hourly partition update routines.

4.3 Query Processing

In this work, we consider predictive time-slice queries [14, 21] which retrieve tentative future locations of the moving objects. We evaluate both predictive range queries and predictive k nearest neighbor (kNN) queries in the experiments (Sect. 5). Specifically, a predictive range query is associated with two coordinates (bottom-left point and upper-right point of the range query window), while a predictive kNN query is associated with a coordinate (center of the kNN query) and kNN-k. Both of the two kinds of queries are associated with a query predict (future) time, which indicates that the queries are performed on the objects' predicted locations at that time.

Query processing for SP is straightforward. The original queries are duplicated (with modifications if necessary) and processed within each partition either concurrently or sequentially. In order to compare the performance between partitioned indexes and their unpartitioned counterparts, in this paper, we conduct the duplicated queries sequentially. Within each index partition, queries are performed using the algorithm associated with the basic indexing structure (e.g. the B^x-tree or the TPR*-tree).

5 Experimental Study

In this section, we conduct extensive experiments to evaluate the performance of our speed partitioning technique with both main memory indexes and disk indexes. Both simulated traffic data and real world GPS tracking data are used in the experiments. We evaluate both update throughput (average number of updates performed in a second) and query response time. Query response time consists of I/O latency and CPU time for disk indexes while only CPU time for main memory indexes.

We use the B^x-tree and the TPR*-tree as the basic indexing structures. We compare our approach of speed partitioning (SP-B^x and SP-TPR*) with the state-of-the-art approaches of DVA-based partitioning [12] (dVP-B^x and dVP-TPR*) and VMBR-based partitioning [23] (mVP-B^x and mVP-TPR*) as well as the baseline approaches (B^x and TPR*). We set the number of partitions k in DVA and VMBR-based partitioning techniques as 3 and 5, respectively, which is

(a) Part of Seoul road network (b) Part of London road network (c) Part of Boston road network (d) Part of Shenzhen road network

Fig. 5. City road networks for traffic simulation

Table 1. Experimental settings

Parameter	Setting
Space domain (m×m)	**10,000×10,000**
Number of objects	**100K**, 200K, ..., 500K
Query window size (m×m)	200×200, **400×400**, ..., 1000×1000
kNN - k	10, 20, **30**, ..., 50
Query predict time (ts)	0, 30, **60**, ...,120
Node size (byte)	1K, 2K, **4K**,···, 16K
datasets	SEO, **LD**, BOS, SZ

consistent with the experimental settings in the original papers. All algorithms are implemented with C++ language and all experiments are performed with 2.93 GHz Intel Xeon CPU and 1TB RAM in CentOS Linux. The experimental settings are displayed in Table 1 where the default settings are boldfaced.

5.1 Datasets

In this subsection, we introduce datasets used in the experiments. Figure 5 shows city road networks corresponding to the datasets in the experiments.

Simulated Traffic Data. The simulation of city traffic consists of two parts: road network generation and traffic generation. City road networks are generated from the XML map data downloaded from http://www.openstreetmap.org. Our traffic generator is based on the digital representation of real road networks and the network-based moving object generator of Brinkhoff [2]. A road is a polyline consisting of a sequence of connected line segments. The initial location of a moving object is randomly selected on the road segments. The object then moves along this segment in either direction until reaching crossroads, where it has a 25 % chance to stop for several seconds due to the traffic and then continues moving along another randomly selected connected segment.

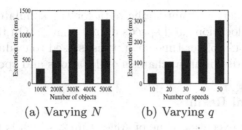

(a) Varying N (b) Varying q

Fig. 6. Overhead for partition update

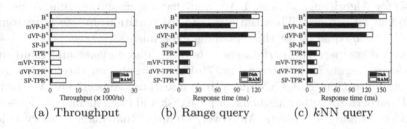

(a) Throughput   (b) Range query   (c) kNN query

Fig. 7. Disk indexes v s. RAM indexes

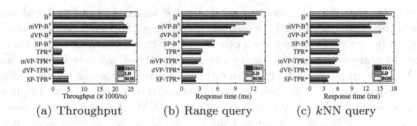

(a) Throughput   (b) Range query   (c) kNN query

Fig. 8. Varying datasets

We assume speed values of the moving vehicles in each road segment follow a random variable X and $X \sim \mathcal{N}(\mu, \sigma^2)$, where $\mathcal{N}$ is the normal distribution, μ and σ are set according to categories of the road segments. We divide the road segments into three categories: (C1) freeways/motorways with fastest traffic, (C2) primary roads with secondary fastest traffic, and (C3) street ways or residential roads with slowest traffic. We randomly select the normal distribution parameter μ from a range in terms of m/s for each category: (C1) [25, 40], (C2) [5, 25], (C3) [0, 15]. We set σ=10 m/s for all road segments.

GPS Tracking Data. The SZ dataset contains 100 K trajectories of taxis within the urban area of Shenzhen, China. Each trajectory contains a sequence of GPS tracking data with timestamps in a single day. The trajectories are not sampled with equal time intervals and the smallest sampling interval is 15 s. The dataset can be accessed at http://mathcs.emory.edu/aims/spindex/taxi.dat.zip.

Note that we generate 2 min traffic data for the simulated datasets, where the distributions of locations and speeds do not change. On the other hand, the SZ dataset has a much longer time span, thus is used to evaluated the temporal factor that leads to distribution changes on locations and speeds.

5.2 Experimental Results

Firstly, we show the execution time of Algorithm 2, which is the main overhead for partition updates, with different number of objects (N) and number of speed values (q). Figure 6 shows the results, which are consistent with the complexity analysis for Algorithm 2 in Sect. 4. We find that the execution time is less then 2 second in all settings. Thus the overhead for partition update is reasonably small. We set q equal 50 for the remaining experiments.

Next we compare the performances between disk indexes and main memory indexes. Figure 7(a) through Fig. 7(c) show results on throughput, range query response time and kNN query response time, respectively. We can see that SP outperforms other methods for both disk and main memory indexes with both B^x-trees and TPR*-trees. Moreover, we found that main memory indexes enjoy much better performance than disk indexes on both throughput and query response time. In the remaining experiments, we report only the results of main memory indexes since we have limited space.

Next we compare the experimental results across three simulated traffic datasets (SEO, LD, and BOS), which are summarized in Fig. 8. We can see that SP enjoys better performance than other velocity-based partitioning methods as well as the non-partitioning counterparts on a variety of datasets (road networks from Asian, European, and American cities). This is because, as shown in Fig. 5, road networks for large space domain ($10,000 \times 10,000$ m^2) usually implies no explicit velocity seeds or DVAs which are used in VMBR-based partitioning and DVA-based partitioning techniques, respectively. Moreover, Boston road network has more high speed roads than other city road networks thus nodes in the corresponding indexes expand faster, which makes the BOS dataset has higher query costs than other datasets.

In the next experiment, we vary the number of moving objects from 100 K to 500K. Figure 9 shows the results about throughput, range query and kNN query. We can see that when the number of objects increases, throughput decreases, query response time increases for both range queries and kNN queries. Moreover, B^x-trees enjoy higher throughput due to the simple update process of B^+-tree but lower query utility due to the "false hits" caused by the space-filling curves [8,22]. On the contrary, TPR*-trees have more complicated update operations which makes query more efficient at a sacrifice of throughput. Finally, SP indexes consistently outperform other indexes in all settings.

Next we vary the node size from 1 KB to 16 KB. Figure 10(a) through Fig. (c) show the experimental results. Generally speaking, performance decreases when node size increases, since index nodes with larger sizes require more maintaining and retrieving efforts. However, query performance of B^x-trees is not significantly affected by node size. This is because nodes of B^x-trees store

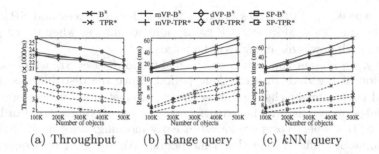

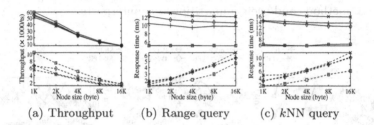

(a) Throughput   (b) Range query   (c) kNN query

Fig. 9. Varying number of objects

(a) Throughput   (b) Range query   (c) kNN query

Fig. 10. Varying node size

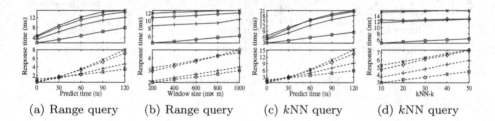

(a) Range query   (b) Range query   (c) kNN query   (d) kNN query

Fig. 11. Varying query parameters

the values computed from space-filling curves, which makes the spatial areas of B^x-tree nodes insensitive to their storage sizes. Note that the experimental results are different from both those for disk indexes, where disk I/O latency dominates the performance [3], and those for main memory indexes with secondary index on object IDs, which enables constant time locating the objects for updates [16]. Finally, SP significantly outperforms other methods in this experiment.

Next we study the impact of query parameters including query predict time, range query window size and kNN-k. The experimental results are summarized in Fig. 11. Figure 11(a) and (b) show the results about range queries while Fig. 11(c) and (d) show those about kNN queries. We can conclude from the figures that, generally speaking, TPR*-trees perform better than B^x-trees and SP outperforms other methods. Moreover, SP gains more advantages when query predict time, query window size, and kNN-k increase.

Finally, we present the results on the real world dataset SZ, which contains information of the taxis in a day long period. Since the distributions of locations and speeds might change during the experiment time, we perform partition updates every 1 hour. The experimental results are summarized in Fig. 12. We can see that query costs are lowest at early morning, since most cities have least volume of traffic during that time period. We also find that query costs raise at noon and night. This is because the taxis drive faster resulting in higher expanding speeds of the index nodes. The variation of throughput during the day is relatively small. Again, SP significantly and consistently outperforms other partitioning methods and their unpartitioned counterparts.

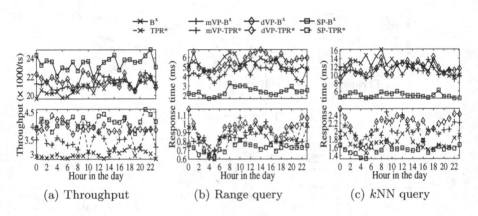

Fig. 12. Varying hour of the day

6   Conclusions and Future Work

In this paper, we proposed a novel and generic speed partitioning technique (SP) for indexing moving objects and implemented SP with the state-of-the-art indexing structures including the B^x-tree and the TPR*-tree. We empirically evaluated the performance of SP through extensive experiments on both simulated traffic data and real world GPS tracking data.

There are several future works which can further improve the performance of SP. Firstly, seeking more accurate estimations on search space expansion can always help finetune the optimal partitioning. Secondly, analytic methods such as *kernel density estimation* (KDE), instead of empirical methods, can be used to estimate the speed distribution. Moreover, sophisticated partition update algorithms might further improve performance in highly dynamic scenarios, where the distributions of locations and speeds change frequently. Finally, we will extend our method to grid-based indexing structures.

Acknowledgments. This research is supported by the AFOSR DDDAS program (Grant No. FA9550-12-1-0240) and National Natural Science Foundation of China (Grant No. 11271351).

References

1. Beckmann, N., Kriegel, H.-P., Schneider, R., Seeger, B.: The r*-tree: an efficient and robust access method for points and rectangles. In: SIGMOD, pp. 322–331 (1990)
2. Brinkhoff, T.: A framework for generating network-based moving objects. GeoInformatica **6**(2), 153–180 (2002)
3. Chen, S., Jensen, C.S., Lin, D.: A benchmark for evaluating moving object indexes. PVLDB **1**(2), 1574–1585 (2008)
4. Chen, S., Ooi, B.C., Tan, K.-L., Nascimento, M.A.: St^{2}b-tree: a self-tunable spatio-temporal b$^+$-tree index for moving objects. In: SIGMOD, pages 29–42, 2008
5. Dittrich, J., Blunschi, L., Vaz Salles, M.A.: Indexing moving objects using short-lived throwaway indexes. In: Mamoulis, N., Seidl, T., Pedersen, T.B., Torp, K., Assent, I. (eds.) SSTD 2009. LNCS, vol. 5644, pp. 189–207. Springer, Heidelberg (2009)
6. Finkel, R.A., Bentley, J.L.: Quad trees: a data structure for retrieval on composite keys. Acta Inf. **4**, 1–9 (1974)
7. Guttman, A.: R-trees: A dynamic index structure for spatial searching. In: SIGMOD, pp. 47–57 (1984)
8. Jensen, C.S., Lin, D., Ooi, B.C.: Query and update efficient b$^+$-tree based indexing of moving objects. In: VLDB, pp. 768–779 (2004)
9. Jensen, C.S., Lu, H., Yang, B.: Indexing the trajectories of moving objects in symbolic indoor space. In: Mamoulis, N., Seidl, T., Pedersen, T.B., Torp, K., Assent, I. (eds.) SSTD 2009. LNCS, vol. 5644, pp. 208–227. Springer, Heidelberg (2009)
10. Jensen, C.S., Pakalnis, S.: Trax - real-world tracking of moving objects. In: VLDB, pp. 1362–1365 (2007)
11. Nascimento, M.A., Silva, J.R.O., Theodoridis, Y.: Evaluation of access structures for discretely moving points. In: Böhlen, M.H., Jensen, C.S., Scholl, M.O. (eds.) STDBM 1999. LNCS, vol. 1678, pp. 171–188. Springer, Heidelberg (1999)
12. Nguyen, T., He, Z., Zhang, R., Ward, P.: Boosting moving object indexing through velocity partitioning. PVLDB **5**(9), 860–871 (2012)
13. Patel, J.M., Chen, Y., Chakka, V.P.: Stripes: an efficient index for predicted trajectories. In: SIGMOD, pp. 637–646 (2004)
14. Saltenis, S., Jensen, C.S., Leutenegger, S.T., Lopez, M.A.: Indexing the positions of continuously moving objects. In: SIGMOD, pp. 331–342 (2000)
15. Schiller, J., Voisard, A.: Location-Based Services. Elsevier, Amsterdam (2004)
16. Sidlauskas, D., Saltenis, S., Christiansen, C.W., Johansen, J.M., Saulys, D.: Trees or grids?: indexing moving objects in main memory. In: SIGSPATIAL, pp. 236–245 (2009)
17. Sidlauskas, D., Saltenis, S., Jensen, C.S.: Parallel main-memory indexing for moving-object query and update workloads. In: SIGMOD, pp. 37–48 (2012)
18. Sidlauskas, D., Saltenis, S., Jensen, C.S.: Processing of extreme moving-object update and query workloads in main memory. VLDB J. **23**(5), 817–841 (2014)
19. Silva, Y.N., Xiong, X., Aref, W.G.: The rum-tree: supporting frequent updates in r-trees using memos. VLDB J. **18**(3), 719–738 (2009)

20. Sistla, A.P., Wolfson, O., Chamberlain, S., Dao, S.: Modeling and querying moving objects. In: ICDE, pp. 422–432 (1997)
21. Tao, Y., Papadias, D., Sun, J.: The tpr*-tree: an optimized spatio-temporal access method for predictive queries. In: VLDB, pp. 790–801 (2003)
22. Yiu, M.L., Tao, Y., Mamoulis, N.: The b^{dual}-tree: indexing moving objects by space filling curves in the dual space. VLDB J. **17**(3), 379–400 (2008)
23. Zhang, M., Chen, S., Jensen, C.S., Ooi, B.C., Zhang, Z.: Effectively indexing uncertain moving objects for predictive queries. PVLDB **2**(1), 1198–1209 (2009)
24. Zhu, Y., Wang, S., Zhou, X., Zhang, Y.: RUM+-tree: a new multidimensional index supporting frequent updates. In: Wang, J., Xiong, H., Ishikawa, Y., Xu, J., Zhou, J. (eds.) WAIM 2013. LNCS, vol. 7923, pp. 235–240. Springer, Heidelberg (2013)

Spatio-Temporal Approaches

Using Lowly Correlated Time Series to Recover Missing Values in Time Series: A Comparison Between SVD and CD

Mourad Khayati[1,2]([✉]), Michael H. Böhlen[1], and Philippe Cudré Mauroux[2]

[1] Department of Computer Science, University of Zurich, Zurich, Switzerland
{mkhayati,boehlen}@ifi.uzh.ch
[2] EXascale Infolab, University of Fribourg, Fribourg, Switzerland
{mkhayati,phil}@exascale.info

Abstract. The Singular Value Decomposition (SVD) is a matrix decomposition technique that has been successfully applied for the recovery of blocks of missing values in time series. In order to perform an accurate block recovery, SVD requires the use of highly correlated time series. However, using lowly correlated time series that exhibit shape and/or trend similarities could increase the recovery accuracy. Thus, the latter time series could also be exploited by including them in the recovery process.

In this paper, we compare the accuracy of the Centroid Decomposition (CD) against SVD for the recovery of blocks of missing values using highly and lowly correlated time series. We show that the CD technique better exploits the trend and shape similarity to lowly correlated time series and yields a better recovery accuracy. We run experiments on real world hydrological and synthetic time series to validate our results.

1 Introduction

In real world applications sensors are used to measure time series data of different types, which are then collected, processed and stored in central stations. In the hydrological field, for instance, weather stations collect measurements that describe meteorological phenomena, e.g., temperature, humidity, air pressure, precipitation, etc. These time series contain blocks of missing values due to many reasons, e.g., sensor failure, power outage, sensor to central server transmission problem, etc. In order to recover these missing values, existing recovery techniques use the (base) time series that contains the missing values in addition to highly correlated (reference) time series. However, these recovery techniques can not learn from the trend and shape similarity of lowly correlated reference time series. Thus, the latter are not included in the recovery process.

The Foehn, for instance, is a warm wind that reaches weather stations at different time points. This environmental phenomenon yields time series with shape and trend similarities, but shifted in time. For example, the Foehn yields shifted temperature time series with similar shapes, e.g., peaks that contain

© Springer International Publishing Switzerland 2015
C. Claramunt et al. (Eds.): SSTD 2015, LNCS 9239, pp. 237–254, 2015.
DOI: 10.1007/978-3-319-22363-6_13

similar spikes. These shifted time series are lowly correlated. It is of interest to benefit from Foehn based time series and include them, in addition to the highly correlated time series, in the recovery process. In this paper, we consider the category of lowly correlated reference time series, e.g., Fohen based time series, that exhibit shape and/or trend similarities to the base time series.

Matrix decomposition techniques decompose an input matrix into the product of k matrices where $k \in [2, 3]$. The truncated Singular Value Decomposition (SVD) has been successfully applied to recover missing values in time series [1]. The truncated SVD performs a decorrelation of vectors and subsequently an unweighted relative reduction of the Mean Squared Error (MSE) to the reference time series. The unweighted MSE reduction yields a recovery that ignores the correlation difference between the input time series. Thus, this recovery technique is not suitable to apply in case of using highly and lowly correlated reference time series (cf. Sect. 5). To the best of our knowledge, there does not exist any technique that introduces different weights in the decomposition process of SVD. In [2–4] fast approximations of the truncated SVD have been proposed. Similarly to SVD, the latter approximations perform a decorrelation of vectors and thus, produce an unweighted MSE relative reduction.

In this work, we are interested in the case of using highly and lowly correlated time series for the recovery of blocks of missing values. Intuitively, in such cases, an accurate recovery technique should give different weights to the used time series. In contrast of the truncated SVD, the truncated Centroid Decomposition (CD) technique gives a weight proportional to the correlation between the base and the reference time series (cf. Sect. 5). Consequently, the obtained recovery produces a relative reduction of the MSE to the highly correlated reference time series more than to the lowly correlated one yielding a block recovery better than the one produced by the truncated SVD. We assume that the lowly correlated time series that exhibit trend and/or shape similarity are given as input. Searching for these time series is beyond the scope of this paper.

The main contributions of this paper are:

- We prove that CD technique produces correlated output vectors while SVD technique produces uncorrelated output vectors.
- We empirically show that CD performs a weighted MSE relative reduction that is proportional to the correlation of the input time series. The resulting recovery of missing values takes into account the correlation difference between the input time series.
- We empirically show that SVD performs an unweighted MSE relative reduction. The resulting recovery of missing values ignores the correlation difference between the input time series.
- We present the results of an experimental evaluation of the recovery accuracy of the CD and SVD techniques. The iterated truncated CD produces a better recovery accuracy in case of using a similar number of highly and lowly correlated time series.

The rest of this paper is organized as follows. Section 2 discusses related work. Section 3 describes the recovery process using SVD and CD techniques. Section 4 defines the unweighted recovery and the correlation based recovery respectively performed by SVD and CD. Section 5 reports the evaluation results. Section 6 concludes the paper and points to future work.

2   Related Work

The Singular Value Decomposition (SVD) is a commonly used matrix decomposition technique. It computes the singular values with their corresponding right and left singular vectors. The truncated SVD, which is computed out of SVD by nullifying the smallest singular values, has been extensively used in many fields, e.g., compression, noise reduction, etc. Khayati et al. [1] applied the truncated SVD for the recovery of missing values in time series. The basic idea is as follows: the truncated SVD is iteratively applied to a matrix that has as columns the time series for which the missing values have been initialized through linear interpolation. The iterative process refines only the initialized missing values and terminates when the difference between the updated values before and after the refinement is smaller than a small threshold value, e.g., 10^{-5}. The Mean Squared Error (MSE), between the real values and the recovered ones, is used to evaluate the recovery accuracy [5].

The Centroid Decomposition (CD) is a matrix decomposition technique that decomposes an input matrix into the product of two matrices. Chu et al. [6] introduce an algorithm that computes the CD of an input matrix in quadratic run time, but requires the construction of a correlation square matrix that has a quadratic space complexity. Khayati et al. [7] propose an algorithm to compute the CD out of the input matrix using a weight vector instead of the construction of the correlation matrix. They prove the correctness of the proposed solution. The space complexity is thus reduced from quadratic to linear while keeping the same run time complexity.

The Semi Discrete Decomposition (SDD) [8] is a matrix decomposition technique that decomposes an input matrix into three matrices such that their product approximates the input matrix, i.e., $\mathbf{X} \approx \mathbf{X}' \cdot \mathbf{D} \cdot \mathbf{Y}^T$. The resulting $\mathbf{D}$ is a diagonal matrix and the values of $\mathbf{X}'$ and $\mathbf{Y}$ are restricted to belong to the set $\{-1, 0, 1\}$. The truncated SDD has been used as clustering method [9]. The non-zero elements of the matrix obtained from the product $d_{ii} \times X'_{*i} \cdot Y_{*i}^T$ are the elements of the input matrix $\mathbf{X}$ which have the closest values and thus can be clustered together. Due to the set restriction of the elements of $\mathbf{X}'$ and $\mathbf{Y}$, the application of SDD for the recovery of blocks of missing values does not produce accurate results.

In addition to matrix decomposition techniques, matrix factorization techniques have been also applied for the recovery of missing values. The latter techniques start from k random matrices in order to approximate the input matrix. Stochastic Gradient Descent (SGD) [10] is a matrix factorization technique that approximates an input matrix $\mathbf{X}$ by the product of two matrices $\mathbf{P}$ and $\mathbf{Q}$, i.e., $\mathbf{X} \approx \mathbf{P} \cdot \mathbf{Q}$. SGD iteratively minimizes an error function by computing the gradient. At each iteration, the gradient is computed using random sample square blocks of the input matrix. The accuracy of the gradient increases with the size and the number of the used blocks [11]. Thus, using an input matrix with high number of rows and columns yields an accurate gradient's computation and subsequently a good approximation of the input matrix. In [12], SGD has been successfully applied to predict ratings in recommender systems for a matrix of items as rows and users as columns. Balzano et al. [13] propose an SGD-based solution, called GROUSE, for the recovery of blocks of missing values in an input matrix. GROUSE performs an accurate recovery for matrices of a high number of rows and columns. The recovery accuracy of the proposed solution deteriorates if the number of columns is much smaller than the number of rows such as in the hydrology field where the number of time series is much smaller than the number of observations.

3   Preliminaries

3.1   Notation

Bold upper-case letters refer to matrices, regular font upper-case letters to vectors (rows and columns of matrices) and lower-case letters to elements of vectors/matrices. For example, $\mathbf{X}$ is a matrix, $\mathbf{X}^T$ is the transpose of $\mathbf{X}$, X_{i*} is the i-th row of $\mathbf{X}$, X_{*i} is the i-th column of $\mathbf{X}$ and x_{ij} is the j-th element of X_{i*}.

In multiplication operations we use the sign $\times$ for scalar multiplication and the sign $\cdot$ otherwise. The symbol $\|\|$ refers to the l-2 norm of a vector. Assume $X = [x_1, \ldots, x_n]$, then $\|X\| = \sqrt{\sum_i^n (x_i)^2}$.

3.2   Background

Time Series. A time series $X_{i*} = \{(t_1, v_1), (t_2, v_2), \ldots, (t_n, v_n)\}$ is a set of n temporal values v_i ordered with respect to their timestamps t_i. We consider time series that have the same granularity of values. Thus, we omit the timestamps and we write time series using only their ordered values, e.g., time series $X_{1*} = \{(1, 4), (2, 5), (3, 1)\}$ is written as $X_{1*} = \{4, 5, 1\}$. Time series are inserted as columns of the input matrix $\mathbf{X}$.

Pearson Correlation Coefficient. Given two vectors X and Y of equal length n, with respective averages $\bar{x}$ and $\bar{y}$, the Pearson correlation coefficient is defined as,

$$r(X, Y) = \frac{\sum_{i=1}^{n}(x_i - \bar{x})(y_i - \bar{y})}{\sqrt{\sum_{i=1}^{n}(x_i - \bar{x})^2}\sqrt{\sum_{i=1}^{n}(y_i - \bar{y})^2}} \tag{1}$$

The absolute value of r ranges between 0 and 1 where $r \in [0.7, 1]$ stands for highly correlated vectors. The value of r is undefined if all x_i (and/or y_i) are equal.

Initialization Strategy. The missing values of each time series are initialized as a preprocessing step before the application of the recovery process. A missing value is initialized with a linear interpolation between the predecessor and the successor values. If the missing value occurs as the first or the last elements of the time series, we use the nearest neighbor initialization. Thus, the missing values of a time series X_{*1} are initialized as follows:

$$(t_i, v_i) = \begin{cases} (t_i, v) \text{ if } (s(t_i), \_) \notin X_{*1}, \\ \quad (p(t_i), v) \in X_{*1} \\ [2pt](t_i, v) \text{ if } (p(t_i), \_) \notin X_{*1}, \\ \quad (s(t_i), v) \in X_{*1} \\ [2pt](t_i, \frac{(t_i - p(t_i))(s(v_i) - p(v_i))}{s(t_i) - p(t_i)} + s(v_i)) \\ \quad \textbf{otherwise} \end{cases}$$

where $p(t_i) = max\{t_j \mid (t_j, \_) \in X_{*1} \wedge t_j < t_i\}$ is the predecessor of timestamp t_i in X_{*1} and $s(t_i) = min\{t_j \mid (t_j, \_) \in X_{*1} \wedge t_j > t_i\}$ is the successor timestamp of t_i in X_{*1}. Similarly, $p(v_i) = \{v_j \mid (t_j, \_) \in X_{*1} \wedge t_j = p(t_i)\}$ is the predecessor of value v_i in X_{*1} and $s(v_i) = \{t_j \mid (t_j, \_) \in X_{*1} \wedge t_j = s(t_i)\}$ is the successor value of v_i in X_{*1}.

3.3   Matrix Decomposition

Singular Value Decomposition. The *Singular Value Decomposition (SVD)* is a matrix decomposition technique that decomposes an $n \times m$ matrix, $\mathbf{X} = [X_{*1}| \ldots |X_{*m}]$, into an $n \times p$ matrix, $\mathbf{U}$, a $p \times m$ matrix, $\mathbf{\Sigma}$, and an $m \times m$ matrix $\mathbf{V}$, i.e.,

$$\mathbf{X} = \mathbf{U} \cdot \mathbf{\Sigma} \cdot \mathbf{V}^T \tag{2}$$

$$= \sum_{i=1}^{p} \sigma_i \times U_{*i} \cdot (V_{*i})^T,$$

where $p = min(n, m)$, the columns of $\mathbf{U}$ and $\mathbf{V}$ are respectively called left and right singular vectors, and $\mathbf{\Sigma}$ is a matrix whose diagonal elements, σ_i, are called singular values and are arranged in decreasing order, i.e., $\sigma_1 \geq \sigma_2 \geq \ldots \geq \sigma_p \geq 0$. The obtained columns of $\mathbf{U}$ are orthogonal to each other, i.e., $U_{*1} \perp U_{*2} \perp \ldots \perp U_{*p}$. Similarly, the columns of $\mathbf{V}^T$ are orthogonal to each other. In order to guarantee the orthogonality of columns of respectively $\mathbf{U}$ and $\mathbf{V}$, SVD requires the use of the same input matrix [14]. Since SVD decomposition is performed based on the same input matrix $\mathbf{X}$, we refer to SVD as a *flat* decomposition method.

Figure 1 illustrates the SVD decomposition of an input matrix $\mathbf{X}$.

Centroid Decomposition. The *Centroid Decomposition (CD)* is a matrix decomposition technique that decomposes an $n \times m$ matrix, $\mathbf{X} = [X_{*1}| \ldots |X_{*m}]$,

$$\mathbf{X} = \begin{bmatrix} -4 & 0 \\ 2 & 1 \\ 3 & -2 \end{bmatrix}; \; \mathrm{SVD}(\mathbf{X}) = \underbrace{\begin{bmatrix} -0.725 & -0.307 \\ 0.333 & 0.627 \\ 0.603 & -0.716 \end{bmatrix}}_{\mathbf{U}}, \underbrace{\begin{bmatrix} 5.445 & 0 \\ 0 & 2.086 \end{bmatrix}}_{\Sigma}, \underbrace{\begin{bmatrix} 0.987 & 0.16 \\ -0.16 & 0.987 \end{bmatrix}}_{\mathbf{V}}$$

such that

$$\mathbf{X} = \underbrace{\begin{bmatrix} -0.725 & -0.307 \\ 0.333 & 0.627 \\ 0.603 & -0.716 \end{bmatrix}}_{\mathbf{U}} \times \underbrace{\begin{bmatrix} 5.445 & 0 \\ 0 & 2.086 \end{bmatrix}}_{\Sigma} \times \underbrace{\begin{bmatrix} 0.987 & -0.16 \\ 0.16 & 0.987 \end{bmatrix}}_{\mathbf{V}^T}$$

Fig. 1. Example of Singular Value Decomposition.

into an $n \times m$ loading matrix, $\mathbf{L}$, and an $m \times m$ relevance matrix, $\mathbf{R}$, i.e.,

$$\mathbf{X} = \mathbf{L} \cdot \mathbf{R}^T = \sum_{i=1}^{m} L_{*i} \cdot (R_{*i})^T, \tag{3}$$

where $\|L_{*1}\| > \|L_{*2}\| > \ldots > \|L_{*m}\| \geq 0$. Figure 2 illustrates the CD of matrix $\mathbf{X}$.

$$\mathbf{X} = \begin{bmatrix} -4 & 0 \\ 2 & 1 \\ 3 & -2 \end{bmatrix}; \; \mathrm{CD}(\mathbf{X}) = \underbrace{\begin{bmatrix} -3.977 & -0.43 \\ 1.878 & 1.214 \\ 3.202 & -1.658 \end{bmatrix}}_{\mathbf{L}}, \underbrace{\begin{bmatrix} 0.994 & 0.11 \\ -0.11 & 0.994 \end{bmatrix}}_{\mathbf{R}}$$

such that

$$\mathbf{X} = \begin{bmatrix} -4 & 0 \\ 2 & 1 \\ 3 & -2 \end{bmatrix} = \underbrace{\begin{bmatrix} -3.977 & -0.43 \\ 1.878 & 1.214 \\ 3.202 & -1.658 \end{bmatrix}}_{\mathbf{L}} \times \underbrace{\begin{bmatrix} 0.994 & -0.11 \\ 0.11 & 0.994 \end{bmatrix}}_{\mathbf{R}^T}$$

Fig. 2. Example of Centroid Decomposition.

The CD technique applies an iterative process to compute matrices $\mathbf{L}$ and $\mathbf{R}$. At each iteration i, the input matrix $\mathbf{X}$ is updated by subtracting the product $L_{*i} \cdot R_{*i}^T$ from it. The columns of $\mathbf{L}$ (and $\mathbf{R}$) are not orthogonal to each other. Since CD decomposition is performed by hierarchically updating $\mathbf{X}$, we refer to CD as a *hierarchical* decomposition method.

Chu et al. [6] prove that the decomposition performed by CD best approximates the one produced by SVD, i.e., $\mathbf{L}$ approximates the product $\mathbf{U} \cdot \mathbf{\Sigma}$ and $\mathbf{R}$ approximates $\mathbf{V}$.

Truncation. The *truncated SVD* computes a matrix $\mathbf{X}_k$ out of the SVD of $\mathbf{X}$. It takes only the k first columns of $\mathbf{U}$ and $\mathbf{V}$ and the k largest elements of $\mathbf{\Sigma}$ such that $k < p$, i.e.,

$$\mathbf{X}_k = \sum_{i=1}^{k} \sigma_i \times U_{*i} \cdot (V_{*i})^T. \tag{4}$$

Equation (4) is equivalent to $\mathbf{X}_k = \mathbf{U} \cdot \mathbf{\Sigma}_k \cdot \mathbf{V}^T$ where $\mathbf{\Sigma}_k$ is obtained by setting the $r - k$ smallest (non zero) singular values of $\mathbf{\Sigma}$ to 0. Let rank p be the maximal number of linearly independent rows or columns of $\mathbf{X}$. Then, among all matrices with rank $k < p$, $\mathbf{X}_k$ is proven to be the optimal approximation to the input matrix $\mathbf{X}$ in the Frobenius norm [15].

The *truncated CD* computes a matrix $\mathbf{X}_k$ out of the CD of $\mathbf{X}$ by setting to 0 the $m - k$ (non zero) last columns of $\mathbf{L}$, with $k < m$, in order to respectively get $\mathbf{L}_k$ and $\mathbf{X}_k = \mathbf{L}_k \cdot \mathbf{R}^T$.

4   Decomposition Comparison

In this section, we compare the decomposition produced by the truncated SVD against the one produced by the truncated CD using the Mean Squared Error ($MSE = \frac{1}{k}\sum_{i=1}^{k}(\tilde{x}_i - x_i)^2$; initialized value x_i; recovered value $\tilde{x}_i$; number of missing observations k) between the initialized values and the recovered ones.

4.1   Recovery Process

Algorithm 1 describes the pseudo code of function *RecM()* that applies truncated SVD and truncated CD to recover missing values. The algorithm takes an input matrix $\mathbf{X}$ where the missing values have been initialized, and returns a matrix $\widetilde{\mathbf{X}}$ with recovered values. Different initialization techniques would lead to the same result but with a higher number of iteration [7]. *RecM()* iteratively replaces the initialized missing values by the result of the truncation of a given matrix decomposition technique. The algorithm terminates if the difference in Frobenius norm ($\|\mathbf{X} - \widetilde{\mathbf{X}}\|_F = \sqrt{\sum_{i=1}^{n}\sum_{j=1}^{m}(x_{ij} - \tilde{x}_{ij})^2}$; x_{ij}: element of $\mathbf{X}$; $\tilde{x}_{ij}$: element of $\widetilde{\mathbf{X}}$) between the matrix before the update of missing values, $\mathbf{X}$, and the one after, $\widetilde{\mathbf{X}}$, is less than a small threshold value, e.g., $\epsilon = 10^{-5}$.

In what follows we describe the recovery properties using respectively SVD and CD. We assume the case where the correlation ranking between time series does not change over the entire history, i.e., the most correlated reference time series has the highest correlation value to the base time series all over the entire history. In case where the correlation ranking changes over the history, then a segmentation of the time series has to be applied.

Algorithm 1. RecM($\mathbf{X}$, n, m, T_j^m)

    Input: $n \times m$ matrix $\mathbf{X}$; set of missing time stamps T_j^m in X_{*j}
    Output: $n \times m$ matrix $\widetilde{\mathbf{X}}$ of recovered values

1 **repeat**
2     $\widetilde{\mathbf{X}} = \mathbf{X}$;
      // Apply truncated SVD or truncated CD
3     $\mathbf{X}_k = Truncate(\widetilde{\mathbf{X}})$;
      // Update missing values
4   **foreach** $t \in T_j^m$ **do**
5        $x_{tj} = w_{tj}$;
        // w_{tj} element of $\mathbf{X}_k$
6 **until** $\|\mathbf{X} - \widetilde{\mathbf{X}}\|_F < \epsilon$;
7 **return** $\widetilde{\mathbf{X}}$

4.2 SVD Recovery

Lemma 1. *Given an input matrix $\mathbf{X}$ of m correlated columns. SVD($\mathbf{X}$) produces non correlated vectors.*

Proof 1. *By definition of SVD, we have that $U_{*1} \perp U_{*2} \perp \ldots \perp U_{*p}$. This implies that the pairwise dot product of columns of $\mathbf{U}$ is equal to 0 and thus, $\forall a, b \in [1, p] \wedge a \neq b$, we get $(U_{*a})^T \cdot U_{*b} = 0$. Using the fact that all input time series have been normalized to have mean equal to 0 (cf. Sect. 5), we assume u and u' to be the i-th elements of respectively U_{*a} and U_{*b}, and get from Eq. (1) the following*

$$r(U_{*a}, U_{*b}) = \frac{\sum_{i=1}^{n}(u_i \times u_i')}{\sqrt{\sum_{i=1}^{n}(u_i - \bar{u})^2}\sqrt{\sum_{i=1}^{n}(u_i' - \bar{u'})^2}}$$

$$= \frac{(U_{*a})^T \cdot U_{*b}}{\sqrt{\sum_{i=1}^{n}(u_i - \bar{u})^2}\sqrt{\sum_{i=1}^{n}(u_i' - \bar{u'})^2}} = 0$$

As a result, the pairwise correlation between all columns of $\mathbf{U}$ is equal to 0. The previous property holds also for the columns of $\mathbf{V}^T$.

Definition 1 (Unweighted Recovery). *Let $\mathbf{X}$ be an input matrix that contains a base time series B and $k > 2$ reference time series each with a correlation r_i to B. An unweighted recovery of B produces a similar relative reduction of the MSE between B and the reference time series.*

Proposition 1. *Assume an $n \times m$ matrix $\mathbf{X} = [B, R_1, \ldots, R_{m-1}]$. A truncated matrix decomposition of $\mathbf{X}$ that produces uncorrelated vectors performs an unweighted recovery of B.*

Based on Lemma 1 and Proposition 1, we get that the truncated SVD performs an unweighted recovery.

Example 1. *Let's take the example of a matrix* $\mathbf{X} = [B, R_1, R_2]$ *where initialized missing values are marked in bold.*

$$\mathbf{X} = \begin{bmatrix} -4 & 1 & 3 \\ -1 & 3 & -1 \\ \mathbf{2} & 6 & 6 \\ \mathbf{5} & 5 & 3 \end{bmatrix}$$

R_1 *is a highly correlated reference time series to B with $r(B, R_1) = 0.88$ and R_2 is a lowly correlated reference time series to B with $r(B, R_2) = 0.32$. The computation of the MSE before the recovery gives $MSE(B, R_1) = 16$ and $MSE(B, R_2) = 8$.*

The following matrix $\widetilde{\mathbf{X}} = [\widetilde{B}, R_1, R_2]$ is an example of an SVD based recovery of B.

$$\widetilde{\mathbf{X}} = \begin{bmatrix} -4 & 1 & 3 \\ 0 & 3 & -1 \\ 4 & 6 & 6 \\ 5 & 5 & 3 \end{bmatrix}$$

The computation of the MSE after the recovery gives $MSE(\widetilde{B}, R_1) = 6.5$ and $MSE(\widetilde{B}, R_2) = 2.5$. The percentage of the MSE relative reduction between B and R_1 is $red(R_1) = \frac{16-6.5}{16} \times 100 = 60\%$. Similarly, the percentage of the MSE relative reduction between B and R_2 is $red(R_2) = 69\%$. As a result, we have $red(R_1) \approx red(R_2)$.

4.3 CD Recovery

Lemma 2. *Given an input matrix $\mathbf{X}$ of m correlated columns. CD($\mathbf{X}$) produces correlated vectors.*

Proof 2. *This proof follows directly from the proof of Lemma 1. On the contrary of SVD, the columns of $\mathbf{L}$ and $\mathbf{R}^T$ computed by the truncated CD are not orthogonal and thus, the pairwise dot product and consequently the pairwise correlation values are different from 0.*

Definition 2 (Correlation Weighted Recovery). *Let $\mathbf{X}$ be an input matrix that contains a base time series B and $k > 2$ reference time series each with a correlation r_i to B. A correlation weighted recovery of B performs a relative reduction of the MSE between B and the reference time series proportionally to $|r_i|$.*

Proposition 2. *Assume an $n \times m$ matrix $\mathbf{X} = [B, R_1, \ldots, R_{m-1}]$. A truncated matrix decomposition of $\mathbf{X}$ that produces correlated vectors performs a correlation weighted recovery of B.*

Based on Lemma 2 and Proposition 2 we get that the truncated CD performs a correlation weighted recovery.

Example 2. *Let's take the example of a matrix $\mathbf{X} = [B, R_1, R_2]$ used in Example 1. The following matrix $\widetilde{\mathbf{X}} = [\widetilde{B}, R_1, R_2]$ is an example of a CD based recovery of B.*

$$
\widetilde{\mathbf{X}} = \begin{bmatrix} -4 & 1 & 3 \\ 2 & 3 & -1 \\ 5 & 6 & 6 \\ 5 & 5 & 3 \end{bmatrix}
$$

The computation of the MSE after the recovery gives $MSE(\widetilde{B}, R_1) = 1$ and $MSE(\widetilde{B}, R_2) = 5$. The percentage of the MSE relative reduction between B and R_1 is $red(R_1) = 94\%$. The percentage of the MSE relative reduction between B and R_2 is $red(R_2) = 37.5\%$. As a result, we have $red(R_1) \gg red(R_2)$.

4.4 Complexity

We compare the runtime and space complexity of CD based recovery against SVD based recovery. We use the algorithm that computes the exact decomposition for each technique.

Run Time. Consider an input matrix $\mathbf{X}$ with n rows and m columns. The number of arithmetic operations to compute SVD of $\mathbf{X}$, using Golub and Reinsch algorithm [14], is $4n^2m + 8mn^2 + 9m^3$. The number of arithmetic operations to compute CD of $\mathbf{X}$ is $2pnm$ where p is the number of iterations [7]. At each iteration of CD, the input matrix is subtracted yielding an updated matrix that contains negative elements. Thus, the value of p depends on the distribution of the minus sign across the updated matrix. In practice, the value of p ranges between $\frac{n}{2}$ and $\frac{n}{3}$ (cf. Sect. 5.5).

Space. SVD technique requires the storage of nm values of $\mathbf{X}$, nm values of $\mathbf{U}$, m values of Σ and m^2 values of $\mathbf{V}$. Additionally, SVD has to transform $\mathbf{X}$ to a bidiagonal matrix using Householder reduction [16] which requires the storage of three additional matrices, i.e., the first matrix contains nm values and the two others contain m^2 values each. The total number values stored by SVD is thus equal to $m(3n + 3m + 1)$ values. CD technique requires the storage of nm values of $\mathbf{X}$, nm values of $\mathbf{L}$ and m^2 values of $\mathbf{R}$. No data structure other than the input and the two output matrices is stored. Thus, the total number values stored by CD is equal to $m(2n + m)$ values.

5 Experiments

The experiments are performed using real world datasets that describe hydrological time series where each tuple records a timestamp and a value of a specific

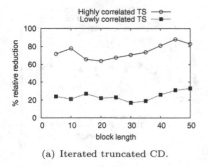

(a) Iterated truncated CD.

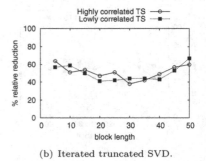

(b) Iterated truncated SVD.

Fig. 3. MSE relative reduction of CD and SVD using highly and lowly correlated time series: case 1.

observation. Hydrological time series with shifted peaks and/or valleys are lowly correlated. Our first set of time series, HYD^1, contains 200 time series of six years length each, where measurements are recorded every five minutes. The second set of time series we refer to, SBR^2, contains 120 time series of twelve years length each, where measurements are recorded every 30 min. The hydrological time series have been normalized with the z-score normalization technique [17]. We consider hydrological time series where the correlation ranking does not change all over the history. We use also synthetic time series, where the correlation is constant all over the entire history. To measure the recovery accuracy, we compute the Mean Squared Error (MSE) between the original and the recovered blocks (cf. Sect. 4).

5.1 Recovery Using Real World TS

MSE Relative Reduction. In this experiment we compute the MSE relative reduction between a base time series B and two reference time series. In Fig. 3 we choose one highly and one lowly correlated reference time series with the respective correlation values $r(B, R_1) = 0.83$ and $r(B, R_2) = 0.18$. The result of this experiment shows that the iterated truncated CD produces a correlation weighted recovery that reduces the relative MSE more to the highly correlated time series than the lowly correlated time series. The iterated truncated SVD performs an unweighted recovery that produces an almost equal reduction of the relative MSE to both reference time series.

In Fig. 4 we consider one highly correlated reference time series with a correlation value $r(B, R_1) = 0.76$. We add also a lowly correlated time series with a correlation value $r(B, R_2) = 0.62$ that is higher than the one used in the experiment of Fig. 3. As expected, the MSE relative reduction of the iterated truncated CD is slightly higher to R_1 than to R_2. The MSE relative reduction of the iterated truncated SVD remains similar to both reference time series.

---

[1] The data was kindly provided by HydroloGIS (http://www.hydrologis.edu).

[2] The data was kindly provided by Südtiroler Beratungsring (http://www.beratungsring.org).

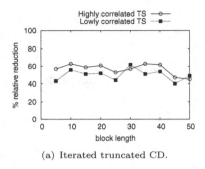

(a) Iterated truncated CD.

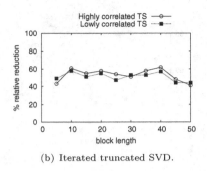

(b) Iterated truncated SVD.

Fig. 4. MSE relative reduction of CD and SVD using highly and lowly correlated time series: case 2.

Recovery Accuracy. In this section we compare the recovery accuracy of the iterated truncated SVD against the iterated truncated CD using highly and lowly correlated time series.

In the experiment of Fig. 5, we use three temperature time series from HYD measured respectively in Aria Borgo (B), Ponte Adige (R_1) and Aria La Villa (R_2) in the region of South Tyrol, Italy. B is highly correlated to R_1 with $r(B, R_1) = 0.75$. B is lowly correlated to R_2 with $r(B, R_2) = 0.32$. However, the peaks of B and R_2 exhibit shape similarity, i.e., the peaks contain similar spikes. The time shift is caused by the Foehn phenomenon (cf. Sect. 1). We drop from the base time series, B, a block for ts $\in [45, 95]$ and recover it using two reference time series, R_1 and R_2. The result of this experiment shows that the iterated truncated CD gives a weight to the reference time series proportional to their correlation with B, yielding a good block recovery accuracy, i.e., the amplitude and the shape of the missing block are accurately recovered. On the contrary, the iterated truncated SVD performs a block recovery that gives the same weight to both time series R_1 and R_2 at a time yielding a bad block recovery accuracy.

Figure 6 shows the MSE for removed blocks of values of increasing length from a base time series: starting from the middle of a block we increase the length of the removed block in both directions and we compute the MSE for each block. We run the experiment on five different base time series from HYD and we take the average of the MSE. For each run we use, in addition to the base time series, one highly correlated and one lowly correlated time series. As expected, the iterative truncated CD learns from the highly and lowly correlated time series at a time and thus, produces a small recovery error that slightly increases with the length of the missing block to recover. However, the recovery accuracy of the iterated truncated SVD considerably deteriorates with the length of the missing block to recover.

Impact of the Time Shift. In Fig. 7 we evaluate the impact of a varying time shift, denoted as s, on the recovery accuracy of the iterated truncated CD and the iterated truncated SVD. We show that for a high value of time shift, the

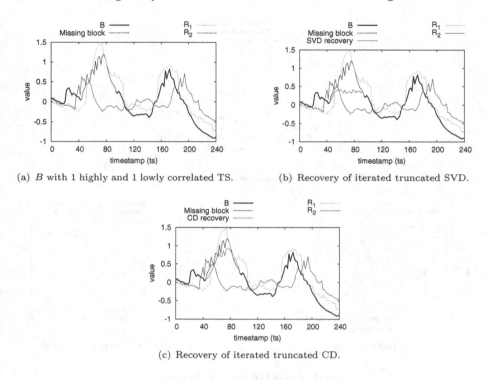

(a) B with 1 highly and 1 lowly correlated TS.   (b) Recovery of iterated truncated SVD.

(c) Recovery of iterated truncated CD.

Fig. 5. Recovery using highly and lowly correlated hydrological TS.

two techniques produce similar block recoveries. In Fig. 7(a) we take three time series from SBR measured respectively in Kaltern (B), Kollman (R_1) and Ritten (R_2) in the region of South Tyrol, Italy. The peaks of B and R_2 have a similar shape, but with a time shift. We drop one peak from B, we shift backwards R_2 with a value s and we compute the MSE recovery accuracy. The result of the experiment shows that starting from $s = 30$, the iterated truncated CD is not able anymore to exploit the lowly correlated time series and produces a block recovery similar to the one produced by the iterated truncated SVD.

5.2   Recovery Using Synthetic TS

For the following experiments, we consider a time series $sin(t)$ that has a small valley at each of the peaks, denoted as B, from which we drop a block of values for $t \in [70, 110]$ that we recover using both techniques.

Recovery Accuracy. In Fig. 8 we add to B one highly correlated time series $-0.5 * sin(t)$ denoted as R_1 such that $r(B, R_1) = 0.84$. We add also a lowly correlated time series by shifting B and we denote it as R_2 such that $r(B, R_2) = 0.16$. As expected, by giving a higher weight to R_2, the iterated truncated CD is able to perform a good recovery of the shape and the amplitude of the missing

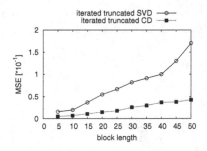

Fig. 6. MSE for successive removed blocks.

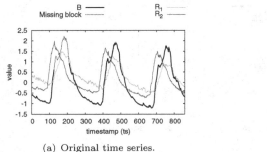

(a) Original time series.

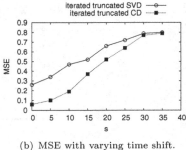

(b) MSE with varying time shift.

Fig. 7. Impact of varying time shift

block. The iterated truncated SVD fails to recover the shape and the amplitude of the missing block.

Impact of Number of Input Time Series. In Fig. 9 we evaluate the robustness of the recovery produced by both techniques using a varying number of highly and lowly correlated time series. In Fig. 9(a) we take B from the experiment of Fig. 8 and one highly correlated time series with $r = 0.9$ to which we add a varying number of lowly correlated time series, by shifting B, such that $r \in [0.2, 0.6]$. The latter time series are added in the decreasing order of their correlation. This experiment shows that for $p_1 < 4$, the iterated truncated CD is able to use the most correlated time series yielding a smaller MSE than the iterated truncated SVD. For $p_1 \geq 4$, the MSE of both techniques converges towards similar value. In the experiment of Fig. 9(b) we take B and one lowly correlated time series with $r = 0.2$ to which we add a varying number of highly correlated time series such that $r \in [0.7, 0.9]$. The latter time series are added in the increasing order of their correlation. In the presence of one lowly correlated time series, the iterated truncated SVD requires at least three additional highly correlated time series in order to reach the same MSE as one of the iterated truncated CD.

The experiment of Fig. 9 shows that, for a close number of highly and lowly correlated time series, the correlation weighted recovery helps the iterated

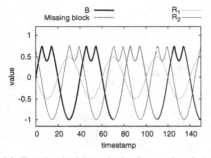

(a) R with 1 highly and 1 lowly correlated time series

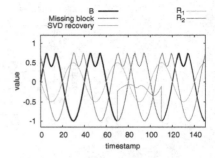

(b) Recovery of iterated truncated SVD.

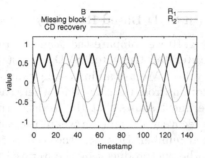

(c) Recovery of iterated truncated CD.

Fig. 8. Recovery using highly and lowly correlated synthetic TS.

truncated CD to produce a better recovery than the one produced by the iterated truncated SVD. Otherwise, the two techniques produce similar recovery of missing values. However, the iterated truncated CD technique is computationally more efficient than the iterated truncated SVD, i.e., CD is linear with the number of input time series while SVD is cubic with the number of input time series.

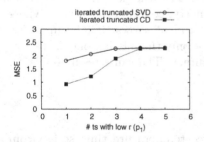

(a) MSE using varying # of lowly correlated time series.

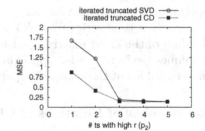

(b) MSE using varying # of highly correlated time series

Fig. 9. Recovery accuracy using varying number of input TS.

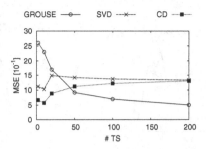

Fig. 10. Recovery accuracy of CD, SVD and GROUSE.

5.3   Comparison with SGD Based Recovery

In the experiment of Fig. 10 we compare the accuracy recovery of the iterated truncated CD and the iterated truncated SVD against GROUSE [12] for the recovery of 20 missing values. We use an increasing number of segments of time series of different correlations from the same type where each contains 200 values. The result of this experiment shows that the iterated truncated CD produces a more accurate block recovery in the case where the length of the input time series is bigger than their number. However, the recovery accuracy produced by GROUSE outperforms the one produced by the iterated truncated CD and the iterated truncated SVD as the number of time series approaches the number of observations (cf. Sect. 2). This is explained by the fact that the segments have different correlation values and GROUSE selects only blocks out of these segments. In real world applications such as hydrology, the length of time series is much bigger than their number and thus, CD based recovery outperforms GROUSE recovery.

5.4   Approximation Accuracy

Figure 11 compares the approximation accuracy of the iterated truncated CD and the iterated truncated SVD to the input matrix. We use the Frobenius norm between the input matrix and the one obtained after the decomposition as an approximation error (cf. Sect. 4.1). The input matrix contains 10 columns where each one is a time series from HYD. This experiment shows that by updating all values of the input matrix at a time (and not only the missing ones), the two techniques perform similar approximation accuracy. The same result holds for different values of the truncation parameter k.

5.5   Number of Iterations of CD

In the experiment of Fig. 12 we consider three temperature time series from HYD: a base time series, one highly correlated reference time series and one lowly correlated time series. We compute the number of iterations p required by the CD technique with an increasing number of rows n. The result of this experiment shows that p ranges between $\frac{n}{2}$ and $\frac{n}{3}$.

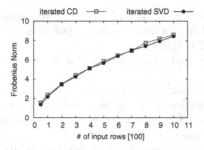

Fig. 11. Approximation error.

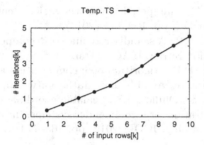

Fig. 12. number of iterations performed by CD.

6 Conclusion

In this paper, we compare the CD and SVD techniques for the recovery of missing values using time series with mixed correlation values. We empirically show that CD produces a weighted relative reduction of MSE that is proportional to the correlation of the input time series, while SVD produces an unweighted relative reduction of MSE. Our experiments on real world hydrological and synthetic time series also show that the iterated truncated CD performs a better recovery in case of similar number of highly and lowly correlated time series.

In future work, it would be of interest to compare the segmentation techniques that are applied in the case where the correlation ranking varies along the time series history. Another promising direction is to refine the definition of highly and lowly correlated time series.

References

1. Khayati, M., Böhlen, M.: Rebom: Recovery of blocks of missing values in time series. In: Proceedings of the 2012 ACM International Conference on Management of Data. COMAD 2012, pp. 44–55. Computer Society of India (2012)
2. Li, M., Bi, W., Kwok, J.T., Lu, B.: Large-scale nyström kernel matrix approximation using randomized SVD. IEEE Trans. Neural Netw. Learn. Syst. **26**, 152–164 (2015)

3. Halko, N., Martinsson, P.G., Tropp, J.A.: Finding structure with randomness: probabilistic algorithms for constructing approximate matrix decompositions. SIAM Rev. **53**, 217–288 (2011)

4. Achlioptas, D., McSherry, F.: Fast computation of low-rank matrix approximations. J. ACM 54 (2007)

5. Li, L., McCann, J., Pollard, N.S., Faloutsos, C.: Dynammo: mining and summarization of coevolving sequences with missing values. In: Proceedings of the 15th ACM SIGKDD International Conference on Knowledge Discovery and Data Mining, pp. 507–516. Paris, France, 28 June–1 July 2009

6. Chu, M., Funderlic, R.: The centroid decomposition: relationships between discrete variational decompositions and svds. SIAM J. Matrix Anal. Appl. **23**, 1025–1044 (2001)

7. Khayati, M., Böhlen, M., Gamper, J.: Memory-efficient centroid decomposition for long time series. In: ICDE. pp. 100–111 (2014)

8. Kolda, T.G., O'Leary, D.P.: A semidiscrete matrix decomposition for latent semantic indexing information retrieval. ACM Trans. Inf. Syst. **16**, 322–346 (1998)

9. Kolda, T.G., O'Leary, D.P.: Algorithm 805: computation and uses of the semidiscretematrix decomposition. ACM Trans. Math. Softw. **26**, 415–435 (2000)

10. Yu, H., Hsieh, C., Si, S., Dhillon, I.S.: Scalable coordinate descent approaches to parallel matrix factorization for recommender systems. In: 12th IEEE International Conference on Data Mining, ICDM 2012, pp. 765–774. Brussels, Belgium, 10–13 December 2012

11. Gemulla, R., Nijkamp, E., Haas, P.J., Sismanis, Y.: Large-scale matrix factorization with distributed stochastic gradient descent. In: KDD, pp. 69–77 (2011)

12. Koren, Y., Bell, R.M., Volinsky, C.: Matrix factorization techniques for recommender systems. IEEE Comput. **42**, 30–37 (2009)

13. Balzano, L., Nowak, R., Recht, B.: Online identification and tracking of subspaces from highly incomplete information. CoRR abs/1006.4046 (2010)

14. Golub, G.H., van Loan, C.F.: Matrix computations, 3rd edn. Johns Hopkins University Press, Baltimore (1996)

15. Björck, A.: Numerical methods for least squares problems. SIAM (1996)

16. Griffiths, D.V., Smith, I.M.: Numerical Methods for Engineers. CRC Press, Boca Raton (2006)

17. Jain, A., Nandakumar, K., Ross, A.: Score normalization in multimodal biometric systems. Pattern Recogn. **38**, 2270–2285 (2005)

Minimal Spatio-Temporal Database Repairs

Markus Mauder[1], Markus Reisinger[1], Tobias Emrich[1], Andreas Züfle[1]([✉]),
Matthias Renz[1], Goce Trajcevski[2], and Roberto Tamassia[3]

[1] Ludwig-Maximilians-Universität München, Munich, Germany
{mauder,reisinger,zuefle,renz}@dbs.ifi.lmu.de
[2] Northwestern University, Evanston, USA
goce@eecs.northwestern.edu
[3] Brown University, Providence, USA
rt@cs.brown.edu

Abstract. This work addresses the problem of efficient detection and fix-ing of inconsistencies in spatio-temporal databases. In contrast to tradi-tional database settings, where integrity constraints pertain to explicitly stored values and values defined via views and aggregates, spatio-temporal data may exhibit other types of constraint violations that cannot be tied to stored or aggregated values. The main reason is that spatio-temporal phenomena are continuous but their database representations are discrete. Thus, the constraints are semantic in nature, as opposed to being depen-dent on the actual stored data. We give a general definition of semantic constraints of a trajectory database and define rules to repair violations of these constraints. In order to minimize the distortion of the state of the database, we aim at minimizing the changes needed for repairing viola-tions of such semantic constraints. Towards this goal, we define a measure of dissimilarity between the initial database and its repaired state. Also, to minimize dissimilarity, we propose several simple rules of space- and time-distortion that shift inconsistent observations in space and time to remove inconsistencies. Our evaluation shows that these rules often run into local minima, and thus may not be able to repair a database. To remedy this problem, we propose a hybrid approach that chooses between several pos-sible space and time distortions. We show that a greedy approach which always chooses the locally best repair may still run into local minima and propose a simulated-annealing approach that combines greedy and ran-dom repairs to avoid these local minima.

1 Introduction

By the end of 2014, there were nearly 7 billion mobile subscriptions world-wide [1]. This fact, along with miniaturization of computing and sensing devices and GPS and RFID technologies, has provided a foundation for generating extremely large volumes of location-in-time data: petabytes of location-based (i.e., spatio-temporal) data are generated every day [11]. The management of (*location, time*) information about mobile entities is essential for a variety of application domains, ranging from navigation and efficient traffic management to emergency/disaster rescue management, environmental monitoring, fly-through

© Springer International Publishing Switzerland 2015
C. Claramunt et al. (Eds.): SSTD 2015, LNCS 9239, pp. 255–273, 2015.
DOI: 10.1007/978-3-319-22363-6_14

visualization, and various military applications (e.g., radar data, troops tracking) [14]. Essentially, every application requiring some form of Location Based Services (LBS) [16] needs efficient techniques for storage, retrieval and query processing of spatio-temporal data—topics studied in the field of Moving Objects Databases (MOD) [10].

Physical factors, such as the imprecision of sensing devices and communication links, often cause location data to be inaccurate and noisy. In addition to this problem—even with perfect sampling accuracy—the data intended to capture a continuous motion can be represented only at discrete time-instances. Moreover, data records can be obsolete as users may update their location infrequently, e.g., due to bad connectivity or to preserve battery power. Thus, one has to cater to the uncertainty as a natural factor when considering the representation of spatio-temporal data (cf. [6]). A complementary observation is that data sources may be various heterogeneous devices: roadside-sensors, weather stations, satellite imagery, (mobile) weather radar, crowd sourced observations, ground and aerial LIDAR—to name but a few. Having multiple sources may yield not only cause type-mismatch issue, but also generate conflicting location information about the same object and cause problems in reconciling the data [18]. Complementary to uncertainty, the above contexts may cause other types of semantic inconsistencies that have not been addressed so far. Namely, a user posing a continuous k-Nearest Neighbor (k-NN) query, may be presented with an answer containing two (or more) vehicles that "have collided." This is an example of violating the following basic semantic constraint: "two objects cannot be at the same place at the same time." Such a violation may be due to imprecise location-samples. Also, it often arises from the use of interpolation (linear, Bezier, etc.) in-between observed samples [9].

The main objective this work is to provide techniques detecting and fixing' such inconsistencies. The focus of this paper is not on removing the inherent uncertainty, which follows from the imprecision of location detection and can be an additional cause of inconsistencies. Rather, we aim at repairing the "symptoms" of the "inevitable uncertainty". As an example, the interpolation of GPS signals may lead to the consequence of having a trajectory of a given car going through a lake. Fixing this problem by having the trajectory going around the lake may still yield the wrong trajectory, as the true trajectory may look different. Hence, while we cannot claim to have alleviated the root-causes for errors in spatio-temporal databases, we take a first step towards fixing the symptoms based on semantic constraints. Clearly, a method for database repairs should aim at minimizing the distortion between the original database and the repaired database. The main contribution of this work can be summarized as follows:

- We identify and formalize the problem of semantic inconsistencies in spatio-temporal data. This formalization identifies a wide class of problems, that have been largely neglected in the moving object and trajectory database literature.
- Since the problem of finding an optimal database repair is NP-hard, we propose a number of heuristics to repair a spatio-temporal database, which are

organized into three general categories of solution, including time-distortion, space-distortion and hybrid approaches.
- We present experimental observations quantifying different trade-offs among the proposed methods.

The rest of this paper is organized as follows. Section 2 presents a review related work. In Sect. 3, we formalize the problem of Moving Object Database (MOD) inconsistencies along with metrics to measure the quality of a database repair. Section 5 gives the details of the proposed algorithmic solutions and the experimental observations are presented in Sect. 6. Finally, in Sect. 7, we conclude the paper and outline directions for future work.

2   Related Work

We now overview the literature on several different topics related to the problems addressed in this paper. However, as we will argue, although each body of work has yielded interesting and relevant results, none has addressed the specific problems tackled by our work, nor has provided a readily applicable "tool-chain."

Relational Database Repairs: Traditional database approaches *repair* [3,4,17] the identified inconsistencies by removing objects or by changing attribute values. Such approaches however, can not be applied directly to spatio-temporal data. Arbitrarily changing a (location, time) pair is likely to yield new inconsistencies, as the changed trajectory may reach an unreachable state, or may have an unrealistically high speed in the repaired version of the database. The main challenge in spatio-temporal data is to incorporate repair rules to span a space of semantically meaningful repairs.

Probabilistic Spatio-Temporal Database Repairs: The recently published approach of [12] aims at repairing probabilistic spatio-temporal databases as defined in [13]. In this setting, each mobile object is assigned a set of spatial regions and a probability interval defining the likelihood to be within this region. In an interpretation of such a database, the probability of a region must be within its interval and the probabilities of all regions of an object must sum up to one. Such a database is inconsistent if no interpretation exists. The approach of [12] shows how to minimally change probability intervals in order to obtain an interpretation. The problem setting in this work can not be extended to trajectory databases.

A recent approach presented in [8] models the motion of a spatio-temporal object by a stochastic process, such that each possible world is indeed associated with a probability. Constraints such as "Object x must not be in state s at time t" can be incorporated into this model by adapting the corresponding probabilities. More complex constraints, such as inter-object constraints that prohibit objects from being at in same state at the same time, can not be incorporated into such models as easily.

Linear Temporal Logic: Our problem of removing inconsistencies from a trajectory database can be cast in the realm of temporal logic. For instance, using

Propositional Linear Temporal Logic (LTL) [7], a trajectory $T = s_1, s_2, ..., s_{|T|}$ can be described using the *eventually* operator $\diamond$ by $\diamond s_1 \diamond s_2 ... \diamond s_{|T|}$. Semantically, this LTL formulation induces a trajectory where eventually state s_1 must be visited after any number of intermediate states, then s_2 must eventually be visited after possible more intermediate states and so on. Further constraints can be formulated, e.g. to constrain the database such that no two objects may be at the same location at the same time, by applying the always operator $\square$ to express the rule $\forall T_1, T_2 \in \mathcal{D}, t \in T : \square T_1(t) \neq T_2(t)$. Using logical solvers for LTL [15], we can efficiently find an interpretation[1] for each trajectory such that all constraints are satisfied, if any such interpretation exists. While LTL allows to formulate any semantic constraint, the main problem of LTL is that, being a logic rather than a function, it does not allow to find any optimal solution. Thus, LTL allows to check if there exists a model that satisfies all given constraints, which any way of formulating a cost function that can be optimized. In most applications, the problem of finding such a model is trivia. For example, the solution of using a *serial schedule*, which avoids any inconsistency between objects by simple removing any temporal overlap between trajectories, does always work.

While the solution based on serially scheduling each trajectory is valid, it is prohibitively expensive, since "repaired" trajectories may be extensively distorted in time. We are looking for a solution which minimizes the changes to the database performed by the repair.

3   Problem Definition

This section presents the details of the novel types of inconsistencies and desirable properties of (methodologies for) enforcing the semantic constraints in a given MOD.

A spatio-temporal database $\mathcal{D}^{ST}$ stores triples (*oid, location, time*), where $oid \in \{o_1, ..., o_N\}$ is a unique object identifier, $location \in \mathcal{S}$ is a spatial position in space and $time \in \mathcal{T}$ is a point in time. Semantically, each such triple corresponds to the location of object o_i at some time. In $\mathcal{D}$, an object can be described by a function $tr_{o_i} : \mathcal{T} \rightarrow \mathcal{S}$ that maps each point in time to a location in space[2] $\mathcal{S}$; this function is called *trajectory*. The corresponding trajectory database is denoted as $\mathcal{D} = \{tr_{o_1}, ..., tr_{o_N}\}$.

Assuming that the location of an object o_i is known for any point in time is unrealistic as the location of object o_i can only be determined at discrete time-instants. The frequency of location-samplings is also bounded by physical constraints, such as the availability of a GPS signal. Between discrete observations, the position of a moving object has to be estimated via some type of a *dead reckoning*. These estimations are based on incomplete information, and thus, may be imprecise.

---

[1] In LTL, an interpretation is a Kripke structure which, in our case, maps each trajectory and each point in time to a state.

[2] Most often the Euclidian 2D space is considered, however, extension to 3 (or higher) dimensions as well as road-network constraints have been commonly considered in the literature.

3.1   Spatio-Temporal Constraints

The violation of a constraint in a trajectory database $\mathcal{D}$ indicates that $\mathcal{D}$ contains erroneous trajectories, possibly incurred due to faulty dead reckoning, or due to deficiency of measuring devices used to capture trajectories. Since we are considering historical data, we lack the option of improving the available information, e.g., by requesting the objects to give a more accurate position update. Since the cause for the inconsistency cannot be removed, the only viable approach is to repair the trajectories in order to mitigate the symptoms of this lack of information.

In contrast to traditional database settings, where integrity constraints pertain to explicitly stored values and values defined via views and aggregates, spatio-temporal data may exhibit other types of constraint violations that cannot be tied to stored or aggregated values. The main reason is that spatio-temporal phenomena are continuous but their database representations are discrete. Thus, the constraints are semantic in nature, as opposed to being dependent on the actual stored data.

Definition 1. *Let $\mathcal{C}$ be a set of constraints. A database $\mathcal{D}$ is said to satisfy $\mathcal{C}$, noted as $\mathcal{D} \vDash \mathcal{C}$, if all constraints are satisfied in $\mathcal{D}$. If $\mathcal{D} \nvDash \mathcal{C}$, then $\mathcal{D}$ is said to be inconsistent.*

Loosely speaking, a *spatio-temporal constraint* can be thought of as any rule describing some semantic constraint related to the trajectories in $\mathcal{D}$. A constraint $c \in \mathcal{C}$ may pertain to an individual object. An example of such an *Object Constraint* is the constraint "An object must not enter a specified area R on Sunday between 2am and 5am." This constraint can be formally expressed as

$$\forall(tr_o \in \mathcal{D}), \forall(t \in [\text{Sunday 2am}, \text{Sunday 5am}]) : tr_o(t) \notin R.$$

In contrast, some constraints may be defined between trajectories, such as "two objects must not be in the same place at the same time" which can be expressed as

$$\forall(tr_{o_i}, tr_{o_j}, i \neq j), \forall t : tr_{o_i}(t) \neq tr_{o_j}(t).$$

In practice, constraints involving more than one object lead to hard optimization problems, as a single repair of one trajectory may have a large number of consequences on the constraints involving other objects. Section 4 will show that such constraints lead to NP-hard optimization problems. Since we are considering the general case, we will be considering such hard inter-object constraints in our experimental evaluation in Sect. 6.

3.2   Database Repair Rules

In this work, we will propose a number of trajectory database repair rules. These rules define for a given trajectory $T \in \mathcal{D}$ the set of possible repairs $\overline{T^\mathcal{R}}$. Before we propose these rules in Sect. 5, we first formally define the purpose of a repair rule.

Definition 2 (Trajectory Database Repair Rule). *Let $\overline{\mathcal{D}}$ denote the set of all possible trajectory databases. A trajectory database repair rule $R : \overline{\mathcal{D}} \mapsto \overline{\mathcal{D}}^*$ is a function, which maps a trajectory database $\mathcal{D}$ to a set of possible repairs.*

As an example, a repair rule R may allow to simply remove a trajectory from $\mathcal{D}$. This repair rule can be specified by

$$R(\mathcal{D}) = \{\mathcal{D}'|\mathcal{D}' \subset \mathcal{D}\}.$$

Definition 3 (Database Repair). *Let $\mathcal{D}$ be a trajectory database inconsistent with respect to a set of semantic constraints $\mathcal{C}$ and let $\mathcal{R}$ be a set of repair rules. Let $\mathcal{D}^R \in \mathcal{R}^*(\mathcal{D})$ be a trajectory database derived by iteratively applying repair rules $R \in \mathcal{R}$ to $\mathcal{D}$. If $\mathcal{D}^R \models \mathcal{C}$ holds, then the trajectory database $\mathcal{D}^R$ is called a database repair of $\mathcal{D}$.*

In many cases, such as the aforementioned exemplary repair rule that allows to discard trajectories, one trivial way of obtaining a database repair $\mathcal{D}^R$ which satisfies all given constraints $c \in C$ is, for example, the empty database $\mathcal{D}^R = \{\}$. Given the lack of any actual trajectory, it trivially satisfies many constraints. Hence, strictly speaking, the challenge is not only to find just any database repair, but to find a database repair having the *minimal difference* from the initial database $\mathcal{D}$.

Definition 4 (Minimal Database Repair). *Let $\mathcal{D}$ be a trajectory database inconsistent with respect to a set of semantic constraints C. Let $dist(\mathcal{D}, \mathcal{D}^R)$ be a dissimilarity function between databases. A minimal repair $\mathcal{D}^R_{min}$ is defined as*

$$\mathcal{D}^R_{min} = argMin_{\mathcal{D}^R \in \overline{\mathcal{D}^R}, \mathcal{D}^R \models C} dist(\mathcal{D}, \mathcal{D}^R),$$

where $\overline{\mathcal{D}^R}$ represents the set of all possible repairs of $\mathcal{D}$.

The goal of this work is to efficiently compute, for a given trajectory database $\mathcal{D}$ and a set of semantic constraints C, a minimal repair $\mathcal{D}^R_{min}$ of $\mathcal{D}$. This problem falls into the class of constraint satisfaction problems and we show in Sect. 4 that it is NP-hard. We will relax the problem to find heuristic solutions that yield a database repair having sufficiently low dissimilarity to the initial database.

3.3   Quality of a Repair

To measure the quality of a repair, a dissimilarity function $dist(\mathcal{D}, \mathcal{D}^R)$ is needed. In accordance with Definition 4, this function will be minimized. Thus, this function defines the semantic of a "good" database repair, which is expected to minimize the total number of changes of the database $\mathcal{D}$, and should guarantee that changes are divided fairly over all trajectories. To measure the total dissimilarity between $\mathcal{D}$ and $\mathcal{D}^R$, we can simply aggregate the dissimilarity of individual trajectories:

$$dist(\mathcal{D}, \mathcal{D}^R) = \sum_{T \in \mathcal{D}} dist(T, T^R),$$

where $dist(T, T^R)$ is a dissimilarity function defined on trajectories such as average Euclidean-distance or edit distance. In addition, changes in $\mathcal{D}^R$ should be divided fairly among trajectories, in order to avoid starvation of single trajectories in the repaired database. Such fairness can be enforced as follows

$$dist(\mathcal{D}, \mathcal{D}^R) = \sum_{T \in \mathcal{D}} g(dist(T, T^R)),$$

where $g(x)$ is a function that monotonically increases in $\mathbb{R}^+$, such as the square function, to take into account the distances of individual trajectories.

In the remainder of this work, we propose solutions to remove inconsistencies from a trajectory database $\mathcal{D}$. For this purpose, in Sect. 4, we provide a formal proof of the NP-hardness of fixing inconsistencies in a trajectory database. Heuristic solutions are presented in Sect. 5. In Sect. 6, we perform an experimental analysis of the quality of these solutions on real data sets, evaluating both run-time and quality of the resulting repair.

4   Complexity Analysis

In the following, we show that the problem of finding the optimal repair $\mathcal{D}^R$ of a trajectory database $\mathcal{D}$ is generally hard. For this purpose, we show that the simpler problem of finding *any* repair is already NP-complete.

Lemma 1. *Given a trajectory database $\mathcal{D}$, a set of constraints C and a set of repair actions A, the problem of deciding whether there exists a repair $\mathcal{D}^R$ which is derived from $\mathcal{D}$ using rules in A, such that $\mathcal{D}^R \models C$ is NP-complete.*

Proof. Let $\mathcal{D}$ be a database of arbitrary trajectories, and let A be repair action such that for each trajectory $T_i \in \mathcal{D}$ there exists exactly one possible repair. For each $T_i \in \mathcal{D}$, let p_i denote the unrepaired trajectory T_i, and let $\not{p}_i$ denote the repaired trajectory which is derived by applying the only possible repair in A to T_i. Furthermore, let C be a set of inter-object constraints such that each constraint $c_{s,t} \in C$ requires that at least one object must be in state s at time t. Let $c_{s,t}(\mathcal{D}, A) \subseteq \bigcap_{1 \leq i \leq N}\{p_i, \not{p}_i\}$ denote the set of all possible trajectories that satisfy constraint $c_{s,t}$, i.e., all possible trajectories that are located in state s at time t. Since each constraint $s_{s,t}$ requires at least one trajectory to be in state s at time t, the constraint $s_{s,t}$ can be rewritten as the disjunction of all trajectories satisfying this constraint:

$$c_{s,t} = \bigvee_{p \in c_{s,t}(\mathcal{D},A)} p.$$

This boolean formula returns true if and only if the constraint $c_{s,t}$ is satisfied. For all constraints to be satisfied, the conjunction of all these disjunctions yields the following boolean formula:

$$\bigwedge_{c_{s,t} \in C} \bigvee_{p \in c_{s,t}(\mathcal{D},A)} p.$$

This formula returns true, if and only if, for a given database repair $\mathcal{D}^R \in \{p_1, \not{p}_1\} \times \{p_N, \not{p}_N\}$ satisfies all constraints in C. Consequently, the problem of finding a valid repair of $\mathcal{D}$ is equivalent to the satisfiability problem of the above boolean formula. This satisfiability problem, known as k-SAT, is known to be NP-complete.   □

Due to the hard nature of the problem, will omit an exact algorithm to find an optimal database repair, i.e., a repair that minimizes the amount of database distortion. It should be noted that such an algorithm can be specified using integer linear programming, yet such a solution may have unbearable run-times even for toy databases. Instead, in the next section, we will propose approximate algorithms, which return a database repair $\mathcal{D}^R$ which may not be minimal in terms of distortion of the original database $\mathcal{D}$, or which may fail to satisfy some constraints.

5   Algorithms

Before discussing our algorithmic solutions for spatio-temporal database repairs in Sect. 5.2, we specify the following components in Sect. 5.1:

1. Spatio-temporal constraints and techniques for their detection
2. Allowed repair rules
3. Dissimilarity function to measure the quality of a database repair

5.1   Component Specifications

Spatio-Temporal Constraints. There are several alternatives for spatio-temporal constraints. In this paper, we consider the following very general constraint: "Two objects must not be within a threshold of ε meters of each other at any time." This constraint is formally expressed as follows:

$$\forall (tr_{o_i}, tr_{o_j}, i \neq j), \forall t : dist(tr_{o_i}(t), tr_{o_j}(t)) > \varepsilon.$$

This constraint is able to ensure that objects with a spatial extent of ε never occupy the same space at the same time, or that objects do not get too close to each other.

As the next step, it is important to be able to quickly find violations of the above constraint in the database. To detect these violations, we use a spatio-temporal R^*-tree to index the set $\mathcal{S}$ of all trajectory segments defined by two successive GPS signals of the same object, using time as a third dimension. Each trajectory segment s is minimally bounded by a rectangle $\square(s)$ and added to the tree. Thus, each leaf of this R^*-tree is a single rectangle pointing to the exact representation of the approximated trajectory segment. To find all initial collisions, we perform a similarity-join [5] joining the indexed database with itself (ignoring identity) and using ε as the similarity threshold that is only applied on the spatial dimensions (and not on the time). The result is a set of intersection pairs (s, c) where s and c are segments of two different trajectories.

Once the initial collisions have been found, future collisions caused by database repairs can be found very efficiently, by querying against the tree only segments that have been changed by a repair.

Repair Rules. In our problem setting (Sect. 3), a database repair is still unspecified. In the following, we focus on the manipulation of the vertices of the trajectories in order to obtain a countable number of possible repairs. To identify the vertex to be repaired to remedy a constraint violation, we always consider the vertex closest to the violation point of both corresponding trajectories. The vertex can then be manipulated in one of the following ways:

- *Time domain:* The manipulation of a vertex v back in time implies that the movement from the previous vertex to v is slowed down and the movement from v to its subsequent vertex is sped up. The manipulation of v forward in time has the opposite effect. Note that the time manipulation of a vertex is constrained by its predecessor and its successor. Manipulating the time of v beyond the times of its predecessor or its successor yields anomalous movement in the spatial domain.
- *Spatial domain:* Manipulating the spatial position of a trajectory vertex has also impact on the speed of the movement.
- *Time and spatial domains:* Obviously, the spatial and temporal manipulation can be combined. A special case of spatio-temporal manipulation is the manipulation of v along the spatio-temporal path to its predecessor or its successor.

Based on these observations, we define the following three rules named after the cardinality of the set provided. Throughout this section, the input to a rule is the repair triple v_p, v, v_f, where v is the vertex to repair, v_p is the predecessor of v, and v_f is the successor of v in the trajectory. Furthermore, a vertex v is characterized by the triple $(v.t, v.x, v.y)$ representing the time, the x position and the y position of v, respectively.

Definition 5 (Two-Rule). *Given repair triple v_p, v, v_f, the Two-Rule returns two vertices v_1' and v_2', where*

$$v_1' = \frac{v_p + v}{2} \qquad v_2' = \frac{v_f + v}{2} \qquad (1)$$

Note that the two vertices returned by the Two-Rule are located in time and space half the way forward and backward around vertex v.
The Four-Rule adds temporal repairs.

Definition 6 (Four-Rule). *Given repair triple v_p, v, v_f and time distortion Δt, the Four-Rule returns the two vertices returned by the Two-Rule plus the two vertices v_3' and v_4', where*

$$v_3' = (v.t - \Delta t, v.x, v.y)$$
$$v_4' = (v.t + \Delta t, v.x, v.y) \qquad (2)$$

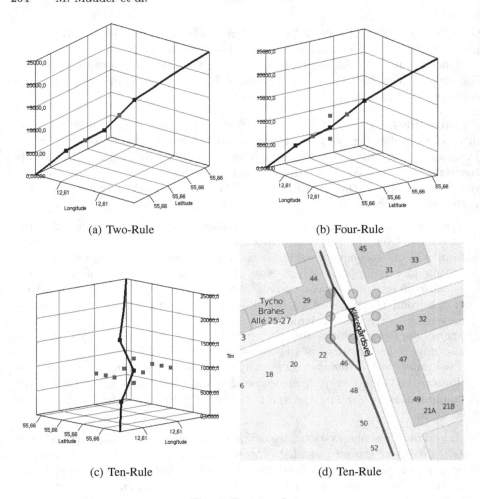

(a) Two-Rule

(b) Four-Rule

(c) Ten-Rule

(d) Ten-Rule

Fig. 1. Repair rules

ensuring that $v.t - \Delta t \geq v_p.t$ and $v.t + \Delta t \leq v_f.t$

The Ten-Rule adds eight absolute spatial distortions.

Definition 7 (Ten-Rule). *Given repair triple* v_p, v, v_f, *time distortion* Δt, *and space distortion* Δs, *the Ten-Rule returns the following ten vertices:*

$$v_3' = (v.t - \Delta t, v.x, v.y), v_4' = (v.t + \Delta t, v.x, v.y),$$
$$v_5' = (v.t, v.x - \Delta s, v.y), v_6' = (v.t, v.x + \Delta s, v.y),$$
$$v_7' = (v.t, v.x, v.y - \Delta s), v_8' = (v.t, v.x, v.y + \Delta s), \tag{3}$$
$$v_9' = (v.t, v.x - \Delta s, v.y - \Delta s), v_{10}' = (v.t, v.x + \Delta s, v.y + \Delta s),$$
$$v_{11}' = (v.t, v.x - \Delta s, v.y + \Delta s), v_{12}' = (v.t, v.x + \Delta s, v.y - \Delta s)$$

ensuring that $v.t - \Delta t \geq v_p.t$ and $v.t + \Delta t \leq v_f.t$

Figure 1 gives an overview of these three rules, where we show in 1(a) the shift on the segment and we show in 1(b) the additional time shift. Figures 1(c) and 1(d) show all ten options, while the Southwest option was chosen in Fig. 1(d). The effectiveness of these rules will be evaluated later on, but obviously the Ten-Rule should perform best, as it offers the most possibilities and so the algorithms will typically choose one of the ten vertices output by the Ten-Rule as the best repair. Besides that, the Two-Rule and Four-Rule are again working relative to the surrounding vertices which brings the already discussed disadvantages.

Quality Measure. As the cause of a constraint violation is unknown, the only sensible approach is to limit the changes to the database as much as possible. Accordingly, the quality of a repair is assessed by the magnitude of its effect on $\mathcal{D}$. A heuristic solution will generally generate a number of possible solutions, one of which will be chosen as the best one after a finite amount of processing time. For this purpose, a quality-measurement function $dist(\mathcal{D}, \mathcal{D}^R)$ for repairs is used for ranking.

$$dist(\mathcal{D}, \mathcal{D}^R) = \sum_{i \in [1, |\mathcal{D}|]} dist(tr_i, tr_i^R). \tag{4}$$

For the purpose of measuring the distance between the original trajectories and the repaired trajectories, we propose the following three dissimilarity functions.

The first function, $dist_{euclid}$, intuitively yields the spatial difference of two trajectories.

$$dist_{euclid}(tr, tr^R) = \sum_{i \in [1, |tr|]} \left((v_{i_\phi} - v_{i_\phi}^R)^2 + (v_{i_\lambda} - v_{i_\lambda}^R)^2 + (v_{i_t} - v_{i_t}^R)^2 \right)^{\frac{1}{2}}. \tag{5}$$

Utilizing weights for every dimension (w_ϕ, w_λ, w_t) it is easy to provide a Weighted Euclidean Distance function that takes into account the weighting of the time in contrast to the spatial dimensions.

$$dist_{weighted}(tr, tr^R) = \sum_{i \in [1, |tr|]} \left(w_\phi(v_{i_\phi} - v_{i_\phi}^R)^2 + w_\lambda(v_{i_\lambda} - v_{i_\lambda}^R)^2 + w_t(v_{i_t} - v_{i_t}^R)^2 \right)^{\frac{1}{2}}. \tag{6}$$

Finally, in order to provide an alternative to the Euclidean Distance, the third function is based on the Maximum Distance:

$$dist_{max}(tr, tr^R) = \sum_{i \in [1, |tr|]} max\{(v_{i_\phi} - v_{i_\phi}^R), (v_{i_\lambda} - v_{i_\lambda}^R), (v_{i_t} - v_{i_t}^R)\}. \tag{7}$$

In order to improve efficiency, the implementation of the above dissimilarity functions does not compute the complete $dist(\mathcal{D}, \mathcal{D}^R)$ after every repair step, but rather compares only trajectories that have been changed during the repair step and sums up the differences for every repair step.

5.2 Generate Database Repairs

The components outlined above can be combined to create an algorithm to generate a database repair. As finding a minimal database repair is NP-hard (Sect. 4), any resulting algorithm should employ heuristics to find a good (but not necessarily optimal) repair.

We have identified the following paradigms as possible approaches: Random, Greedy, and Simulated Annealing.

In our description of these algorithms we use the following functions:

- $c : \mathcal{D} \to \mathcal{V}$ returns the set of vertices that are part of any conflict in $\mathcal{D}$.
- $R_v : \mathcal{D} \to \mathcal{D}$ is the repair function R (as defined in Sect. 3), but limited to manipulations of the conflicting vertex v.

Random. The simplest approach does not try to choose a good repair function at all. Instead it applies a random instance of a set of possible repair functions to a random conflicting vertex in the database. See Algorithm 1 for a detailed description.

Algorithm 1. Random($\mathcal{D}$, $\mathcal{R}$)

1: **while** $c(\mathcal{D}) \neq \emptyset$ **do**
2:    $V \leftarrow c(\mathcal{D})$
3:    $v \leftarrow rnd(V)$
4:    $R \leftarrow rnd(\mathcal{R})$
5:    $\mathcal{D} \leftarrow R_v(\mathcal{D})$
6: **end while**

Applying a random repair function does not necessarily reduce the number of conflicts. As a consequence, the algorithm might not converge on a solution.

Greedy. The more sophisticated *Greedy* algorithm uses the number of remaining constraints after applying each function to make a better choice. The Random algorithm's weak spot is its unguided choice of repair function. The Greedy algorithm considers only the local improvement of each repair. The repair yielding the lowest number of remaining constraint violations is picked and applied to $\mathcal{D}$. See Algorithm 2 for details.

Compared to the Random algorithm, Greedy's locally optimal repairs yield a much faster convergence on a (possibly local) optimum. However, the increase of complexity leads to an increase in runtime. To find a repair that is closer to the minimal database repair, an algorithm must avoid the local minimum Greedy is prone to converge on. The following algorithm addresses this problem by combining random and greedy elements.

Algorithm 2. Greedy($\mathcal{D}$, $\mathcal{R}$)

1: **while** $c(\mathcal{D}) \neq \emptyset$ **do**
2:      $V \leftarrow c(\mathcal{D})$
3:      $v \leftarrow V[0]$
4:      $R_{opt} \leftarrow argmin_{R \in \mathcal{R}} \|R(\mathcal{D})\|$
5:      $\mathcal{D} \leftarrow R_{opt}(\mathcal{D})$
6: **end while**

Simulated Annealing. The deterministic nature of greedy algorithms makes them prone to local minima. For this reason, algorithm Greedy presented above is likely to return a valid database quickly, but this result is unlikely to be minimal (or close to minimal). To increase the likelihood of finding a global minimum, we now describe an algorithm based on simulated annealing. See Algorithm 3 for a detailed description.

Algorithm 3. SA($\mathcal{D}$, $\mathcal{R}$)

1: $\delta = 1$
2: **while** $c(\mathcal{D}) \neq \emptyset$ **do**
3:      **if** $random(]0; 1]) < \delta$ **then**
4:          $\mathcal{D} \leftarrow Random(\mathcal{D}, \mathcal{R})$
5:      **else**
6:          $\mathcal{D} \leftarrow Greedy(\mathcal{D}, \mathcal{R})$
7:      **end if**
8:      $\delta \leftarrow \delta - \Delta_\delta$
9: **end while**

By consolidating the Random and Greedy algorithms we counter the overly deterministic nature of greedy algorithms by introducing some randomness in a directed way. This algorithm avoids local minima by initially choosing random repairs, then trying to improve on the best random result using more and more greedy approaches. In each iteration, this algorithm first decides to either perform a Random repair or a Greedy repair with increasing bias toward greediness. In the first iteration, the probability δ of performing a Greedy repair is zero. In each subsequent iteration, this probability increases by a parameter $\Delta_\delta \in [0, 1]$.

The Simulated Annealing algorithm is expected to be slower than the Greedy algorithm, but more flexible and able to find a more global minimum. Compared to the Random algorithm, Simulated Annealing is faster and more directed. This claim will be evaluated in the following.

6 Experiments

The experimental evaluation presented in this section was conducted using a desktop computer having an Intel i7-870 CPU at 2.93 GHz and 8 GB of RAM.

Table 1. Abbreviations for experiments

tdra Absolute Time-Distortion Repair	**tdrr** Relative Time-Distortion Repair
ldra Absolute Location-Distortion Repair	**ldrr** Relative Location-Distortion Repair
ra Random	**gr** Greedy
sa Simulated Annealing	**2** Two-Repair-Rules
4 Four-Repair-Rules	**10** Ten-Repair-Rules

The spatio-temporal dataset that we are using consists of workout GPS data, i.e., running and hiking GPS-traces obtained from Endomondo (https://www.endomondo.com). For each GPS-position of a workout a trajectory is stored in $\mathcal{D}$ using linear interpolation, which is the main source of inconsistencies. The service is most popular in Scandinavia, so most workouts are located in cities there. The dataset we used was from the area of Copenhagen, which has a number of vertices between 2567 and 652854. In a data cleaning step, we removed: (1) trajectories that do not have an absolute time-stamp, to avoid having a huge number of runners at the beginning of time; (2) outlier GPS signals yielding a run-speed of more than 50 km per hour. The constraint is that two objects must not be closer than ϵ to each other, where ϵ is a parameter that we can vary in order to alter the number of inconsistencies, called *collisions*. Unless otherwise specified, the default value is $\epsilon = 3$ m. In this evaluation, we use four straightforward algorithms as a baseline. These four algorithms randomly pick a conflicting GPS-signal p that is adjacent to a conflicting trajectory segment. The, p is distorted by (i) moving its time-stamp one second towards the time of the next GPS-signal (*Absolute Time-Distortion*), or (ii) by moving its time-stamp half-way to the time of the next GPS-signal, or (iii) moving its location one meter towards the location of the next GPS-signal, or (iv) by moving its location half-way to the location of the next GPS-signal. Table 1 lists the various repair heuristics that we presented in Sect. 5 and shows the respective abbreviations that we will use in the following evaluation.

6.1   Collision Detection

In Sect. 5.1 we describe how we can find collisions in a trajectory database. This is performed by querying individual trajectory segments at an R^*-tree. For our experiments, we use the R^*-tree implementation of the ELKI-framework [2]. The average time required for a single intersection query, depends on the capacity of the R^*-tree. For leaf capacities of 10, 100, and 1000, we measured an average query time of 0.1283, 0.3390, and 7.2444, respectively. We are using a leaf capacity of 10 in the following. Figure 2(a) shows the total time required to find all initial collisions, which requires a large number of intersection queries. The number of collisions is also influenced by the intersection pipe radius ϵ, and Fig. 2(a) illustrates the effect on the Endomondo dataset. It is notable that the time required to find collisions seems independent of ϵ. This is attributed to the fact that even for a ϵ of 50 m, the number of collision candidates that have to be evaluated is too small to significantly

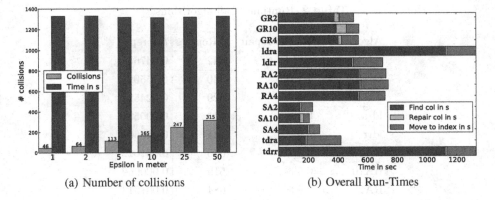

(a) Number of collisions   (b) Overall Run-Times

Fig. 2. Runtime experiments

impact the run-time. Thus, the vast majority of time is lost in the collision candidate generation step.

Figure 2(b) shows the time required to repair the found collisions. In each iteration of each algorithm, three steps are required: (i) Repairing a collision, (ii) then updating the index with the new distorted trajectory, and (iii) finding new collisions involving the distorted trajectory. The times required for these three steps are shown in Fig. 2(b). We note that despite the use of an efficient index structure, the time needed to repair two colliding trajectories lasts only a fraction of the time needed to find the collision and update/move the trajectory.

6.2   Run Time

The time to repair a collision is further shown in Table 2, along with the number of repair iterations – which varies depending on the heuristic used (likely that the collision has not been fixed, or new collisions may have been incurred). In Table 2, stars next to run-times imply that in at least one case, the repair algorithm did not terminate. Non-terminating cases are ignored for the computation of run-times in this experiments. We can make the following observations: We see that purely time distorting heuristics (tdra and tdrr) and purely location distorting heuristics (ldra and ldrr) are able to repair a database quickly. However, due to the simple rules that these approaches follow, they are unable to handle some cases which may occur in trajectory databases.

- In the case of relative repairs (tdrr and ldrr) this is caused be the fact that if two trajectory segments completely falls into their ε-range, then no distortion on these segments can yield a successful repair.
- In the case of absolute repairs (tdra and ldra), some special cases can not be handled. For instance, in the case where two trajectories remain at the same location for multiple GPS-signals: in this case GPS-signals are shifted, but the likelihood of reaching a state where all signals are collision free becomes minimal.

Table 2. Runtime of all algorithms

Algorithm	Time to repair	# Repairs	$t/\#$rep
GR4	16.294	342	0.047643
GR10	51.181	330	0.155094
GR2	10.522	429	0.024527
ldra	0.198*	341	0.000581
ldrr	20.92*	1341	0.015600
RA4	0.557	545	0.001022
RA10	0.503	519	0.000969
RA2	0.898	684	0.001313
SA4	16.046	343	0.046781
SA10	47.673	332	0.143593
SA2	11.708	464	0.025233
tdra	18.02*	5506	0.003273
tdrr	17.823*	1340	0.013301

When omitting the cases where these approaches do not terminate, we note that the fastest repair is achieved by the ldra heuristics, which distort observed GPS-signals in space. Furthermore, we can see that among the heuristics to choose a possible repair, the Random-heuristics (RA2, RA4, RA10) perform best, which is expected as these heuristics are not required to make any expensive greedy probing steps. The Greedy (GR2, GR4, GR10) and the Simulated Annealing (SA2, SA4, SA10) require approximately the same time to apply their repairs, but require significantly more time than the pure random approach. Finally, we can see that an increase of the number of repair rules does not affect the random approach, since the time to pick a rule at random can be neglected. For the Simulated Annealing and Greedy approaches, the run-time increases sub-linearly in the number of repair rules: Firstly, each greedy-step requires probing all possible repair rules to pick the most promising one. Yet, this greedy choice is rewarded by reducing the number of total repair iterations that are required to fix the database, thus lowering the run-time.

6.3   Experiments on Quality of Repair

In Sect. 5.1 we established three different dissimilarity functions. The results of the experiments are shown in Fig. 3. The larger the dissimilarity, the lower the quality of the corresponding repair. The Euclidean and Maximum Distances almost always return the same values, as in most cases, a single trajectory is distorted at one segment only. For the Euclidean distance, we set the weights to $(1, 1, 0.5)$ in order to weight the location stronger, because the domain of latitude and longitude is smaller than the time domain of a workout in seconds. At first glance, it appears that the purely time distorting heuristics (**tdra** and

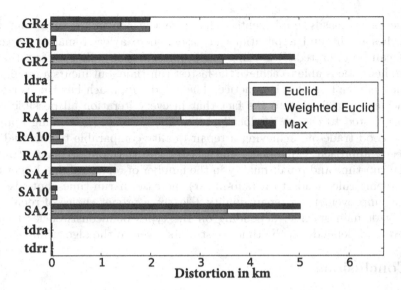

Fig. 3. Quality of repairs.

tdrr) and purely location distorting heuristics (ldra and ldrr) seem to yield a nearly perfect repair quality. However, this conclusion is misleading, since for this experiment, we were not able to consider the cases where ldra, ldrr, tdra and tdrr do not terminate. These cases however, are the interesting and hard cases, where the most distortion is required to repair the database. Despite this bias, which arises from the fact that ldra, ldrr, tdra and tdrr can not repair some collisions, we decided to keep the quality experiments for completeness. Another important observation that we can make in Fig. 3, is that for the repair quality of approaches utilizing several repair rules (Random, Greedy and Simulated Annealing), the repair quality improves significantly as the number of possible repair rules increases. In particular, the approach that allows to dodge collisions by distorting space in one of eight directions or by distorting time in one of two directions (the ten-repair-rule case) achieves an extremely high repair quality. When we compare the three heuristics to choose a repair rule, we see that the random heuristic performs by far the worst, thus leading to a large number of needless distortions. The greedy heuristic and the simulated annealing heuristic show comparable results. In fact, the simulated annealing approach yields a better quality in some cases. This is possible, as our greedy approach only selects the locally best next repair rule, which may not lead to the global best repair. In contrast, the simulated annealing allows to initially do quick random decisions to get rid of the majority of collisions, and then fix the remaining ones by using greedy decisions.

To summarize, our initial proposed repair rules using only spatial distortion (ldra and ldrr) and our proposed repair rules using only time distortion (tdra and tdrr) are not able to repair complex inconsistencies. Nevertheless, these approaches are easily implemented and have low run-times, such that these approaches might find applications in cases where a few remaining inconsistencies can be tolerated. Regarding our proposed repair rules, we saw that the random heuristic is able to achieve the fastest run-time, but incurs a repair-error that may not be tolerable in practice. The greedy approach has the worst run-time, which is attributed to the fact that in every iteration all possible repair rules are tested to choose the locally best. The simulated annealing approach yields a good trade-off, achieving a repair quality comparable to the quality of the greedy approach, while being much faster. Furthermore, we saw a trade-off between run-time and repair quality in the number of repair rules: a larger number of repair rules leads to a (sub-linear) increase in run-times but also to a (drastic) improvement of repair quality. Clearly, a proper choice of repair rules is highly domain specific, depending on the types of inconsistencies that are repaired, and depends on the time-constraints given to the algorithm.

7   Conclusions

In this work, we have formalized a category of problems that has been largely neglected in moving object literature – repairing inconsistencies in historical trajectory databases. This is an important problem since such databases are inherently uncertain for a number of reasons and, in addition, attempt to capture continuous phenomena via discrete values. We have shown that this problem is NP-hard, such that we aim at finding heuristics that find a good repair rather than finding the optimal repair. For this purpose, we presented a number of initial solutions, including a time-distortion algorithm, a space-distortion algorithm, as well as a set of generic algorithms that apply pre-defined repair rules, including a random algorithm, a greedy algorithm and a simulated annealing algorithm. Our experimental setting is aimed at one specific type of inconsistency, namely collisions. The results show that the simple approaches fail to find any repair at all. In contrast, our proposed repair-rule based solutions are able to find a good repair in acceptable time. We believe that this work will spur many challenges in identifying different domain-properties and corresponding heuristics to speed up the "fixings" for different constraints. While finding an optimal repair is a hard problem, we feel that a combination of the techniques presented in this work, as well as the consideration of new ideas, may yield a new solution that combines the best these worlds.

The problem of fixing inconsistencies in moving objects database becomes even more challenging when moving regions are involved. The removal of inconsistencies in such setting may have the potential of existing prediction models that are used in geo-sciences. Addressing the context of mobile regions is one part of our future work. We are also planning to investigate the trade-offs between fixing the inconsistencies in the data vs. fixing inconsistencies in the (answers to) pending queries – which can be challenging in the context of streaming (*location,time*) data. Another challenge that we plan to address is to investigate the impact – and efficient removal – of the inconsistencies in various spatio-temporal data mining tasks.

References

1. Mobile subscribers 2014: ITU World Telecommunication/ICT Indicators-database. http://www.itu.int/en/ITU-D/Statistics/Documents/facts/ICTFacts Figures2014-e.pdf
2. Achtert, E., Kriegel, H.-P., Schubert, E., Zimek, A.: Interactive data mining with 3d-parallel-coordinate-trees. In: Proceedings of the 2013 ACM SIGMOD International Conference on Management of Data, pp. 1009–1012. ACM (2013)
3. Arenas, M., Bertossi, L., Chomicki, J.: Consistent query answers in inconsistent databases. In: Proceedings of the Eighteenth ACM SIGMOD-SIGACT-SIGART Symposium on Principles of Database Systems, PODS 1999, pp. 68–79 (1999)
4. Bohannon, P., Fan, W., Flaster, M., Rastogi, R.: A cost-based model and effective heuristic for repairing constraints by value modification. In: Proceedings of SIGMOD, pp. 143–154 (2005)
5. Brinkhoff, T., Kriegel, T., Seeger, B.: Efficient processing of spatial joins using R-trees. In: Proceedings of the 1993 ACM SIGMOD International Conference on Management of Data, Washington, D.C., pp. 237–246, 26–28 May 1993
6. Cheng, R., Emrich, R., Kriegel, H., Mamoulis, N., Renz, M., Trajcevski, G., Züfle, A.: Managing uncertainty in spatial and spatio-temporal data. In: IEEE 30th International Conference on Data Engineering, Chicago, ICDE 2014, IL, USA, March 31 - April 4, 2014, pp. 1302–1305 (2014)
7. Emerson, E.: Temporal and modal logic. In: Handbook of Theoretical Computer Science, Volume B: Formal Models and Sematics (B) (1990)
8. Emrich, T., Kriegel, H.-P., Mamoulis, N., Renz, M., Züfle, A.: Querying uncertain spatio-temporal data. In: Kementsietsidis, A., Salles, M.A.V. (eds) ICDE, pp. 354–365. IEEE Computer Society (2012)
9. Gindele, T., Brechtel, S., Dillmann, R.: Learning driver behavior models from traffic observations for decision making and planning. IEEE Intell. Transport. Syst. Mag. **7**(1), 69–79 (2015)
10. Güting, R.H., Schneider, M.: Moving Objects Databases. Morgan Kaufmann, Amsterdam (2005)
11. Manyika, J., Chui, M., Brown, B., Bughin, J., Dobbs, R., Roxburgh, C., Byers, A.H.: Big data: The next frontier for innovation, competition, and productivity. McKinsey and Company Report, May 2009
12. Parisi, F., Grant, J.: Repairs and consistent answers for inconsistent probabilistic spatio-temporal databases. In: Straccia, U., Calì, A. (eds.) SUM 2014. LNCS, vol. 8720, pp. 265–279. Springer, Heidelberg (2014)
13. Parker, A., Subrahmanian, V., Grant, J.: A logical formulation of probabilistic spatial databases. IEEE Trans. Knowl. Data Eng. **19**(11), 1541–1556 (2007)
14. Pitoura, E., Samaras, G.: Locating objects in mobile computing. IEEE Trans. Knowl. Data Eng. (TKDE) **13**(4), 571–592 (2001)
15. Rozier, K., Vardi, M.: LTL satisfiability checking. In: Automated Technology for Verification and Analysis (2011)
16. Schiller, J., Voisard, A.: Location-Based Services. The Morgan Kaufmann Series in Data Management Systems. Morgan Kaufmann, San Francisco (2004)
17. Wijsen, J.: Database repairing using updates. ACM Trans. Database Syst. **30**(3), 722–768 (2005)
18. Zhang, B., Trajcevski, G.: The tale of (fusing) two uncertainties. In: Proceedings of the 22nd ACM SIGSPATIAL International Conference on Advances in Geographic Information Systems, Dallas/Fort Worth, TX, USA, pp. 521–524, 4–7 November 2014

A Spatio-Temporally Opportunistic Approach to Best-Start-Time Lagrangian Shortest Path

Sarnath Ramnath[2]([✉]), Zhe Jiang[1], Hsuan-Heng Wu[3],
Venkata M.V. Gunturi[1], and Shashi Shekhar[1]

[1] Computer Science and Engineering, University of Minnesota,
Minneapolis, USA
{zhe,shekhar,gunturi}@cs.umn.edu
[2] Computer Science and Information Technology, St. Cloud State University,
St. Cloud, MN, USA
rsarnath@stcloudstate.edu
[3] Department of Computer Science, National Tsing Hua University,
Hsinchu, Taiwan
wuxx1279@umn.edu

Abstract. The Best-start-time Lagrangian Shortest Path (BLSP) problem requires choosing the start time that yields the shortest path in a time-dependent graph. The inputs to the problem are a spatio-temporal network, an origin, o, a destination, d, and a discrete interval of possible start times. The solution is a path, P, and a start time, t, such that the total time taken to travel along P, starting at t, is no greater than the time taken to travel along any path from o to d, if we start in the given interval. The problem is important when the traveler is flexible about the start time, but would like to select a start time that minimizes the travel time. Its computational challenges arise from the large number of start time instants, and the manner in which the length of the shortest lagrangian path can vary from one start time instant to the next. Earlier work focused largely on finding the shortest path for a single start time. Researchers recently considered the BLSP problem, and proposed an approach based on finding the shortest lagrangian path for each start time, and then picking the best. Such an approach performs redundant evaluation of common sub-expressions, because time is explored in a sequential manner. We present an algorithm, BESTIMES, and propose an implementation that uses a Temporally Expanded priority queue. Our algorithm is built on the idea of "spatio-temporal opportunism", which allows us to navigate both space and time simultaneously in a non-sequential manner and appropriately combine sub-paths. Theoretical analysis and experiments on real data show that there is a well-defined range of inputs over which this approach performs significantly better than previous approaches.

1 Introduction

Given a spatio-temporal (ST) network, an origin(o), a destination(d), and a start-time interval, $[T_s \ldots T_f]$, the Best-start-time Lagrangian Shortest Paths

© Springer International Publishing Switzerland 2015
C. Claramunt et al. (Eds.): SSTD 2015, LNCS 9239, pp. 274–291, 2015.
DOI: 10.1007/978-3-319-22363-6_15

problem (BLSP) determines a path, P, and a start time t, $t \in [T_s \ldots T_f]$, such that for all paths from o to d, that leave o during the interval $[T_s \ldots T_f]$, P has the least travel time. BLSP tells us the *best start time*, i.e., the time at which we must start so that the total travel time is minimized, and also the path along which we must travel in order to attain this travel time.

Problem Example: The following example illustrates how the Lagrangian Shortest Path can vary with start time, in a time-dependent network. Figure 1(a) shows a spatio-temporal network, with nine possible start times. Our objective is to find the time instant at which we should leave O in order to minimize the travel time for reaching D.

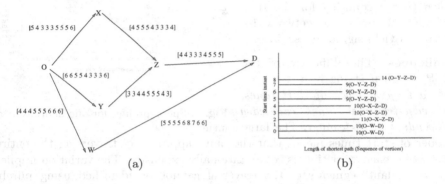

(a)   (b)

Fig. 1. Problem example: lagrangian shortest paths for different start times

Associated with each edge is a vector that lists the travel times for all possible start times. The edge (X, Z) has the vector $[4\,5\,5\,5\,4\,3\,3\,3\,4]$. The first value in the vector is 4, which means that if we start from X at time instant 0, it takes us 4 time steps to reach Z. The next three values are 5, which means that if we start at time instants 1, 2, or 3, it takes us 5 time steps to reach Z. To compute the length of a lagrangian path, consider the path $O - W - D$, when we start from O at time instant 0. The first value in each vector corresponds to time instant 0, and this value for edge (O, W) is 4. This means that we if we leave from O at time instant 0, we reach W at time instant 4. We then traverse the edge (W, D) starting at time instant 4. The vector for (W, D) is $[5\,5\,5\,5\,6\,8\,7\,6\,6]$, and the fifth value (which corresponds to a time instant of 4) in this vector is 6. Thus it will take us 6 time steps to traverse (W, D), and the total time to reach D along this path is 10 time steps. As it turns out, this is in fact the shortest lagrangian path from O to D, starting at time instant 0.

Figure 1(b) shows the shortest paths from O to D for each start time. As we can see, both the path length and the path vary with the start time. Here, the best start times are 5, 6 or 7, when the path $O - Y - Z - D$ has length 9. It should be noted that not only does the shortest path change from one time instant to the next, but the length of a given path can change significantly as well. For instance, the length of path $O - Y - Z - D$ is 9 if we start at time instant 7, and 14 if we start at time instant 8.

Application Domain: Determining the shortest paths is important in many applications related to air travel, road travel, and other spatio-temporal networks [13]. Several examples can be found in the literature that show how the length of the shortest path between two points varies with time in a travel network [11,12,16]. Figure 2 shows the speed profiles for a particular highway segment in our dataset over a period of 30 days. As can be seen, the speed varies with the time of day.

The problem becomes more interesting in the context of multi-modal networks, which involve several modes of transport [19,23]. Studies have shown that accounting for the time dependent nature of the network [10, 23,24] can yield significant savings.

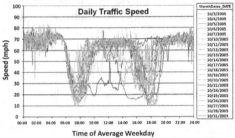

Fig. 2. Speed profiles for one highway segment

Challenges: The difficulty of the BLSP problem stems from two factors: *a large number of start times,* and *edge-traversal times can vary independent of each other.* The large number of start times means that the naive approach of traversing the entire network for each start time is computationally expensive. The variation implies that redundant segments of the traversal cannot be identified using purely local computation. In addition, we have to deal with the challenge of *non-stationarity*, which appears in all routing problems in spatio-temporal networks. Non-stationarity refers to the fact that the ranking of alternate paths between any particular source and destination pair in the network is not stationary. In other words, the optimal path between a source and destination for one start time may not be optimal for other start times.

Related work and their limitations: Time-dependent networks have been extensively modeled and studied to solve a wide variety of problems [8,20,21]. A significant body of research exists for the problem of finding the shortest lagrangian path for a single start time. The problem was

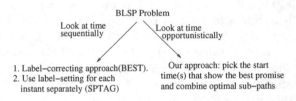

Fig. 3. Related work.

first posed in [6], and a label correcting algorithm was proposed. More recently, a variety of techniques such as $A^\star$ search [11], bi-directional search [22], use of landmarks [2], use of arc-flags [4], and hierarchical partitioning [17] have been applied to obtain faster algorithms for computing the shortest path in a time-dependent networks. A survey and a systematic analysis of such techniques can be found in [3].

The problem of finding the best start time in a time-dependent network was first considered by George et. al. [15]. They proposed an algorithm, BEST, which uses a label-correcting approach to find the shortest path for each start time in the interval, and then picks the best solution. Their research also evaluated the performance of an algorithm for finding the shortest path for a single start time. This algorithm, named SPTAG, used a label-setting approach. We can also use the label-setting approach to solve BLSP by trying out SPTAG for every instant in the given start-time interval. However, applying this method for each of a possibly large number of start times is expected to be computationally expensive. As part of this research we have done a comparison between all three approaches. Figure 3 shows the relationship between these approaches as applicable to the BLSP problem.

Contributions: This paper proposes the concepts of spatio-temporally opportunistic algorithms and temporally expanded priority queues. These concepts allow for a simultaneous exploration of both the spatial and temporal dimensions of time-dependent networks in a non-sequential manner. We apply these to design a spatio-temporal search algorithm, BESTIMES, for finding the Best Lagrangian Shortest Path and the best start time in a time-dependent travel network. The correctness of BESTIMES is proved and theoretical bounds on its complexity are established. The algorithm is experimentally evaluated using real data sets [1], and the results are compared against earlier approaches. Experimental results correlate well with our analysis, and show that the proposed algorithm outperforms earlier approaches for a well-defined range of inputs.

Scope: The scope of our research is limited to evaluating the performance of our algorithm, as compared to previous approaches for the BLSP problem. We restrict ourselves to the discrete time domain. We do not explore any other techniques such as bi-directional search or network partitioning [3].

Outline: The rest of the paper is organized as follows. Section 2 presents a formal problem definition and briefly reviews some basic concepts. Section 3 gives a formal description of the algorithm with a sample trace. In Sect. 4, we prove the correctness and analyze the complexity of our algorithm. Section 5 describes the results of our experimental evaluation. Section 6 presents a discussion of these results and some issues related to the use of specialized data structures for time-dependent algorithms. Section 7 concludes the paper.

2   Problem Definition and Basic Concepts

The Model Being Used: There are two models commonly used to represent a time-dependent travel network [16]: the Time Expanded Graph, or TEG, and The Time Aggregated Graph, or TAG. Both models are built around the notion of a finite time sequence, T, of all possible time instants. The TEG has a vertex for each node at every time instant. Thus, corresponding to node u, we have vertices $(u, 0)$, $(u, 1)$, $(u, 2)$,..., $(u, T^L - 1)$, where T^L is the length of the time sequence. Suppose it takes five time steps to travel from node u to node v, if we leave u at time instant 3. The TEG then contains and edge from vertex $(u, 3)$ to vertex $(v, 8)$. The TAG provides a more compact notation. Associated with each edge is a vector of length T^L, which gives the travel time corresponding to each start time. It was shown in [15] that the TAG model provided better performance than the TEG model for traversal algorithms. We will therefore use the TAG model throughout this paper. An example of a TAG was seen in Fig. 1(a).

Problem Definition:

Inputs:

(a) Spatio-temporal(ST) network $G = (V, E)$, where V is the set of vertices, and E is the set of edges;
(b) An origin o and a destination d $o, d \in V$;
(c) A discrete time interval, $T_{in} = [T_s \ldots T_f]$ over which best start time for the path between o and d is to be determined;
(d) The network has an associated edge cost function, denoted as *len*, and $len(u, v, t)$ represents the time taken to traverse (u, v), if we start from u at time instant t. The cost of an edge represents the time required to travel on that edge. The cost function of an edge repeats after every T time instances, i.e., $len(u, v, t)$ equals $len(u, v, t+T)$. It is therefore a *cyclic time series* with integer values.

Output: A route, P, from o to d, and a start time $\alpha \in T_{in}$.
Objective function: P is the shortest lagrangian path between o and d, if the travel starts at a time instant in T_{in}.
Constraints on the Input: The time horizon of the ST network is finite. The edge cost function is a cyclic, integer time series, with positive (non-zero) values.

How Our Approach Yields Efficiency: Before we formalize our process and design experiments, it is useful to have an intuitive under- standing of how computational savings can be realized by using our approach. Consider again the example shown in Fig. 1(a). Our objective is to find the time instant at which we should leave O in order to minimize the travel time for reaching D. We present a short synopsis of how spatio-temporal opportunism works in this case. (See details at http://web.stcloudstate.edu/rsarnath/bestimes.pdf.) The travel-time series for (O, X) is [5 4 3 3 3 5 5 5 6]. The quickest option for this edge is therefore, to leave O at time instant 2, 3 or 4. The travel-time series for (O, Y) is [6 6 5 5 4 3 3 3 6], which means the quickest option is time instant 5, 6 or 7. If we expand these best choices further (as would be dictated by an opportunistic approach), we see that the path $O - X - Z$ has length 6 for arrival time interval $[8 \ldots 10]$ and the path $O - Y - Z$ has length 6 for the arrival time interval $[11 \ldots 13]$. This means that these two sub-paths can be combined to tell us that we have a set of paths from O to Z of length 6 for the arrival time interval $[8 \ldots 13]$. Combining this information with the travel times for the edge (Z, D), we see that there is a path of length 9 arriving at D during the interval $[14 \ldots 16]$. We have therefore avoided traversing the entire network for those start times from O, when the travel times for edges (O, X) and (O, Y)

were 4, 5 or 6. The spatial aspect of the opportunism is demonstrated by picking the vertex with the most promising path, and temporal aspect is demonstrated by picking the best set of time instances for each vertex. The difficulty is that multiple distinct temporal routes can arrive at the same vertex and these must be appropriately merged to manage the complexity.

3   A Spatio-Temporally Opportunistic Algorithm

Notation:
T: the time series associated with each edge.
T^L: length of T.
T_{in}: set (range) of possible start times, $T_s, \ldots, T_f$.
T_{in}^L: length of T_{in}.
o: vertex of origin.
d: destination vertex.
L: length of shortest path from o to d that starts at o at some time instant in T_{in}.
$len(u, v, t)$: time to traverse the edge (u, v) starting at t.
   The algorithm keeps track of the following:

1. *Spatio-temporal labels.* The algorithm creates **spatio-temporal labels** (or simply **labels**). Each label is associated with a vertex, v, and tells us that a path of a certain length, starting at the source(o), and arriving at v a certain time(or a set of times) has been discovered in our traversal. For some such path, $o, x_1, x_2, \ldots, x_k, v$, the penultimate vertex, x_k, is also stored in the label.
2. A_v: When we process a label, we are looking at a vertex, v, and a set of time instants. It may so happen, that after the label was enqueued, other shorter paths to v were discovered. Therefore shorter paths to v may be known for some of the time instants. A_v helps us track this by storing a collection of all the lengths of shortest paths arriving at different time instants at the vertex v. Let t be a time instant, such that a path has been found from o to v, arriving at v at time t. Let l be the length of the shortest such path. Then $A_v[t]$ stores the value l. For all vertices v(except o), and time instants t, $A_v[t]$ is initialized to ∞. A_o is initialized to zero for all t.
3. S_v^l: Several distinct paths, arriving at vertex v at different time instants, could have the same length. To handle this, we keep a collection of arrival times when a path of length l arrives at v. With each arrival instant, t, we store the penultimate vertex on the corresponding path. Formally,
   $S_v^l = \{(t, u) \mid a \ path \ of \ length \ l, \ reaches \ v \ at \ time \ t, \ through \ (u, v)\}$.
4. *A priority queue.* The priority contains all the spatio-temporal labels generated by the process. Each label consists of a vertex, v, the length, l, of the paths reaching v, and the set of times, S_v^l, at which these paths arrive at v. The priority queue is a collection of 3-tuples of the form (v, l, S_v^l).

   Algorithm 1 formally describes the algorithm BESTIMES. Figure 4 shows a simple network with four nodes and traces the algorithm for the first few steps:

Algorithm 1. BESTIMES Algorithm

1: **Inputs:** A graph G with a set of nodes V and edges E, where each edge has a travel-time series ($len(u, v, t)$ is the time taken to traverse edge (u, v) starting at time t), an origin o, a destination d, and a set of possible start times, T_{in}. We assume that for all u, v, and t, $len(u, v, t) > 0$.

2: **Output:** The shortest path from o to d that starts at some instant in T_{in}.

3: Initialize the collections A_v and S_v^l. /*INITIALIZE*/

4: Insert the label $(o, 0, S_o^0)$ in priority queue. /*INITIALIZE*/

5: **while** the shortest path has not been found /*MAIN LOOP*/ **do**

6:     (u, l, S_u^l) = Dequeue()

7:     **if** $u == d$ **then**

8:         For any pair (t, x), $(t, x) \in S_u^l$ report that shortest path to d departs from o at time $t - l$. Extract the shortest path recursively by extracting the shortest path to x. **EXIT**.

9:     **end if**

10:    **for** each t, such that $(t, x) \in S_u^l$ /*PROCESS EACH ARRIVAL TIME*/ **do**

11:        **if** $A_u[t] == l$ /*l is the least time needed to reach u at t*/ **then**

12:            **for** each edge (u, v) /*PROCESS EACH OUTGOING EDGE*/ **do**

13:                $t_1 = (t + len(u, v, t))$

14:                $l_1 = A_v[t_1]$

15:                **if** $l_1 > l + len(u, v, t)$ **then**

16:                    /*a shorter path is found for reaching v at t_1*/

17:                    $A_v[t_1] = (l + len(u, v, t))$ /*$A_v[t_1]$ is updated*/

18:                    Insert($S_v^{l+len(u,v,t)}$, (t_1, u))

19:                    Enqueue($v, l + len(u, v, t), S_v^{l+len(u,v,t)}$)

20:                **end if**

21:            **end for**

22:        **end if**

23:    **end for**

24: **end while**

---

1. Initially we have the label [o1: (o, 0,{})], which represents that we are at vertex o, and the shortest path arriving here is of length zero.

2. When we process this label, we see from the travel time sequence for the edge (o, x) that we can reach x in 3 time steps if we leave at any of the time instants 0, 1 or 2, and arrive at time instants 3, 4 or 5, respectively. This is captured by the label [x1: (x, 3, {(3, o), (4, o), (5, o)})]. The information associated with this label is a 3-tuple which gives us the vertex, x, the path length, 3, and the arrival time instants, 3, 4, or 5. Along with each arrival time, we store the penultimate vertex on the path. This information is useful to retrieve the path after the computation is complete. Likewise, we see from the travel time sequence for the edge (o, x) that we can reach x in 4 time steps if we leave at any of the time instants 3, 4, 5, 6, 7 or 8, and arrive at time instants 7, 8 9, 10, 11 or 12, respectively. However, our time series is of length 9, 12 is the same as 3, modulo 9; as we already have a path of length 3 arriving at time instant 3,

we do not record the path of length 4. Accordingly, we get the label
[x2: (x, 4, {(7, o), (8, o), (9, o), (10, o), (11, o)})]

3. Next we look at the edge (o, y), and from the travel time sequence, we generate three labels: [y1: (y, 3, {(9, o), (10, o) (11, o)})],
[y2: (y, 4, {(7, o), (8, o)})],
and [y3: (y, 5, {(5, o), (6, o)})] This completes the processing of label o1, and leaves us with five labels in our priority queue.

4. In the next step, we pick the label with the shortest path from the priority queue and process it. This could be either x1 or y1, since both have length 3. Say we choose y1. When we process this, two more labels are added, [d1: (d, 8, {(16, y)})], and [d2: (d, 9, {(15, y)})].

5. The next choice has to be x1, which adds the labels [d3: (d, 6, {(6, x)})] and [d4: (d, 10, {(11, x), (12, x)})].

6. Note that the label d3 captures the shortest path. However, the algorithm must first process x2, y2, and y3 before it picks d3.

4   Analysis of the Algorithm

For a complete version with details of proofs see full paper at: http://web.stcloudstate.edu/rsarnath/bestimes.pdf.

4.1   Correctness

Since the label (u, l, S_u^l) is associated with a path of length l from o to u, and the queue is prioritized by l, we have the following lemma.

Lemma 1. *Suppose that in the algorithm BESTIMES, the label (u, l, S_u^l) is dequeued before $(u', l', S_{u'}^{l'})$. Then $l \leq l'$.*

Lemma 2 follows from the manner in which $A_v[t]$ is updated.

Lemma 2. *The values stored in the arrays A_v, for any vertes v, can only decrease or remain the same during the course of the algorithm.*

Lemma 3. *Suppose that in the algorithm BESTIMES, the label (u, l, S_u^l) is dequeued, and inner FOR loop is executed for some arrival time t. Then, the shortest path from o to u, starting at some instant in $T = [T_s \dots T_f]$, and reaching u at time t, has length l.*

From the above lemmas, we get the following result.

Theorem 1. *Of all the paths from o to d that leave d at some instant in $T = [T_s \dots T_f]$, the algorithm BESTIMES returns a path from o to d that has the least travel time.*

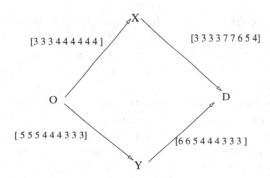

Execution Trace, showing the set of labels after each label is expanded:

Initial set of labels: [o1: (o, 0,{ })]

After expanding o1: [x1: (x, 3, {(3, o), (4, o), (5, o)})], [x2: (x, 4, {(7, o), (8, o), (9, o), (10, o), (11, o)})],
[y1: (y, 3, {(9, o), (10, o) (11, o)})], [y2: (y, 4, {(7, o), (8, o)})], [y3: (y, 5, {(5, o), (6, o)})]

After expanding y1: [x1: (x, 3, {(3, o), (4, o), (5, o)})], [x2: (x, 4, {(7, o), (8, o), (9, o), (10, o), (11, o)})],
[y2: (y, 4, {(7, o), (8, o)})], [y3: (y, 5, {(5, o), (6, o)})], [d1: (d, 8, {(16, y)})],
[d2: (d, 9, {(15, y)})]

After expanding x1: [x2: (x, 4, {(7, o), (8, o), (9, o), (10, o), (11, o)})],[y2: (y, 4, {(7, o), (8, o)})],
[y3: (y, 5, {(5, o), (6, o)})], [d1: (d, 8, {(16, y)})], [d2: (d, 9, {(15, y)})]
[d3: (d, 6, {(6, x)})], [d4: (d, 10, {(11, x), (12, x)})]

Fig. 4. Execution trace of BESTIMES for a simple example

4.2   Complexity

A naive analysis (i.e., multiplying the maximum number of iterations of each loop) gives an incorrect (possibly exponential) result. This can be improved, since the FOR loops may not be executed for each iteration of the WHILE loop. A more careful analysis, placing bounds on the number of iterations is provided here.

Lemma 4. *If n is the number of nodes, and L is the length of the shortest path from o to d returned by the algorithm, then the WHILE loop executes at most nL times.*

In a worst case situation, we could have nT^L iterations of the inner FOR loop in a single iteration of the WHILE loop. This would give us a loose bound of n^2LT^L on the complexity of the algorithm. This bound can be significantly reduced by placing independent bounds on the total number of times that the inner loops can execute in the course of the entire algorithm. These bounds are described in the following lemmas.

Lemma 5. *Let m be the number of edges, and T^L be the length of the time series associated with each edge. Algorithm BESTIMES can be implemented so that the inner FOR loop executes at most mT^L times.*

Lemma 6. *Let m be the number of edges, and L be the length of the shortest path returned. In the Algorithm BESTIMES, the inner FOR loop executes at most $m(L + T_{in}^L)$ times.*

The enqueue operation is invoked only when the inner FOR loop is executed. The following lemma follows immediately from Lemma 6.

Lemma 7. *Let m be the number of edges, and L be the length of the shortest path returned. Algorithm BESTIMES performs at most $m(L + T_{in}^L)$ enqueue operations.*

The number of dequeue operations is obviously bounded by the number of enqueue operations. We can therefore combine Lemmas 4 through 7 to get the following:

Lemma 8. *Let $P = min(T^L, L + T_{in}^L)$. Algorithm BESTIMES performs at most mP dequeue operations and at most mP executions of the inner FOR loop.*

Theorem 2. *Let $P = min(T^L, L + T_{in}^L)$. The Algorithm BESTIMES can be implemented such that the total number of operations performed by the algorithm is $O(mP(log\,m + log\,P))$.*

Proof. From Lemma 8 we know that the WHILE loop and the inner FOR loop both execute at most mP times. To deal with the case where the outer FOR loop executes but the inner FOR loop does not, note that this happens when there is a pair, $(t, x) \in S_u^l$, such that $(A_u[t] \neq l)$. However, for any such pair, such an event will happen exactly once. Hence the total number times that the outer FOR loop executes, but the inner FOR loop does not, cannot exceed the number total number of Insert operations, i.e., the total number of iterations of the inner FOR loop. All the steps in the algorithm can be done in $O(1)$ time or in logarithmic time using standard data structures [7] and our result follows.

Note that in the case where T^L equals 1, we get a complexity of $O(|E| \, log \, |E|)$. For road networks, this is asymptotically the same as that of Dijkstra's algorithm (which is $O(|E| + |V| \, log \, |V|)$), since the nodes have bounded degree.

5 Experimental Evaluation

The goal of the experimental evaluations was twofold. The first goal was to compare the computational performance of the proposed BLSP algorithm, BESTIMES, with the existing BEST and SPTAG algorithms, as different parameters were varied. Theoretical analysis tells us that the performance should depend on the network size(n and m), the length of the travel time sequence(T^L), the length of the start time interval(T_{in}^L), and the travel time, i.e., the length of the shortest path(L). Our second goal was to see how well our experimental results correlate with the theoretical analysis.

5.1   Experiment Setup

Experiment Design: The experiment design is illustrated in Figure 5. Since we ran the experiments on real data, the network size(n and m) and the length of the travel time sequence(T^L) were fixed. In the comparative experiments, we therefore varied three parameters: the start time point of query time interval (T_s), the length of query time interval (T_{in}^L), and the approximate travel time between the source node and the destination node (roughly equal to L). All the algorithms were implemented in C++. Experiments were conducted on a iMac with 8 GB memory and Intel Core i7 CPU with 4 cores. The time costs are the sums of 50 runs.

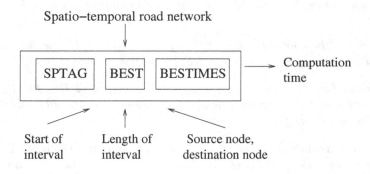

Fig. 5. Experiment design

Dataset Description: The experiments were carried out on a real data set containing the highway road network of Hennepin county, Minnesota, provided by NAVTEQ [1]. The dataset contained 1417 nodes and 3754 edges. It also contained travel times for each edge at time quanta of 15 min. For experimental purposes, the travel times were converted into time quanta of 1mins by linear interpolation.

5.2   Experiment Results

Effect of Length of Start Time Interval (T_{in}^L): We fixed the start time $(T_s$ as non-rush hour at 0:00 a.m.) and the travel times. We increased the length of the query time interval from 100, 200, 300, 400, 500, to 1000, 2000, 3000, 4000, 5000. Figure 6(a–c) shows the results with travel times fixed at 30, 50, and 150 min respectively. As can be seen, the BESTIMES algorithm has lower costs than the SPTAG and BEST algorithms when the travel times are small, e.g., 30 min and 50 min. The reason why BESTIMES is faster than SPTAG is that it considers multiple time instances at the same time while SPTAG computes shortest path repeatedly for every start time. The BEST algorithm has significantly larger costs than the BESTIMES algorithm when the travel time is small. The reason is that BEST uses a label correcting approach and iterates through

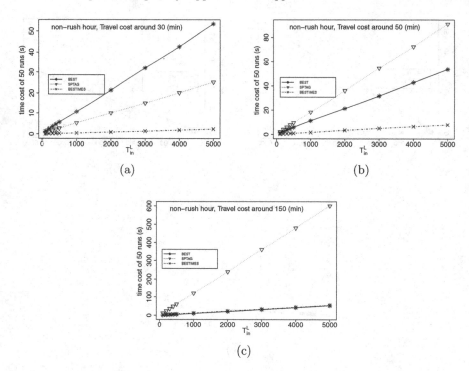

Fig. 6. Computational performance comparison with different T_{in}^L starting at non-rush hour

entire network, whereas BESTIMES is label-setting algorithm which stops as soon as destination is reached. When the travel time from source and destination is large, e.g., 150 min, BESTIMES and BEST seem to have the same cost. This result is somewhat surprising, since the theoretical complexity of BEST is $O(n^2mT)$ [15], which is much worse than that of BESTIMES. However, it has been pointed out [9] that label-correcting approaches on road networks converge much faster than what the graph-theoretic bounds predict for general graphs.

Effect of Network Traffic: To test the effect of traffic volume, we re-conducted the analysis shown above with a different start time of 420th minute (7 a.m. in the morning, which is during the rush hour). This comparison is useful to test whether the cost reduction achieved by BESTIMES is significantly affected when the traffic pattern changes. The results are shown in Fig. 7. Comparing Figs. 6 and 7 together, we can see that this change does not seem to influence the computational performance significantly in the experiment.

Effect of Different Travel Times: We fixed a non-rush hour (0 min, midnight at 0:00 a.m.) time for the start of the query interval(T_s) and varied the length of the query time interval (T_{in}^L was set to 100, 1000, and 5000 respectively). We increased the travel times, i.e., length of the shortest path, from 10, 20, to 150 by changing the destination node. Figure 8(a–c) shows the results of fixing T_{in}^L

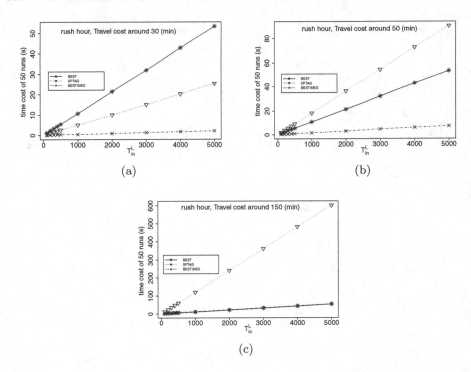

Fig. 7. Computational performance comparison with different T_{in}^L starting at non-rush hour

at 100, 1000, and 5000 respectively. As can be seen, the cost of BEST algorithm does not change with the travel time. This is consistent with the fact that BEST is a label correcting algorithm, updating the entire network. In contrast, the computational costs of both SPTAG and BESTIMES increase with travel times, with SPTAG being much higher. When the travel times are small, e.g., less than 40 min, the computational costs of SPTAG and BESTIMES are smaller than BEST algorithm. When the travel times increase, the computational cost of BESTIMES approaches that of BEST, due to the small network size (i.e., around 1400 nodes).

6   Discussion

The results in Sect. 5 clearly show that when the travel time (L) is small compared to the size of the network (n), the proposed BESTIMES algorithm is significantly faster than both the SPTAG and the BEST algorithms. As the travel time increases the performances of BEST and BESTIMES converge. This clearly shows that the spatio-temporally opportunistic approach yields an improved algorithm. A few other issues are discussed below:

Improving the Efficiency of the BESTIMES Implementation: Instead of a set of time instants for each label we could treat them as sub-intervals of

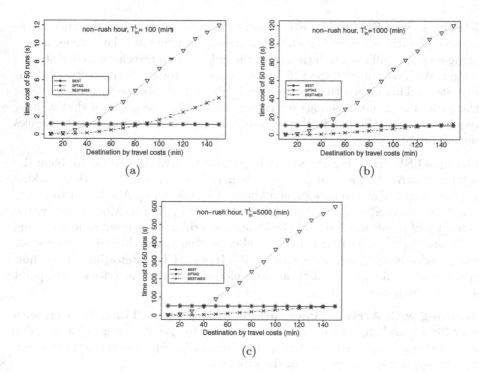

Fig. 8. Computational performance comparison with different travel times starting at non-rush hour

time instants. This could save us some redundancy, but will incur an additional overhead for maintaining data structures that deal with interval operations.

Finding Multiple Solutions: The algorithm can be easily adapted to find, say, the ten best start times. To do this, we do not terminate the process upon reaching d, but continue until we have dequeued the labels necessary for locating the first ten paths. Such an adaptation is useful for a system that provides the traveler with several choices in response to a query.

Enhancing Performance of BESTIMES: Since BESTIMES is a label-setting algorithm, it seems possible to enhance it by employing techniques like A*, partial pre-computation and bi-directional search. This would make it a useful search technique to build practical tools.

Discrete vs Continuous Model: Some work has been done on using a continuous model of time-dependent networks, in which we have a continuous time series. An algorithm for the continuous model was presented in [18], which reports all the shortest paths between two query vertices for a continuous time interval. It has been pointed out that for some cases their algorithm does not converge [11,14], and some of the issues with [18] were addressed in [12]. Since the actual data that we are using is discrete, and the models are very different, it is very difficult to compare our results with these.

FIFO vs Non-FIFO: The FIFO assumption [16] states that the time series associated with any edge is such that a traveler starting at a later time instant along an edge, will not reach the end of the link before a traveler who had started earlier. We have not used the FIFO assumption in proving the correctness of our algorithm. This means that no waiting is allowed at intermediate nodes along the path. This is not a serious restriction, since it has been shown that waiting requirements can be handled by suitably modifying the travel time sequences during pre-processing [5, 16].

Using ALSP to Solve BLSP: It is possible to solve the BLSP problem, by solving the All Lagrangian Shortest Path (ALSP) problem and then picking the best time. This requires some additional bookkeeping. A solution to ALSP was first proposed in [5], using a label-correcting approach. More recently, the efficiency of ALSP was improved by identifying critical time points and removing redundancy [16]. Since BEST uses a label-correcting approach, we have in essence compared our approach with that of [5]. It would be interesting to see how spatio-temporally opportunistic algorithms compare with the critical time point approach presented in [16].

Working with Arrival Times Instead of Departure Time: Travelers with very little knowledge of the network may prefer to specify a range of acceptable arrival times. It appears that the spatio-temporally opportunistic approach can also be applied to this version of the problem.

6.1   Temporally-Expanded Priority Queues

The role of priority queues can be generalized in the context of algorithms for time-dependent networks.

Table 1 illustrates the properties and operations of a temporally-expanded (TE) priority queue in contrast to a regular priority queue. Unlike the scalar elements in a regular priority queue, an element of a TE priority queue is a time-series of values. A TE priority queue is ordered on the value of the component time series at a particular time index. This time instant, t_{arg}, could be the same across all the keys or be different for each key. Apart from the standard priority queue operations such as insert, extract-min, and update-key, we have two new operations called "Retrieve-keys" and "forecast-critical-time-point." "Retrieve-keys" returns all the keys whose key-label matches the input id. "forecast-critical-time-point" can be called at the end of each extract-min, when it compares the time-series corresponding to the extract-min with the other time-series in the queue to determine the maximum time duration (beyond t_{arg}) for which the chosen time-series has the lowest value. Temporally-expanded priority queues can be used both for our BESTIMES algorithm and for the CTAS algorithm for the ALSP problem proposed in [16], as follows:

Implementation for BESTIMES: Temporally-expanded priority queues can be used implement the S_v^l's which are inserted into the priority queue of the BESTIMES algorithm. Here, a key in the TE priority queue would contain contiguous segments of arrival times at the node v, all of which have the same travel

Table 1. Priority queue vs temporally-expanded priority queue.

Properties/ operations	Priority Queue	Temporally-expanded Priority Queue
Keys	Scalar values	a tuple containing (key-label, t_{arg}, time-series of values)
Ordering	Ascending or Descending	Ascending or Descending on the value of time series in the keys at time index t_{arg}
Insert operation	Inserts scalar value	Inserts a time-series
Update-Key operation	Updates a scalar value	Updates an entire time-series
Extract-Min	Returns a scalar with lowest cost	Returns a time-series with lowest cost
Retrieve-keys(id)	--Not Available--	Retrieves all keys with certain key-id
Forecast Critical Time-Point	--Not Available--	Returns Maximum time $t_{ctp}(>$ t_{arg}) for which the previous extract-min has least value among all keys in the queue

time. The queue is ordered on this travel time. Here, t_{arg} could be set to any time index within this contiguous segment of arrival time series at node v.

Implementation for CTAS: Temporally-expanded priority queues can be used to fulfill priority queue requirements (line number 5 in Algorithm 1 of [16]) of the CTAS algorithm. For this implementation, all the keys in the TE priority queue would be ordered on the same time index. In other words, t_{arg} would be the same for all the keys. After each extract-min (line number 7 in Algorithm 1 of [16]), the CTAS algorithm could determine a potential critical-time-point using the "forecast-critical-time-point" operation (line number 8 in Algorithm 1 of [16]).

7   Conclusion and Future Work

In this paper we proposed the concept of spatio-temporally opportunistic algorithms, and used this concept to design an algorithm that finds the best lagrangian start time. This problem is an important component of applications in transportation networks. Traditional approaches have treated time in a sequential manner, which results in redundancy. Our experimental results have shown a well-defined range of inputs on which our algorithm performs better than earlier algorithms.

There are two ways in which we could enhance the performance of our algorithm. One way is to develop efficient TE data structures that can be fine-tuned to the mix of operations performed in a spatio-temporal approach. The other way is to apply some of the techniques cataloged in [3]. Bi-directional search could significantly reduce the region of the network to be searched. It would also be useful to incorporate the algorithm into other existing systems to study the performance.

Acknowledgment. We are grateful to the members of the Spatial Database Research Group at the University of Minnesota and Dr Betsy George for their valuable feedback, and Kim Koffolt for editing help. This material is based upon work supported by the National Science Foundation under Grant No. 1029711, IIS-1320580, 0940818 and IIS-1218168, the USDOD under Grant No. HM1582-08-1-0017 and HM0210-13-1-0005, and the University of Minnesota under the OVPR U-Spatial.

References

1. Oracle spatial and graph. http://www.oracle.com/technetwork/database/options/spatialandgraph/downloads/navteq-lic-168395.html
2. Delling, D., Wagner, D.: Landmark-based routing in dynamic graphs. In: Demetrescu, C. (ed.) WEA 2007. LNCS, vol. 4525, pp. 52–65. Springer, Heidelberg (2007)
3. Delling, D., Wagner, D.: Time-dependent route planning. In: Ahuja, R.K., Möhring, R.H., Zaroliagis, C.D. (eds.) Robust and Online Large-Scale Optimization. LNCS, vol. 5868, pp. 207–230. Springer, Heidelberg (2009)
4. Bauer, R., Delling, D.: Sharc: fast and robust unidirectional routing. J. Exp. Algorithmics **4:2.14**, 4:2.4–29 (2010)
5. Chabini, I.: Discrete dynamic shortest path problems in transportation applications: complexity and algorithms with optimal run time. Transp. Res. Rec. **1645**, 170–175 (1998)
6. Cooke, K.L., Halsey, E.: The shortest route through a network with time-dependent internodal transit times. J. Math. Anal. App. **14**, 493–498 (1966)
7. Cormen, T.H., Leiserson, C.E., Rivest, R.L., Stein, C.: Introduction to Algorithms. MIT Press, Cambridge (2001)
8. Costa, C.F., Nascimento, M.A., de Macêdo, J.A.F., Machado, J.C.: Nearest neighbor queries with service time constraints in time-dependent road networks. In: Proceedings of 2nd ACM SIGSPATIAL MobiGIS 2013, Orlando, Florida, USA, pp. 22–29 (2013)
9. Dean, B.C.: Shortest paths in fifo time-dependent networks: theory and algorithms. Technical report (2004)
10. Demiryurek, U., Banaei-Kashani, F., Shahabi, C.: A case for time-dependent shortest path computation in spatial networks. In: Proceedings of 18th SIGSPATIAL International Conference on Advances in Geographic Information Systems, GIS 2010, pp. 474–477 (2010)
11. Demiryurek, U., Banaei-Kashani, F., Shahabi, C., Ranganathan, A.: Online computation of fastest path in time-dependent spatial networks. In: Pfoser, D., Tao, Y., Mouratidis, K., Nascimento, M.A., Mokbel, M., Shekhar, S., Huang, Y. (eds.) SSTD 2011. LNCS, vol. 6849, pp. 92–111. Springer, Heidelberg (2011)

12. Ding, B., Yu, J.X., Qin, L.: Finding time-dependent shortest paths over large graphs. In: Proceedings of 11th International Conference on Extending Database Technology (EDBT), pp. 205–216 (2008)
13. Evans, M.R., Yang, K., Kang, J.M., Shekhar, S.: A lagrangian approach for storage of spatio-temporal network datasets: a summary of results. In: Proceedings of 18th SIGSPATIAL International Conference on Advances in GIS, GIS 2010, pp. 212–221 (2010)
14. Foschini, L., Hershberger, J., Suri, S.: On the complexity of time-dependent shortest paths. In: Proceedings of the Twenty-Second Annual ACM-SIAM Symposium on Discrete Algorithms, SODA 2011, pp. 327–341 (2011)
15. George, B., Kim, S., Shekhar, S.: Spatio-temporal network databases and routing algorithms: a summary of results. In: Papadias, D., Zhang, D., Kollios, G. (eds.) SSTD 2007. LNCS, vol. 4605, pp. 460–477. Springer, Heidelberg (2007)
16. Gunturi, V.M.V., Nunes, E., Yang, K.S., Shekhar, S.: A critical-time-point approach to all-start-time lagrangian shortest paths: a summary of results. In: Pfoser, D., Tao, Y., Mouratidis, K., Nascimento, M.A., Mokbel, M., Shekhar, S., Huang, Y. (eds.) SSTD 2011. LNCS, vol. 6849, pp. 74–91. Springer, Heidelberg (2011)
17. Jagadeesh, G., Srikanthan, T., Quek, K.: Heuristic techniques for accelerating hierarchical routing on road networks. IEEE Trans. Intell. Transp. Syst. **3**(4), 301–309 (2002)
18. Kanoulas, E., Du, Y., Xia, T., Zhang, D.: Finding fastest paths on a road network with speed patterns. In: Proceedings of the 22nd International Conference on Data Engineering (ICDE), p. 10 (2006)
19. Kirchler, D., Liberti, L., Calvo, R.W.: Efficient computation of shortest paths in time-dependent multi-modal networks. J. Exp. Algorithmics **19**, 1–29 (2015)
20. Ma, Y., Yang, B., Jensen, C.S.: Enabling time-dependent uncertain eco-weights for road networks. In: Proceedings of Workshop on Managing and Mining Enriched Geo-Spatial Data, SIGMOD 2014, p. 1 (2014)
21. Mouratidis, K., Yiu, M.L., Papadias, D., Mamoulis, N.: Continuous nearest neighbor monitoring in road networks. In: Proceedings of 32nd International Conference on Very Large Data Bases, pp. 43–54, September 2006
22. Nannicini, G., Delling, D., Liberti, L., Schultes, D.: Bidirectional A^* search for time-dependent fast paths. In: McGeoch, C.C. (ed.) WEA 2008. LNCS, vol. 5038, pp. 334–346. Springer, Heidelberg (2008)
23. Pinelli, F., Hou, A., Calabrese, F., Nanni, M., Zegras, C., Ratti, C.: Space and time-dependant bus accessibility: a case study in Rome. In: 12th International IEEE Conference on Intelligent Transportation Systems (2009)
24. Yuan, J., Zheng, Y., Zhang, C., Xie, W., Xie, X., Sun, G., Huang, Y.: T-drive: driving directions based on taxi trajectories. In: Proceedings of 18th SIGSPATIAL International Conference on Advances in GIS, GIS 2010 (2010)

Privacy and Matching

Combining Differential Privacy and PIR for Efficient Strong Location Privacy

Eric Fung, Georgios Kellaris, and Dimitris Papadias[✉]

Hong Kong University of Science and Technology, Hong Kong, China
{khfungac,gkellaris,dimitris}@cse.ust.hk

Abstract. Data privacy is a huge concern nowadays. In the context of location based services, a very important issue regards protecting the position of users issuing queries. Strong location privacy renders the user position indistinguishable from any other location. This necessitates that every query, independently of its location, should retrieve the same amount of information, determined by the query with the maximum requirements. Consequently, the processing cost and the response time are prohibitively high for datasets of realistic sizes. In this paper, we propose a novel solution that offers both strong location privacy and efficiency by adjusting the accuracy of the query results. Our framework seamlessly combines the concepts of ϵ-*differential privacy* and *private information retrieval* (PIR), exploiting query statistics to increase efficiency without sacrificing privacy. We experimentally show that the proposed approach outperforms the current state-of-the-art by orders of magnitude, while introducing only a small bounded error.

1 Introduction

Mobile devices enable the use of location based services (LBS) in order to facilitate everyday tasks. An LBS allows users to issue queries along with their locations to a server, which in turn replies with the results. For example, a user may ask for the closest gas station to his current location, the shortest path from his home to a shopping mall, real-time traffic condition in his area, and so on. Each of the queried locations, e.g., gas station, shopping mall, is called a Point of Interest (POI). However, location based queries raise privacy concerns, as they can reveal the sensitive location of the user. For example, a user may wish to find the nearest bar without revealing his presence in the specific area. In this work we focus on private k-Nearest Neighbor queries (kNN), which ask for the k nearest POIs to the user.

Numerous algorithms have been proposed for private kNN queries. Strong Location Privacy for kNN [23] is currently the only solution which renders the position of the user truly indistinguishable from all other possible locations. It leverages hardware PIR and a query plan. Hardware PIR ensures that the server is oblivious to the data acquired by the users, while the query plan requires that every user receives the same amount of information independent of the data size needed. By combining these two properties, the query process for any user from

© Springer International Publishing Switzerland 2015
C. Claramunt et al. (Eds.): SSTD 2015, LNCS 9239, pp. 295–312, 2015.
DOI: 10.1007/978-3-319-22363-6_16

any location appears exactly the same to the server. In order to guarantee that all users receive accurate answers, the algorithm of [23] sets the query plan as the maximum data size required by any possible query. Although this solution is viable for small databases, it becomes prohibitively expensive for a large number of POIs because the result size required to satisfy any query may be enormous.

In order to overcome this problem, we propose the *adaptive query plan*, which relaxes the need for answering all queries accurately. Instead, it computes a minimum data size, which guarantees that at least a predefined percentage of queries are answered correctly. The adaptive query plan depends on the actual user behavior and changes periodically based on statistics of previously issued queries. However, utilizing statistics on sensitive location data may reveal the whereabouts of a user [7]. In order to avoid this type of privacy breaches, we employ the notion of ϵ-differential privacy [8]. This concept offers theoretical privacy guarantees when publishing statistics on sensitive data. Our solution is applicable to [23], and in general to PIR techniques based on similar principles.

We demonstrate the efficiency and effectiveness of our approach by using rigorous secure hardware simulations on two real datasets consisting of millions of POIs. Compared to [23], it offers up to orders of magnitude better efficiency, while retaining high levels of accuracy, rendering it practical for large datasets.

2   Related Work

Section 2.1 describes the notion of ϵ-differential privacy. Section 2.2 reviews location privacy techniques in general and presents the implementation of the state-of-the-art method of [23].

2.1 ϵ-Differential Privacy

Differential privacy hides sensitive information about individual users when publishing statistics. Specifically, the published results are produced in a random way, so that the presence of any individual in the data has negligible impact. Let $\mathcal{D}$ be a set of finite databases with d attributes. Each $D \in \mathcal{D}$ is a set of rows. For example, each row of D represents a user, and each column a location. A cell $c_{i,j}$ of D is 1 if user i has visited location j, and 0 otherwise. Two databases $D, D' \in \mathcal{D}$ are considered **neighboring** if they differ in at most one row; essentially, D and D' differ in the locations of one user. A mechanism $\mathcal{M}$ is a randomized algorithm performed by the publisher; given a database D, $\mathcal{M}$ applies some functionality and outputs a *transcript* o.

Definition 1. *A mechanism* $\mathcal{M} : \mathcal{D} \to \mathcal{O}$ *satisfies* ϵ **-differential privacy** *if for all sets* $O \subseteq \mathcal{O}$, *and every pair* $D, D' \in \mathcal{D}$ *of neighboring databases*

$$Pr[\mathcal{M}(D) \in O] \leq e^{\epsilon} \cdot Pr[\mathcal{M}(D') \in O] \tag{1}$$

The smaller the value of ϵ, the stronger the privacy guarantees. Intuitively, $\mathcal{M}$ satisfies ϵ-differential privacy, if changing the attributes of one individual in the database has a negligible effect on the distribution of the output of $\mathcal{M}$.

A common differential privacy technique adds Laplace noise to the outputs using the Laplace Perturbation Algorithm (LPA [7,8]). Before presenting LPA, we formulate the notion of *sensitivity*. We view the release of statistical information as a query performed on the data. For example the query asks for the counts on each column of D, i.e., the number of users who visited each location. We model the query as a function $\mathbf{Q} : \mathcal{D} \to \mathbb{N}^d$, where d is the number of elements in the output. For $D, D' \in \mathcal{D}$, $\mathbf{Q}(D), \mathbf{Q}(D')$ are two d-dimensional vectors. Let $\|\mathbf{Q}(D) - \mathbf{Q}(D')\|$ be the L_1 norm of $\mathbf{Q}(D), \mathbf{Q}(D')$. Then, the *sensitivity* of $\mathbf{Q}$ is $\Delta(\mathbf{Q}) = \max_{D,D' \in \mathcal{D}} \|\mathbf{Q}(D) - \mathbf{Q}(D')\|$ for all neighboring $D, D' \in \mathcal{D}$.

Let $Lap(\lambda)$ be a random variable drawn from Laplace distribution with mean zero and scale parameter λ. LPA achieves ϵ-differential privacy through the mechanism outlined in the following theorem [7].

Theorem 1. *Let* $\mathbf{Q} : \mathcal{D} \to \mathbb{N}^d$, *and define* $\mathbf{c} \stackrel{def}{=} \mathbf{Q}(D)$. *A mechanism* $\mathcal{M}$ *that adds independently generated noise from a zero-mean Laplace distribution with scale parameter* $\lambda = \Delta(\mathbf{Q})/\epsilon$ *to each of the d output values of* $\mathbf{Q}$, *i.e., which produces transcript*

$$\mathbf{o} = \mathbf{c} + \langle Lap(\Delta(\mathbf{Q})/\epsilon)\rangle^d$$

achieves ϵ-*differential privacy. The error introduced in the* i^{th} *element of* $\mathbf{o}$ *by* LPA *is*

$$error^i_{\mathsf{LPA}} = \mathbb{E}|\mathbf{o}[i] - \mathbf{c}[i]| = \mathbb{E}|Lap(\lambda)| = \sqrt{2}\lambda = \sqrt{2}\Delta(\mathbf{Q})/\epsilon$$

The higher the error the more the published results deviate from their actual values, reducing their *utility*. Next, we include a *composition* theorem [19] that is useful for our proofs. It concerns successive executions of differentially private mechanisms on the same input, and allows us to view ϵ as a *privacy budget*, distributed among these mechanisms.

Theorem 2. *Let* $\mathcal{M}_1, \ldots, \mathcal{M}_r$ *be a set of mechanisms, where each* $\mathcal{M}_i$ *provides* ϵ_i-*differential privacy. Let* $\mathcal{M}$ *be another mechanism that executes* $\mathcal{M}_1(D), \ldots,$ $\mathcal{M}_r(D)$ *using independent randomness for each* $\mathcal{M}_i$, *and returns the vector of the outputs of these mechanisms. Then,* $\mathcal{M}$ *satisfies* $(\sum_{i=1}^r \epsilon_i)$-*differential privacy.*

An interesting problem concerns computing differentially private range-sums over sensitive histograms. Applying LPA in this scenario would result in high error due to the noise accumulation [3,9] reduce the error by building a full binary tree over the input values. Every node stores the sum of the values of its children, plus noise with logarithmic scale to the input size. Then, each sum of consecutive values is computed by summing the root values of the maximal subtrees covering the values. We exploit this technique in our solution in order to maximize utility.

2.2   Location Privacy for kNN Queries

The setting of private kNN queries assumes a data owner, who provides the POIs to an LBS, and users, who issue queries to the LBS. The goal is to conceal the user locations from the LBS. There exist two notions of privacy in the literature: (i) weak location privacy, where the LBS can derive that the query issuer lies in some general area, without, however, being able to pinpoint his exact position, and (ii) strong location privacy, where the LBS cannot infer anything about the position of the user issuing the query. Weak location privacy solutions adopt three general methodologies, namely K-anonymity, location obfuscation, and data transformation.

K-anonymity [14,20] methods assume the existence of a trusted third party that receives and anonymizes the queries before sending them to the LBS. Specifically, a trusted *anonymizer* that has the locations of all users, generalizes each query so that the LBS cannot distinguish who among K users issued the query. Location obfuscation [5,6,16] substitutes the exact user location with a cloaking region, which is sent to the LBS instead of the exact location. In some obfuscation methods (e.g., [28]), the user sends a fake location and keeps obtaining results until it acquires all k nearest neighbors. In all the above techniques, the LBS can restrict the position of the querying users in some area within the data space, without however being able to pinpoint their exact location.

Data transformation techniques [16,25] assume that the owner encodes the data before sending them to the LBS. Subsequently, the users send encoded queries to the server. The latter cannot determine either the queried data or the user query location. Data transformation methods conceal the user locations better than K-anonymity and obfuscation. However, they are expensive due to the encoding/decoding operations. Additionally, they are prone to *access pattern attacks* [27] because the same query always returns the same encoded results. For example, the LBS may use the query frequency and data density to infer the position of a user (e.g., queries at a city center are much more frequent than those from the suburbs).

Strong location privacy is based on private information retrieval (PIR) [10,11], which allows the users to retrieve data from a database obliviously. There are three categories of PIR, namely information-theoretical, computational, and hardware-based. Information-theoretical PIR [4] offers privacy with theoretical guarantees, while computational PIR [13,18,21,24,26] assumes computationally bounded adversaries. They are both infeasible even for databases of moderate sizes [25]. Hardware-based PIR relies on a tamper-resistant CPU, trusted by the clients, attached to the server. This CPU receives client block requests, which are unreadable by the server, obliviously extracts the requested blocks from the server disk, and returns them to the client. Hardware-based methods are the only viable PIR solutions for large datasets.

[15] applies hardware-based PIR for kNN processing, allowing, however, a variable number of PIR requests for different queries. Consequently, although each PIR retrieval is private, the cardinality of these retrievals may allow access pattern attacks. On the other hand, the method of [12] achieves strong location privacy because every query involves a single PIR request and, hence, all queries are indistinguishable. Nevertheless, this scheme focuses on single NN processing ($k = 1$), and relies on a prohibitively expensive computational PIR protocol [18].

Currently, the only viable technique that guarantees strong location privacy for kNN queries is AHG [23]. AHG is a hardware-based PIR algorithm that utilizes a query plan to avoid access pattern attacks.

Specifically, the setting of [23] assumes that the LBS maintains the data as sequential blocks[1]. AHG initially imposes a Hilbert index grid G on the POIs, grouping them into cells. The Hilbert index grid is a mapping which defines an ordering among the cells according to their unique Hilbert values, e.g., Hilbert values of cell c_{21} and cell c_{11} are defined by $H(2,1) = 1$ and $H(1,1) = 0$, respectively. To preserve locality, the LBS stores the cells ordered by the Hilbert values along with their POI counts in multiple PIR blocks of a database $\mathcal{DB}_1$. It also keeps the blocks of individual POIs in two databases, namely $\mathcal{DB}_2$ and $\mathcal{DB}_3$. Essentially, $\mathcal{DB}_1$ maintains an index of the POIs stored in $\mathcal{DB}_2$. $\mathcal{DB}_2$ holds the actual locations of the POIs, i.e., the longitudes and latitudes, and pointers to $\mathcal{DB}_3$. Finally, $\mathcal{DB}_3$ stores the tail records of the POIs, i.e., other data related to the POIs, such as street addresses, phone numbers, and detailed information.

The users issue kNN queries to the trusted CPU (attached to the LBS), using a fixed query plan to ensure that each query retrieves the same number of blocks from each database, independent of the query location. The query plan is defined as $QP = ((\mathcal{DB}_1, cnt_1), (\mathcal{DB}_2, cnt_2), (\mathcal{DB}_3, k))$. Specifically, when a user asks a kNN query, he first obtains cnt_1 index blocks from $\mathcal{DB}_1$. Then, using the index, he retrieves cnt_2 blocks with POI coordinates from $\mathcal{DB}_2$, and locates the k nearest POIs using these coordinates. Finally, he issues a query to $\mathcal{DB}_3$ and obtains the k corresponding blocks from $\mathcal{DB}_3$. In order to guarantee that every user receives enough blocks for an accurate answer, cnt_1 and cnt_2 constitute upper bounds for the number of blocks needed by *any possible* query location[2].

3   Adaptive Query Plan

We assume the same setting as [23], where a curious, but not malicious LBS maintains the data as sequential blocks. Users issue kNN queries in the form of block requests to a trusted CPU attached to the LBS. This CPU obliviously extracts the requested blocks from the server, and returns them to the client. According to the fixed query plan of AHG, every user receives the maximum number of blocks required to accurately answer all possible queries. Consequently, most users obtain numerous *redundant* blocks since the vast majority of queries need relatively few blocks due to the fact that the query distribution usually follows the distribution of the POIs [2,17,22]. This has a negative impact on the LBS (in terms of processing cost) and the users (in terms of response time), rendering AHG too slow for large spatial datasets commonly found in practice.

To overcome this problem, we propose an adaptive query plan (AQP) that yields the exact kNN set for the majority of the queries, but may lead to inaccurate results for a pre-defined percentage $(1 - \alpha)$ of queries at sparse areas of the

---

[1] The size of each block depends on the PIR hardware.

[2] There is a distinct query plan for every allowed value of k. For ease of presentation, we focus on a single value of k.

data space. The value of α adjusts the trade-off between accuracy and efficiency. In order to derive the size of AQP, we utilize differentially private statistics about previous queries. Strong location privacy is always preserved independently of the value of α and the size of AQP.

Our framework involves two stages: (1) The *query stage* has a fixed period (e.g., a day), in which users issue kNN queries to the LBS. Every user can ask up to q_{max} private kNN queries, where q_{max} is a system parameter. For each query, he records the number of redundant blocks during the current period. (2) At the query plan *re-computation stage*, the LBS obtains the redundancy data from users in a differentially private manner, and computes the distribution of redundant blocks along with the number of blocks necessary to answer the issued queries. Finally, it generates an AQP for the next query stage, so that at least a percentage α of the queries receive enough PIR blocks for accurate results, according to the current statistics. Section 3.1 describes the query stage, Sect. 3.2 elaborates the re-computation stage, and Sect. 3.3 proves the correctness and analyzes the utility of our approach.

3.1   Query Stage

During this stage, each user u maintains a vector R_u of length $cnt_1 + 1$ that stores the number of redundant blocks received from $\mathcal{DB}_1$. Each element j of R_u holds the number of times that u received j redundant blocks. The last element $R_u[cnt_1]$ indicates the number of queries with insufficient blocks (i.e., those with potentially inaccurate results). For example, if $cnt_1 = 30$, $R_u[0] = 1$, $R_u[2] = 5$, and $R_u[30] = 4$, then u had 1 query for which he received the exact number of required blocks, 5 queries with 2 redundant blocks, and 4 queries without enough blocks. A similar vector S_u of length $cnt_2 + 1$ is maintained for database $\mathcal{DB}_2$.

Figure 1 depicts an example for 20 POIs (P_1 to P_{20}), a 6×6 grid, and a query location Q. Let c_{yx} be the cell of the yth row and x^{th} column. All the c_{yx}'s are ordered according to their unique Hilbert values, and stored in $\mathcal{DB}_1$ as pairs of numbers. The first value of c_{yx} indicates the sum of the POIs contained in all preceding cells in the Hilbert order, while the second one is the number of POIs in c_{yx}. For example, (3,0) for c_{13} denotes 3 POIs lying in $c_{11}, c_{21}, c_{22}, c_{12}$, and no POIs in c_{13}. Each block in $\mathcal{DB}_1$ holds up to 8 cells, and it is denoted as $B_{1,id}$, where id is the block id. $\mathcal{DB}_2$ holds a tuple $< P.id, P.x, P.y, P.ptr >$ for each POI, where $P.id$ is the POI id, $P.x$, $P.y$ its coordinates, and $P.ptr$ a pointer to $\mathcal{DB}_3$. Each $\mathcal{DB}_2$ block, denoted as $B_{2,id}$, consists of up to 4 POIs, which are sorted on the Hilbert values of their cells. Finally, $\mathcal{DB}_3$ contains tuples of the form $< P.id, P.tail >$, where $P.id$ is the POI id, and $P.tail$ is the tail information of the POI. Each block in $\mathcal{DB}_3$ is denoted as $B_{3,id}$.

Let the query plan be $((\mathcal{DB}_1, 2), (\mathcal{DB}_2, 4), (\mathcal{DB}_3, 2))$. Assume that a user u in cell c_{45} issues a 2NN query from location Q. Initially, u discovers the block in $\mathcal{DB}_1$ that contains his residing cell as follows. He computes the Hilbert value $H(4,5) = 29$ of c_{45}, and determines the required block as $(29+1)/8+1 = 4$. Then, he finds the position of c_{45} by computing $(29 + 1) \mod 8 = 6$. Consequently, u derives that the information of c_{45} is at position 6 of the 4th block in $\mathcal{DB}_1$,

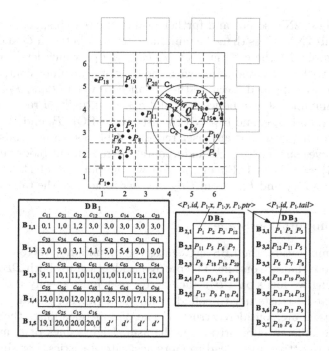

Fig. 1. Query stage example

and instructs the secure CPU to retrieve block $B_{1,4}$. The pair of values in c_{45} is $(17, 0)$, denoting that there are no POIs in u's cell. Thus, u proceeds with the next closest cell to Q, which is c_{35}. The latter belongs to the already retrieved block $B_{1,4}$ and hence, u does not need to obtain another block. Moreover, c_{35} contains only one POI and as such u needs to explore more cells by following the same procedure. After retrieving c_{44} from $B_{1,2}$, u has gathered 2 POIs, which he uses for pruning as follows. He computes the maximum possible distance $maxdist$ between Q and the retrieved cells, and draws a circle C_1 centered at Q, with radius of $maxdist$. All the cells that have no common area with the disk of C_1 cannot contain any POIs comprising the 2NN, and hence, they can be ignored. Therefore, u needs only the information of cells c_{34}, c_{46}, c_{36}, c_{55}, c_{25}, c_{54}, c_{24}, c_{43}, c_{33}, c_{56}, and c_{26}, which are distributed among all the 5 blocks of $\mathcal{DB}_1$. However, he cannot acquire the information for all these cells, since his $\mathcal{DB}_1$ block retrieval is limited to 2 blocks by the query plan. As such, he has insufficient blocks from $\mathcal{DB}_1$ in order to ensure an accurate 2NN, and hence, he updates the redundancy vector for $\mathcal{DB}_1$ by setting $R_u[2] := R_u[2] + 1$.

Next, u requests the coordinates of the POIs from $\mathcal{DB}_2$ in ascending order of the minimum distances between Q and each retrieved cell from $\mathcal{DB}_1$. Specifically, he obtained $(17, 1)$ for c_{35} from the previous step. He computes $(17+1)/4+1 = 5$, and $(17 + 1) \mod 4 = 2$, and derives that the POI is at the 2nd position of the 5th block of $\mathcal{DB}_2$. Thus, he requests $B_{2,5}$ from $\mathcal{DB}_2$, and repeats the process for all of the potential NNs. Each time he retrieves the coordinates of a POI, he updates the best 2NN so far, and further prunes the search space if the current distance of this 2NN is less than the minimum distance between Q and any other cell. For example, let the current 2NN be P_9 and P_{12}, which lets u draw circle C_2. Any cell with no common area with the disk of C_2 cannot contain a better NN, e.g., cell c_{43} is pruned, because the distance between Q and P_{12}, is smaller than the minimum distance between Q and c_{43}. Eventually, u receives 3 blocks, i.e., $B_{2,1}$, $B_{2,4}$, and $B_{2,5}$, and his 2NN consists of the POIs P_9 and P_{12}. However, the query plan requires that any user retrieves 4 blocks, so he sends a random request, retrieves one more block, and updates his redundant block vector S_u for $\mathcal{DB}_2$, to $S_u[1] := S_u[1] + 1$. Finally, he follows the pointers of P_9 and P_{12}, and retrieves blocks $B_{3,2}$ and $B_{3,6}$ from $\mathcal{DB}_3$ in order to acquire the tail information.

3.2   Re-Computation Stage

During re-computation, the LBS aggregates the redundancy vectors of all active users, i.e., those that are on-line. The process is identical for both $\mathcal{DB}_1$ and $\mathcal{DB}_2$; in the following we focus on $\mathcal{DB}_1$. Each active user must specify the percentage ℓ of users that he trusts not to collude with the LBS. For instance, if $\ell = 10$, 10 % of the users are considered trusted. The parameter adjusts the noise scale added by the differentially private mechanism. High values of ℓ, as well as a large number of active users, lead to more accurate statistics. For simplicity, we assume that every user chooses the same ℓ value.

When a user u registers with the service, he sends to the LBS a self-signed Diffie-Hellman (DH) component for computing pairwise keys, and a certificate to authenticate himself. At the beginning of re-computation, the LBS distributes the self-signed DH components and certificates to all active users. Every user computes the pairwise keys shared with the other users using the DH components. Then, each user u performs computations on R_u, before forwarding a noisy and encrypted version $\hat{R}_u$ of R_u to the LBS. Having collected $\hat{R}_u$ from the active users, the LBS derives a differentially private vector $\hat{R}$ of the aggregate statistics and uses it to generate the new AQP.

The process constitutes a combination of secure multiparty computation and distributed differential privacy, for which we adopt the method of [1], originally proposed for computing differentially private sums[3]. Let n be the number of active users, $K_{u,w}$ be the pairwise key shared by users u and w, r_1 and r_2 be two random numbers published by the LBS, and K_u a symmetric key between user u and the LBS. For every element $R_u[v]$, user u spends privacy budget $\lambda = \epsilon/(4 \cdot q_{\max})$ by drawing two noise values from the Gamma distribution, and computes $\hat{R}_u[v] = R_u[v] + G_{u,1}(n \cdot \ell, \lambda) - G_{u,2}(n \cdot \ell, \lambda)$. Subsequently, u selects approximately $n \cdot \ell$ other users randomly by using a secure pseudo random function (PRF), so that if u selects w, then w selects u as well. The PRF works as follows: u chooses w, if $PRF(K_{u,w}, r_1) \le n \cdot \ell/(n-1)$; for each chosen w,

---

[3] In this setting, there are several users, each holding a value, and they wish to publish the total sum, so that the value of any user is not revealed, even if an adversary has complete knowledge of all the remaining users.

u computes $dkey_{u,w} = (u - w)/|u - w| \cdot PRF(K_{u,w}, r_2)$. Note that $dkey_{u,w} = -dkey_{w,u}$. User u then encrypts $\hat{R}_u[v]$ as $Enc(\hat{R}_u[v]) = \hat{R}_u[v] + K_u + \sum_w dkey_{u,w}$ and sends it to the server.

The LBS aggregates all the numbers sent from the users. By doing this, all the $dkey$s are canceled out, and the server decrypts the sum by $\hat{R}[v] = \sum_u Enc(\hat{R}_u[v]) - \sum_u K_u = \sum_u R_u[v] + \sum_u (G_{u,1}(n \cdot \ell, \lambda) - G_{u,2}(n \cdot \ell, \lambda))$. [1] shows that $\hat{R}[v]$ is equivalent to $\hat{R}[v] = \sum_u R_u[v] + Lap(\lambda/\ell)$ because the Laplace noise can be approximated from identically distributed (i.i.d.) gamma distributions. As such, after computing all the elements of $\hat{R}$, the final result satisfies differential privacy.

The LBS calculates cnt_1 by executing method $recomputeQP(\hat{R})$, shown in Fig. 2. Lines 4–5 determine if the percentage of queries without enough blocks exceeds $(1 - \alpha)$ by checking $ratio = \hat{R}[cnt_1]/\sum_{i=0}^{cnt_1} \hat{R}[i]$. In this case, the value of cnt_1 increases, so that queries receive more blocks. In case the percentage of queries without enough blocks does not exceed $(1 - \alpha)$ (lines 6–11), the server iteratively computes $ratio+ = \hat{R}[l]/\sum_{i=0}^{cnt_1} \hat{R}[i]$ by incrementing l until the ratio exceeds $(1 - \alpha)$. Essentially, the ratio for a certain l value represents the percentage of queries that will not receive enough blocks if we decrease the previous query plan by l. Thus, the new query plan is computed as the previous one reduced by $(l - 1)$.

In lines 2 and 9 of $recomputeQP$ (Fig. 2), the LBS sums consecutive noisy values of $\hat{R}$ in order to compute differentially private range-sums over private data. As discussed in the related work session, this would result in high error due to the noise accumulation. Therefore we adapt the technique of [3,9] as follows. Each user u creates a binary tree, such that the original redundancy vector R_u (without noise) represents the leaf nodes and each parent node is the sum of its direct children, i.e., the root node is the sum of all leaf nodes. Moreover, the privacy budget spent by the users decreases from $\epsilon/(4q_{max})$ to $\epsilon/(4q_{max} \cdot \log_2 |\hat{R}|)$, in order to satisfy the same level of privacy as suggested by [3,9]. Then, aggregation is executed for the tree nodes instead for the elements of

Function recomputeQP($\hat{R}$)	
1. **for** $i = 0$ to cnt_1 **do**	
2. $sum+ = \hat{R}[i]$	//count the total number of queries
3. $ratio = \hat{R}[cnt_1]/sum$	//ratio of queries without enough blocks
4. **if** $ratio > (1 - \alpha)$ **then**	//percentage of inaccurate queries exceeds $(1 - \alpha)$
5. $cnt_1 := cnt_1 + 1$	//increase AQP size by 1
6. **else**	
7. $l = 0$	
8. **while** $ratio \leq (1 - \alpha)$	//percentage of inaccurate queries below $(1 - \alpha)$
9. $ratio+ = \hat{R}[l]/sum$	//decrease AQP size
10. $l = l + 1$	
11. $cnt_1 := cnt_1 - (l - 1)$	

Fig. 2. Pseudocode of $recomputeQP$

R_u. Finally, the LBS acquires an aggregate tree with leaf nodes representing the noisy values of $\hat{R}$, and re-computes the query plan as before. The only difference is that, instead of adding the values of $\hat{R}$ one by one in order to compute *ratio*, the LBS utilizes the noisy tree structure, resulting in more accurate sums.

3.3   Correctness and Utility Analysis

The next theorem shows that our adaptive solution, when applied on both $\mathcal{DB}_1$ and $\mathcal{DB}_2$, satisfies ϵ-differential privacy.

Theorem 3. *The AQP algorithm satisfies ϵ-differential privacy for at most q_{max} queries per user.*

Proof. We refer to the query stage as mechanism M_1, and the re-computation stage as mechanism M_2. Due to the fact that the query stage satisfies strong location privacy [23], an adversary cannot distinguish if the user asks a query from a location j or any other location j'. In other words, the probability for the user to be at location j, and receive cnt_1 blocks from $\mathcal{DB}_1$ and cnt_2 blocks from $\mathcal{DB}_2$ is the same with the probability he is at j', and receives exactly the same number of blocks from the two databases. Thus, from Definition 1, M_1 satisfies 0-differential privacy.

In the case of M_2, we further split the mechanism into two mechanisms $M_{2.1}$ and $M_{2.2}$. $M_{2.1}$ computes cnt_1 for $\mathcal{DB}_1$ and $M_{2.2}$ computes cnt_2 for $\mathcal{DB}_2$. Thus, mechanism M_2 comprises of $cnt_1 + 1$ ($M_{2.1}$) and $cnt_2 + 1$ ($M_{2.2}$) mechanisms of [1], each utilizing privacy budget $\epsilon/(4 \cdot q_{max})$.

Let D_1 be a table where each row represents a user, and each column the received redundant blocks from $\mathcal{DB}_1$, i.e., each row u of D_1 corresponds to R_u of user u for $\mathcal{DB}_1$. Each cell i, j of D_1 holds how many times user i received j redundant blocks from $\mathcal{DB}_1$ during the query stage. Similarly, we define a table D_2 for $\mathcal{DB}_2$, i.e., each row u of D_2 corresponds to S_u of user u for $\mathcal{DB}_2$.

Mechanism $M_{2.1}$ (resp. $M_{2.2}$) essentially executes the method of [1] on each column of D_1 (resp. D_2), and returns vector $\hat{R}$ (resp. $\hat{S}$), which holds the noisy sums on the columns. A neighboring database D_1' (resp. D_2') differs at most by $2 \cdot q_{max}$ to D_1 (resp. D_2): A user issues at most q_{max} private queries in D_1 (resp. D_2), and there can be at most q_{max} different private queries in D_1' (resp. D_2'). As such, he can change at most $2 \cdot q_{max}$ values of each database D_1 and D_2.

In order to determine the achieved privacy level we work as follows. For any D_1 (resp. D_2), w.r.t a D_1' (resp. D_2'), we create two databases; D_a which contains only the columns of D_1 (resp. D_2) that differ from D_1' (resp. D_2'), and D_b which contains the columns that are exactly the same as those of D_1' (resp. D_2'). Note that D_a has at most $2 \cdot q_{max}$ columns due to the sensitivity of D_1 (resp. D_2). From D_a we compute the noisy sums of the columns $\hat{R}_a$, and from D_b we compute $\hat{R}_b$, where $\hat{R}_a \cup \hat{R}_b = \hat{R}_{D1}$ (resp. $\hat{R}_a \cup \hat{R}_b = \hat{R}_{D2}$) and $\hat{R}_a \cap \hat{R}_b = \varnothing$. Any mechanism of [1] when applied on D_a with privacy budget $\epsilon/(4 \cdot q_{max})$, it satisfies $\epsilon/(4 \cdot q_{max})$-differential privacy, as shown in [1]. On the other hand, when it is applied on D_b, it satisfies 0-differential privacy, due to Definition 1, since $D_b = D_b'$. Moreover, we have at most $2 \cdot q_{max}$ sub-mechanisms of $M_{2.1}$ (resp. $M_{2.2}$) applied on D_a, since D_a has at most $2 \cdot q_{max}$ columns. Hence, from Theorem 2, $M_{2.1}$ satisfies $(2 \cdot q_{max} \cdot \epsilon/(4 \cdot q_{max}) = \epsilon/2)$-differential privacy, and equivalently $M_{2.2}$ satisfies $\epsilon/2$-differential privacy, while M_1 satisfies 0-differential privacy. Thus, the whole procedure satisfies $(0 + \epsilon/2 + \epsilon/2 = \epsilon)$-differential privacy due to Theorem 2.

Next, we quantify the expected error. We focus on $\mathcal{DB}_1$ since the analysis for $\mathcal{DB}_2$ is the same. Let q be the total number of queries performed at query stage, and R be the actual redundancy vector without the noise required for differential privacy. Due the the noise addition, we expect that the computed (at the LBS) vector $\hat{R}$ deviates from R, yielding an error during the query plan re-computation.

Let l_{real} be the position of R representing the α percentile, i.e., the minimum l_{real} value such that $R[cnt_1] + \sum_{i=0}^{l_{real}} R[i] > (1-\alpha) \cdot q$. Then, $cnt_1 - l_{real} - 1$ is the number of necessary blocks for accuracy α. On the other hand, the corresponding number in Fig. 2 is computed as the minimum value of l for which it holds that $\hat{R}[cnt_1] + \sum_{i=0}^{l} \hat{R}[i] > (1-\alpha) \cdot q$. As such, $recomputeQP$ may stop at an $l \leq l_{real}$ (or $l > l_{real}$), and return $|l_{real} - l|$ more blocks (resp. fewer blocks) for each query than required for accuracy α. We define $|l_{real} - l|$ as the error due to the noise perturbation.

As an example, let $n = 10$, $\alpha = 70\%$, $q = 10$, $cnt_1 = 3$, $|\hat{R}| = 4$, and R and $\hat{R}$ as shown in Table 1. Then, $(1 - \alpha) \cdot q = 3$, $R[3] + R[0] = 3$, and $l_{real} = 2$. Consequently, $cnt_1 - (l_{real} - 1) = 2$, and hence, $cnt_1 = 2$ for the next period. However, the server knows only the noisy $\hat{R}$. It computes $\hat{R}[3] + \hat{R}[0] + \hat{R}[1] > 3$, and sets $l = 3$. As a result, $cnt_1 - (l-1) = 1$, or $cnt_1 = 1$, and due to noise each query receives $|l_{real} - l| = 1$ block less than required for 70 % accuracy.

In the worst case, R is highly skewed, and $(\alpha + \delta)q$ queries, for any small $\delta > 0$, receive $cnt_1 - 1$ redundant blocks (i.e., $R[cnt_1 - 1] = (\alpha + \delta)q$), while the rest receive insufficient blocks (i.e., $R[cnt_1] = (1-\alpha-\delta)q$). Therefore, l_{real} should be $|R| - 1$, i.e., the new query plan should be set as $cnt_1 = 1$. In this case, if the Laplace noise added while computing $\hat{R}[cnt_1]$ is positive, $recomputeQP$ will compute the new query plan as $cnt_1 := cnt_1 + 1 = |R|$, resulting in the maximum error of $|R| - 1$. Due to the fact that the Laplace distribution is symmetric about its mean 0, the probability for $recomputeQP$ to stop at position 0 is 50 %, at position 1 is 25 %, and so on. This is equivalent to multiple Bernoulli trials and hence, the probability to stop at position i can be described with the Binomial distribution with $p = 0.5$. Thus, with probability 50 %, we get the worst possible error $|R| - 1$, while the probability for the error to be reduced by l (i.e. to stop at position l) is equal to the probability we receive l heads in l coin flips.

Table 1. An example illustrating variables

Redundancy	0	1	2	Insufficient
R	2	2	5	1
$\hat{R}$	1	3	5	1

In order to better quantify the expected error, we assume a uniform distribution of the queries in R, i.e., $q/|R|$ queries receive 0 redundant blocks, $q/|R|$ queries receive 1 redundant block, and so on. In this case, in order to achieve α accuracy, we need to set $l_{real} = (1 - \alpha)|R|$. Then, it suffices to compute the expected value of l (returned by $recomputeQP$) in order to find the expected error $|l_{real} - l|$. Initially, we calculate the expected error of each sum $\sum_{i=0}^{i} \hat{R}[i]$, for any $0 \leq i \leq |R|$. The sum is computed utilizing the aggregate tree at the server side with noisy node values. The noise at each node is equivalent to Laplace noise with scale $\lambda = \frac{4q_{max} \cdot \log |R|}{\epsilon \cdot \ell}$. In order to calculate each sum i we use the technique of [3], which results in error err less than $\lambda \cdot \sqrt{\log(i + 1)} \cdot \log \frac{1}{\delta_i}$ with probability $(1 - \delta_i)$.

$RecomputeQP$ checks the value of each sum for $i = 0$ to $|R|$, and returns the first i that results in a noisy sum which is higher to $(1 - \alpha) \cdot q$. Thus, $i = l$ if it does not stop at $i = 0 \ldots l - 1$, with probability higher than $1 - \delta$. The probability for the algorithm to stop at a position l is equal to the probability it does not stop until position $l - 1$ or $\prod_{i=0}^{l} \left(1 - \Pr\left[(i + 1)\frac{q}{|R|} + err \geq (1 - \alpha) \cdot q\right]\right)$. Thus, the value of l can be computed as the first value for which the following inequality does not hold.

$$\prod_{i=0}^{l} \left(1 - \Pr\left[(i + 1)\frac{q}{|R|} + err \geq (1 - \alpha) \cdot q\right]\right) \geq (1 - \delta)$$

$$\prod_{i=0}^{l} \left(1 - \Pr\left[err \geq (1 - \alpha) \cdot q - (i + 1)\frac{q}{|R|}\right]\right) \geq (1 - \delta)$$

$$\prod_{i=0}^{l} (1 - \delta_i) \geq (1 - \delta)$$

where $\delta_i = 0.5^{\frac{(1-\alpha) \cdot q - (i+1) \cdot q/|R|}{\lambda \sqrt{\log(i+1)}}}$.

4   Experimental Evaluation

In this section we compare our adaptive query plan AQP with the fixed query plan of AHG [23]. We implemented the methods in C++ on a Linux server with Intel Core i7-4770 and 32 GB of RAM. Since the re-computation takes only a few seconds and is performed once after each query stage, it is excluded from the evaluation, and all experiments focus on query processing. In order to evaluate efficiency, we measure the query response time, which directly affects the user, and the number of block accesses, which determines the processing cost at the LBS. We used two datasets with real POIs from Germany (denoted as *Germany*) and the United States (denoted as *USA*)[4]. The former consists of 1.9 million POIs, while the latter has 12.8 million POIs.

---

[4] SimpleGeo's Places, available at http://freegisdata.rtwilson.com/.

Table 2. Parameter values

Parameter	Values	Default
# of active users n	2000, 4000, 6000, 8000, 10000	6000
% of trusted users ℓ	5, 10, 15, 20, 25	10
Accuracy α	0.75, 0.8, 0.85, 0.9, 0.95	0.95

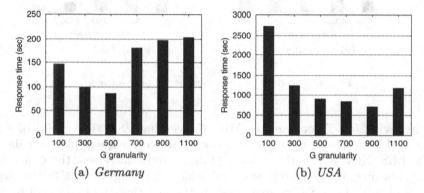

(a) *Germany* (b) *USA*

Fig. 3. Response time vs. G granularity

Table 2 illustrates the examined parameters, along with their default values. The number n of active users denotes those participating in the re-computation stage. The percentage ℓ of trusted users corresponds to those trusted not to collude with others. Accuracy α is the percentage of queries that should be answered correctly. In all experiments, we set $\epsilon = 1$ for differential privacy, and $k = 10$ as the number of returned nearest neighbors. Every user issues $q_{max} = 10$ private queries that follow the same distribution as the POIs. Since all queries retrieve the same number of blocks they incur the same cost and response time.

Following [23], we first fine-tune the granularity of the grid G, used by the basic AHG and AQP. Figure 3 shows the query response time, i.e. the total elapsed time until a user receives a query answer, as a function of the granularity of G, assuming fixed query plans (as in [23]). Coarse granularity leads to high cost because there are numerous POIs in each cell, leading to many $\mathcal{DB}_2$ PIR retrievals. The response time is also high when the granularity is too fine because there are numerous empty cells, yielding many $\mathcal{DB}_1$ PIR retrievals. In the remaining experiments, we set the grid granularity to the optimal configuration, which is 500×500 for *Germany*, and 900×900 for *USA*. Note that a single query by basic AHG requires more than 10 min in *USA*, even for the best granularity, motivating the need for AQP.

Figure 4 plots the response time versus the number of active users, setting $\alpha = 95\%$ and $\ell = 10\%$. The cost drops as the active users increase because the accuracy of statistics estimation improves, leading to a smaller query plan. The basic solution needs 93.3 s for *Germany*, and 743.1 s for *USA*, in order to answer a single query. For *Germany*, AQP reaches the lowest value (about 20 s)

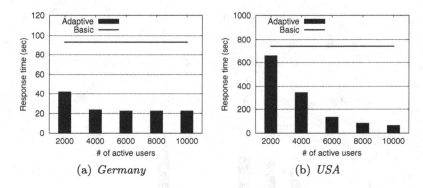

Fig. 4. Response time vs. number of active users

quickly, saving 78.6 % compared to AHG. Concerning *USA*, the benefits of AQP are limited for a small number of users due to inaccurate query plan calculation by the LBS. However, as the number of active users increases, the cost drops quickly, reaching 65 s for 10,000 users and achieving savings of 91.3 %. The larger benefits of AQP for *USA* are explained by the fact that the basic plan is very expensive leaving more space for optimization.

To better elaborate performance, we investigate the size (in blocks) of the query plans. The fixed query plan for *Germany* is $((\mathcal{DB}_1, 83), (\mathcal{DB}_2, 53), (\mathcal{DB}_3, 10))$ under all settings. This implies that any 10NN query retrieves $cnt_1 = 83$ blocks from $\mathcal{DB}_1$, $cnt_2 = 53$ from $\mathcal{DB}_2$, and $cnt_3 = 10$ from $\mathcal{DB}_3$. Recall that the cnt_1 blocks correspond to cell retrievals, whereas the cnt_2 blocks represent POI retrievals. The cnt_3 blocks refer to detail information about the 10NNs, and cannot be avoided by AQP or any other method. The fixed query plan for *USA* is $((\mathcal{DB}_1, 299), (\mathcal{DB}_2, 485), (\mathcal{DB}_3, 10))$. Note that cnt_1 (cnt_2) is higher for *USA* because of the finer grid granularity (larger number of POIs).

Figures 5 and 6 show the number of blocks cnt_1 and cnt_2, in both fixed and adaptive plans, as a function of the number of active users, setting $\alpha = 95\%$ and

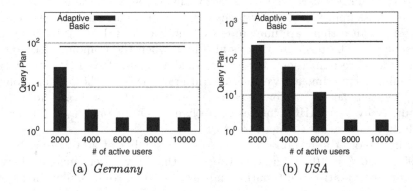

Fig. 5. Query plan vs. number of active users for $\mathcal{DB}_1$

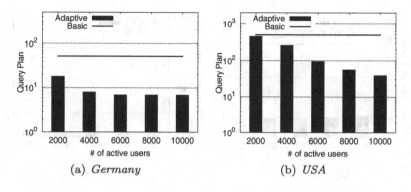

Fig. 6. Query plan vs. number of active users for $\mathcal{DB}_2$

$\ell = 10\,\%$. Comparing Figs. 5 and 6, the benefits of AQP are more pronounced in $\mathcal{DB}_1$. This is explained by the fact that the pruned cells are likely to contain few POIs; therefore the cell reduction in $\mathcal{DB}_1$ does not directly translate to an equivalent POI reduction in $\mathcal{DB}_2$. In general, the results are consistent with those on response time in Fig. 4; i.e., for *Germany* a small number of active users suffices, while for *USA* more users are necessary to substantially reduce the number of block retrievals.

Figure 7 shows the response time as a function of the percentage of trusted users ℓ, fixing $\alpha = 95\,\%$ and $n = 6{,}000$. The time drops as ℓ increases because the resulting statistics have smaller noise scale. For *Germany*, even $\ell = 5\,\%$ (i.e., 300 trusted users) yields the lowest cost. Similar to Fig. 4, for *USA* the number of users necessary for convergence is larger, but the savings with respect to the basic plan are more substantial.

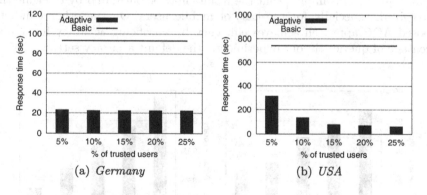

Fig. 7. Response time vs. percentage of trusted users

Figure 8 illustrates the response time as a function of the percentage of accuracy α, setting $\ell = 10\,\%$ and $n = 6{,}000$. As expected, the cost drops with the required accuracy, but the effect is more pronounced in *USA* where reducing

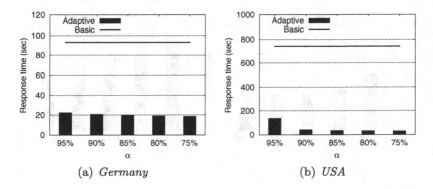

Fig. 8. Response time vs. accuracy α

the accuracy from 95 % to 75 % decreases the time from 135 to 30 s. The same reduction in *Germany* gains only about 3 s. It is worth pointing out that even for the inaccurate queries, the retrieved 10NN set is similar to the real one; i.e., 80 %–90 % of the actual nearest neighbors are in the query result.

In the next experiment, we evaluate the actual versus the expected accuracy of queries. Specifically, we first executed $n \cdot q_{max} = 60,000$ queries based on which the LBS generated the AQP. Then, we performed another 100,000 queries and measured the percentage for which the users obtain accurate results, i.e., the retrieved and the actual 10NN sets are identical. Note that the query distributions in both cases are the same as that of the POIs. As shown in Fig. 9, the accurate queries always exceed the desired accuracy level because the error incurred by the re-computation stage corresponds to a conservative estimation.

Summarizing the experimental evaluation, even the current state-of-the-art method may take several minutes (more than 10 for *USA*) to answer a nearest neighbor query. This implies that the results may be out-dated by the time they are received, especially for the case of mobile users. On the other hand, the proposed AQP approach achieves efficiency by sacrificing accuracy for a small percentage of queries (in our experiments, the default accuracy setting is 95 %).

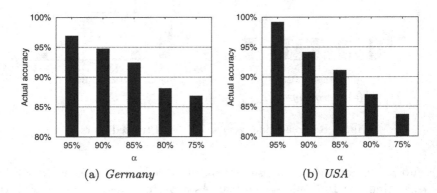

Fig. 9. Actual accuracy vs. α

5 Conclusion

Strong location privacy requires that every query retrieves the same number of blocks in order to protect users from access pattern attacks. This has serious performance implications for both the LBS (in terms of processing cost) and the users (in terms of response time). To overcome the problem, we propose a novel approach that utilizes query statistics and ensures privacy by adopting the concept of ϵ-differential privacy. The trade-off is that accuracy is sacrificed for a small predefined percentage of queries. As shown in a comprehensive experimental evaluation with real POIs, ours is the first practical approach for strong location privacy in large datasets.

Acknowledgments. This work was supported by GRF grant 618011 from Hong Kong RGC.

References

1. Ács, G., Castelluccia, C.: I have a DREAM! (DiffeRentially privatE smArt Metering). In: Filler, T., Pevný, T., Craver, S., Ker, A. (eds.) IH 2011. LNCS, vol. 6958, pp. 118–132. Springer, Heidelberg (2011)
2. Böhm, C.: A cost model for query processing in high dimensional data spaces. TODS **25**(2), 129–178 (2000)
3. Chan, T.-H.H., Shi, E., Song, D.: Private and continual release of statistics. TISSEC **14**(3), 26:1–26:24 (2011)
4. Chor, B., Kushilevitz, E., Goldreich, O., Sudan, M.: Private information retrieval. J. ACM (JACM) **45**(6), 965–981 (1998)
5. Duckham, M., Kulik, L.: A formal model of obfuscation and negotiation for location privacy. In: Gellersen, H.-W., Want, R., Schmidt, A. (eds.) PERVASIVE 2005. LNCS, vol. 3468, pp. 152–170. Springer, Heidelberg (2005)
6. Duckham, M., Kulik, L.: Simulation of obfuscation and negotiation for location privacy. In: Cohn, A.G., Mark, D.M. (eds.) COSIT 2005. LNCS, vol. 3693, pp. 31–48. Springer, Heidelberg (2005)
7. Dwork, C.: A firm foundation for private data analysis. CACM **54**(1), 86–95 (2011)
8. Dwork, C., Kenthapadi, K., McSherry, F., Mironov, I., Naor, M.: Our data, ourselves: privacy via distributed noise generation. In: Vaudenay, S. (ed.) EUROCRYPT 2006. LNCS, vol. 4004, pp. 486–503. Springer, Heidelberg (2006)
9. Dwork, C., Naor, M., Pitassi, T., Rothblum, G.N.: Differential privacy under continual observation. In: STOC (2010)
10. Gasarch, W.: A survey on private information retrieval. Bull. EATCS **82**, 72–107 (2004)
11. Ghinita, G.: Privacy for location-based services. Synth. Lect. Inf. Secur. Priv. Trust **4**(1), 1–85 (2013)
12. Ghinita, G., Kalnis, P., Khoshgozaran, A., Shahabi, C., Tan, K.-L.: Private queries in location based services: anonymizers are not necessary. In: SIGMOD (2008)
13. Ghinita, G., Kalnis, P., Skiadopoulos, S.: PRIVE: anonymous location-based queries in distributed mobile systems. In: WWW (2007)

14. Kalnis, P., Ghinita, G., Mouratidis, K., Papadias, D.: Preventing location-based identity inference in anonymous spatial queries. IEEE Trans. Knowl. Data Eng. **19**(12), 1719–1733 (2007)
15. Khoshgozaran, A., Shahabi, C., Shirani-Mehr, H.: Location privacy; moving beyond k-anonymity, cloaking and anonymizers. KAIS **26**, 435–465 (2010)
16. Kido, H., Yanagisawa, Y., Satoh, T.: An anonymous communication technique using dummies for location-based services. In: ICPS (2005)
17. Korn, F., Pagel, B.-U., Faloutsos, C.: On the 'dimensionality curse' and the 'self-similarity blessing'. TKDE **13**(1), 96–111 (2001)
18. Kushilevitz, E., Ostróvsky, R.: Replication is not needed: single database, computationally-private information retrieval. In: 2013 IEEE 54th Annual Symposium on Foundations of Computer Science, pp. 364–364. IEEE Computer Society (1997)
19. McSherry, F.D.: Privacy integrated queries: an extensible platform for privacy-preserving data analysis. In: Proceedings of the 2009 ACM SIGMOD International Conference on Management of Data, pp. 19–30. ACM (2009)
20. Mokbel, M.F., Chow, C.-Y., Aref, W.G.: The new casper: query processing for location services without compromising privacy. In: Proceedings of the 32nd International Conference on Very Large Data Bases, pp. 763–774. VLDB Endowment (2006)
21. Ostrovsky, R., Skeith III, W.E.: A survey of single-database private information retrieval: techniques and applications. In: Okamoto, T., Wang, X. (eds.) PKC 2007. LNCS, vol. 4450, pp. 393–411. Springer, Heidelberg (2007)
22. Pagel, B.-U., Korn, F., Faloutsos, C.: Deflating the dimensionality curse using multiple fractal dimensions. In: ICDE (2000)
23. Papadopoulos, S., Bakiras, S., Papadias, D.: Nearest neighbor search with strong location privacy. Proc. VLDB Endow. **3**(1–2), 619–629 (2010)
24. Shang, N., Ghinita, G., Zhou, Y., Bertino, E.: Controlling data disclosure in computational pir protocols. In: Proceedings of the 5th ACM Symposium on Information, Computer and Communications Security, pp. 310–313. ACM (2010)
25. Sion, R., Carbunar, B.: On the computational practicality of private information retrieval. In: Proceedings of the Network and Distributed Systems Security Symposium (2007)
26. Wang, S., Agrawal, D., El Abbadi, A.: Generalizing PIR for practical private retrieval of public data. In: Foresti, S., Jajodia, S. (eds.) Data and Applications Security and Privacy XXIV. LNCS, vol. 6166, pp. 1–16. Springer, Heidelberg (2010)
27. Williams, P., Sion, R.: Usable PIR. In: NDSS (2008)
28. Yiu, M.L., Jensen, C., Huang, X., Lu, H.: SpaceTwist: managing the trade-offs among location privacy, query performance, and query accuracy in mobile systems. In: ICDE (2008)

Privacy-Preserving Detection of Anomalous Phenomena in Crowdsourced Environmental Sensing

Mihai Maruseac[1], Gabriel Ghinita[1]([✉]), Besim Avci[2], Goce Trajcevski[2], and Peter Scheuermann[2]

[1] University of Massachusetts, Boston, MA 02125, USA
{mmarusea,gghinita}@cs.umb.edu
[2] Northwestern University, Evanston, IL 60208, USA
{besim,goce,peters}@eecs.northwestern.edu

Abstract. Crowdsourced environmental sensing is made possible by the wide-spread availability of powerful mobile devices with a broad array of features, such as temperature, location, velocity, and acceleration sensors. Mobile users can contribute measured data for a variety of purposes, such as environmental monitoring, traffic analysis, or emergency response. One important application scenario is that of detecting anomalous phenomena, where sensed data is crucial to quickly acquire data about forest fires, environmental accidents or dangerous weather events. Such cases typically require the construction of a *heatmap* that captures the distribution of a certain parameter over a geospatial domain (e.g., temperature, CO_2 concentration, water polluting agents, etc.).

However, contributing data can leak sensitive private details about an individual, as an adversary may be able to infer the presence of a person in a certain location at a given time. In turn, such information may reveal information about an individual's health, lifestyle choices, and may even impact the physical safety of a person. In this paper, we propose a technique for privacy-preserving detection of anomalous phenomena, where the privacy of the individuals participating in collaborative environmental sensing is protected according to the powerful semantic model of *differential privacy*. Our techniques allow accurate detection of phenomena, without an adversary being able to infer whether an individual provided input data in the sensing process or not. We build a differentially-private index structure that is carefully customized to address the specific needs of anomalous phenomenon detection, and we derive privacy-preserving query strategies that judiciously allocate the privacy budget to maintain high data accuracy. Extensive experimental results show that the proposed approach achieves high precision of identifying anomalies, and incurs low computational overhead.

1 Introduction

Environmental sensing using crowdsourcing is a promising direction due to the widespread availability of mobile devices with positioning capabilities and a

© Springer International Publishing Switzerland 2015
C. Claramunt et al. (Eds.): SSTD 2015, LNCS 9239, pp. 313–332, 2015.
DOI: 10.1007/978-3-319-22363-6_17

broad array of sensing features, e.g., audio and video capture, temperature, velocity, acceleration, etc. In addition, mobile devices can easily interface with external sensors and upload readings for many other environmental parameters (e.g., CO_2, water pollution levels, atmospheric pressure). The growing trend towards crowdsourcing environmental sensing is beneficial for a wide range of applications, such as pollution levels monitoring or emergency response. In such a setting, authorities can quickly and inexpensively acquire data about forest fires, environmental accidents or dangerous weather events.

One particular task that is relevant to many application domains is that of detecting anomalous phenomena. Such cases typically require to determine a *heatmap* capturing the distribution of a certain sensed parameter (e.g., temperature, CO_2 level) over a geospatial domain. When the parameter value in a certain region reaches a predefined threshold, then an alarm should be triggered, signaling the occurrence of an anomaly. Furthermore, the alarm should identify with good accuracy the region where the dangerous event occurred, so that countering measures can be deployed to that region.

However, there are important privacy concerns related to crowdsourced sensing. Contributed data may reveal sensitive private details about an individual's health, lifestyle choices, and may even impact the physical safety of a person. To protect against such disclosure, the state-of-the-art model of *differential privacy (DP)* adds noise to data in a way that prevents an adversary from learning whether the contribution of an individual is present in a dataset or not. Several DP-compliant techniques for protecting location data have been proposed in [1,16,17]. However, these approaches consider only simple, general-purpose count queries, and rely on simplifying assumptions that make them unsuitable for our considered problem of anomalous phenomenon detection.

Consider the example of a forest fire, where mobile users report air temperature in various regions. To model the fire spread, one needs to plot the temperature distribution, which depends on the values reported by individual users, and the users' reported locations. With existing techniques, one could partition the dataspace according to a regular grid and split the available privacy budget between two aggregate query types, one counting user locations in each grid cell, and the other summing reported values. Next, a temperature heatmap is obtained by averaging the temperature for each cell. As we show in our experimental evaluation, this approach results to useless data, due to the high amount of noise injected. This is the result of a more fundamental limitation of existing approaches that are designed only for general-purpose queries, and do not take into account correlations that are specific to more complex data processing algorithms.

In this paper, we propose an accurate technique for privacy-preserving detection of anomalous phenomena in crowdsourced sensing. We also adopt the powerful semantic model of *differential privacy*, but we devise a tailored solution, specifically designed for privacy-preserving heatmap construction. Our technique builds a flexible data indexing structure that can provide query results at arbitrary levels of granularity. Furthermore, the sanitization process fuses together distinct types of information (e.g., user count, placement and reported value scale) to obtain an effective privacy-preserving data representation that can help decide with high accuracy whether the sensed value in a certain geographical region exceeds the threshold or not. To the best of our knowledge, this is the first work that addresses the problem of value heatmap construction within the differential privacy framework. Our specific contributions are:

1. We introduce a hierarchical differentially-private structure for representing sensed data collected by mobile users. The structure is customized to address the specific requirements of value heatmap construction, and accurately supports queries at variable levels of granularity.
2. We examine the impact of structure parameters and privacy budget allocation on data accuracy, and devise algorithms for parameter selection and tuning.
3. We investigate techniques for reducing the impact of DP-injected noise, and devise effective voting strategies during data processing that increase accuracy of anomalous phenomenon detection.
4. We perform an extensive experimental evaluation which shows that the proposed techniques accurately detect anomalous phenomena, and clearly outperform existing general-purpose sanitization methods that fare poorly when applied to the studied problem.

The paper is organized as follows: Sect. 2 provides background information on differential privacy. In Sect. 3, we introduce the system model, and the metrics used to characterize anomalous phenomenon detection accuracy. Section 4 presents the proposed privacy-preserving data indexing structure and analytical models for characterizing query accuracy. We introduce strategies for anomaly detection in Sect. 5, followed by experimental evaluation results in Sect. 6. We present related work in Sect. 7, and conclude with directions for future work in Sect. 8.

2 Background

2.1 Differential Privacy

Differential privacy (DP) [2,3] addresses the limitation of syntactic privacy models (e.g., k-anonymity [19], ℓ-diversity [12], t-closeness [9]) which are vulnerable against background knowledge attacks. DP is a semantic model which argues that one should minimize the risk of disclosure that arises from an individual's participation in a dataset.

Two datasets $\mathcal{D}$ and $\mathcal{D}'$ are said to be *siblings* if they differ in a single record r, i.e., $\mathcal{D}' = \mathcal{D} \cup \{r\}$ or $\mathcal{D}' = \mathcal{D} \setminus \{r\}$. An algorithm $\mathcal{A}$ is said to satisfy differential privacy with parameter ε (called *privacy budget*) if the following condition is satisfied [2]:

Definition 1 (ε-indistinguishability). *Consider algorithm $\mathcal{A}$ that produces output $\mathcal{O}$ and let $\epsilon > 0$ be an arbitrarily-small real constant. Algorithm $\mathcal{A}$ satisfies ε-indistingui-shability if for every pair of sibling datasets $\mathcal{D}, \mathcal{D}'$ it holds that*

$$\left| \ln \frac{Pr[\mathcal{A}(\mathcal{D}) = \mathcal{O}]}{Pr[\mathcal{A}(\mathcal{D}') = \mathcal{O}]} \right| \leq \varepsilon \tag{1}$$

In other words, an attacker is not able to learn, with significant probability, whether output $\mathcal{O}$ was obtained by executing $\mathcal{A}$ on input $\mathcal{D}$ or $\mathcal{D}'$. To date, two prominent techniques have been proposed to achieve ε-indistinguishability [3,13]: the *Laplace mechanism* (and the closely related geometric mechanism for integer-valued data) and the *exponential mechanism*. Both mechanisms are closely related to the concept of *sensitivity*.

Definition 2 (*L_1-sensitivity* [3]). *Given any two sibling datasets $\mathcal{D}$, $\mathcal{D}'$ and a set of real-valued functions $\mathcal{F} = \{f_1, \ldots, f_m\}$, the L_1-sensitivity of $\mathcal{F}$ is measured as $\Delta_{\mathcal{F}} = \max_{\forall \mathcal{D}, \mathcal{D}'} \sum_{i=1}^{m} |f_i(\mathcal{D}) - f_i(\mathcal{D}')|$.*

The *Laplace mechanism* is used to publish the results to a set of statistical queries. A statistical query set $\mathcal{Q} = \{Q_1, \ldots, Q_m\}$ is the equivalent of a set of real-valued functions, hence the sensitivity definition immediately extends to such queries. According to [3], to achieve DP with parameter ε it is sufficient to add to each query result random noise generated according to a Laplace distribution with mean $\Delta_{\mathcal{Q}}/\varepsilon$. For COUNT queries that do not overlap in the data domain (e.g., finding the counts of users enclosed in disjoint grid cells), the sensitivity is 1.

An important property of differentially-private algorithms is *sequential composability* [13]. Specifically, if two algorithms $\mathcal{A}_1$ and $\mathcal{A}_2$ executing in isolation on dataset $\mathcal{D}$ achieve DP with privacy parameters ε_1 and ε_2 respectively, then executing both $\mathcal{A}_1$ and $\mathcal{A}_2$ on $\mathcal{D}$ in sequence achieves DP with parameter $(\varepsilon_1 + \varepsilon_2)$. In contrast, *parallel composability* specifies that executing $\mathcal{A}_1$ and $\mathcal{A}_2$ on disjoint partitions of the dataset achieves DP with parameter $\max(\varepsilon_1, \varepsilon_2)$.

2.2   Private Spatial Decompositions (PSD)

The work in [1] introduced the concept of *Private Spatial Decompositions (PSD)* to release spatial datasets in a DP-compliant manner. A PSD is a spatial index transformed according to DP, where each index node is obtained by releasing a noisy count of the data points enclosed by that node's extent. Various index types such as grids, quad-trees or k-d trees [18] can be used as a basis for PSD.

Accuracy of PSD is heavily influenced by the type of PSD structure and its parameters (e.g., height, fan-out). With space-based partitioning PSD, the split position for a node does not depend on data point locations. This category includes flat structures such as grids, or hierarchical ones such as BSP-trees (Binary Space Partitioning) and quad-trees [18]. The privacy budget ϵ needs to be consumed only when counting the users in each index node. Typically, all nodes at same index level have non-overlapping extents, which yields a constant and low sensitivity of 1 per level (i.e., adding/removing a single location in the data may affect at most one partition in a level). The budget ϵ is best distributed across levels according to the *geometric allocation* [1], where leaf nodes receive more budget than higher levels. The sequential composition theorem applies across nodes on the same root-to-leaf path, whereas parallel composition applies to disjoint paths in the hierarchy. Space-based PSD are simple to construct, but can become unbalanced.

Object-based structures such as k-d trees and R-trees [1] perform splits of nodes based on the placement of data points. To ensure privacy, split decisions must also be done according to DP, and significant budget may be used in the process. Typically, the exponential mechanism [1] is used to assign a merit score to each candidate split point according to some cost function (e.g., distance from median in case of k-d trees), and one value is randomly picked based on its noisy score. The budget must be split between protecting node counts and building the index structure. Object-based PSD are more balanced in theory, but they are not very robust, in the sense that accuracy can decrease abruptly with only slight changes of the PSD parameters, or for certain input dataset distributions.

The recent work in [16] compares tree-based methods with multi-level grids, and shows that two-level grids tend to perform better than recursive partitioning counterparts. The paper also proposes an *Adaptive Grid (AG)* approach, where the granularity of the second-level grid is chosen based on the noisy counts obtained in the first-level (sequential composition is applied). AG is a hybrid which inherits the simplicity and robustness of space-based PSD, but still uses a small amount of data-dependent information in choosing the granularity for the second level.

All these methods assume general-purpose and homogeneous queries (i.e., find counts of users in various regions of the dataspace), and, as we show later in this paper, are not suitable for the problem of anomalous phenomenon detection. We compare against state-of-the-art PSD techniques in our experimental evaluation (Sect. 6).

3   System Model and Evaluation Metrics

We consider a two-dimensional geographical region and a phenomenon characterized by a scalar value (e.g., temperature, CO_2 concentration) within domain $[0, M]$. A number of N mobile users measure and report phenomenon values recorded at their location. If a regular grid is super-imposed on top of the data domain, then the histogram obtained by averaging the values reported within each grid cell provides a *heatmap* of the observed phenomenon. Since our focus is on detecting anomalous phenomena, the actual value in each grid cell is not important; instead, what we are concerned with is whether a cell value is above or below a given threshold T, $0 < T < M$.

Mobile users report sensed values to a trusted data collector, as illustrated in Fig. 1. The collector sanitizes the set of reported values according to differential privacy with parameter ε, and outputs as result a data structure representing a noisy index of the data domain, i.e., a PSD. This PSD is then released to data recipients (i.e., general public) for processing. Based on the PSD, data recipients are able to answer queries with arbitrary granularity that is suitable for their specific data uses. Furthermore, each data recipient has flexibility to choose a different threshold value T in their analysis. In practice, the trusted collector role can be fulfilled by cell phone companies, which already know the locations of mobile users, and may be bound by contractual obligations to protect users' location privacy. The collector may charge a small fee to run the sanitization process, or can perform this service free of charge, and benefit from a tax break, e.g., for supporting environmental causes.

According to differential privacy, the goal of the protection mechanism is to hide whether a certain individual contributed to the set of sensed values or not. To achieve protection, noise is added to the values of individual value reports. Furthermore, fake value reports may have to be inserted, and some actual readings may have to be deleted from the dataset. Inherently, protection decreases data accuracy.

To measure the accuracy of sanitization, we need to quantify the extent to which the outcome for certain regions changes from above the threshold to below, or vice-versa. Given an arbitrary-granularity regular grid, we define the following metrics:

ϕ_{both}: number of grid cells above the threshold according to *both* actual and sanitized readings.
ϕ_{either}: number of grid cells above the threshold according to *either* actual or sanitized readings.
ϕ_{flip}: number of grid cells above the threshold in one dataset and below in the other.
ϕ_{all}: total number of grid cells.

It results immediately from the metric definitions that $\phi_{either} = \phi_{flip} + \phi_{both}$. Hence, we can define two additional metrics with domain $[0, 1]$ and ideal value of 1 (i.e., perfect accuracy). **FlipRatio (FR)** quantifies the proportion of cells that change their outcome due to sanitization:

$$FR = 1 - \frac{\phi_{flip}}{\phi_{all}}$$

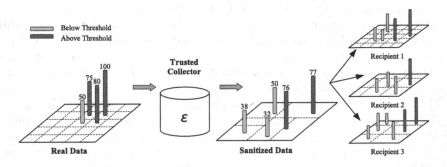

Fig. 1. System model

The **Jaccard (J)** metric, derived from *the Jaccard similarity coefficient* [2], measures the dissimilarity between the real and sanitized datasets:

$$J = \frac{\phi_{both}}{\phi_{either}}$$

The FR and J metrics have the advantage of being less dependent on the grid granularity, i.e., the ϕ_{all} values, so they maintain their relevance across a broad range of query granularities. However, only the J metric captures the local impact of the sanitization method. Interchanging the state of two random cells will not change the values of any other metrics than J, so they are not sufficient to determine the accuracy of the heatmap. Therefore, in the rest of the paper, we focus on the J metric. Formally, our problem statement is:

Problem 1. Given N users moving within a two-dimensional space, a phenomenon characterized by a scalar value with domain range $[0, M]$, an anomaly threshold T, $0 < T < M$ and privacy budget ε, determine an ε-differentially-private release such that the Jaccard metric between the real and sanitized dataset is maximized.

4   PSD for Anomalous Phenomenon Detection

Constructing an appropriate PSD is an essential step, since the accuracy of the entire solution depends on the structure properties. Furthermore, due to the specific requirements of our problem, general-purpose PSDs such as the ones optimized for count queries [1,16,17] are not suitable.

The anomalous phenomenon detection may be performed with respect to a regular grid of arbitrarily fine-grained granularity. On the other hand, creating a PSD that is too fine-grained is not a suitable approach. According to the Laplace mechanism, each cell's query result is added with random noise of magnitude independent of the actual value. Therefore, PSDs with small cells and PSDs that do not adapt to data density are not appropriate, as the resulting inaccuracy is high. Instead, we construct a flexible structure, based on which the threshold condition can be answered for arbitrary regular grids, as illustrated on the right side of Fig. 1.

The PSD must keep track of two measures necessary to determine phenomena heatmaps: sensor counts[1] and phenomenon value sums, which together provide average values for each cell. We denote the actual values for sensor count and value sum in a cell by n and s, respectively (we use subscript indices to distinguish the n and s values across cells). We denote the sanitized counts and sums by n^* and s^*. The sensitivity of n is 1, whereas the sensitivity of s is M (adding a new sensor in a cell can increase n by 1 and s by M). Hence, if n is answered using privacy budget ε_n and s is answered using privacy budget ε_s, the variance of n^* is $\frac{2}{\varepsilon_n^2}$, whereas the variance of s^* is $\frac{2M^2}{\varepsilon_s^2}$.

---

[1] In the rest of the paper, the terms *mobile user* and *sensor* are used interchangeably.

To simplify presentation, we introduce our PSD in incremental fashion: first, we outline the main concepts and parameters for a single-level regular grid. Next, we extend our findings to a two-level structure, and then generalize to a multiple-level structure. Table 1 summarizes the notations used.

Single-level Grid. Assume a regular grid of $N_0 \times N_0$ cells spanning over a data domain of size $w \times w$. Similar to other work on PSD [10,16], we assume that a negligible fraction of the privacy budget is spent to estimate n_0^*, the total number of sensors, and s_0^*, the sum of all sensed values. Granularity N_0 must be chosen to minimize the expected error over all rectangular queries (since any query can be decomposed into non-overlapping rectangular regions). The error has two sources:

– *Laplace error* within a single cell due to noise addition by the Laplace mechanism. These errors are added for all cells covered by the query.
– *Non-uniformity error* caused by non-uniformity of sensor distribution within a grid cell. These errors occur only for cells which are partially covered by the query rectangle. In such a case, we output a value proportional to the fraction of the cell that overlaps the query.

Furthermore, errors occur for both sensor counts and sensed values. Since the threshold T is expected to be proportional to scale M, we normalize the error for sensed values to account for the skew introduced by M. The error expression subject to minimization becomes the sum of all count errors plus $\frac{1}{M}$ of the sum of all value sum errors.

Table 1. Symbols and notations used in the paper.

Symbol	Description
n, s	Real count and sum of values of sensors in a cell
n^*, s^*	Noisy count and sum of values of sensors in a cell
n', s'	Count and sum of values of sensors in a cell after weighted averaging
$\bar{n}, \bar{s}$	Count and sum of values of sensors in a cell after mean consistency step
ε	Privacy budget
$\varepsilon_n, \varepsilon_s$	Privacy budget used for answering count and, respectively, sum queries in the cell
α	Proportion of available privacy budget to use at current PSD level
β	Proportion of privacy budget for the current level used for answering count queries
N_u	Split factor for cell u
M	Maximum value of a sensor's scale
T	Threshold for the anomalous heatmap
N_t	Threshold for minimum (noisy) number of sensors in a cell
K	Non-uniformity constant

Consider an arbitrary rectangle query of size rw^2, $r \in (0,1)$. The query will cover approximately rN_0^2 cells. The total variance of the query result is $\frac{2rN_0^2}{\varepsilon_n^2}$ for n and $\frac{2M^2rN_0^2}{\varepsilon_s^2}$ for s. Hence, the count error is expressed as $\sqrt{2r}\frac{N_0}{\varepsilon_n}$, and the sum error as $\sqrt{2r}\frac{MN_0}{\varepsilon_s}$. The total Laplace error is $\sqrt{2r}N_0\left(\frac{1}{\varepsilon_n} + \frac{1}{\varepsilon_s}\right)$.

The query rectangle might partially cover some cells. The number of such cells is of the order $\mathcal{O}(\sqrt{r}N_0)$ (determined by the perimeter of the query rectangle). Hence, we can assume that the number of points in partially covered cells is of the order $\mathcal{O}(\sqrt{r}N_0\frac{n_0^*}{N_0^2}) = K\sqrt{r}\frac{n_0^*}{N_0}$, where K is a constant. Assuming uniform sensor density, the error for value sum in partially covered cells is $K\sqrt{r}\frac{s_0^*}{N_0}$. Hence, the non-uniformity error is $K\frac{\sqrt{r}}{N_0}\left(n_0^* + \frac{s_0^*}{M}\right)$.

Thus, we must minimize the expression:

$$\sqrt{2r}N_0\left(\frac{1}{\varepsilon_n} + \frac{1}{\varepsilon_s}\right) + K\frac{\sqrt{r}}{N_0}\left(n_0^* + \frac{s_0^*}{M}\right) \tag{2}$$

According to the sequential composition property (Sect. 2), the available privacy budget ε must be split between ε_n and ε_s. We capture this split with parameter $\beta \in (0,1)$, defined as the fraction used by the count sanitization: $\varepsilon_n = \beta\varepsilon$ and $\varepsilon_s = (1-\beta)\varepsilon$. Minimizing Eq. (2) with respect to N_0, we obtain the optimal single-level granularity

$$N_0 = \sqrt{\varepsilon \times \frac{K}{\sqrt{2}} \times \beta(1-\beta)\left(n_0^* + \frac{s_0^*}{M}\right)} \tag{3}$$

Two-level Grid. Starting with the optimal single-level N_0 setting, we further divide each cell according to its noisy n^* and s^*. The privacy budget must be split between the two levels according to sequential composition. We model this split with parameter $\alpha \in (0,1)$, which quantifies the budget fraction allocated to the level 1 grid. Levels 1 and 2 receive respectively budgets $\varepsilon_1 = \alpha\varepsilon$ and $\varepsilon_2 = (1-\alpha)\varepsilon$. Each level budget is further divided between counts and sums using parameter $\beta \in (0,1)$:

$$\varepsilon_{n1} = \beta\varepsilon_1, \varepsilon_{s1} = (1-\beta)\varepsilon_1, \varepsilon_{n2} = \beta\varepsilon_2, \varepsilon_{s2} = (1-\beta)\varepsilon_2 \tag{4}$$

Since each level-1 cell is further divided, we define N_0 as a fraction of the value in Eq. (3) (later in this section, Eq. (11) shows how to choose η):

$$N_0 = \frac{1}{\eta}\sqrt{\varepsilon \times \frac{K}{\sqrt{2}} \times \beta(1-\beta)\left(n_0^* + \frac{s_0^*}{M}\right)} \tag{5}$$

For each cell u in the first level we use budgets ε_{n1} and ε_{s1} to determine n_{u1}^* and, respectively, s_{u1}^*. Based on these values, we split cell u into N_u^2 cells. For each cell $v \in child(u)$, we use ε_{n2} and ε_{s2} to determine n_{v2}^* and, respectively, s_{v2}^* (the subscript indicates the level of the grid where the value is computed).

Since the actual sensor count in a cell at level 1 is the same as the sum of the sensor counts in all of its children at level 2 (and the same holds for the sums), we perform a constrained inference procedure with the purpose of improving accuracy. Based on the values n_{u1}^*, s_{u1}^*, n_{v2}^*, s_{v2}^* we determine $\overline{n_{u1}}$, $\overline{s_{u1}}$, $\overline{n_{v2}}$ and $\overline{s_{v2}}$ such that

$$\overline{n_{u1}} = \sum_{v \in child(u)} \overline{n_{v2}}$$

$$\overline{s_{u1}} = \sum_{v \in child(u)} \overline{s_{v2}}$$

and $\forall u$, the variances of $\overline{n_{u1}}$ and $\overline{s_{u1}}$ are minimized. Note that, since all input values are already sanitized, no budget is consumed in the constrained inference step, and differential privacy is still enforced.

We determine these values in two steps:

1. We determine the **weighted average** estimators n_{u1}' and s_{u1}' with minimal variance. We average the values of n_{u1}^* and $\sum_{v \in child(u)} n_{v2}^*$ to determine n_{u1}' and the corresponding ones for s_{u1}'. To do so, we are using the fact that the variance of the weighted average of two random variables X and Y with variances $Var(X)$ and $Var(Y)$ is minimized by the value

$$\frac{Var(Y)}{Var(X) + Var(Y)} \times X + \frac{Var(X)}{Var(X) + Var(Y)} \times Y \tag{6}$$

In our case, X is n_{u1}' (s_{u1}') and Y is $\sum_{v \in child(u)} n_{v2}^*$ (respectively $\sum_{v \in child(u)} s_{v2}^*$).

2. We update the values to ensure **mean consistency** according to:

$$\overline{n_{u1}} = n_{u1}' \quad \overline{n_{v2}} = n_{v2}' + \frac{1}{N_u^2} \left(\overline{n_{u1}} - \sum_{v \in child(u)} n_{v2}' \right) \tag{7}$$

$$\overline{s_{u1}} = s_{u1}' \quad \overline{s_{v2}} = s_{v2}' + \frac{1}{N_u^2} \left(\overline{s_{u1}} - \sum_{v \in child(u)} s_{v2}' \right) \tag{8}$$

The effects of the constrained inference so far concern only queries which partially cover level-1 cells. Suppose that a query covers $i \times j$ sub-cells of cell u, where $i, j \in \{1, 2, \ldots N_u\}$. Then, the effect of the constrained inference is that $\min(i \times j, N_u^2 - i \times j)$ level-2 cells will be used to answer the query. On average, the number of level-2 cells required to answer a query is:

$$\frac{1}{N_u^2 - 1} \sum_{i=1}^{N_u} \sum_{j=1}^{N_u} \min(i \times j, N_u^2 - i \times j) \approx \frac{N_u^2}{5} + \mathcal{O}(N_u)$$

Hence, the total variances are $\frac{2N_u^2}{5\varepsilon_{n2}^2}$ and $\frac{2M^2 N_u^2}{5\varepsilon_{s2}^2}$, and the resulting total Laplace error is $\frac{\sqrt{10} N_u}{5} \left(\frac{1}{\varepsilon_{n2}} + \frac{1}{\varepsilon_{s2}} \right)$.

For non-uniformity errors, assume r is the ratio between the area used to answer the query and the total area of the cell. We know from the single-level case that the non-uniformity errors are $K\sqrt{r}\frac{n_u^*}{N_u}$ and $K\sqrt{r}\frac{s_u^*}{N_u}$. To eliminate the $\sqrt{r}$ factor, we integrate over its domain $((0, 0.5])$ and compute the expected value of the total non-uniformity error. Since $\frac{\int_0^{0.5}\sqrt{r}dr}{\int_0^{0.5}dr} = \frac{\sqrt{2}}{3}$ we get that the total non-uniformity error is $\frac{\sqrt{2}K}{3N_u}\left(n_u^* + \frac{s_u^*}{M}\right)$.

Thus, we must minimize the expression

$$\frac{\sqrt{10}N_u}{5}\left(\frac{1}{\varepsilon_{n2}} + \frac{1}{\varepsilon_{s2}}\right) + \frac{\sqrt{2}K}{3N_u}\left(n_u^* + \frac{s_u^*}{M}\right)$$

and we obtain

$$N_u = \sqrt{\frac{\sqrt{5}}{3}\varepsilon K\beta(1-\beta)(1-\alpha)\left(n_u^* + \frac{s_u^*}{M}\right)} \tag{9}$$

where we can approximate $\frac{\sqrt{10}}{3}$ by 1. This also provides a value for η (Eq.(5)), such that:

$$N_0 = \sqrt{\varepsilon \times \frac{K}{\sqrt{2}} \times \beta(1-\beta)\alpha\left(n_0^* + \frac{s_0^*}{M}\right)} \tag{10}$$

$$N_u = \sqrt{\varepsilon \times \frac{K}{\sqrt{2}} \times \beta(1-\beta)(1-\alpha)\left(n_u^* + \frac{s_u^*}{M}\right)} \tag{11}$$

Generalization to Multiple Levels. The analysis for two levels can be extended to a multiple-level structure, where the privacy budget is split across levels (keeping $\alpha\varepsilon$ for the current level and dividing privacy budget between count and sum using β, as before), and the granularity for each new level is determined based on the sanitized data and variance analysis at the previous level. However, we must carefully decide when to end the recursion, as having too many levels will decrease the budget per level, and consequently decrease accuracy. Because of this, we implement two stopping mechanisms: first, we introduce a maximum depth of the PSD, max_depth, to prevent excessive reduction of per-level privacy budget. Second, we introduce a threshold, N_t such that a cell u is divided only if its estimated sensor count satisfies inequality $n_u^* > N_t$.

The number N_u of children nodes of u is given by:

$$N_u = \sqrt{\varepsilon_u \times \frac{K}{\sqrt{2}} \times \beta(1-\beta)(1-\alpha)\left(n_u^* + \frac{s_u^*}{M}\right)} \tag{12}$$

We illustrate the proposed multiple-level PSD approach with a running example, in parallel with the description of the pseudocode provided in Algorithm 1. The PSD is built in three phases. First, the PSD structure is determined (i.e., the spatial extent of each index node), by splitting cells according to Eq. (12), and noisy values are computed for sensor counts and value sums. This is the only step that accesses the real dataset of readings, and hence the only step that consumes privacy budget. The recursive procedure `buildPSD` (Algorithm 1) summarizes this process.

Algorithm 1. Splitting a PSD cell u at depth $depth$, with privacy budget ε

1: **function** BUILDPSD(ε, u, $depth$)
2:     **if** $depth == max\_depth$ **then**
3:         $\varepsilon_{crt} \leftarrow \varepsilon$
4:     **else**
5:         $\varepsilon_{crt} \leftarrow \alpha\varepsilon$
6:     **end if**
7:     $\varepsilon_n \leftarrow \beta\varepsilon_{crt}$
8:     $\varepsilon_s \leftarrow (1 - \beta)\varepsilon_{crt}$
9:     $(n, s) \leftarrow$ GETREALVALUES(u)
10:     $n^* \leftarrow n +$ LAPLACE($1/\epsilon_n$)
11:     $s^* \leftarrow s +$ LAPLACE(M/ϵ_s)
12:     $N_u \leftarrow$ COMPUTESPLIT(ε, n^*, s^*)
13:     **if** $N_u < Nt$ **then**
14:         $\varepsilon_n \leftarrow \beta(1 - \alpha)\varepsilon$
15:         $\varepsilon_s \leftarrow (1 - \beta)(1 - \alpha)\varepsilon$
16:         $n'^* \leftarrow n +$ LAPLACE($1/\epsilon_n$)
17:         $s'^* \leftarrow s +$ LAPLACE(M/ϵ_s)
18:         $n' \leftarrow$ AVERAGE(n^*, n'^*)
19:         $s' \leftarrow$ AVERAGE(s^*, s'^*)
20:     **end if**
21:     **for all** $v \in$ SPLITCELL($u, N_u, depth$) **do**
22:         BUILDPSD($(1 - \alpha)\varepsilon, v, depth + 1$)
23:     **end for**
24: **end function**

Figure 2 illustrates PSD construction with $\alpha = 0.2$, $\beta = 0.5$ and $\varepsilon = 1.6$. The root node will receive a budget of $\varepsilon_{n,root} = 0.5 \times 0.2 \times 16 = 0.16$ (lines 2–8 of Algorithm 1). Line 9 computes the real values for the count and sum of sensor values inside the cell (the sensor counts for the running example are presented in Fig. 2(d)). Lines 10–11 add Laplace noise, resulting in a value of $n^*_{root} = 14$. The split granularity for next level is determined as in Eq. (12). Assume we obtain $N_u = 4$, larger than the threshold $N_t = 2$. The root is split into four cells, and the procedure is recursively applied to each of them with $\varepsilon_1 = (1 - \alpha)\varepsilon = 0.8 \times 1.6 = 1.28$.

The budget for level 1 is further split between sum and count values, to obtain $\varepsilon_{n,1} = 0.128$ (lines 2–8). Adding the corresponding Laplace noise to the real values of 2, 1, 2 and 3 (Fig. 2(d)) (lines 10–11), results in noisy counts 9, 2, 6 and, respectively, -2 (Fig. 2(a)).

The cells with values 9 and 6 are further split, while the one with $n^*_1 = -2$ is not, due to the value of N_t. In case no further splits are performed, the remaining

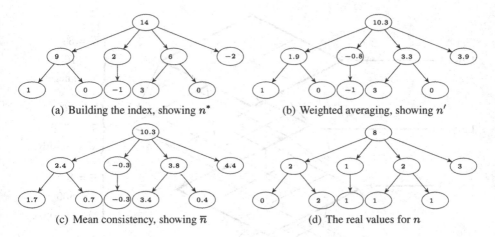

(a) Building the index, showing n^*

(b) Weighted averaging, showing n'

(c) Mean consistency, showing $\bar{n}$

(d) The real values for n

Fig. 2. Representation of PSD construction, including weighted averaging and mean consistency.

budget is used by running lines 13–20 of Algorithm 1, which compute new noisy estimates which are averaged to determine n' and, respectively, s'.

Since the remaining cells are at the maximal depth allowed by the method, the remaining privacy budget of $\varepsilon_{n,2} = 0.512$ is used to compute the remaining noisy values. The result of the algorithm is shown in Fig. 2(a).

The second phase of the index building method is **weighted averaging**. We average for each internal node the two estimates and compute n' and s' according to Eq. (6). For each node, we keep track of the variance of the noisy variables and the averaged values, since they will be needed in the higher levels of the tree. The resulting tree at the end of this phase is shown in Fig. 2(b).

Finally, the last phase performs **mean consistency**, which ensures that the estimate from one node is the same as the sum of the estimates from its children. We use Eqs. (7) and (8) in a top-down traversal of the tree, the result of which is shown in Fig. 2(c).

5 PSD Processing and Heatmap Construction

As illustrated in Fig. 1 (Sect. 3), after the PSD is finalized at the trusted collector, it is distributed to data recipients who process it according to their own granularity and threshold requirements. The objective of the data recipient is to obtain a binary heatmap that captures areas with anomalous phenomena, i.e., regions of the geographical domain where the measured values are above the recipient-specified threshold.

We assume that the recipient is interested in building a heatmap according to a *recipient resolution grid* (rrg). Recall that our solution is designed to be flexible with respect to recipient requirements, and each recipient may have its own rrg of arbitrary granularity. In this section, we show how a recipient is able to accurately determine a phenomenon heatmap given as input the PSD, the

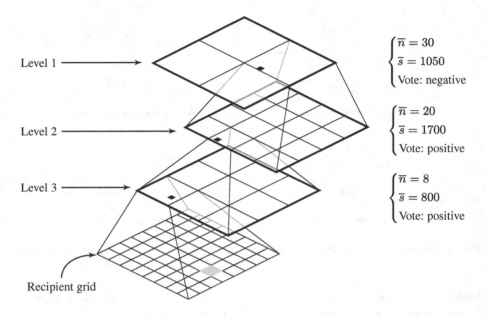

Level 1 $\longrightarrow$

$\begin{cases} \overline{n} = 30 \\ \overline{s} = 1050 \\ \text{Vote: negative} \end{cases}$

Level 2 $\longrightarrow$

$\begin{cases} \overline{n} = 20 \\ \overline{s} = 1700 \\ \text{Vote: positive} \end{cases}$

Level 3 $\longrightarrow$

$\begin{cases} \overline{n} = 8 \\ \overline{s} = 800 \\ \text{Vote: positive} \end{cases}$

Recipient grid

Fig. 3. Construction of heatmap at the data recipient site

recipient-defined rrg and threshold T. The objective of heatmap construction is to determine for each rrg cell a binary outcome: *positive* if the value derived for the cell is above T, and *negative* otherwise.

Figure 3 shows an example of rrg superimposed on the PSD index. The PSD has four levels, out of which only three are shown (the root is split into four cells, and it is omitted from the diagram due to space considerations). The bottom layer in the diagram represents the rrg. The shaded cell in the rrg layer represents the cell for which we are currently determining the outcome. In this example, we illustrated a high-resolution rrg, so most rrg cells are completely enclosed within a PSD cell at each index level. However, in general, there may be cases when a rrg cell overlaps with several PSD cells. We consider both cases below.

Since the recipient has no other information other than the PSD, we assume that the count and sum values inside a PSD cell are uniformly distributed over the cell's extent. Hence, for each rrg cell we compute n and s in proportion to the overlap between the rrg and PSD cells, normalized by the PSD cell area. If one rrg cell overlaps two or more PSD cells, the values for n and s are determined as the weighted sum of the values corresponding to each PSD cell, where the weight is represented by the overlap amount.

Note that, even if the above procedure may result in values for n and s for each rrg cell which are not too far apart from the actual values, there is another important source of inaccuracy due to the fact that the outcome for an rrg cell is obtained by dividing the noisy s and n values. The ratio can be significantly affected even if the noise is not very high. Furthermore, even though the leaf cells of the PSD are likely to be closer in resolution to the rrg grid, considering solely leaf nodes in the outcome evaluation may have undesirable effects, due to the fact that the noise added to leaf nodes is more significant compared to their actual values compared to PSD nodes that are higher in the hierarchy (i.e., relative errors are higher closer to the leaf level).

In our solution, we account for these factors. Instead of naïvely dividing estimates for n and s in each rrg grid cell (which may have low accuracy), we evaluate individually the outcome based on information at each PSD level, and then combine the outcomes through a voting process in order to determine the outcome for each individual rrg cell. Returning to the example in Fig. 3, assume that threshold $T = 80$. We determine the outcome of the gray cell at the rrg layer by using the outcomes for all the marked PSD cells on the three levels shown (cells are marked using a small black square). Specifically, the Level 1 PSD cell containing the shaded grid cell has $\bar{n} = 30$ and $\bar{s} = 1050$, resulting in a phenomenon value $\bar{p} = \frac{\bar{s}}{\bar{n}} = 35$, below the threshold $T = 80$. Hence, the root cell's vote would be negative, meaning that with the information from that layer, the grayed grid cell does not present an anomalous reading.

However, at Level 2 of the PSD, we have $\bar{n} = 20$ and $\bar{s} = 1700$, resulting in a value of 85, greater than the threshold. Hence, this layer will contribute a positive vote. Similarly, at Level 3, $\bar{n} = 8$ and $\bar{s} = 800$ which also results in a positive vote.

The resulting outcome for any rrg cell depends on the distribution of the votes it has received. We could use the difference between positive and negative votes, but this will report a biased result for grid cells overlapping multiple PSD cells at the same level. A better solution is to use the ratio of positive votes to the total votes. In our example, the grayed cell got two positive votes and a single negative one, hence it would be marked as anomalous.

An alternative approach is to use only the number of positive votes that have been received. For instance, a rrg cell would receive a positive outcome if at least two PSD cells vote positively. This approach has two advantages: first, it captures locality better than the previous strategy. If the region where the phenomenon has an anomalous value is small, majority voting would tend to flatten the heatmap at higher levels, and the sharp spike may be missed. The two-vote strategy, however, may correctly identify the spike if both the leaf level PSD and another level above vote positively. Second, the two-vote strategy may prevent false alarms, caused by small PSD cells that may receive a high amount of random noise. By having a second level confirm the reading, many of the false negatives are eliminated, as it is unlikely that two PSD cells at different levels that overlap each other both receive very high noise due to the Laplace mechanism.

6   Experiments

We evaluate experimentally the proposed technique for privacy-preserving detection of anomalous phenomena. We implemented a C prototype, and we ran our experiments on an Intel Core i7-3770 3.4 GHz CPU machine with 8 GB of RAM running Linux OS. We first provide a description of the experimental settings used. Next, we evaluate the accuracy of our technique in comparison with benchmarks. Finally, we investigate the performance of our technique when varying fundamental system parameters.

Experimental Settings. We consider a square two-dimensional location space with size 100×100, and a phenomenon with range $M = 100$ and threshold $T = 80$. We consider between $10,000$ and $50,000$ mobile users (i.e., sensors), uniformly distributed over the location domain. The average non-anomalous phenomenon value is 20, and to simulate an anomaly we generate a Gaussian distribution of values with scale parameter 20, centered at a random focus point within the location domain.

We consider two benchmark techniques for comparison. The first method, denoted as *Uniform Grid (U)*, considers a single-level fixed-granularity regular grid. The parameters of the grid are chosen according to the calculations presented in the first part of Sect. 4. The second method, *Adaptive Grid (AG)*, implements the state-of-the-art technique for PSDs as introduced in [16]. Specifically, it uses a two-level grid, where the first grid granularity is chosen according to a fixed split as indicated in [16], whereas the second-level granularity is determined based on the data density in the first level.

Comparison with Competitor Methods. We measure the accuracy in detecting anomalous phenomena for the proposed tree-based technique (denoted as t) and the benchmarks U and AG when varying privacy budget ε. For fairness, we consider the *1-vote* decision variant, which is supported by all methods. Figure 4(a) shows that our technique (presented with two distinct depth settings) clearly outperforms both benchmarks with respect to the Jacard metric. The U and AG method are only able to achieve values around 0.1 or less. Furthermore, they are not able to make proper use of the available privacy budget, and sometimes accuracy decreases when ε increases. The reason for this behavior is that the procedure for grid granularity estimation proposed in [16] has some built-in constants that are only appropriate for specific datasets and query types. In our problem setting, the granularity of these choices increases when ε increases, and the noise injected offsets the useful information in each cell.

To validate the superiority of the proposed technique beyond the J metric, Fig. 4(b) and (c) provide visualization of the heatmap obtained for the U method and our technique, respectively (the heatmap obtained for AG is similar to that of U). The anomalous phenomenon in the real data is shown using the circle area (i.e., points inside the circle are above the threshold). The heatmap produced by the U method is dominated by noise, and indicates that there are small regions with above-the-threshold values randomly scattered over the data domain. In contrast, our technique accurately identifies a compact region that overlaps almost completely with the actual anomalous region. Furthermore, for the t technique we consider two distinct maximum depth settings, $d = 3$ and $d = 4$. We observe that, although both variants outperform the benchmarks, as the height of the structure increases, a potentially negative effect occurs due to the fact that the privacy budget per level decreases. Hence, it is not advisable to increase too much the PSD depth.

Both the UG and the AG method are unable to maintain data accuracy, and return virtually unusable data, without the ability to detect the occurrence of anomalous phenomena. In the rest of the experiments, we no longer consider competitor methods, and we focus on the effect of varying system parameters on the accuracy of the proposed technique. We also note that our method incurs low performance overhead, similar to that of the U method (between 2 and 4 s to sanitize and process the entire dataset). The AG method requires slightly longer, in the range of 15–20 s.

Effect of Varying System Parameters. We perform experiments to measure the accuracy of the proposed technique when varying fundamental system parameters, such as budget split parameters α, β and sensor count N.

Figure 5 shows the accuracy of our method when varying α, the budget split fraction across levels. Each graph illustrates several distinct combinations of budget ε and count-sum budget split β. For smaller α values, a smaller fraction of the budget is kept for the current level, with the rest being transferred for the children cells. Since the root node and the high levels of the tree have large spans, a smaller budget does not have a significant effect on accuracy, so it is best when a larger fraction is used in the lower-levels. For $\alpha = 0.2$, the proposed method reaches close to perfect J metric value.

We also illustrate the effect of the various decision variants based on voting. Comparing Fig. 5(a) and (b), we can see that the accuracy increases slightly for the 2-vote scenario. This confirms that the 2-vote approach is able to filter out cases where some large outlier noise in one of the lower-level cells creates a false positive. On the other hand, the majority-voting strategy from Fig. 5(c) obtains lower accuracy, as it suffers from a relatively high false negative rate. Even if some of the levels signal an alarm, it is possible that a large amount of noise on several levels flips the outcome to "below the threshold". We conclude that the 2-vote strategy is the best available option.

Figure 6 shows the effect of varying parameter β, which decides the privacy budget split between the counts and sums in the PSD. Similar to previous results, we observe that the majority voting strategy has lower accuracy, due to the

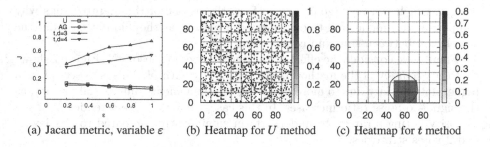

(a) Jacard metric, variable ε (b) Heatmap for U method   (c) Heatmap for t method

Fig. 4. Accuracy evaluation in comparison with U and AG benchmarks.

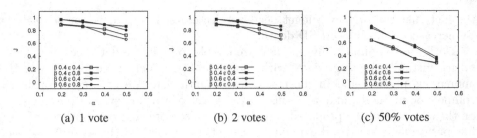

(a) 1 vote   (b) 2 votes   (c) 50% votes

Fig. 5. Impact of cross-level privacy budget split parameter α, $d = 3$

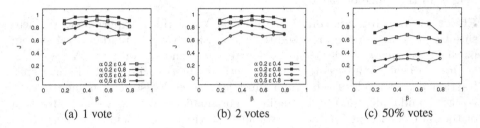

(a) 1 vote   (b) 2 votes   (c) 50% votes

Fig. 6. Impact of "count vs sum" privacy budget split parameter β, $d = 3$

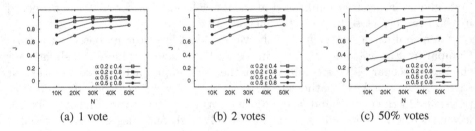

(a) 1 vote   (b) 2 votes   (c) 50% votes

Fig. 7. Impact of number of mobile users N, $\beta = 0.5$, $d = 3$

increased occurrence of false negatives. The results also show that an equal split between counts and sums yields good results. As long as the β split is not severely skewed, the parameter does not significantly influence accuracy. However, when β is excessively low or high, one of the sum or count components gets very little budget, which causes large errors. In fact, this is one of the main reasons why competitor techniques fail to obtain good accuracy, as they do not consider the correlation between sum and count errors.

Finally, we consider the effect of varying number of sensors N. Figure 7 shows that the accuracy of the method increases slightly with N. This is expected, as a higher data density due to more reporting sensors benefits differential privacy, as the signal-to-noise ratio increases. In this case, we also notice a tendency of the majority voting strategy to underperform significantly compared to the 1-vote and 2-votes strategies.

7   Related Work

Collaborative sensing enables information extraction from a large number of wireless devices, spanning from smart phones to motes in a WSN. We focus on personal devices which are carried by users and may be used in sensing applications – from tracking to shapes-detection – in settings in which there are no WSNs available [11,15]. Such settings occur in many real-life applications in which the deployment of a WSN is either not possible or the WSN approach is not sustainable. We note that collaborative sensing is, in some sense, a broader paradigm than *participatory sensing* and *opportunistic sensing*, and when it comes to issues related to privacy protection, it subsumes the ones from the latter two paradigms in the risk of leaking personal/sensitive information [8]. While privacy-preserving computation has its history in domains such as cryptography and data mining, the existing methodologies cannot be straightforwardly mapped into the collaborative sensing applications.

Existing work addressed different aspects of the problem of detecting and representing spatial features of a particular monitored phenomenon [4,5]. Spatial summaries (e.g., isocontours [5]) may be constructed for energy-efficient querying. A natural trade-off is the precision of the aggregated representation vs the energy efficiency.

Location privacy has been studied extensively. Some techniques make use of cryptographic protocols such as private information retrieval [6]. Another category of methods focuses on location cloaking, e.g., using spatial k-anonymity [7,14], where a user hides among k other users. As discussed in Sect. 2, such techniques have serious security drawbacks. Closest to our work are the PSD construction techniques in [1,16,17]. As discussed in Sect. 4, these techniques are general-purpose, and our experimental evaluation shows that they are not suitable for anomalous phenomenon detection.

8   Conclusions

We proposed an accurate differentially-private technique for detection of anomalous phenomena in crowdsourced environmental sensing. Our solution consists of a PSD specifically-tailored to the requirements of phenomenon heatmap data, and strategies for flexible processing of sanitized datasets with values collected from mobile users. Experimental results show that the proposed technique is accurate, and clearly outperforms existing state-of-the-art in private spatial decompositions. In the future, we plan to extend our solution to continuous monitoring of phenomena, where multiple rounds of reporting are performed. This scenario is more challenging, as an adversary may correlate readings from multiple rounds to breach individual privacy.

References

1. Cormode, G., Procopiuc, C., Srivastava, D., Shen, E., Yu, T.: Differentially private spatial decompositions. In: ICDE, pp. 20–31 (2012)

2. Dwork, C.: Differential privacy. In: Bugliesi, M., Preneel, B., Sassone, V., Wegener, I. (eds.) ICALP 2006. LNCS, vol. 4052, pp. 1–12. Springer, Heidelberg (2006)
3. Dwork, C., McSherry, F., Nissim, K., Smith, A.: Calibrating noise to sensitivity in private data analysis. In: Halevi, S., Rabin, T. (eds.) TCC 2006. LNCS, vol. 3876, pp. 265–284. Springer, Heidelberg (2006)
4. Fayed, M., Mouftah, H.T.: Localised alpha-shape computations for boundary recognition in sensor networks. Ad Hoc Netw. **7**(6), 1259–1269 (2009)
5. Gandhi, S., Kumar, R., Suri, S.: Target counting under minimal sensing: complexity and approximations. In: Fekete, S.P. (ed.) ALGOSENSORS 2008. LNCS, vol. 5389, pp. 30–42. Springer, Heidelberg (2008)
6. Ghinita, G., Kalnis, P., Khoshgozaran, A., Shahabi, C., Tan, K.L.: Private queries in location based services: anonymizers are not necessary. In: SIGMOD, pp. 121–132 (2008)
7. Gruteser, M., Grunwald, D.: Anonymous usage of location-based services through spatial and temporal cloaking. In: USENIX MobiSys (2003)
8. He, W., Liu, X., Nguyen, H.V., Nahrstedt, K., Abdelzaher, T.F.: PDA: privacy-preserving data aggregation for information collection. TOSN **8**(1), 6 (2011)
9. Li, N., Li, T., Venkatasubramanian, S.: T-closeness: privacy beyond k-anonymity and l-diversity. In: ICDE 2007, pp. 106–115. IEEE, Istanbul, Turkey (2007)
10. Li, N., Qardaji, W., Su, D., Cao, J.: Privbasis: frequent itemset mining with differential privacy. Proc. VLDB Endow. **5**(11), 1340–1351 (2012)
11. Li, W., Bao, J., Shen, W.: Collaborative wireless sensor networks: a survey. In: Proceedings of the IEEE International Conference on Systems, Man and Cybernetics, Anchorage, Alaska, USA, 9–12 October 2011, pp. 2614–2619 (2011)
12. Machanavajjhala, A., Gehrke, J., Kifer, D., Venkitasubramaniam, M.: l-diversity: privacy beyond k-anonymity. In: Proceedings of International Conference on Data Engineering (ICDE) (2006)
13. McSherry, F., Talwar, K.: Mechanism design via differential privacy. In: Proceedings of Annual IEEE Symposium on Foundations of Computer Science (FOCS), pp. 94–103 (2007)
14. Mokbel, M.F., Chow, C.Y., Aref, W.G.: The new casper: query processing for location services without compromising privacy. In: Proceedings of VLDB (2006)
15. Peralta, L.M.R., de Brito, L.M.P.L., Santos, J.F.F.: Improving users' manipulation and control on wsns through collaborative sessions. I. J. Knowl. Web Intell. **3**(3), 287–311 (2012)
16. Qardaji, W., Yang, W., Li, N.: Differentially private grids for geospatial data. In: Proceedings of IEEE Intlernational Conference on Data Engineering (ICDE) (2013)
17. Qardaji, W., Yang, W., Li, N.: Priview: practical differentially private release of marginal contingency tables. In: Proceedings of ACM SIGMOD (2014)
18. Samet, H.: The Design and Analysis of Spatial Data Structures. Addison-Wesley, Reading (1990)
19. Sweeney, L.: K-anonymity: a model for protecting privacy. Int. J. Uncertainty Fuzziness Knowl. Based Syst. **10**(5), 557–570 (2002)

Efficient Top-k Subscription Matching for Location-Aware Publish/Subscribe

Jiafeng Hu[1]($\boxtimes$), Reynold Cheng[1], Dingming Wu[1], and Beihong Jin[2]

[1] Department of Computer Science, The University of Hong Kong,
Pokfulam Road, Hong Kong, China
{jhu,ckcheng,dmwu}@cs.hku.hk

[2] State Key Laboratory of Computer Science,Institute of Software,
Chinese Academy of Sciences, Beijing, China
beihong@iscas.ac.cn

Abstract. The dissemination of messages to a vast number of mobile users has raised a lot of attention. This issue is inherent in emerging applications, such as location-based targeted advertising, selective information disseminating, and ride sharing. In this paper, we examine how to support location-based message dissemination in an effective and efficient manner. Our main idea is to develop a location-aware version of the *Pub/Sub* model, which was designed for message dissemination. While a lot of studies have successfully used this model to match the interest of *subscriptions* (e.g., the properties of potential customers) and *events* (e.g., information of casual users), the issues of incorporating the location information of subscribers and publishers have not been well addressed. We propose to model subscriptions and events by boolean expressions and location data. This allows complex information to be specified. However, since the number of publishers and subscribers can be enormous, the time cost for matching subscriptions and events can be prohibitive. To address this problem, we have developed the R^I-tree. This data structure is an integration of the R-tree and the dynamic interval-tree. Together with our novel pruning strategy on R^I-tree, our solution can effectively and efficiently return the top-k subscriptions with respect to an event. We have performed extensive evaluations to verify our approach.

1 Introduction

Due to the advance of telecommunications and Internet technologies, tremendous amounts of location information can now be obtained easily. For instance, a user's location is often tracked by base stations in a cellular network; a vehicle's position can be obtained through GPS receivers or sensors on roads; a user reveals her location when she "checks in" (e.g., through Facebook and Twitter). The availability of location information stimulates the development of location-based messaging services, which disseminates interesting messages to users based on their positions and other information. Taking location-based targeted advertising as an example, advertisements are sent to users selected in terms of age,

© Springer International Publishing Switzerland 2015
C. Claramunt et al. (Eds.): SSTD 2015, LNCS 9239, pp. 333–351, 2015.
DOI: 10.1007/978-3-319-22363-6_18

gender, interest, and location[1]. Another example is the Location-Based App Recommendation (LBAR), where software or "apps" are suggested to a user based on where she is. An LBAR feature recently appears in Apple's iOS 8[2], which shows the picture of an app (e.g., "Starbucks") in the lock screen based on the user's context (e.g., she is close to a Starbucks coffee shop). As pointed out by Verve Mobile in 2013, the LBTA outperforms non-location-targeted advertising by a factor of two, and the usage of the LBTA exceeds the industry average click-through rate (CTR) of 0.4 %[3].

Those applications can be built on the top of Publish/Subscribe (Pub/Sub) systems [6] which can provide large-scale matching and information dissemination. In a Pub/Sub system, there are two kinds of clients, *subscriber* and *publisher*. A subscriber, typically an information provider such as an advertising company, specifies the properties of users in which it is interested. These properties, or constraints, are collectively known as a subscription. For instance, an advertising company A (e.g., a restaurant), acting as a subscriber, posts the following subscription to the Pub/Sub system: ($15 < age < 30, interest = \{barbecue, sushi\}, gender = male, visited\_time \geq 3$). The constraints specified in this subscription are used to match the "events" published by a publisher; once a matching is found, information from a subscriber is sent to the publisher. A publisher can be a casual mobile phone user. When a publisher, say, U, browses a homepage (say, Facebook), an event, containing information about this user (e.g., *age=25, interest=barbecue, gender=male, visited_time=5*), is sent to the system.

Since one or more constraints may be specified in a subscription, it may not be possible for the values of an event to match all the constraints. Hence, researchers have proposed to allow more flexibility in the matching process by allowing matching between subscribers and publishers to be inexact or partial. This variant of Pub/Sub systems, called ranked Pub/Sub systems [14,17,18], return the k best subscriptions (or top-k subscriptions) to a publisher based on some scoring functions.

We notice that the current work only focuses on normal boolean expressions including strings and numbers. However, the newly emerging applications bring the new technical challenges. Continuing the example of a location-based targeted advertising application, when a publisher opens an app, his geographic coordinates will be sent to the system. A subscriber can also specify this kind of location, for instance, by saying that his shop is at a specific location. On the other hand, to push advertisements to a mobile user, due to the factors like limited network bandwidth and the screen size of user's mobile phone, only the advertisements whose distribution scopes are near to the user's current location may become the candidate advertisements. Since subscriptions contain both complex boolean expressions and location information, it is rather costly to retrieve top-k relevant subscriptions from millions of subscriptions for an event.

---

[1] http://www.google.com/ads/admob/.
[2] http://goo.gl/wZeSg5.
[3] http://goo.gl/kJpQPG.

Different from the ranked Pub/Sub systems, the location information has been integrated into the so-called keyword Boolean matching Pub/Sub systems [11,13,19] where the numeric attribute matching is not supported. Their methods of incorporating the location information are either not applicable or inefficient for the ranked Pub/Sub systems. In the empirical studies, we extend the existing work related to the location-aware Pub/Sub systems as the competitors to support the numeric attribute matching. The experimental results demonstrate that our approach significantly outperforms the competitors.

In this paper, we explore the issues of incorporating location information into a ranked Pub/Sub system. To support top-k subscription matching for location-aware Pub/Sub systems, we propose a novel R-tree based index, the R^I-tree, by integrating the dynamic interval tree into the R-tree nodes. When an event with location information arrives, our algorithm can quickly report the top-k subscriptions most relevant to the event. To summarize, our main contributions are:

1. We formalize a new variant of top-k subscription matching, permitting location data to be a part of a subscription or an event;
2. We propose an index structure, called the R^I-tree;
3. We design an efficient matching algorithm; and
4. Our experimental evaluation validates the feasibility of our R^I-tree based solution.

The rest of the paper is organized as follows. Section 2 overviews the related work. Section 3 formulates the top-k subscription matching. Section 4 gives the threshold algorithm based solution as a baseline solution. In Sect. 5, we present the R^I-tree index and describe the matching algorithm. Finally, we evaluate the performance of the R^I-tree based solution by extensive experiments in Sect. 6 and conclude the paper in Sect. 7.

2 Related Work

The related work can be categorized into two main areas: ranked Pub/Sub systems and location-aware Pub/Sub systems. The differences between our R^I-tree solution and the existing solutions are summarized in Table 1 (BE denotes Boolean Expression).

Table 1. Comparison of existing location-aware Pub/Sub systems

	Pub/Sub					spatial keyword		R^I-tree
	SOPT-R-tree [14]	k-index [18]	BE*-tree [17]	R^t-tree [13]	OpIndex [20]	IR-tree [4]	I^3 [21]	
matching semantics	BE	BE	BE	keyword	BE	keyword	keyword	BE
location data	✗	✗	✗	✓	✗	✓	✓	✓
top-k	✓	✓	✓	✗	✗	✓	✓	✓

2.1   Ranked Pub/Sub Systems

Since the issue of the top-k subscription matching is posed in [14], there have existed some researches on top-k subscription matching [8,14,17,18]. [14] designates a subscription to be a set of intervals over a multi-dimensional space, in which each dimension is associated with a weight, and an event to be a point over the same multi-dimensional space. Based on this, [14] builds scored interval indexes for each dimension of the subscriptions. As thus, while an event arrives, the matching is carried out on every dimension of subscriptions, returning a corresponding subscription list sorted in a descending order of the scores. Then, the threshold algorithm [7] (TA for short) is employed to merge multiple sorted lists to obtain the subscriptions whose scores are ranked among the top k. Moreover, [14] presents two novel index structures: the IR-tree and the SOPT-R-tree to support top-k subscription matching. However, these index structures cannot support dynamic insertion and deletion of subscriptions. So they are not applicable to the scenarios where the subscriptions are updated frequently.

Whang et al. [18] turn the top-k subscription matching into another problem: how to efficiently index Disjunctive Normal Form (DNF) and Conjunctive Normal Form (CNF) boolean expressions over a high-dimensional space so as to quickly find the boolean expressions that evaluate to true for a given assignment of values to attributes. [18] presents the k-index, an inverted list based index for DNF and CNF boolean expressions, and then finds the top-k matched boolean expressions by virtue of the k-index. As an extension of [18], the methods presented in [8] are not restricted to the normal form expressions, but can deal with arbitrarily complex boolean expressions. [8] leverages existing techniques for evaluating leaf-level conjunctions, and then develops two bottom-up evaluation techniques, Dewey ID matching and Interval ID matching, to reduce unnecessary evaluation.

For the hierarchical top-k subscription matching, [17] presents a novel index structure named BE*-tree, which permits the values of attributes to be a continuous or discrete domain and combines a bi-directional tree expansion mechanism and an overlap-free splitting strategy to adapt to different workloads. For very high dimensional subscription matching, [20] proposes an in-memory index OpIndex, which builds an inverted index on the pivot attributes of subscriptions and designs a two-level partitioning scheme. However, OpIndex is not yet available to support location data and ranking.

2.2   Location-Aware Pub/Sub Systems

Recently, there have been many researches on location-aware Pub/Sub systems from a database perspective [2,9,11,13,19]. [13] proposes the R^t-tree which can efficiently filter geo-textual data. [19] extends R^t-tree to support ranking semantics, i.e., return all subscriptions whose similarities with the query event are not smaller than a given threshold θ. But the pruning algorithm proposed in [19] can not be used for top-k search. Further, [11] studies the location-aware Pub/Sub problem for parameterized spatio-textual subscriptions and presents a filter-verification framework by integrating prefix filtering and spatial pruning techniques. However, only keywords are considered in [11,13,19], and they can not support to retrieve top-k subscription matching. Note that Pub/Sub systems which only consider keywords cannot support numeric attribute matching such as the example of the subscription described in Sect. 1. Compared to [11,13,19], the R^I-tree proposed in this paper has two distinguishing features. First, it allows users to specify their interests with boolean expressions, which is more expressive than keywords. Second, it focuses on the top-k semantics which is frequently used in many emerging applications (e.g., location-based targeted advertising).

Chen et al. [2] considers the temporal spatial-keyword top-k subscription query. They present an efficient solution which can continuously maintain up-to-date top-k most relevant results (events) over a stream of geo-textual objects for each subscription. [9] proposes a new location-aware Pub/Sub system, i.e., *Elaps*, that focuses on continuously monitoring moving users subscribing to dynamic event streams. However, their problems are different from ours. Our work is also different from spatial keyword search [1,4,21]. The main reason is that they focus on keywords while we adopt boolean expressions to express subscriber's requirements in subscriptions, which is more expressive than keywords.

3 Problem Formalization

3.1 Data Model

Definition 1. *Subscription: A subscription s contains a boolean expression Ω, a location loc, and a tuning parameter α, i.e., $s : \Omega \wedge loc \wedge \alpha$. The boolean expression is a conjunction of predicates, i.e., $\Omega = \{p_1 \wedge \cdots \wedge p_n\}$. A predicate is a quadruple, i.e., $p =< attr, op, val, \omega >$, with attr being an attribute id that uniquely represents a dimension, op being an operator (e.g., from the relational operators ($<, \leq, =, \neq, \geq, >$), the set operators ($\in_s, \notin_s$) and the interval operator ($\in_i$)), val being a value, a set of values in discrete domains or a range of values in continuous domains, and ω being an assigned predicate weight, where $\sum_{i=1}^{n} p_i.\omega = 1$.*
The predicate weight signifies the relevance between the predicate and the event and can be given by subscribers (e.g., the advertiser assigns higher weights to more relevant predicates). The location loc represents the spatial dimension and is denoted as a conjunction of two triples, i.e., $(lat = val_{lat}) \wedge (lon = val_{lon})$, where lat (lon) denotes the latitude (longitude) of the object. The parameter α is used to balance the relative importance of non-spatial and spatial similarity.

For example, in the targeted advertising, the subscription for an advertisement from a restaurant can be:$\{(age \in_i [15, 30], 0.3) \wedge (income > 5000, 0.4) \wedge (credit\_score > 80, 0.3) \wedge (lat = 22.27) \wedge (lon = 114.17) \wedge \alpha = 0.5\}$.

As explained in [16], predicates with different types of operators can be converted into one-dimensional intervals, as shown in Table 2, where v_{min} and v_{max} are the smallest and the largest possible values in the corresponding domain, and $\{v_1, \cdots, v_k\}$ is sorted in an ascending order. The way of converting $\neq$ and $\notin_s$ to an interval that spans the entire domain is built on the following reasonable speculation: these kinds of predicates are satisfied with a high probability by an event having a predicate on the corresponding attribute. Thus, the given transformation can help the early pruning during the matching. Therefore, in this paper, we focus on predicates in the form of intervals.

Table 2. Predicate Conversion

Predicates	Interval	Predicates	Interval
$i < v_1$	$[v_{min}, v_1)$	$i \geq v_1$	$[v_1, v_{max}]$
$i \leq v_1$	$[v_{min}, v_1]$	$i \in_s \{v_1, \cdots, v_k\}$	$[v_1, v_k]$
$i = v_1$	$[v_1, v_1]$	$i \notin_s \{v_1, \cdots, v_k\}$	$[v_{min}, v_{max}]$
$i \neq v_1$	$[v_{min}, v_{max}]$	$i \in_i [v_1, v_2]$	$[v_1, v_2]$
$i > v_1$	$(v_1, v_{max}]$		

Definition 2. *Event: An event e includes a non-spatial set of attribute name and value pairs and a location loc, i.e., $e : (attr_1 = val_1) \wedge (attr_2 = val_2) \wedge \cdots \wedge (attr_{|e|} = val_{|e|}) \wedge (lat = val_{lat}) \wedge (lon = val_{lon})$, where $attr_i$ is the attribute identifier, val_i is the associated value, val_{lat} and val_{lon} are the publisher's current latitude and longitude, respectively.*

Here is an example of an event: $\{age = 25 \wedge income = 5000 \wedge credit\_score = 1000 \wedge lat = 22.27 \wedge lon = 114.17\}$.

Definition 3. *Similarity Function ψ: Given event $e : (attr_1 = val_1) \wedge (attr_2 = val_2) \wedge \cdots \wedge (attr_{|e|} = val_{|e|}) \wedge loc$ and subscription $s : (\Omega \wedge loc \wedge \alpha) = (p_1 \wedge p_2 \wedge \cdots \wedge p_n \wedge loc \wedge \alpha)$, the similarity function $\psi(e, s)$ is defined as follows.*

$$\psi(e, s) = (1 - s.\alpha) \cdot \psi_t(e, s) + s.\alpha \cdot \psi_s(e, s), \tag{1}$$

where ψ_t is a non-spatial similarity function and ψ_s is a spatial similarity function.

Further, the non-spatial similarity[4] is given by:

$$\psi_t(e, s) = \sum_{e.attr_i \in e, s.p_j \in s.\Omega, e.attr_i = s.p_j.attr} s.p_j.\omega \cdot check(e.val_i, s.p_j) \tag{2}$$

where $e.val_i$ denotes i^{th} attribute value of event e, $s.p_j$ $(j = 1, \cdots, n)$ denotes the jth predicate of subscription s and function $check(e.val_i, s.p_j)$ is defined in Eq. 3 to check whether the constraint $s.p_j$ is satisfied by $e.val_i$.

$$check(e.val_i, s.p_j) = \begin{cases} 1 & e.val_i \in_i s.p_j.val \\ 0 & otherwise \end{cases} \tag{3}$$

---

[4] More generally, the non-spatial similarity can be any monotonic function of the weights.

The spatial similarity is given by:

$$\psi_s(e, s) = 1 - \frac{dist(e.loc, s.loc)}{MaxDist}, \qquad (4)$$

where $dist(e.loc, s.loc)$ is the Euclidian distance between $e.loc$ and $s.loc$, and $MaxDist$ is the maximum Euclidian distance between subscriptions.

3.2   Problem Definition

Based on the above definitions, we formulate the problem we will solve as follows. Given a set of subscriptions $\mathcal{S}$, an event e, and a parameter k, the *Top-k Subscription Matching problem* (SM-k problem for short) finds the top-k best matching set $\mathcal{S}_k \subseteq \mathcal{S}$ which is defined as $\mathcal{S}_k = \{s | \psi(e, s) \geq \psi(e, s'), \forall s' \in \mathcal{S} \backslash \mathcal{S}_k\}$ and $|\mathcal{S}_k| = k$.

Example 1. Figure 1 shows 9 subscriptions $s_0 \cdots s_8$ and an event e. The similarities between subscriptions and the event e are shown in Table 3. For event e, subscription s_0 is the result of the top-1 matching according to Eq. 1, i.e., $\psi(e, s_0) = \mathbf{0.925}$.

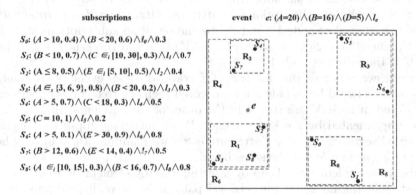

Fig. 1. Example of subscriptions and an event

Table 3. Similarities between event e and subscriptions in Fig. 1

s	s_0	s_1	s_2	s_3	s_4	s_5	s_6	s_7	s_8
$\psi_s(s, e)$	0.75	0.5	0.875	0.75	0.75	0.5	0.5	0.8	0.8
$\psi_t(s, e)$	1.0	0.0	0.0	0.2	0.7	0.0	0.1	0.6	0.0
$\psi(s, e)$	**0.925**	0.35	0.35	0.365	0.725	0.1	0.42	0.7	0.64

4   Baseline Solution

In this section, we present the *Threshold Algorithm-based Solution* for SM-k problem as a baseline solution. Considering that top-k subscription matching belongs to the relaxed matching and the subscription is not required to match with the event on each attribute exactly, we build an index for each attribute, i.e., build a scored segment-tree [14] for every non-spatial attribute and R-tree [10] for the spatial attribute. Based on that, we propose the *Threshold Algorithm based Solution (TAS)*.

TAS builds a two level index for subscriptions. At the root level, a hashmap is used to map attribute names to the sub-level data structures. At the sub-level, an index is built for each attribute to index those subscriptions contain that attribute.

For each non-spatial attribute, we build a scored segment-tree [14] for all subscription intervals in that attribute. The scored segment-tree is a variant of segment trees. A segment-tree [5] is a binary-tree structure to index intervals (segments). It partitions the intervals into a collection of disjoint, atomic intervals. Each atomic interval corresponds to a leaf node in the tree. If the length of the whole interval is n, then a segment tree is a balanced binary tree with n nodes as leaves and $\log n$ as the height of the tree.

Let $\mathcal{I}$ be the set of all subscription intervals in attribute $attr_i$, I be the interval constraint of subscription s in attribute $attr_i$ where $I \in \mathcal{I}$, $interval(V)$ denote the interval of node V, and $\mathcal{S}_V$ be the set of all subscriptions stored on node V, where $\forall I \in \mathcal{S}_V$, we have $interval(V) \subseteq I$ and $interval(U) \nsubseteq I$, here, node U is the father of node V.

To retrieve top-k scoring interval of $\mathcal{I}$ stabbed by an event point $e.val_i$, a segment-tree should be modified into a scored segment tree [14], i.e., subscriptions stored in node V are sorted in the order of their weights (in practice, it can be implemented by a priority queue). When retrieving top-k subscriptions in a scored segment-tree $\mathcal{T}$ for attribute $attr_i$, we maintain a global max-heap of size $O(\log n)$ and follow the steps below.

Firstly, we replenish the heap by inserting the top element of each subscription list from the nodes on the retrieval path in $\mathcal{T}$. Secondly, we pop out the top element of the heap as a candidate subscription (assuming that the element popped out is stored on node V) and insert the next element of the subscription list from node V. Run the second step in turns, until k elements are picked up. As thus, we can get top-k subscriptions (elements) on $\mathcal{T}$.

For the spatial attribute, we build an R-tree for all locations of subscriptions. On each node N of the R-tree, an extra information α_{max} is stored. If the node N is a leaf node, the value of α_{max} is the α of the corresponding subscription. Otherwise, if the node N is a non-leaf node, the value of α_{max} is the maximum of all its children's α_{max}. Thus, the new R-tree can support incremental nearest neighbor (NN) search. When an event location $e.loc$ arrives, for each non-leaf node N on the R-tree, the upper bound is

$$N.\alpha_{max} \cdot (1 - \frac{MinDist(e.loc, N.rectangle)}{MaxDist}),$$

where $MinDist(e.loc, N.rectangle)$ denotes the minimum Euclidian distance between $e.loc$ and any point on the *Minimum Bounding Box* of node N.

Now we describe the matching process. When an event e arrives, for each attribute in e, use the hashmap to get the related index structure, and incrementally return the subscriptions matching with e on that attribute in the order of their weight on that attribute (The weight of a subscription s on each attribute is multiplied by $(1 - s.\alpha)$ when s is inserted into the scored segment-tree). Then we use the Threshold Algorithm (TA) [7] to merge multiple ranked lists. It is proved that TA is correct and instance optimal in [7].

Thus, the steps of *TAS* are as follows:

1. At the beginning, for each attribute of the event, retrieve the best candidate subscription using the corresponding index structure. Then go to 3.
2. For each attribute of the event, retrieve the next best candidate subscription.
3. Merge those candidates using TA. If the terminal criterion in TA cannot be satisfied, then go to 2. Otherwise, go to 4.
4. Return the top-k subscriptions got in TA.

TAS will search on each attribute separately. If there exists a subscription matched with an event on many attributes with small weight for each predicate (the total similarity is very large), the searching list for each attribute can be very long. Thus, TAS is not very efficient. In order to avoid this problem, we design a novel index which combines all attributes on one tree index.

5 R^I-tree Based Solution

In this section, we present a framework that integrates the R-tree and the interval-tree into a new index, named R^I-tree and that includes an algorithm for processing SM-k problem using the R^I-tree.

5.1 R^I-tree Index Structure

The R-tree [10] is a widely used index for spatial queries and the interval-tree [15] is the "standard" known solution for efficiently processing simple stabbing queries. They are designed separately for different kinds of queries.

The R^I-tree is essentially an R-tree, each node of which is enriched with reference to a set of dynamic interval trees for objects contained in its sub-tree.

In the R^I-tree, if node N is a leaf node, it contains a number of entries of the form $(sid, \Omega, loc, \alpha)$, where sid is the identifier of an subscription, Ω, loc and α are the boolean expression, the location and the tuning parameter of the subscription s_{sid}, respectively. Here, it is important to note that the weight of a subscription s_{sid} on each attribute is multiplied by $(1 - s_{sid}.\alpha)$ when s_{sid} is inserted into the R^I-tree. A leaf node also contains some metadata. The metadata includes *rectangle*, which is the Minimum Bounding Rectangle of all constituent entries, α_{min} and α_{max} which are the minimum and maximum value of α among all constituent entries, and the aggregated information Γ for each attribute of the form $(attr_i, range, \omega_{max})$. In addition, a leaf node also contains a pointer to a dynamic interval tree forest $\mathcal{F}$, i.e., a set of dynamic interval trees organized by a hashmap, shown as Fig. 2a. Let $\mathcal{S}_{attr}$ be the set of all attributes stored on the leaf node. The hashmap manages all attributes in $\mathcal{S}_{attr}$. For each attribute $attr_i$, the hashmap maps it to a dynamic interval tree $\mathcal{T}_{attr_i}$. The tree $\mathcal{T}_{attr_i}$ stores all intervals of the leaf node N's entries on attribute $attr_i$. The form of intervals stored on the tree is $(range, sid, \omega)$, where $range$ denotes the range of the interval, sid is the id of the corresponding subscription and ω is the weight of that subscription on $attr_i$.

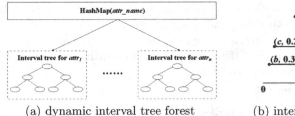
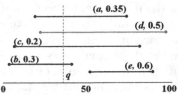

(a) dynamic interval tree forest   (b) interval stabbing-max problem

Fig. 2. Example of the dynamic interval tree forest and the interval stabbing-max problem

The dynamic interval tree $\mathcal{T}_{attr_i}$ dynamically maintains a set of intervals $\mathcal{I}$, where each interval $I \in \mathcal{I}$ has a weight $I.\omega$ such that the interval with the maximum weight containing an event point can be found efficiently. This structure can solve the *interval stabbing-max* problem. For instance, as shown in Fig. 2b, for the query point q, it stabs four intervals (a, b, c, d). Since interval d has the greatest weight 0.5, it will be returned. There exists several solutions which can solve this problem. In this paper, we use the modified interval tree structure mentioned in [12] which can support queries in $O(\log^2 n)$ time, updates in $O(\log n)$ time and only requires $O(n)$ space.

On the other hand, if node N is a non-leaf node, it contains a number of entries cp, which points to the corresponding child node. Being same as the type of leaf nodes, node N also maintains the metadata and a dynamic interval tree forest organized by a two-level index structure which contains the interval information for each associated attributes. For each child node U of node N, the interval of $U.\Gamma_{attr_i}$ will be stored on the dynamic interval tree $\mathcal{T}_{attr_i}$ of the node N. Thus, the number of intervals in $\mathcal{T}_{attr_i}$ will not exceed the number of entries in node N.

Example 2. Figure 3 illustrates the R^I-tree index for the subscriptions in Fig. 1. Figure 4 is an example of $Metadata_1$ and $Forest_1$.

After describing the R^I-tree index, now we introduce an important metric, the *Upper Bound* (UB) of the similarity. Given an event e and a node N in the R^I-tree, the metric UB provides an upper bound of the similarity between the

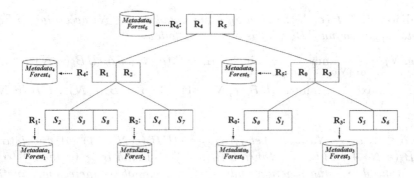

Fig. 3. R^I-tree index for subscriptions in Figure

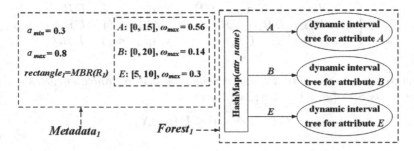

Fig. 4. Example of $Metadata_1$ and $Forest_1$

event e and all subscriptions located at the rectangle of node N. It can be used to order and efficiently prune the paths of the search space in the R^I-tree. To get the value of UB, we first calculate the upper bound of non-spatial similarity UB_t, and then calculate the upper bound of spatial similarity UB_s.

Definition 4. $UB_t(e, N)$: *Given an event e and a node N, the upper bound of non-spatial similarity $UB_t(e, N)$ is defined as follows:*

$$UB_t(e, N) = \sum_{e.attr_i = N.\Gamma.attr_j} N.\mathcal{T}_{attr_j}(e.val_i), \qquad (5)$$

where $\mathcal{T}_{attr_j}(e.val_i)$ returns the maximum weight of the interval containing $e.val_i$ in attribute $N.\Gamma.attr_j$.

Definition 5. $UB_s(e, N)$: *Given an event e and a node N, the upper bound of spatial similarity $UB_s(e, N)$ is defined as follows:*

$$UB_s(e, N) = 1 - \frac{MinDist(e.loc, N.rectangle)}{MaxDist}, \qquad (6)$$

where $MaxDist$ is the same as in Eq. 4, and $MinDist(e.loc, N.rectangle)$ is the minimum Euclidian distance between $e.loc$ and any point on $N.rectangle$.

Definition 6. $UB(e, N)$**:** *Given an event e and a node N, according to Eq. 1, the total upper bound $UB(e, N)$ is defined as follows:*

$$UB(e, N) = \max_{\alpha \in [N.\alpha_{min}, N.\alpha_{max}]} \min(1 - \alpha, \ UB_t(e, N)) + \alpha \cdot UB_s(e, N)$$

$$= \max_{\alpha \in [N.\alpha_{min}, N.\alpha_{max}]} \min(\alpha \cdot (UB_s(e, N) - 1) + 1, \ \alpha \cdot UB_s(e, N) + UB_t(e, N))$$

(7)

where $\alpha \in [N.\alpha_{min}, N.\alpha_{max}]$. Let $f_1(\alpha) = \alpha \cdot (UB_s(e, N) - 1) + 1$ and $f_2(\alpha) = \alpha \cdot UB_s(e, N) + UB_t(e, N)$. Since $UB_s(e, N) \leq 1$ and $\alpha \geq 0$, then $f_1(\alpha)$ is a monotone decreasing function and $f_2(\alpha)$ is a monotone increasing function. Thus, we can get a more succinct formula to calculate $UB(e, N)$.

$$UB(e, N) = \begin{cases} 1 - N.\alpha_{min} + N.\alpha_{min} \cdot UB_s & N.\alpha_{min} \geq (1 - UB_t) \\ (1 - UB_t) \cdot UB_s + UB_t & N.\alpha_{min} < (1 - UB_t) < N.\alpha_{max} \\ N.\alpha_{max} \cdot UB_s + UB_t & N.\alpha_{max} \leq (1 - UB_t) \end{cases}$$

(8)

Theorem 1. *Given an event e and a node N whose rectangle encloses a set of subscriptions $S = \{s_i, 1 \leq i \leq n\}$, we have:*

$$\forall s \in S, \ \psi(e, s) \leq UB(e, N)$$

(9)

Proof. Since subscription s is enclosed in the rectangle of node N, the minimum Euclidian distance between $e.loc$ and any point on $N.rectangle$ is no larger than the Euclidian distance between $e.loc$ and $s.loc$, i.e.:

$$MinDist(e.loc, N.rectangle) \leq dist(e.loc, s.loc)$$

(10)

Thus, the spatial similarity between e and s is no larger than the upper bound of spatial similarity between e and node N according to the Eqs. 4 and 6, i.e.:

$$\psi_s(e, s) \leq UB_s(e, N)$$

(11)

Meanwhile, the set of attributes in $N.\Gamma$ is the union of all subscriptions in node N. And for each attribute appears in both event e and $N.\Gamma$, let $e.attr_i = N.\Gamma.attr_j$, the value $N.\mathcal{T}_{attr_j}(e.val_i)$ is the maximum weight of all subscriptions in node N on that attribute. Thus:

$$(1 - s.\alpha) \cdot \psi_t(e, s) \leq \min(1 - s.\alpha, UB_t(e, N))$$

(12)

Since $s.\alpha \in [N.\alpha_{min}, N.\alpha_{max}]$, according to Eqs. 1, 7, 11 and 12, we can get:

$$\begin{aligned} \psi(e, s) &= (1 - s.\alpha) \cdot \psi_t(e, s) + s.\alpha \cdot \psi_s(e, s) \\ &\leq \min(1 - s.\alpha, UB_t(e, N)) + s.\alpha \cdot UB_s(e, N) \\ &\leq \max_{\alpha \in [N.\alpha_{min}, N.\alpha_{max}]} \min(1 - \alpha, \ UB_t(e, N)) + \alpha \cdot UB_s(e, N) \\ &= UB(e, N) \end{aligned}$$

(13)

□

Algorithm 1. R^I-treeMatch(e, *tree*, k)

1: *queue* ← new PriorityQueue(); /*The higher the value of UB, the greater the priority*/
2: *heap* ← new Min-Heap(); /*Store candidate top-k subscriptions*/
3: *queue*.push(*tree.root*, 1);
4: **while not** *queue*.empty() **do**
5:     *element* ← *queue*.top();
6:     *queue*.pop();
7:     **if** *element.Node* is a leaf node **then**
8:         **for** each entry *sub* ∈ *element.Node* **do**
9:             **if** *heap.size*() < k **then**
10:                *heap*.insert(*sub*, $\psi(e, sub)$);
11:            **else if** $\psi(e, sub) > heap$.begin().*key* **then**
12:                *heap*.erase(*heap*.begin());
13:                *heap*.insert(*sub*, $\psi(e, sub)$)
14:        **if** *heap*.size()==k **and**
            (*queue*.empty() **or** *heap*.begin().*key* ≥ *queue*.top().*key*) **then**
15:            break;
16:    **else**
17:        **for** each entry *node* ∈ *element.Node* **do**
18:            *queue*.push(*node*, $UB(e, node)$);
19: reverse all elements in *heap* and return;

5.2 Matching

Now, we discuss how to use the R^I-tree to solve SM-k problem when an event e arrives. The best-first traversal algorithm (e.g., [15]) is used to retrieve the top-k best matched subscriptions. A priority queue is used to store the nodes that have yet to be visited (i.e., the node with higher UB has a greater priority). And a global min-heap of size $O(k)$ is maintained and is used to store top-k subscriptions among all subscription objects visited.

Algorithm 1 shows the pseudocode of retrieving top-k subscriptions on the R^I-tree. The algorithm always picks the node N with the largest $UB(e, N)$ value in the priority queue. The algorithm terminates when k subscriptions have been found and the similarity of the smallest one is not smaller than the largest one in the priority queue (or the priority queue is empty).

6 Evaluation

In this section, we evaluate the R^I-tree based solution by conducting extensive experiments on a very large data set. All algorithms are implemented in C++ and compiled using g++ 4.2.1 and the experiments are run on a 2.3 GHz Intel(R) Core(TM) core i7 processor with 16 GB of *RAM*.

Table 4. Experimental Parameters

Param	Description	Value
N_e	# Events	1000
N_s	# Subscriptions	1M, **2M**, 3M, 4M, 5M
d	# Dimensions	50, 100, **400**, 600, 800
c	Dimension cardinality	10, 50,**100**,250,500
L_e	Avg. event length	6, 8, **10**, 12, 14
L_s	Avg. subscription length	2, 3, **4**, 5, 6
k	Top-k parameter	1, 3, **5**, 7, 9
f_α	Distribution of α	$\mathcal{N}(0.1, 0.05), \mathcal{N}(0.5, 0.05), \mathcal{N}(0.9, 0.05), \boldsymbol{\mathcal{U}(0,1)}$

6.1   Experimental Setup

We evaluate the following algorithms: (1) SCAN(a sequential scan); (2) R^t-tree (we extend R^t-tree [19] to support our model, i.e., constructing an R^t-tree, changing its *TokenSet* to the attribute set which includes attribute id, range and maximum weight and traversing the R^t-tree from the root to leaves); (3) TAS (the threshold algorithm based solution); (4) R^I-tree (the R^I-tree based solution). In the experiments, all subscriptions are first loaded into the memory and indexed by corresponding index structures, and then events are read as input continuously. We record and analyze the average time and space cost of matching top-k matched subscriptions with an event. The maximum number of children of a node in the R^I-tree and the R^t-tree is 50 in our experiments.

Since there is no suitable public real subscription and event dataset for top-k subscription matching, we use a synthetic dataset by combining non-spatial data generated by BE-Gen[5] [16] and spatial data selected from real twitter data with location information in USA [3] for both events and subscriptions. For each boolean expression generated by BE-Gen, we randomly assign a location to it from the twitter dataset.

As to the weight for each predicate of subscriptions, we adopt the weight generation technique proposed in [18] (this way is also used in [17]): (i) For each unique attribute $attr$, first compute its reciprocal of frequency, denoted by ξ_{attr}, based on the concept that popular attribute should be assigned a low weight while an infrequent attribute should be assigned a high weight. (ii) For each predicate $p_i = (attr, op, val, \omega)$, its weight is computed as follows: $p_i.\omega = \max(\xi_{p_i.attr}, \ x)$, where x is randomly generated from a Gaussian distribution: $\mathcal{N}(0.8 \times \xi_{p_i.attr}, 0.05 \times \xi_{p_i.attr})$.

In our experiments, we evaluate the performance of those algorithms under different data distributions(Uniform and Zipf). Table 4 summarizes the main parameters used in experiments (default values are in bold).

---

[5] http://msrg.org/datasets/BEGen.

6.2 Experimental Results and Analyses

We conduct 6 groups of experiments, observing the effect of the number of subscriptions, the space dimensionality, the dimension cardinality, the average subscription/event length, the top-k parameter and the distribution of α on matching time. In each group of experiments, we conduct the experiments under different distributions of choosing predicates' attributes (Uniform and Zipf).

Varying the Number of Subscriptions. In the first group of experiments, we observe the effect of the number of subscriptions. Figure 5 show the performance of 4 algorithms when the number of subscriptions is varied from 1 million to 5 millions under different workload distributions. In general, with the increase of the number of subscriptions, the matching times of R^t-tree, TAS, and SCAN all increase quickly while our solution is very smooth. In these experiments, the matching time of R^I-tree on average is 4.8 and 4.3 times faster than the next best algorithm for the uniform distribution and the Zipf distribution, respectively.

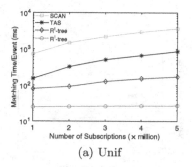
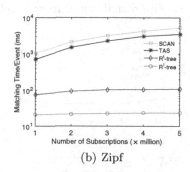

| (a) Unif | (b) Zipf |

Fig. 5. Varying the number of subscriptions

Varying Space Dimensionality. Compared with the effect of varying the number of subscriptions, the effect of space dimensionality is slight. All algorithms with the exception of TAS are almost unchanged as the dimensionality varies, as shown in Fig. 6. In Fig. 6a, under the uniform distribution the matching time of TAS decreases as the dimensionality increases, since subscriptions tend to share less common predicates when the dimensionality is large. However, under the Zipf distribution the matching time of TAS does not increase as the dimensionality increases, it is because there are a few popular dimensions among all subscriptions, leading to a large of overlap among subscriptions. Overall, on average R^I-tree is 3.7 and 4.4 times faster than the next best algorithm for the uniform distribution and the Zipf distribution, respectively.

Varying the Dimension Cardinality. In this group of experiments, we observe the effect of the dimension cardinality. When the dimension cardinality increases, the matching rate between an event and all subscriptions will decrease.

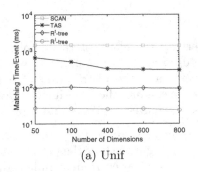
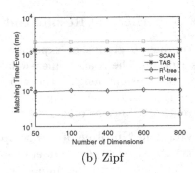

(a) Unif   (b) Zipf

Fig. 6. Varying space dimensionality

Facing such data set, the R-tree based algorithms will visit more nodes, resulting in the increasing of matching time. However, as shown in Fig. 7, on average, R^I-tree is still 4.8 and 4.2 times faster than R^t-tree for the uniform distribution and the Zipf distribution, respectively.

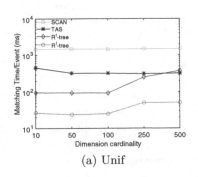
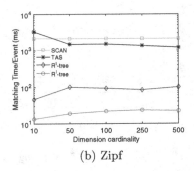

(a) Unif   (b) Zipf

Fig. 7. Varying dimension cardinality

Varying Average Subscription/Event Length. Another key factor which can affect the algorithm performance is the average number of predicates per subscription and event. As shown in Fig. 8, all algorithms are sensitive to the average subscription length. It is because the overlap among subscriptions increases. Compared with the average subscription length, 4 algorithms are insensitive to the average event length, as shown in Fig. 9. The matching times of R^I-tree and R^t-tree both increase since the upper bound of non-spatial similarity will increase as the average event length increases. In general, R^I-tree performs best because of its filtering strategy. It is 4.3 and 4.6 times faster than the next best algorithm for the uniform distribution and the Zipf distribution, respectively, as the average subscription length increases. It is also 4 and 4.4 times faster than the next best algorithm for the uniform distribution and the Zipf distribution, respectively, as the average event length increases.

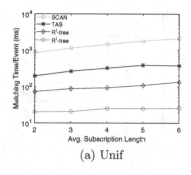

(a) Unif

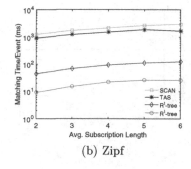

(b) Zipf

Fig. 8. Varying average subscription length

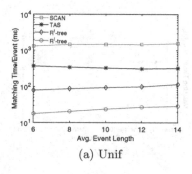

(a) Unif

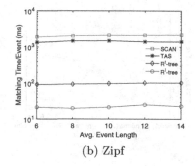

(b) Zipf

Fig. 9. Varying average event length

Varying the Value of k. Now we observe the effect of the top-k parameter, here, the value of k grows from 1 to 9. Figure 10 shows the performance of 4 algorithms when varying k under different workload distributions. In general, with the increase of k, the matching times of R^t-tree and R^I-tree increase marginally. However, R^I-tree still has the best performance. For instance, at $k = 9$, under the uniform distribution, the matching time of R^I-tree is 3, 6, and 30 times better than R^t-tree, TAS, and SCAN, respectively. Similarly, under the Zipf distribution, the speed-up ratios are 4, 39, and 57 times, respectively.

Varying the Distribution of α. Finally, we conduct a group of experiments observing the effect of the distribution of α. Figure 11 shows the performance of 4 algorithms under different distribution of α, i.e., Normal distribution :$\mathcal{N}(0.1, 0.05)$, $\mathcal{N}(0.5, 0.05)$ and $\mathcal{N}(0.9, 0.05)$; Uniform distribution $\mathcal{U}(0, 1)$. Under Normal distribution, the R^I-tree performs much better than all other algorithms and also is better than the situation R^I-tree at $\mathcal{U}(0, 1)$. The reason behind this is that the values of α for all subscriptions are close to the mean and most search paths can be filtered efficiently by the pruning strategy. For example, at $\mathcal{N}(0.5, 0.05)$, the matching time of R^I-tree is 21 and 133 times faster than the next best algorithm (TAS) for the uniform distribution and the Zipf distribution, respectively.

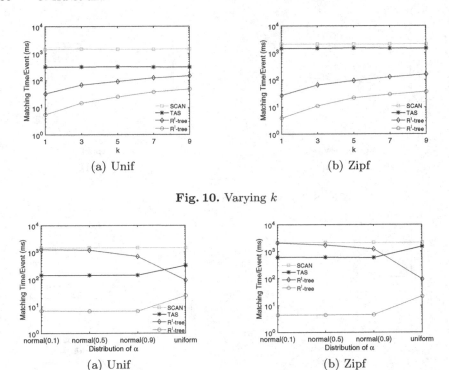

Fig. 10. Varying k

Fig. 11. Varying the distribution of α

7 Conclusions

In this paper, we propose and formalize a variant of top-k subscription matching, i.e., top-k subscription matching for location-aware Pub/Sub systems which supports boolean expressions in subscriptions. We propose a novel index structure R^I-tree, which combines the R-tree and the dynamic interval-tree. In addition, we develop an efficient filtering strategy to reduce the search space. Finally, we evaluate the R^I-tree based solution by experiments on a large-scale dataset. The experimental results convincingly demonstrate the benefits of our algorithm.

Acknowledgments. Jiafeng Hu and Reynold Cheng were supported by the Research Grants Council of Hong Kong (RGC Project (HKU 711110)). Dingming Wu was supported by HKU 714712E. Beihong Jin was supported by the National Natural Science Foundation of China under Grant No. 61472408 and the Opening Foundation of Beijing Key Lab of Intelligent Telecommunications Software and Multimedia, Beijing University of Posts and Telecommunications.

References

1. Chen, L., Cong, G., Jensen, C.S., Wu, D.: Spatial keyword query processing: an experimental evaluation. In: PVLDB'2013, pp. 217–228. VLDB Endowment (2013)
2. Chen, L., Cong, G., Cao, X., Tan, K.-L.: Temporal spatial-keyword top-k publish/subscribe. In: ICDE 2015, pp. 255–266 (2015)
3. Cheng, Z., Caverlee, J., Lee, K., Sui, D.Z.: Exploring millions of footprints in location sharing services. In: ICWSM 2011, pp. 81–88 (2011)
4. Cong, G., Jensen, C.S., Wu, D.: Efficient retrieval of the top-k most relevant spatial web objects. In: VLDB 2009, vol. 2, pp. 337–348 (2009)
5. de Berg, M., van Kreveld, M., Overmars, M., Schwarzkopf, O.: Computational Geometry: Algorithms and Applications. Springer-Verlag, Heidelberg (2000)
6. Eugster, P.T., Felber, P.A., Guerraoui, R., Kermarrec, A.-M.: The many faces of publish/subscribe. ACM Comput. Surv. **35**(2), 114–131 (2003)
7. Fagin, R., Lotem, A., Naor, M.: Optimal aggregation algorithms for middleware. In: PODS 2001, pp. 102–113. ACM, NY, USA (2001)
8. Fontoura, M., et al.: Efficiently evaluating complex boolean expressions. In: SIGMOD 2010, pp. 3–14. ACM, NY, USA (2010)
9. Guo, L., Zhang, D., Li, G., Tan, K.-L., Bao, Z.: Location-aware pub/sub system: When continuous moving queries meet dynamic event streams. In: SIGMOD 2015, pp. 843-857. ACM (2015)
10. Guttman. A.: R-trees: a dynamic index structure for spatial searching. In SIGMOD 1984, pp. 47–57. ACM, NY, USA (1984)
11. Hu, H., Liu, Y., Li, G., Feng, J., Tan, K.-L.: A location-aware publish/subscribe framework for parameterized spatio-textual subscriptions. In: ICDE 2015, pp. 711–722 (2015)
12. Kaplan, H., Molad, E., Tarjan, R.E.: Dynamic rectangular intersection with priorities. In: STOC 2003, p. 639. ACM Press, New York, June 2003
13. Li, G., Wang, Y., Wang, T., Feng, J.: Location-aware publish/subscribe. In: KDD 2013, pp. 802–810. ACM Press, New York (2013)
14. Machanavajjhala, A., Vee, E., Garofalakis, M., Shanmugasundaram, J.: Scalable ranked publish/subscribe. In: VLDB 2008, vol. 1, issue no. 1, pp. 451–462, August 2008
15. Mehlhorn, K.: Data structures and algorithms 3: Multi-dimensional Searching and Computational Geometry. Monographs in Theoretical Computer Science. An EATCS Series. Springer, Heidelberg (1984)
16. Sadoghi, M., Jacobsen, H.-A.: Be-tree: an index structure to efficiently match boolean expressions over high-dimensional discrete space. In: SIGMOD 2011, pp. 637-648. ACM (2011)
17. Sadoghi, M., Jacobsen, H.-A.: Relevance matters: Capitalizing on less (top-k matching in publish/subscribe). In: ICDE 2012, pp. 786–797 (2012)
18. Whang, S.E., Garcia-Molina, H., Brower, C. J. Shanmugasundaram, S. Vassilvit-skii, E. Vee, and R. Yerneni. Indexing boolean expressions. In: VLDB 2009, vol. 2, issue no. 1, pp. 37–48 (2009)
19. Yu, M., Li, G., Wang, T., Feng, J., Gong, Z.: Efficient filtering algorithms for location-aware publish/subscribe. IEEE TKDE **27**(4), 950–963 (2015)
20. Zhang, D., Chan, C.-Y., Tan, K.-L.: An efficient publish/subscribe index for e-commerce databases. In: Proceedings VLDB Endow, vol. 7, issue no. 8, pp. 613–624 (2014)
21. Zhang, D., Tan, K.-L., Tung, A.K.H.: Scalable top-k spatial keyword search. In: EDBT 2013, pp. 359–370. ACM Press, New York, USA (2013)

Similarity Search and Pattern

Similarity Search using Patterns

Spatiotemporal Similarity Search in 3D Motion Capture Gesture Streams

Christian Beecks[1]([✉]), Marwan Hassani[1], Jennifer Hinnell[2], Daniel Schüller[3], Bela Brenger[3], Irene Mittelberg[3], and Thomas Seidl[1]

[1] Data Management and Exploration Group, RWTH Aachen University, Aachen, Germany
{beecks,hassani,seidl}@cs.rwth-aachen.de
[2] Department of Linguistics, University of Alberta, Alberta, Canada
hinnell@ualberta.ca
[3] Natural Media Lab, RWTH Aachen University, Aachen, Germany
{schueller,brenger,mittelberg}@humtec.rwth-aachen.de

Abstract. The question of how to model spatiotemporal similarity between gestures arising in 3D motion capture data streams is of major significance in currently ongoing research in the domain of human communication. While qualitative perceptual analyses of co-speech gestures, which are manual gestures emerging spontaneously and unconsciously during face-to-face conversation, are feasible in a small-to-moderate scale, these analyses are inapplicable to larger scenarios due to the lack of efficient query processing techniques for spatiotemporal similarity search. In order to support qualitative analyses of co-speech gestures, we propose and investigate a simple yet effective distance-based similarity model that leverages the spatial and temporal characteristics of co-speech gestures and enables similarity search in 3D motion capture data streams in a query-by-example manner. Experiments on real conversational 3D motion capture data evidence the appropriateness of the proposal in terms of accuracy and efficiency.

Keywords: Similarity search · Spatiotemporal data · 3D motion capture data · Streams · Co-speech gestures · Gesture matching distance · Gesture signature · Dynamic time warping

1 Introduction

Human communication typically involves multiple modalities such as vocalizations, spoken or signed language, manual gestures, eye gaze, body posture and facial expressions. In face-to-face conversation, gestures are a communication component of the most natural form and accompany over 75 % of all clauses [10,36]. They support us in expressing information in non-verbal form. Gestures that emerge spontaneously during conversation and are associated with speech are denoted as *co-speech gestures*. These gestures are understood as kinetic action involving hand and arm configurations or movements that have some communicative function and are an integral part of utterance formation and communicative

© Springer International Publishing Switzerland 2015
C. Claramunt et al. (Eds.): SSTD 2015, LNCS 9239, pp. 355–372, 2015.
DOI: 10.1007/978-3-319-22363-6_19

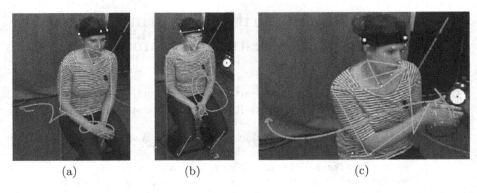

(a)   (b)   (c)

Fig. 1. Three example gestures of different spatiotemporal movement types: (a) gesture of type spiral, (b) gesture of type circle, and (c) gesture of type straight. Blue trajectories indicate the main movement of the gestures (Color figure online).

interaction [29,36,41]. They are contextualized and disambiguated according to the concurrent speech and are multifunctional, i.e., they may fulfill a broad range of cognitive, communicative, performative and interactive functions, and oftentimes they perform several of them at the same time [42]. Examples of co-speech gestures are brief hand motions, denoted as *beats*, pointing gestures referring to physical objects and concepts, denoted as *deictics*, and miming of actions or shapes of concrete and abstract things which are referred to as *iconics* and *metaphorics* [10,36].

Since co-speech gestures appear spontaneously and unconsciously during conversation, they are not bound by well-formedness conditions and imply a high degree of idiosyncrasy. In other words, the spatiotemporal patterns underlying co-speech gestures do not follow a rigorous theoretical model as expected from codified gestures such as *emblems*, i.e. manual signs with fixed culture-dependent meanings, e.g. the victory sign [43], and *sign languages* that are essentially linguistic [10] and exhibit large vocabularies and complex grammars.

Currently ongoing research in the domain of human communication with respect to co-speech gestures mainly focus on qualitative analyses. To this end, co-speech gestures are typically manually compared with respect to their observed similarity and contextualized according to their associated speech. Hypotheses are then advanced or refuted by inspecting the video recordings of single or multiple conversations.

In order to support such qualitative analyses and access similar co-speech gestures automatically, we propose a simple yet effective distance-based similarity model for co-speech gestures arising in 3D motion capture data streams. In contrast to conventional 2D video capture technology, the utilization of 3D motion capture technology has the advantage of measuring and visualizing the spatiotemporal dynamics of otherwise invisible movement traces with the highest possible accuracy. We aim at maintaining this accuracy by aggregating movement traces, i.e. trajectories, into a lossless, spatiotemporal feature representation, namely *gesture signature*, which has the ability of weighting trajectories individually according to their relevance. In order to compare two gesture signatures with each other, we propose the *Gesture Matching Distance* as a distance-based similarity measure that matches individual trajectories of the gesture signatures. Our approach is able to deal with gesture signatures varying in size and structure. Additionally, it effectively takes care of gesture signatures that do not include all trajectories due to the problem of hidden markers. We focus our evaluation on co-speech gestures exhibiting specific spatiotemporal movement patterns which result in different kinds of image-schematic gestalts, e.g. spirals, circles, and straight paths [14,38]. An illustration of co-speech gestures belonging to the aforementioned movement types is depicted in Fig. 1, where blue trajectories indicate the main movement of the gestures. For the sake of convenience we will denote co-speech gestures as gestures in the remainder of this paper.

Our main contributions are summarized as follows:

- We propose to model gestures arising in 3D motion capture data streams by means of gesture signatures maintaining the spatiotemporal characteristics and allowing for individual trajectory weighting.
- We present two different weighting schemes to reflect the inherent characteristics of relevant trajectories.
- We propose the *Gesture Matching Distance* as a distance-based similarity measure for assessing the dissimilarity between two gesture signatures and investigate its theoretical properties.
- We introduce a sequential query processing algorithm for the analysis of 3D motion capture data streams in a query-by-example manner.
- We benchmark the proposed approach with respect to the qualities of accuracy and efficiency on real conversational 3D motion capture data.

The remainder of this paper is structured as follows: Sect. 2 outlines related work with respect to adjacent research fields. Section 3 proposes the concept of gesture signatures for modeling gestures arising in 3D motion capture data streams. The Gesture Matching Distance is introduced in Sect. 4 together with an analysis of its theoretical properties. Section 5 then introduces our query processing algorithm. The results of the experimental evaluation on real conversational 3D motion capture data are presented in Sect. 6. The paper is concluded with an outlook on future work in Sect. 7.

2   Related Work

As pointed out by Campbell [10], conversational gestures have been investigated since ancient times. One of the first academic works which tries to visually classify spontaneous gestures has been carried out by Efron [16]. Following his work, Ekman and Friesen [17], Kendon [27,28], and MacNeil [36] also studied various aspects of spontaneous gestures and tried to classify such gestures according to their discourse functions. The resulting gesture taxonomies lead to the research question of whether co-verbal gestures can be classified automatically.

Campbell [10] investigated this research direction and investigated *Hidden Markov Models* for classifying co-verbal gestures. He came to the conclusion that the recognition rate is not good enough to be useful in a real world working system and compared the situation of gesture recognition with the early days of speech recognition [10].

From the computer science point of view, gestures are mainly researched in terms of gesture recognition which aims at recognizing meaningful expressions of human motion including hand, arm, face, head, and body movements [37]. Many surveys [25,32,34,35,37,47,49,56,57] have been released in the past years, providing an extensive overview of the many facets of gesture recognition. Many approaches either rely on 2D video capture technology and, thus, computer vision techniques, cf. [39,40], or on 3D motion capture technology which provides higher accuracy and thus more potential for precise spatiotemporal similarity search. A recent survey of vision-based gesture recognition approaches can be found in [47].

Frequently encountered approaches for recognizing hand gestures are based on *Hidden Markov Models* [31,44,46,53] or more generally *Dynamic Bayesian Networks* [52]. More recent approaches are based for instance on *Feature Fusion* [13], on *Dynamic Time Warping* [1,9], on *Longest Common Subsequences* [51], or on *Neural Networks* [20].

Since gestures addressed within the scope of this paper are represented in a lossless feature representation by aggregating movement traces, i.e. trajectories, the question of how to measure similarity between 3D motion capture trajectories is of crucial importance. Measuring the similarity between two trajectories can be carried out for instance by *Dynamic Time Warping* [8,58], *Levenshtein Distance* [19], *Minimal Variance Matching* [33], *Longest Common Subsequence* [54,55], *Edit Distance with Real Penalty* [11], *Edit Distance on Real Sequences* [12], or *Mutual Nearest Point Distance* [18]. When considering a real-time management of the trajectories, many streaming solutions have been proposed to track the emerging patterns that are appearing in the trajectory stream and their similarities as they are generated [21,22].

In addition to the aforementioned trajectory similarity measures, there exists also a rich amount of research literature on distance-based similarity measures applicable to feature signatures [2,3,5,48]. Well-known measures are the transformation-based *Earth Mover's Distance* [48], the correlation-based *Signature Quadratic Form Distance* [6], the matching-based *Hausdorff Distance* [23] and its variants [24,45] as well as the *Signature Matching Distance* [4].

Many of the approaches mentioned above are based on complex models or features that are learned or extracted prior to query processing. This assumes that enough training data is available and, in particular, that the gestural patterns to be considered are known to some extent, which is rarely the case for co-speech gestures due to their high degree of idiosyncrasy. For this reason, we propose a distance-based similarity model that is applicable to any type of gestural pattern. In fact, the proposed Gesture Matching Distance on gesture signatures can be utilized in order to model dissimilarity between gestural patterns whose movement types are well-known and between gestural patterns whose inherent structure is completely unknown. In this way, we provide an unsupervised and model-free distance-based approach of modeling gesture similarity.

3 Gesture Signatures

In this section, we introduce *gesture signatures* as lossless, spatiotemporal feature representations of gestures in 3D motion capture data streams. These data streams can be thought of as sequences of points in the three-dimensional Euclidean space $\mathbb{R}^3$. In the scope of this work, these points arise from several reflective markers which are attached to the body and in particular to the hands of a participant. The motion of the markers is triangulated via multiple cameras and finally recorded every 10 milliseconds. In this way, each marker defines a finite trajectory of points in a three-dimensional space. The formal definition of a trajectory is given below.

Definition 1 *(Trajectory). Given a three-dimensional feature space $\mathbb{R}^3$, a trajectory $t : \{1, \ldots, n\} \to \mathbb{R}^3$ is defined for all $1 \leq i \leq n$ as:*

$$t(i) = (x_i, y_i, z_i).$$

A trajectory describes the spatiotemporal motion of a single marker in a three-dimensional space. It is worth noting that the time information is abstracted to integral numbers in order to model trajectories arising from different time intervals and sampling rates. Since a gesture typically arises from multiple markers within a certain period of time, we aggregate several trajectories including their individual relevance by means of a *gesture signature*. For this purpose, we denote the set of all finite trajectories as trajectory space $\mathbb{T} = \bigcup_{k \in \mathbb{N}} \{t | t : \{1, \ldots, k\} \to \mathbb{R}^3\}$ and define a gesture signature as a function from the trajectory space $\mathbb{T}$ into the real numbers $\mathbb{R}$. The formal definition of a gesture signature is given below.

Definition 2 *(Gesture Signature). Let $\mathbb{T}$ be a trajectory space. A gesture signature $S \in \mathbb{R}^{\mathbb{T}}$ is defined as:*

$$S : \mathbb{T} \to \mathbb{R} \ \ subject \ to \ |\{t \in \mathbb{T} | S(t) \neq 0\}| < \infty.$$

A gesture signature formalizes a gesture by assigning a finite number of trajectories non-zero weights reflecting their importances. Negative weights are immaterial in practice but ensure the gesture space $\mathbb{S} = \{S | S \in \mathbb{R}^{\mathbb{T}} \wedge |S^{-1}(\mathbb{R} \setminus \{0\})| < \infty\}$ forms a vector space. While a weight of zero indicates insignificance of a trajectory, a positive weight is utilized to indicate contribution to the corresponding gesture. In this way, a gesture signature allows us to focus on the trajectories arising from those markers which actually form a gesture. For example, if a gesture is expressed by the participant's hands, only the corresponding hand markers and thus trajectories have to be weighted positively.

A gesture signature defines a generic mathematical model but omits a concrete functional implementation. In fact, given a subset of relevant trajectories $\mathcal{T}^+ \subset \mathbb{T}$, the most naive way of defining a gesture signature S consists in assigning relevant trajectories a weight of one and irrelevant trajectories a weight of zero, i.e. by defining $S \in \mathbb{S}$ for all $t \in \mathbb{T}$ as follows:

$$S(t) = \begin{cases} 1 & \text{if } t \in \mathcal{T}^+ \\ 0 & \text{otherwise.} \end{cases}$$

This uniform approach, however, completely ignores the inherent characteristics of the relevant trajectories. We therefore propose to weight each relevant trajectory according to its inherent spatiotemporal properties of *motion distance* and *motion variance*. These properties are defined below.

Definition 3 *(Motion Distance and Motion Variance). Let $\mathbb{T}$ be a trajectory space and $t : \{1, \ldots, n\} \to \mathbb{R}^3$ be a trajectory. The motion distance $m_\delta : \mathbb{T} \to \mathbb{R}$ of trajectory t is defined as:*

$$m_\delta(t) = \sum_{i=1}^{n-1} \|t(i) - t(i+1)\|_2.$$

The motion variance $m_{\sigma^2} : \mathbb{T} \to \mathbb{R}$ *of trajectory t is defined with mean $\mu(t) = \frac{1}{n} \cdot \sum_{i=1}^{n} t(i)$ as:*

$$m_{\sigma^2}(t) = \frac{1}{n} \cdot \sum_{i=1}^{n} \|t(i) - \mu(t)\|_2^2.$$

The intuition behind motion distance and motion variance is to take into account the overall movement and vividness of a trajectory. The higher these qualities, the more information the trajectory may contain and vice versa. Their utilization with respect to a set of relevant trajectories finally leads to the definitions of a *motion distance gesture signature* and a *motion variance gesture signature*, as shown below.

Definition 4 *(Motion Distance Gesture Signature and Motion Variance Gesture Signature). Let $\mathbb{T}$ be a trajectory space and $\mathcal{T}^+ \subset \mathbb{T}$ be a subset of relevant trajectories. A motion distance gesture signature $S_{m_\delta} \in \mathbb{S}$ is defined for all $t \in \mathbb{T}$ as:*

$$S_{m_\delta}(t) = \begin{cases} m_\delta(t) & \text{if } t \in \mathcal{T}^+ \\ 0 & \text{otherwise.} \end{cases}$$

A motion variance gesture signature $S_{m_{\sigma^2}} \in \mathbb{S}$ is defined for all $t \in \mathbb{T}$ as:

$$S_{m_{\sigma^2}}(t) = \begin{cases} m_{\sigma^2}(t) & \text{if } t \in \mathcal{T}^+ \\ 0 & \text{otherwise.} \end{cases}$$

Motion distance and motion variance gesture signatures are able to reflect the spatial and temporal characteristics of the expressed gestures with respect to the corresponding relevant trajectories by adapting the number and weighting of relevant trajectories. As a consequence, the computation of a (dis)similarity value between gesture signatures is frequently based on the (dis)similarity values among the involved trajectories in the trajectory space. Whereas distance functions applicable to trajectories are outlined in Sect. 2, we continue with introducing the *Gesture Matching Distance* in the following section.

4   Gesture Matching Distance

Gesture signatures naturally differ in size and length, i.e., in the number of relevant trajectories that contribute to a certain gesture and in the lengths of those trajectories. In order to quantify the distance between differently structured gesture signatures, we propose the matching-based *Gesture Matching Distance*. The idea of this distance-based similarity measure between gesture signatures is to match the underlying trajectories with respect to their spatial and temporal characteristics. These characteristics are evaluated by means of a trajectory distance function, such as the Dynamic Time Warping Distance. Based on a trajectory distance function, similar trajectories between both gesture signatures are matched according to the principle of the *δ-Nearest-Neighbor Matching*. The formal definition of this matching is given below.

Definition 5 *(δ-Nearest-Neighbor Matching). Let $S_1, S_2 \in \mathbb{S}$ be two gesture signatures and $\delta : \mathbb{T} \times \mathbb{T} \to \mathbb{R}$ be a trajectory distance function. The δ-Nearest-Neighbor Matching $\mathrm{m}^{\delta\text{-NN}}_{S_1 \to S_2} \subseteq \mathbb{T} \times \mathbb{T}$ between S_1 and S_2 is defined as follows:*

$$\mathrm{m}^{\delta\text{-NN}}_{S_1 \to S_2} = \{(t_1, t_2) \in \mathbb{T} \times \mathbb{T} | S_1(t_1) > 0 \wedge S_2(t_2) > 0 \wedge t_2 = \mathrm{argmin}_{t \in \mathbb{T}} \delta(t_1, t)\}.$$

As can be seen in the definition above, the δ-Nearest-Neighbor Matching $\mathrm{m}^{\delta\text{-NN}}_{S_1 \to S_2}$ between two gesture signatures S_1 and S_2 assigns each trajectory t_1 from the first gesture signature S_1 to one or more trajectories t_2 from the second gesture signature S_2. The δ-Nearest-Neighbor Matching $\mathrm{m}^{\delta\text{-NN}}_{S_1 \to S_2}$ satisfies *left totality*, i.e. it holds that $\forall t_1, \exists t_2 : S_1(t_1) > 0 \Rightarrow (t_1, t_2) \in \mathrm{m}^{\delta\text{-NN}}_{S_1 \to S_2}$ but not *right uniqueness*, i.e. it holds that $\forall t_1, t_2, t'_2 : (t_1, t_2) \in \mathrm{m}^{\delta\text{-NN}}_{S_1 \to S_2} \wedge (t_1, t'_2) \in \mathrm{m}^{\delta\text{-NN}}_{S_1 \to S_2} \not\Rightarrow t_2 = t'_2$. Therefore, the size of the δ-Nearest-Neighbor Matching is restricted by the number of trajectories which contribute to both gesture signatures, i.e. it holds that $|\mathrm{m}^{\delta\text{-NN}}_{S_1 \to S_2}| \leq |\{S_1(t) > 0\}_{t \in \mathbb{T}}| \cdot |\{S_2(t) > 0\}_{t \in \mathbb{T}}|$. Given the δ-Nearest-Neighbor Matching between two gesture signatures, we can now introduce the Gesture Matching Distance as shown in the definition below.

Definition 6 *(Gesture Matching Distance). Let $S_1, S_2 \in \mathbb{S}$ be two gesture signatures and $\delta : \mathbb{T} \times \mathbb{T} \to \mathbb{R}$ be a trajectory distance function. The* Gesture Matching Distance $\mathrm{GMD}_\delta : \mathbb{S} \times \mathbb{S} \to \mathbb{R}$ *between S_1 and S_2 is defined as:*

$$\mathrm{GMD}_\delta(S_1, S_2) = \sum_{(t_1, t_2) \in \mathrm{m}^{\delta\text{-NN}}_{S_1 \to S_2}} S_1(t_1) \cdot \delta(t_1, t_2) + \sum_{(t_2, t_1) \in \mathrm{m}^{\delta\text{-NN}}_{S_2 \to S_1}} S_2(t_2) \cdot \delta(t_2, t_1).$$

The Gesture Matching Distance GMD_δ between two gesture signatures is evaluated by adding the distances between matching trajectories and weighting these distances with the corresponding weights. In this way, the spatiotemporal relations of matching trajectories between two gesture signatures are linearly combined by the Gesture Matching Distance. The more similar two gesture signatures in terms of their underlying trajectories, the smaller the corresponding values of the trajectory distance functions between matching trajectories and thus the smaller the value of the Gesture Matching Distance.

The computation time complexity of a single distance computation lies in $\mathcal{O}(|\{S_1(t) > 0\}_{t \in \mathbb{T}}| \cdot |\{S_2(t) > 0\}_{t \in \mathbb{T}}| \cdot \zeta)$ where ζ denotes the computation time complexity of the trajectory distance function δ.

The following theorem formally proves that the Gesture Matching Distance indeed complies with the properties of a distance function [15]: (i) *non-negativity*, (ii) *symmetry*, and (iii) *reflexivity*.

Theorem 1 (*Distance Properties of Gesture Matching Distance*). *Let $\mathbb{S} = \{S | S \in \mathbb{R}^\mathbb{T} \wedge |S^{-1}(\mathbb{R} \setminus \{0\})| < \infty\}$ be a gesture space over the trajectory space $\mathbb{T}$ and $\delta : \mathbb{T} \times \mathbb{T} \to \mathbb{R}$ be a trajectory distance function. The Gesture Matching Distance $\mathrm{GMD}_\delta : \mathbb{S} \times \mathbb{S} \to \mathbb{R}$ satisfies the following distance properties for any gesture signatures $S_1, S_2 \in \mathbb{S}$:*

$$(i) \quad \mathrm{GMD}_\delta(S_1, S_2) \geq 0$$
$$(ii) \quad \mathrm{GMD}_\delta(S_1, S_2) = \mathrm{GMD}_\delta(S_2, S_1)$$
$$(iii) \quad \mathrm{GMD}_\delta(S_1, S_1) = 0$$

Proof. (i): By definition of $\mathrm{m}^{\delta\text{-NN}}_{S_1 \to S_2}$ and the properties of δ it holds that $\forall t_1, t_2 \in \mathbb{T} : (t_1, t_2) \in \mathrm{m}^{\delta\text{-NN}}_{S_1 \to S_2} \Rightarrow S_1(t_1) > 0 \wedge \delta(t_1, t_2) \geq 0$. Therefore, it holds that $\sum_{(t_1, t_2) \in \mathrm{m}^{\delta\text{-NN}}_{S_1 \to S_2}} S_1(t_1) \cdot \delta(t_1, t_2) \geq 0$. It can be shown analogously that $\sum_{(t_2, t_1) \in \mathrm{m}^{\delta\text{-NN}}_{S_2 \to S_1}} S_2(t_2) \cdot \delta(t_2, t_1) \geq 0$ and thus $\mathrm{GMD}_\delta(S_1, S_2) \geq 0$.

(ii): The symmetry of GMD_δ can be shown as follows:

$$\mathrm{GMD}_\delta(S_1, S_2)$$
$$= \sum_{(t_1, t_2) \in \mathrm{m}^{\delta\text{-NN}}_{S_1 \to S_2}} S_1(t_1) \cdot \delta(t_1, t_2) + \sum_{(t_2, t_1) \in \mathrm{m}^{\delta\text{-NN}}_{S_2 \to S_1}} S_2(t_2) \cdot \delta(t_2, t_1)$$
$$= \sum_{(t_2, t_1) \in \mathrm{m}^{\delta\text{-NN}}_{S_2 \to S_1}} S_2(t_2) \cdot \delta(t_2, t_1) + \sum_{(t_1, t_2) \in \mathrm{m}^{\delta\text{-NN}}_{S_1 \to S_2}} S_1(t_1) \cdot \delta(t_1, t_2)$$
$$= \mathrm{GMD}_\delta(S_2, S_1).$$

(iii): By definition of $\mathrm{m}^{\delta\text{-NN}}_{S_1 \to S_1}$ and the properties of δ it holds that $\forall t_1, t_1' \in \mathbb{T} : (t_1, t_1') \in \mathrm{m}^{\delta\text{-NN}}_{S_1 \to S_1} \Rightarrow \delta(t_1, t_1') = 0$. Therefore, it holds that $\sum_{(t_1, t_1') \in \mathrm{m}^{\delta\text{-NN}}_{S_1 \to S_1}} S_1(t_1) \cdot \delta(t_1, t_1') = 0$ and thus $\mathrm{GMD}_\delta(S_1, S_1) = 0$. This gives us the theorem.

Theorem 1 states that the Gesture Matching Distance GMD_δ is a valid distance function between gesture signatures from $\mathbb{S}$ provided that the underlying trajectory distance function δ satisfies the properties of non-negativity and reflexivity. Due to the nature of trajectories whose inherent spatial and temporal properties are rarely expressible in a single figure, trajectories are frequently compared by aligning their coincident similar points with each other. A prominent example is the *Dynamic Time Warping Distance*, which was first introduced in the field of speech recognition by Itakura [26] and Sakoe and Chiba [50] and later brought to the domain of pattern detection in databases by Berndt and Clifford [7]. The idea of this distance is to locally replicate points of the trajectories in order to fit the trajectories to each other. The point-wise distances finally yield the Dynamic Time Warping Distance, whose formal definition is given below.

Definition 7 *(Dynamic Time Warping Distance). Let $t_n : \{1, \ldots, n\} \to \mathbb{R}^3$ and $t_m : \{1, \ldots, m\} \to \mathbb{R}^3$ be two trajectories from $\mathbb{T}$ and $\delta : \mathbb{R}^3 \times \mathbb{R}^3 \to \mathbb{R}$ be a distance function. The* Dynamic Time Warping Distance $DTW_\delta : \mathbb{T} \times \mathbb{T} \to \mathbb{R}$ *between t_n and t_m is recursively defined as:*

$$DTW_\delta(t_n, t_m) = \delta(t_n(n), t_m(m)) + \min \begin{cases} DTW_\delta(t_{n-1}, t_{m-1}) \\ DTW_\delta(t_n, t_{m-1}) \\ DTW_\delta(t_{n-1}, t_m) \end{cases}$$

with

$$DTW_\delta(t_0, t_0) = 0$$
$$DTW_\delta(t_i, t_0) = \infty \quad \forall 1 \le i \le n$$
$$DTW_\delta(t_0, t_j) = \infty \quad \forall 1 \le j \le m.$$

As can be seen in the definition above, the Dynamic Time Warping Distance is defined recursively by minimizing the distances δ between replicated elements of the trajectories. In this way, the distance δ assesses the spatial proximity of two points while the Dynamic Time Warping Distance preserves their temporal order within the trajectories. By utilizing Dynamic Programming, the computation time complexity of the Dynamic Time Warping Distance lies in $\mathcal{O}(n \cdot m)$.

We have decided to utilize the Dynamic Time Warping Distance within the Gesture Matching Distance for the following two reasons: (i) The value of the Dynamic Time Warping Distance is based on all points of the trajectories with respect to their temporal order and is not attributed to partial characteristics of the trajectories and (ii) it provides the ability of efficient query processing by means of lower bounding [30].

Given the proposed spatiotemporal similarity model for gestures arising in 3D motion capture data streams, namely gesture signatures endowed with the Gesture Matching Distance, we will show in the following section how to process queries sequentially in order to provide a means of support for qualitative analyses.

5   Query Processing in 3D Motion Capture Data

In this section, we introduce a query processing algorithm for the query-driven analysis of 3D motion capture data streams. The objective of this algorithm is to produce a dissimilarity plot for the whole 3D motion capture data stream in order to indicate temporal segments comprising high spatiotemporal similarity with respect to a given query signature. To this end, we make use of a sliding window approach in order to average dissimilarity values between the query pattern and patterns within the 3D motion capture data stream. The pseudo code is listed in Algorithm 1.

input : trajectories $t_1, \ldots, t_n \in \mathbb{T}$, signature $Q \in \mathbb{S}$
output: $avgDissimilarity$

```
1  queryLength ← getMaxTrajectoryLength(Q);
2  dataLength ← getMaxTrajectoryLength(t₁,...,tₙ);
3  for i ← 0 to i < (dataLength − queryLength) do
4      signature S ← signaturize(t₁|[i,i+queryLength],...,tₙ|[i,i+queryLength]);
5      S ← weight(S);
6      S ← normalize(S);
7      dist ← GMDδ(Q,S);
8      avgDissimilarity[i,...,i + queryLength] ← update(avgDissimilarity,dist);
9      i ← i + 1;
10 end
```

Algorithm 1. Query processing algorithm

The input of Algorithm 1 consists of a 3D motion capture data stream in form of trajectories $t_1, \ldots, t_n \in \mathbb{T}$ and a query gesture signature $Q \in \mathbb{S}$ reflecting the spatiotemporal pattern of interest. The method `getMaxTrajectoryLength` is invoked twice to determine the maximum length of the query trajectories and of the data trajectories, respectively. The variable $queryLength$ defines the size of the sliding window which is utilized to generate gesture signatures from the 3D motion capture data stream. For this purpose, the method `signaturize` generates a gesture signature from the trajectories $t_1, \ldots, t_n$ restricted to the interval $[i, i + queryLength]$. The resulting gesture signature S is further weighted by means of a weighting scheme, as proposed in Definition 3, via the method `weight`. In addition to this weighting, the algorithm further applies a *min-max-normalization* to the interval $[0, 1]^3 \in \mathbb{R}^3$ by the method `normalize`. This ensures the gesture signatures to be translation invariant. Finally, the Gesture Matching Distance $\text{GMD}_\delta(Q, S)$ between the query gesture signature Q and the generated gesture signature S is evaluated and the variable $avgDissimilarity[i, \ldots, i + queryLength]$ is updated accordingly via the method `update`. The variable $avgDissimilarity$ contains the dissimilarity plot, i.e. the average dissimilarity values as a function of time, and is returned by the algorithm after processing the 3D motion capture data stream.

Unlike query processing algorithms designed for distance-based *range* and *nearest neighbor queries*, whose aims lie in finding the most similar objects with respect to queries, the proposed query processing algorithm aims at supporting domain-specific analyses by providing an average dissimilarity value for each temporal segment within the 3D motion capture data stream. We show the appropriateness of this dissimilarity plot obtained by utilizing the Gesture Matching Distance in the following section.

6 Experimental Evaluation

Evaluating the performance of distance-based similarity models is a highly empirical discipline. It is nearly unforeseeable which approach will provide the best results in terms of accuracy. To this end, we evaluated the Gesture Matching Distance based on the Dynamic Time Warping Distance by using a natural media corpus of 3D motion capture data collected for this project. This dataset comprises three-dimensional motion capture data streams arising from eight participants during a guided conversation. The participants were equipped with a multitude of reflective markers which were attached to the body and in particular to the hands. The motion of the markers has been tracked optically via cameras at a frequency of 100 Hz. In the scope of this work, we used the right wrist marker and two markers attached to the right thumb and right index finger each. The gestures arising within the conversation were classified by domain experts according to the following types of movement: spiral, circle, and straight. Example gestures of these movement types are sketched in Fig. 1. A total of 20 gesture signatures containing five trajectories each was obtained from the 3D motion capture data streams. The trajectories of the gesture signatures have been normalized to the interval $[0,1]^3 \in \mathbb{R}^3$ in order to maintain translation invariance.

	spirals						circles								straights					
spirals	0.00	0.32	0.32	0.22	0.37	0.31	0.64	0.57	0.57	0.44	0.30	0.29	0.46	0.31	0.36	0.22	0.28	0.15	0.30	0.35
	0.32	0.00	0.12	0.19	0.28	0.24	0.66	0.53	0.48	0.45	0.35	0.46	0.30	0.24	0.24	0.33	0.22	0.25	0.23	0.20
	0.32	0.12	0.00	0.14	0.23	0.16	0.63	0.48	0.34	0.29	0.27	0.34	0.24	0.16	0.17	0.36	0.18	0.23	0.17	0.21
	0.22	0.19	0.14	0.00	0.14	0.15	0.75	0.58	0.58	0.30	0.24	0.25	0.30	0.11	0.09	0.22	0.08	0.17	0.10	0.18
	0.37	0.28	0.23	0.14	0.00	0.29	1.00	0.74	0.94	0.60	0.40	0.48	0.44	0.21	0.18	0.28	0.13	0.26	0.16	0.23
	0.31	0.24	0.16	0.15	0.29	0.00	0.53	0.40	0.26	0.15	0.14	0.20	0.17	0.12	0.15	0.43	0.16	0.23	0.15	0.31
circles	0.64	0.66	0.63	0.75	1.00	0.53	0.00	0.08	0.38	0.56	0.34	0.57	0.41	0.73	0.81	0.99	0.79	0.41	0.76	0.88
	0.57	0.53	0.48	0.58	0.74	0.40	0.08	0.00	0.35	0.49	0.32	0.56	0.27	0.56	0.61	0.95	0.60	0.38	0.57	0.77
	0.57	0.48	0.34	0.58	0.94	0.26	0.38	0.35	0.00	0.29	0.20	0.44	0.15	0.45	0.53	0.83	0.61	0.40	0.62	0.78
	0.44	0.45	0.29	0.30	0.60	0.15	0.56	0.49	0.29	0.00	0.17	0.27	0.20	0.22	0.29	0.60	0.29	0.38	0.31	0.54
	0.30	0.35	0.27	0.24	0.40	0.14	0.34	0.32	0.20	0.17	0.00	0.16	0.18	0.21	0.25	0.46	0.24	0.20	0.24	0.43
	0.29	0.46	0.34	0.25	0.48	0.20	0.57	0.56	0.44	0.27	0.16	0.00	0.33	0.24	0.28	0.41	0.26	0.22	0.24	0.45
	0.46	0.30	0.24	0.30	0.44	0.17	0.41	0.27	0.15	0.20	0.18	0.33	0.00	0.26	0.30	0.62	0.29	0.28	0.31	0.52
	0.31	0.24	0.16	0.11	0.21	0.12	0.73	0.56	0.45	0.22	0.21	0.24	0.26	0.00	0.08	0.35	0.10	0.22	0.13	0.30
straights	0.36	0.24	0.17	0.09	0.18	0.15	0.81	0.61	0.58	0.29	0.25	0.28	0.30	0.08	0.00	0.36	0.06	0.22	0.14	0.31
	0.22	0.33	0.36	0.22	0.28	0.43	0.99	0.95	0.83	0.60	0.46	0.41	0.62	0.35	0.36	0.00	0.26	0.23	0.34	0.34
	0.28	0.22	0.18	0.08	0.13	0.16	0.79	0.60	0.61	0.29	0.24	0.26	0.29	0.10	0.06	0.26	0.00	0.18	0.10	0.23
	0.15	0.25	0.23	0.17	0.26	0.23	0.41	0.38	0.40	0.38	0.20	0.22	0.28	0.22	0.22	0.23	0.18	0.00	0.23	0.34
	0.30	0.23	0.17	0.10	0.16	0.15	0.76	0.57	0.62	0.31	0.24	0.24	0.31	0.13	0.14	0.34	0.10	0.23	0.00	0.18
	0.35	0.20	0.21	0.18	0.23	0.31	0.88	0.77	0.78	0.54	0.43	0.45	0.52	0.30	0.31	0.34	0.23	0.34	0.18	0.00

Fig. 2. Distance matrix for the Gesture Matching Distance based on motion variance gesture signatures with respect to different movement types. Bluish and reddish colors indicate small and large distance values, respectively. The distance values are normalized to the interval $[0,1] \in \mathbb{R}$ (Color figure online).

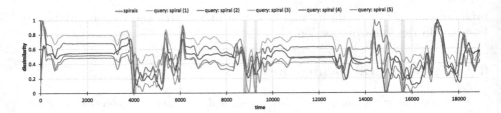

Fig. 3. Average dissimilarity values shown as a function of time with respect to gesture patterns of movement type spiral. Reddish time intervals depict gestural patterns included in the 3D motion capture data streams. Average dissimilarity values for the corresponding queries are shown via bluish and greenish lines (Color figure online).

The resulting distance matrix of the Gesture Matching Distance between all gesture signatures is shown in Fig. 2. Since weighting of relevant trajectories by motion distance and motion variance, cf. Definition 3, approximately shows the same tendency, we include the results regarding motion variance gesture signatures only. We depict small and large distance values by bluish and reddish colors in order to visually indicate the performance of our proposal: gesture signatures from the same movement type should result in bluish colors while gesture signatures from different movement types should result in reddish colors.

As can be seen in Fig. 2, the Gesture Matching Distance is able to distinguish gesture signatures from different movement types. On average, gesture signatures belonging to the same movement type are less dissimilar to each other than gesture signatures from different movement types. We further observed that the distinction between gesture signatures from the movement types spiral and straight are most challenging. This is caused by a similar sequence of movement of these two gestural types. While gesture signatures belonging to the movement type straight follow a certain direction, e.g., movement on the horizontal axis, gesture signatures from the movement type spiral additionally oscillate with respect to a certain direction. Since this oscillation can be dominated by the movement direction, the underlying trajectory distance function is often unable to distinguish oscillating from non-oscillating trajectories and thus gesture signatures of movement type spiral from those of movement type straight.

The question of whether our proposed distance-based similarity model and query processing algorithm are able to find similar spatiotemporal patterns within streams of 3D motion capture data is investigated in the remainder of this section. To this end, we used gestures classified by domain experts according to their movement type as queries and computed the dissimilarity plot for different motion capture data streams by means of Algorithm 1. The resulting average dissimilarity values calculated by the query processing algorithm are visualized as a function of time in Fig. 3 for the movement type spiral, in Figs. 4 and 5 for the movement type straight, and in Figs. 6 and 7 for the movement type circle. We highlight the corresponding gestural patterns included in the 3D motion capture data streams via reddish time intervals. The average dissimilarity values

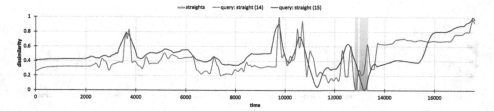

Fig. 4. Average dissimilarity values shown as a function of time with respect to gesture patterns of movement type straight. Reddish time intervals depict gestural patterns included in the 3D motion capture data streams. Average dissimilarity values for the corresponding queries are shown via bluish and greenish lines (Color figure online).

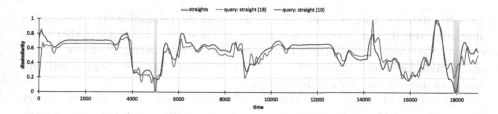

Fig. 5. Average dissimilarity values shown as a function of time with respect to gesture patterns of movement type straight. Reddish time intervals depict gestural patterns included in the 3D motion capture data streams. Average dissimilarity values for the corresponding queries are shown via bluish and greenish lines (Color figure online).

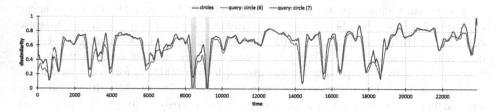

Fig. 6. Average dissimilarity values shown as a function of time with respect to gesture patterns of movement type circle. Reddish time intervals depict gestural patterns included in the 3D motion capture data streams. Average dissimilarity values for the corresponding queries are shown via bluish and greenish lines (Color figure online).

for the respective queries are shown by means of bluish and greenish lines. The better these lines coincide with the corresponding gestural patterns, the better the similarity model.

As can be seen in the figures, more than 90 % of the queries match the corresponding gestural patterns included in the 3D motion capture data streams as indicated by low dissimilarity values. The only exception is the second query of movement type spiral that is visualized in Fig. 3. It can be further observed that similar and dissimilar patterns in the 3D motion capture data streams

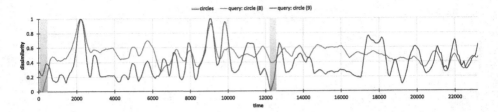

Fig. 7. Average dissimilarity values shown as a function of time with respect to gesture patterns of movement type circle. Reddish time intervals depict gestural patterns included in the 3D motion capture data streams. Average dissimilarity values for the corresponding queries are shown via bluish and greenish lines (Color figure online).

are well separable with respect to a specific query by a dissimilarity threshold of approximately 0.2. In other words, our proposal is able to find gestures that share similar spatiotemporal characteristics with the query pattern. It thus supports qualitative domain-specific analyses by query-driven spatiotemporal similarity search in 3D motion capture data streams.

Apart from the quality of accuracy, efficiency is another important aspect when evaluating the performance of a gesture similarity model. For this purpose, we measured the computation time needed to perform a single distance computation with respect to the gesture signature size, i.e. with respect to the length of relevant trajectories. We implemented the proposed distance-based approach in Java 1.8 and conducted the evaluation on a single-core 3.4 GHz machine. The average time needed to perform a single distance computation strongly depends on the gesture signature size. While the Gesture Matching Distance needs on average 57 ms for a single distance computation between two gesture signatures comprising trajectories of length 100, i.e. for gestures with a duration of 1 second, it needs on average 1.3 sec for a single distance computation between two gesture signatures comprising trajectories of length 500, i.e. for gestures with a duration of 5 sec. Thus, the computation time of the Gesture Matching Distance significantly grows with increasing length of the corresponding trajectories and thus the duration of the gestures.

To sum up, the experimental evaluation reveals that the proposed Gesture Matching Distance is able to model spatiotemporal similarity between gestures arising in 3D motion capture data streams in an unsupervised and model-free way. Without the need for training a complex model or extracting computationally intensive features from the 3D motion capture data streams, the experimental evaluation shows that our approach instantly supports the domain experts' qualitative analyses of gestural patterns.

7 Conclusions and Future Work

In this paper, we have proposed and investigated a distance-based approach to measure spatiotemporal similarity between gestures arising in 3D motion capture data streams. To this end, we have explicated gesture signatures as a way of aggregating the inherent spatial and temporal characteristics of gestures and introduced the Gesture Matching Distance as a novel distance-based approach for quantifying dissimilarity between gesture signatures. The Gesture Matching Distance epitomizes an unsupervised and model-free measure that can be instantly applied to 3D motion capture data streams in order to support the domain experts' qualitative analyses of gestural patterns within multimedia contexts.

In future work, we intend to extend our research on gesture similarity towards indexing and efficient query processing. While the focus of the present paper lies on investigating a distance-based similarity model for gesture signatures and providing an intuitive query processing algorithm for supporting the domain experts' qualitative analyses, we further plan to research lower bounding approaches for incrementally computing distances in a stream-based manner in order to enable real-time query-driven analyses in 3D motion capture gesture streams.

Acknowledgment. This work is partially funded by the Excellence Initiative of the German federal and state governments and by DFG grant SE 1039/7-1.

References

1. Arici, T., Celebi, S., Aydin, A.S., Temiz, T.T.: Robust gesture recognition using feature pre-processing and weighted dynamic time warping. Multimedia Tools Appl. **72**(3), 3045–3062 (2014)
2. Beecks, C.: Distance-based similarity models for content-based multimedia retrieval. PhD thesis, RWTH Aachen University (2013)
3. Beecks, C., Kirchhoff, S., Seidl, T.: On stability of signature-based similarity measures for content-based image retrieval. Multimedia Tools Appl. **71**(1), 349–362 (2014). doi:10.1007/s11042-012-1334-3
4. Beecks, C., Kirchhoff, S., Seidl, T.: Signature matching distance for content-based image retrieval. In: Proceedings of the ACM International Conference on Multimedia Retrieval, pp. 41–48 (2013)
5. Beecks, C., Uysal, M.S., Seidl, T.: A comparative study of similarity measures for content-based multimedia retrieval. In: Proceedings of the IEEE International Conference on Multimedia and Expo, pp. 1552–1557 (2010)
6. Beecks, C., Uysal, M.S., Seidl, T.: Signature quadratic form distance. In: Proceedings of the ACM International Conference on Image and Video Retrieval, pp. 438–445 (2010)
7. Berndt, D., Clifford, J.: Using dynamic time warping to find patterns in time series. In: AAAI 1994 workshop on knowledge discovery in databases, pp. 359–370 (1994)
8. Blackburn, J., Ribeiro, E.: Human motion recognition using isomap and dynamic time warping. In: Elgammal, A., Rosenhahn, B., Klette, R. (eds.) Human Motion 2007. LNCS, vol. 4814, pp. 285–298. Springer, Heidelberg (2007)
9. Bodiroža, S., Doisy, G., Hafner, V.V.: Position-invariant, real-time gesture recognition based on dynamic time warping. In: Proceedings of the International Conference on Human-robot Interaction, pp. 87–88 (2013)
10. Campbell, L.W.: Visual Classification of Co-verbal Gestures for Gesture Understanding. PhD thesis (2001)

11. Chen, L., Ng, R.: On the marriage of Lp-norms and edit distance. In: Proceedings of the International Conference on Very Large Data Bases, pp. 792–803 (2004)
12. Chen, L., Özsu, M.T., Oria, V.: Robust and fast similarity search for moving object trajectories. In: Proceedings of the ACM SIGMOD International Conference on Management of Data, pp. 491–502 (2005)
13. Cheng, J., Xie, C., Bian, W., Tao, D.: Feature fusion for 3D hand gesture recognition by learning a shared hidden space. Pattern Recogn. Lett. **33**(4), 476–484 (2012)
14. Cienki, A.: Cognitive linguistics: Spoken language and gesture as expressions of conceptualization. Body - Language - Communication: An International Handbook on Multimodality in Human Interaction, pp. 182–201 (2013)
15. Deza, M., Deza, E.: Encyclopedia of Distances. Springer, Heidelberg (2009)
16. Efron, D.: Gesture and Environment. Kings Crown Press, New York (1941)
17. Ekman, P., Friesen, W.: The repertoire of nonverbal behavior: Categories, origins, usage, and coding. Semiotica **1**(1), 49–98 (1969)
18. Fang, S., Chan, H.: Human identification by quantifying similarity and dissimilarity in electrocardiogram phase space. Pattern Recogn. **42**(9), 1824–1831 (2009)
19. Hahn, M., Krüger, L., Wöhler, C.: 3D action recognition and long-term prediction of human motion. In: Gasteratos, A., Vincze, M., Tsotsos, J.K. (eds.) ICVS 2008. LNCS, vol. 5008, pp. 23–32. Springer, Heidelberg (2008)
20. Hasan, H., Abdul-Kareem, S.: Static hand gesture recognition using neural networks. Artif. Intell. Rev. **41**(2), 147–181 (2014)
21. Hassani, M., Beecks, C., Töws, D., Serbina, T., Haberstroh, M., Niemietz, P., Jeschke, S., Neumann, S., Seidl, T.: Sequential pattern mining of multimodal streams in the humanities. In: Proceedings of the Conference on Database Systems for Business, Technology, and Web, pp. 683–686 (2015)
22. Hassani, M., Seidl, T.: Towards a mobile health context prediction: Sequential pattern mining in multiple streams. In: Proceedings of the IEEE International Conference on Mobile Data Management, pp. 55–57 (2011)
23. Hausdorff, F.: Grundzüge der Mengenlehre. Von Veit (1914)
24. Huttenlocher, D.P., Klanderman, G.A., Rucklidge, W.: Comparing images using the hausdorff distance. IEEE Trans. Pattern Anal. Mach. Intell. **15**(9), 850–863 (1993)
25. Ibraheem, N.A., Khan, R.Z.: Article: survey on various gesture recognition technologies and techniques. Int. J. Comput. Appl. **50**(7), 38–44 (2012)
26. Itakura, F.: Minimum prediction residual principle applied to speech recognition. IEEE Trans. Acoust. Speech Signal Process. **23**(1), 67–72 (1975)
27. Kendon, A.: Some relationships between body motion and speech. Stud. Dyadic Commun. **7**, 177 (1972)
28. Kendon, A.: Gesticulation and speech: two aspects of the process of utterance. The Relat. Verbal Nonverbal Commun. **25**, 207–227 (1980)
29. Kendon, A.: Gesture: Visible action as utterance. Cambridge University Press (2004)
30. Keogh, E.J.: Exact indexing of dynamic time warping. In: Proceedings of the International Conference on Very Large Data Bases, pp. 406–417 (2002)
31. Keskin, C., Erkan, A., Akarun, L.: Real time hand tracking and 3d gesture recognition for interactive interfaces using hmm. ICANN/ICONIPP **26–29**, 2003 (2003)
32. Khan, R.Z., Ibraheem, N.A.: Survey on gesture recognition for hand image postures. pp. 110–121 (2012)

33. Latecki, L.J., Megalooikonomou, V., Wang, Q., Lakaemper, R., Ratanamahatana, C.A., Keogh, E.: Elastic partial matching of time series. In: European Conference on Principles and Practice of Knowledge Discovery in Databases, pp. 577–584 (2005)
34. LaViola, J.: A survey of hand posture and gesture recognition techniques and technology. Brown University, Providence, RI (1999)
35. Liu, J., Kavakli, M.: A survey of speech-hand gesture recognition for the development of multimodal interfaces in computer games. In: Proceedings of the IEEE International Conference on Multimedia and Expo, pp. 1564–1569 (2010)
36. McNeill, D.: Hand and mind: What gestures reveal about thought. University of Chicago Press (1992)
37. Mitra, S., Acharya, T.: Gesture recognition: a survey. Trans. Sys. Man Cyber Part C **37**(3), 311–324 (2007)
38. Mittelberg, I.: Geometric and image-schematic patterns in gesture space. Equinox Publishing, pp. 351–388 (2010)
39. Moeslund, T.B., Granum, E.: A survey of computer vision-based human motion capture. Comput. Vis. Image Underst. **81**(3), 231–268 (2001)
40. Moeslund, T.B., Hilton, A., Krüger, V.: A survey of advances in vision-based human motion capture and analysis. Comput. Vis. Image Underst. **104**(2), 90–126 (2006)
41. Müller, C.: Redebegleitende Gesten. Berliner Wissenschafts-Verlag, Kulturgeschichte - Theorie - Sprachvergleich (1998)
42. Müller, C., Cienki, A., Fricke, E., Ladewig, S.H., McNeill, D., Teßendorf, S.: Body - Language - Communication: An International Handbook on Multimodality in Human Interaction. (Handbooks of Linguistics and Communication Science 38). De Gruyter Mouton, Berlin/ Boston (2013)
43. Müller, C., Posner, R.: The Semantics and Pragmatics of Everyday Gestures. Kultur. Weidler, Körper, Zeichen (2004)
44. Nam, Y., Wohn, K.: Recognition of hand gestures with 3D, nonlinear arm movement. Pattern Recogn. Lett. **18**(1), 105–113 (1997)
45. Park, B.G., Lee, K.M., Lee, S.U.: Color-based image retrieval using perceptually modified hausdorff distance. EURASIP J. Image Video Process. **2008**, 4:1–4:10 (2008)
46. Psarrou, A., Gong, S., Walter, M.: Recognition of human gestures and behaviour based on motion trajectories. Image Vis. Comput. **20**(5), 349–358 (2002)
47. Rautaray, S.S., Agrawal, A.: Vision based hand gesture recognition for human computer interaction: a survey. Artif. Intell. Rev. **43**(1), 1–54 (2015)
48. Rubner, Y., Tomasi, C., Guibas, L.J.: The earth mover's distance as a metric for image retrieval. Int. J. Comput. Vision **40**(2), 99–121 (2000)
49. Ruffieux, S., Lalanne, D., Mugellini, E., Abou Khaled, O.: A survey of datasets for human gesture recognition. In: Kurosu, M. (ed.) HCI 2014, Part II. LNCS, vol. 8511, pp. 337–348. Springer, Heidelberg (2014)
50. Sakoe, H., Chiba, S.: Dynamic programming algorithm optimization for spoken word recognition. IEEE Trans. Acoust. Speech Signal Process. **26**(1), 43–49 (1978)
51. Stern, H., Shmueli, M., Berman, S.: Most discriminating segment-longest common subsequence (MDSLCS) algorithm for dynamic hand gesture classification. Pattern Recogn. Lett. **34**(15), 1980–1989 (2013)
52. Suk, H.-I., Sin, B.-K., Lee, S.-W.: Recognizing hand gestures using dynamic bayesian network. In: Proceedings of the IEEE International Conference on Automatic Face & Gesture Recognition, pp. 1–6 (2008)

53. Suk, H.-I., Sin, B.-K., Lee, S.-W.: Hand gesture recognition based on dynamic Bayesian network framework. Pattern Recogn. **43**(9), 3059–3072 (2010)
54. Vlachos, M., Hadjieleftheriou, M., Gunopulos, D., Keogh, E.: Indexing multi-dimensional time-series with support for multiple distance measures. In: Proceedings of the ACM SIGKDD International Conference on Knowledge Discovery and Data Mining, pp. 216–225 (2003)
55. Vlachos, M., Kollios, G., Gunopulos, D.: Elastic translation invariant matching of trajectories. Mach. Learn. **58**(2–3), 301–334 (2005)
56. Watson, R.: A survey of gesture recognition techniques. Technical report,Trinity College Dublin, Department of Computer Science (1993)
57. Wu, Y., Huang, T.S.: Vision-based gesture recognition: a review. In: Braffort, A., Gibet, S., Teil, D., Gherbi, R., Richardson, J. (eds.) GW 1999. LNCS (LNAI), vol. 1739, pp. 103–115. Springer, Heidelberg (2000)
58. Yang, J., Li, Y., Wang, K.: A new descriptor for 3D trajectory recognition via modified CDTW. In: Proceedings of the IEEE International Conference on Automation and Logistics, pp. 37–42 (2010)

A Progressive Approach for Similarity Search on Matrix

Tsz Nam Chan[1]([✉]), Man Lung Yiu[1], and Kien A. Hua[2]

[1] Department of Computing, Hong Kong Polytechnic University,
Kowloon, Hong Kong
{cstnchan,csmlyiu}@comp.polyu.edu.hk
[2] College of Engineering and Computer Science, University of Central Florida,
Orlando, USA
kienhua@cs.ucf.edu

Abstract. We study a similarity search problem on a raw image by its pixel values. We call this problem as *matrix similarity search*; it has several applications, e.g., object detection, motion estimation, and super-resolution. Given a data image D and a query q, the best match refers to a sub-window of D that is the most similar to q. The state-of-the-art solution applies a sequence of lower bound functions to filter sub-windows and reduce the response time. Unfortunately, it suffers from two drawbacks: (i) its lower bound functions cannot support arbitrary query size, and (ii) it may invoke a large number of lower bound functions, which may incur high cost in the worst-case. In this paper, we propose an efficient solution that overcomes the above drawbacks. First, we present a generic approach to build lower bound functions that are applicable to arbitrary query size and enable trade-offs between bound tightness and computation time. We provide performance guarantee even in the worst-case. Second, to further reduce the number of calls to lower bound functions, we develop a lower bound function for a group of sub-windows. Experimental results on image data demonstrate the efficiency of our proposed methods.

1 Introduction

Multimedia databases [14,15,21] support similarity search on objects (e.g., images) by their feature vectors. In contrast, we consider a similarity search problem on a raw image by its pixel values. We call this problem as *matrix similarity search*; it has several applications, e.g., object detection [6], motion estimation [17], and super-resolution [7]. For example, we consider a satellite image in which each pixel represents a certain area on Earth (or in the sky). We illustrate a weather satellite image (obtained from [1]) in Fig. 1a and a cloud pattern in Fig. 1b. The matrix similarity search problem has been used for cloud motion estimation on satellite images [4]. This problem takes a data image D and a query image q as inputs (c.f. Fig. 1). A candidate c refers to a sub-window (of D) with the same size as q. The matrix similarity search problem comes in two flavors [19,22]:

© Springer International Publishing Switzerland 2015
C. Claramunt et al. (Eds.): SSTD 2015, LNCS 9239, pp. 373–390, 2015.
DOI: 10.1007/978-3-319-22363-6_20

- **Range search:** given a range τ_{range}, find every candidate c of D such that $dist(q, c) \leq \tau_{range}$.
- **Nearest neighbor (NN) search:** find a candidate c of D such that it has the smallest $dist(q, c)$.

The typical distance function $dist(q, c)$ is the L_p-norm distance (usually L_1 or L_2). In subsequent discussion, we let the size of D be $N_D = L_D \times W_D$, and the size of q be $N_q = L_q \times W_q$.

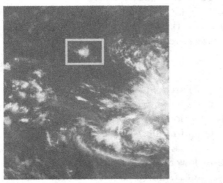

(a) data image D, of size $N_D = L_D \times W_D$ (b) query image q, of size $N_q = L_q \times W_q$
(best match: yellow rectangle)

Fig. 1. The matrix similarity search problem

In this paper, we focus on the NN flavor of matrix similarity problem because some applications [4, 17] require finding the best match. Unlike the range search, the NN search has a fixed result size and does not require the user to supply a range parameter τ_{range} [22].

Schweitzer et al. [22] is the state-of-the-art NN search algorithm for the matrix similarity search problem. It applies a sequence of lower bound functions to filter candidates and reduce the response time. We illustrate this idea in Fig. 2a. It starts with the cheapest lower bound function and then progressively apply tighter lower bound functions when necessary. However, this solution still suffers from two drawbacks. First, the lower bound functions in [22] are based on a Fourier transform on matrix (called the Walsh-Hadamard transform), which can only support query of the size $2^r \times 2^r$. Thus, it cannot support arbitrary query size. Second, in the worst case, it may invoke a large number of lower bound functions on a candidate, which may sum up to a high cost.

To avoid the above drawbacks on matrix similarity search, we contribute two lower bound functions $LB_{level,\ell}$ and LB_{group}, as shown in Fig. 2b.

- When compared to Ref. [22], we present a generic approach to build a sequence of lower bound functions $LB_{level,\ell}$ that are applicable to arbitrary query size. As shown in Fig. 2b, our approach would only call a logarithmic number of functions (in terms of N_q) in the worst-case.

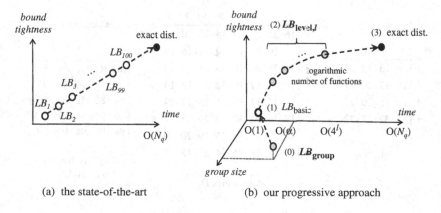

Fig. 2. Intuition

– Existing lower bound functions take a single candidate as input. We develop a lower bound function LB_{group} that can take a group of candidates as input. This significantly reduces the frequency of calling lower bound functions for individual candidates.

The rest of the paper is organized as follows. Section 2 defines our problem and introduces the background information. Section 3 presents our proposed solution. Section 4 discusses our experimental results. Section 5 elaborates on the related work. Section 6 concludes the paper with future research directions.

2 Preliminaries

2.1 Problem Definition

In this paper, we represent each image as a matrix. Let D be the data matrix (of size $N_D = L_D \times W_D$) and q be the query matrix (of size $N_q = L_q \times W_q$). A *candidate* $c_{x,y}$ is a sub-window of D with the same size as q.

$$c_{x,y}[1..L_q, 1..W_q] = D[x..x + L_q - 1, y..y + W_q - 1] \tag{1}$$

The subscript of $c_{x,y}$ denotes the start position in D; we drop it when the context is clear.

Problem 1 (Matrix NN Search). Given a query q and a data matrix D, find the candidate c_{best} such that it has the minimum $dist_p(q, c_{best})$, where the distance is defined as:

$$dist_p(q, c) = (\sum_{i=1}^{L_q} \sum_{j=1}^{W_q} |q[i, j] - c[i, j]|^p)^{\frac{1}{p}} \tag{2}$$

Figure 3 shows a query q of size 4×4 and a data matrix D of size 8×8. There are $5 \times 5 = 25$ candidates in D. For instance, the dotted sub-window refers to

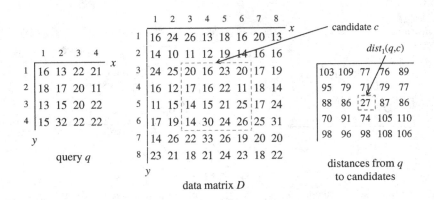

Fig. 3. Example for the problem

the candidate $c_{3,3}$. The right-side of Fig. 3 enumerates the distances from q to each candidate, assuming the L_1 distance (i.e., $p = 1$) is used. In this example, the best match is $c_{3,3}$ because it has the smallest distance $dist_1(q, c_{3,3}) = 27$ from q.

2.2 Background: Prefix-Sum Matrix and Basic Lower Bound Functions

In this section, we first introduce prefix-sum matrix and then discuss how they can be utilized to compute basic lower bound functions.

As we will introduce shortly, lower bound functions require summing the values in a rectangular region in a matrix. We can speedup their computation by using a prefix-sum matrix [11]. It is also called integral image [26] in the computer vision community.

Definition 1 (Prefix-sum matrix). *Given a matrix A (of size $N_A = L_A \times W_A$), we define its prefix-sum matrix P_A (of the same size) with entries:*

$$P_A[x, y] = \sum_{i=1}^{x} \sum_{j=1}^{y} A[i, j] \qquad (3)$$

The prefix-sum matrix occupies $O(N_A)$ space and takes $O(N_A)$ construction time [11]. It enables us to find the sum of values of a rectangular region (say, $[x_1..x_2, y_1..y_2]$) in a matrix A in $O(1)$ time, according to Eq. 4.

$$\sum A[x_1..x_2, y_1..y_2] = \begin{cases} P_A[x_2, y_2] & \text{if } x_1 = 1, y_1 = 1 \\ P_A[x_2, y_2] - P_A[x_1 - 1, y_2] & \text{if } x_1 > 1, y_1 = 1 \\ P_A[x_2, y_2] - P_A[x_2, y_1 - 1] & \text{if } x_1 = 1, y_1 > 1 \\ \begin{aligned} &P_A[x_2, y_2] + P_A[x_1 - 1, y_1 - 1] \\ &-P_A[x_1 - 1, y_2] - P_A[x_2, y_1 - 1] \end{aligned} & \text{otherwise} \end{cases}$$

$$(4)$$

$$\Sigma D[4..7,2..5] = P_D[7,5] - P_D[3,5] - P_D[7,1] + P_D[3,1]$$

	1	2	3	4	5	6	7	8
1	16	24	26	13	18	16	20	13
2	14	10	11	12	19	14	16	16
3	24	25	20	16	23	20	17	19
4	16	12	17	16	22	11	18	14
5	11	15	14	15	21	25	17	24
6	17	19	14	30	24	26	25	31
7	14	26	22	33	26	19	20	20
8	23	21	18	21	24	23	18	22

data matrix D

	1	2	3	4	5	6	7	8
1	16	40	66	79	97	113	133	146
2	30	64	101	126	163	193	229	258
3	54	113	170	211	271	321	374	422
4	70	141	215	272	354	415	486	548
5	81	167	255	327	430	516	604	690
6	98	203	305	407	534	646	759	876
7	112	243	367	502	655	786	919	1056
8	135	287	429	585	762	916	1067	1226

prefix-sum matrix P_D of D

Fig. 4. Example of a prefix-sum matrix

Figure 4 illustrates a data matrix D and its corresponding prefix-sum matrix P_D. The sum of values in the dotted region ($[4..7,2..5]$) in D can be derived from the entries $(7,5)$, $(3,1)$, $(3,5)$, $(7,1)$ in P_D.

We proceed to introduce the basic lower bound function LB_{basic} used in Fig. 2. Since our solution will use LB_{basic} as a building block (cf. Sect. 3), we require that: (i) LB_{basic} can be computed in $O(1)$ time, (ii) $LB_{basic}(q,c) \leq dist_p(q,c)$ always holds, and (iii) LB_{basic} supports arbitrary query size.

In this paper, we use the following lower bound functions as LB_{basic}.

$$LB_{\oplus}(q,c) = \frac{\sqrt[p]{N_q}}{N_q} \cdot \left| \sum_{i=1}^{L_q} \sum_{j=1}^{W_q} q[i,j] - \sum_{i=1}^{L_q} \sum_{j=1}^{W_q} c[i,j] \right| \tag{5}$$

$$LB_{\triangle}(q,c) = \left| \sqrt[p]{\sum_{i=1}^{L_q} \sum_{j=1}^{W_q} |q[i,j]|^p} - \sqrt[p]{\sum_{i=1}^{L_q} \sum_{j=1}^{W_q} |c[i,j]|^p} \right| \tag{6}$$

The first one ($LB_{\oplus}(q,c)$) is given in [28]. The second one ($LB_{\triangle}(q,c)$) is derived from the triangle inequality of the L_p distance [5, 13].

Observe that both of them can be computed in $O(1)$ time, by using a prefix-sum matrix as discussed before. Regarding the summation term for q, we can compute it once and then reuse it for every candidate c. For $LB_{\oplus}(q,c)$, the term $\sum_{i=1}^{L_q} \sum_{j=1}^{W_q} c[i,j]$ can be derived from the prefix-sum matrix P_D (of data matrix D). For $LB_{\triangle}(q,c)$, the term $\sum_{i=1}^{L_q} \sum_{j=1}^{W_q} |c[i,j]|^p$ can be derived from the prefix-sum matrix $P_{D'}$, where the matrix D' is defined with entries: $D'[i,j] = (D[i,j])^p$.

As a remark, we are aware of lower bound functions used in the pattern matching literature [2, 10, 18, 19, 25]. However, since those lower bound functions take more than $O(1)$ time, we choose not to use them as LB_{basic} (the building block) in our solution.

3   Progressive Search Algorithm

We illustrate the flow of our proposed NN search method in Fig. 5. Like [16,23], we employ a min-heap H in order to process entries in ascending order of their lower bound distance. The main difference is that H contains two types of entries: (i) a candidate, (ii) a group of candidates. As discussed before, a *candidate* corresponds to a sub-window of D. On the other hand, a *group* represents a consecutive region of candidates. Initially, H contains a group entry that covers the entire D.

When we deheap an entry from H, we check whether it is a group or a candidate.

1. If it is a group G, then we divide it evenly into 4 groups G_1, G_2, G_3, G_4[1]. For each G_i, we compute the group lower bound $LB(q, G_i)$ and then enheap G_i into H.
2. If it is a candidate c, then we compute the candidate lower bound $LB_{level,\ell}(q, c)$ at the next level ℓ, and then enheap c into H again.

During this process, a group would degenerate into a candidate when it covers exactly one candidate. Similarly, when a candidate reaches the deepest level, we directly apply the exact distance function $dist(q, c)$ on it, and update the best NN distance found so far τ_{best}. The search terminates when the lower bound of a deheaped entry exceeds τ_{best}.

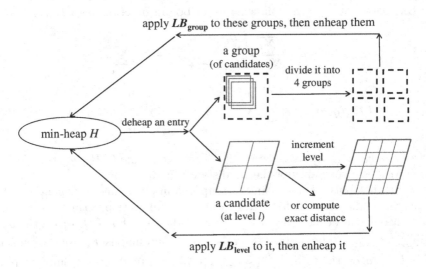

Fig. 5. The flow of our progressive search method

Table 1 lists the lower bound functions to be used in our NN search method. We have introduced LB_{basic} in Sect. 2.2. We will develop $LB_{level,\ell}$ and LB_{group}

---

[1] This is similar to the division of nodes in a quadtree.

in Sects. 3.1 and 3.2, respectively. Section 3.3 explores an efficient technique for computing LB_{group}. Finally, we summarize our proposed NN search algorithm in Sect. 3.4.

Table 1. Types of lower bound functions

Function	Apply to	Cost
LB_{basic} (e.g., $LB_\Delta, LB_\oplus$)	Candidate	$O(1)$
$LB_{level,\ell}$	Candidate	$O(4^\ell)$
LB_{group}	Group	$O(\alpha)$

3.1 Progressive Filtering for Candidates

As discussed in Sect. 1, the lower bound LB_{basic} and the exact distance $dist_p$ have a significant gap in terms of computation time and bound tightness (cf. Fig. 2). In order to save expensive distance computations, we suggest to apply tighter lower bound functions progressively.

In this section, we present a generic idea to construct a parameterized lower bound function $LB_{level,\ell}$ by using LB_{basic} as a building block. The level parameter ℓ controls the trade-offs between the bound tightness and the computation time in $LB_{level,\ell}$. A small ℓ incurs small computation time whereas a large ℓ provides tighter bounds.

Intuitively, we build $LB_{level,\ell}$ by using divide-and-conquer. We can partition the space $[1..L_q, 1..W_q]$ into 4^ℓ disjoint rectangles $\{R_v : 1 \leq v \leq 4^\ell\}$, and then apply LB_{basic} (for q and c) in each rectangle R_v.[2] Then, we combine these 4^ℓ lower bound distances into $LB_{level,\ell}$ in Eq. 7. The time complexity of $LB_{level,\ell}$ is $O(4^\ell)$, as each LB_{basic} takes $O(1)$ time.

$$LB_{level,\ell}(q,c) = \sqrt[p]{\sum_{v=1}^{4^\ell} LB_{basic}(q[R_v], c[R_v])^p} \qquad (7)$$

For example, in Fig. 6, when $\ell = 2$, both the query q and the candidate c are divided into $4^\ell = 16$ rectangles. We apply LB_{basic} on each rectangle in order to compute $LB_{level,\ell}(q,c)$. As a remark, the maximum possible level $\ell_{\max}$ (for ℓ) is:

$$\ell_{\max} = \lceil \log_2(\max\{L_q, W_q\}) \rceil \qquad (8)$$

Next, we show that $LB_{level,\ell}$ satisfies the lower bound property.

Lemma 1. *For any candidate c, we have:* $LB_{level,\ell}(q,c) \leq dist_p(q,c)$.

---

[2] In general, the space $[1..L_q, 1..W_q]$ may have less than $O(4^\ell)$ disjoint rectangles.

Proof. For each region R_v, we have $LB_{basic}(q[R_v], c[R_v])$ $\leq$ $\sqrt[p]{\sum_{(i,j)\in R_v} |q[i,j] - c[i,j]|^p}$, and thus $LB_{basic}(q[R_v], c[R_v])^p \leq \sum_{(i,j)\in R_v} |q[i,j] - c[i,j]|^p$. By summing it over all R_v, we obtain: $\sum_{v=1}^{4^\ell} LB_{basic}(q[R_v], c[R_v])^p \leq \sum_{i=1}^{L_q} \sum_{j=1}^{W_q} |q[i,j] - c[i,j]|^p$, because $\cup_{v=1}^{4^\ell} R_v$ covers all positions in the query matrix q. Thus we have: $LB_{level,\ell}(q,c) \leq dist_p(q,c)$. □

During search, we will apply $LB_{level,\ell}$ on a candidate in the ascending order of ℓ as shown in Fig. 6. If we cannot filter c at level ℓ, then we attempt to filter it with minimal extra effort, i.e., at level $\ell + 1$. We justify this ascending ℓ order in Lemma 2.

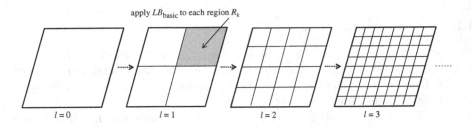

apply LB_{basic} to each region R_k

$l = 0$ $\qquad$ $l = 1$ $\qquad$ $l = 2$ $\qquad$ $l = 3$

Fig. 6. $LB_{level,\ell}$ at different levels

Lemma 2. *Consider a candidate c that is not the nearest neighbor. The ascending level order achieves $cost_{order} \leq \frac{4}{3} \cdot cost_{opt}$, where $cost_{opt}$ is the optimal cost, and $cost_{order}$ is the cost of the order.*

Proof. Recall that the cost of $LB_{level,\ell}(q,c)$ is 4^ℓ. Let ℓ^* be the smallest level such that $LB_{level,\ell^*}(q,c) > dist_{NN}$, where $dist_{NN}$ is the best match distance.

In order to discard c, the optimal way (which knows ℓ^*) is to apply LB_{level,ℓ^*}. Thus, we have: $cost_{opt} = 4^{\ell^*}$.

For the ascending level order, we have: $cost_{order} = \sum_{i=0}^{\ell^*} 4^i = \frac{4^{\ell^*+1}-1}{3}$. Thus, we have: $cost_{order}/cost_{opt} \leq \frac{4^{\ell^*+1}-1}{3 \times 4^{\ell^*}} \leq \frac{4}{3}$.   □

3.2 Progressive Filtering for Groups

We first introduce the concept of a group and then propose a lower bound function for it. A *group* G represents a consecutive region of candidates as shown in Fig. 7. It contains the following attributes: (i) L_g and W_g represent the size of the group, and (ii) x_{start} and y_{start} represent the start position (top-left corner) of the group. In order to cover all candidates in the group (e.g., those at bottom-right corner), we define the *extended region* as $G.R^{ext} = [x_{start}..x_{end}^{ext}, y_{start}..y_{end}^{ext}]$, where $x_{end}^{ext} = \min(x_{start} + L_g + L_q - 1, L_D)$ and $y_{end}^{ext} = \min(y_{start} + L_g + W_q - 1, W_D)$.

Our lower bound functions require the following concepts.

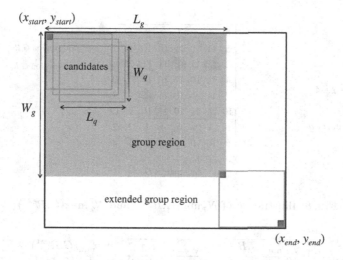

Fig. 7. Illustration of a group with $L_g \times W_g$ consecutive candidates

Definition 2 *(The smallest/largest N_q values). We define $N_q \min(G.R^{ext})$ as the multi-set of the smallest N_q values in the submatrix $D[G.R^{ext}]$, i.e., it satisfies:*

$$\max\{v : v \in N_q \min(G.R^{ext})\} \leq \min\{v : v \in D[G.R^{ext}] - N_q \min(G.R^{ext})\}$$

Then we define the following aggregates:
$$\phi_{\min}(G.R^{ext}) = \sum_{v \in N_q \min(G.R^{ext})} v, \quad \phi^p_{\min}(G.R^{ext}) = \sum_{v \in N_q \min(G.R^{ext})} |v|^p.$$
We define the max *versions (i.e., $N_q \max(G.R^{ext})$, $\phi_{\max}(G.R^{ext})$, $\phi^p_{\max}(G.R^{ext})$) in a similar way.*

We illustrate these concepts in Fig. 8. Assume that $p = 2$ and the query size is $N_q = 2 \times 2 = 4$. Consider the group G with region $G.R = [2..5, 2..5]$ (as dotted square) and the extended region $G.R^{ext} = [2..6, 2..6]$ (as bolded square). In this example, the smallest N_q values $G.R^{ext}$ are: 9, 9, 10, 10. Thus, we have: $\phi_{\min}(G.R^{ext}) = 9 + 9 + 10 + 10 = 38$, $\phi^2_{\min}(G.R^{ext}) = 2 \cdot 9^2 + 2 \cdot 10^2 = 362$.

We then extend basic lower bound functions (e.g., $LB_\oplus, LB_\triangle$) for a group G. We propose the lower bound functions $LB^\oplus_{group}$ and $LB^\triangle_{group}$ for G in Eqs. 9 and 10. They serve as lower bounds of $LB_\oplus(q, c), LB_\triangle(q, c)$ for any candidate c in G (cf. Lemmas 3 and 4).

$$LB^\oplus_{group}(q, G) = \begin{cases} \frac{\sqrt[p]{N_q}}{N_q}(\phi_{\min}(G.R^{ext}) - \sum_* q) & \text{if } \phi_{\min}(G.R^{ext}) > \sum_* q \\ \frac{\sqrt[p]{N_q}}{N_q}(\sum_* q - \phi_{\max}(G.R^{ext})) & \text{if } \phi_{\max}(G.R^{ext}) < \sum_* q \\ 0 & \text{otherwise} \end{cases} \quad (9)$$

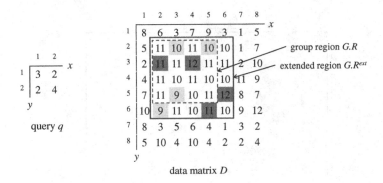

Fig. 8. Illustration of $N_q \min(G.R^{ext})$ and $N_q \max(G.R^{ext})$

$$LB_{group}^{\triangle}(q,G) = \begin{cases} \sqrt[p]{\phi_{\min}^p(G.R^{ext})} - \sqrt[p]{\sum_* |q[i,j]|^p} & \text{if } \phi_{\min}^p(G.R^{ext}) > \sum_* |q[i,j]|^p \\ \sqrt[p]{\sum_* |q[i,j]|^p} - \sqrt[p]{\phi_{\max}^p(G.R^{ext})} & \text{if } \phi_{\max}^p(G.R^{ext}) < \sum_* |q[i,j]|^p \\ 0 & \text{otherwise} \end{cases}$$

$$(10)$$

where $\sum_* q = \sum_{i=1}^{L_q} \sum_{j=1}^{W_q} q[i,j]$ and $\sum_* |q[i,j]|^p = \sum_{i=1}^{L_q} \sum_{j=1}^{W_q} |q[i,j]|^p$.

Lemma 3. *Given a group G, for any candidate c in G, we have:* $LB_{group}^{\oplus}(q,G) \leq LB_{\oplus}(q,c)$.

Proof. First, we focus on the first case of $LB_{group}^{\oplus}(q,G)$, i.e., when $\phi_{\min}(G.R^{ext}) > \sum_* q$.

Consider a candidate c in the group region of G. Since $N_q \min(G.R^{ext})$ contains the least N_q values in the group, we have: $\sum_* c \geq \phi_{\min}(G.R^{ext})$. Combining it with the condition in the first case, i.e., $\phi_{\min}(G.R^{ext}) > \sum_* q)$, we have $\sum_* c \geq \phi_{\min}(G.R^{ext}) > \sum_* q$.

Then we apply the above inequality on $LB_{\oplus}(q,c)$ and derive: $LB_{\oplus}(q,c) = \frac{\sqrt[p]{N_q}}{N_q} \cdot (\sum_* c - \sum_* q) \geq \frac{\sqrt[p]{N_q}}{N_q}(\phi_{\min}(G.R^{ext}) - \sum_* q) = LB_{group}^{\oplus}(q,G)$.

We omit the proof for the second case as it is similar to the above argument. The proof for the third case (i.e., $LB_{group}^{\oplus}(q,G) = 0$) is trivial. $\square$

Lemma 4. *Given a group G, for any candidate c in G, we have:* $LB_{group}^{\triangle}(q,G) \leq LB_{\triangle}(q,c)$.

Proof. First, we focus on the first case of $LB_{group}^{\triangle}(q,G)$, i.e., when $\phi_{\min}^p(G.R^{ext}) > \sum_* |q[i,j]|^p$.

Consider a candidate c in the group region of G. Since $N_q \min(G.R^{ext})$ contains the least N_q values in the group, we have: $\sum_* |c[i,j]|^p \geq \phi_{\min}^p(G.R^{ext})$. Combining it with the condition in the first case, i.e., $\phi_{\min}^p(G.R^{ext}) > \sum_* |q[i,j]|^p$, we have $\sum_* |c[i,j]|^p \geq \phi_{\min}^p(G.R^{ext}) > \sum_* |q[i,j]|^p$.

Then we apply the above inequality on $LB_{\triangle}(q,c)$ and derive: $LB_{\triangle}(q,c) = \sqrt[p]{\sum_* |c[i,j]|^p} - \sqrt[p]{\sum_* |q[i,j]|^p} \geq \sqrt[p]{\phi_{\min}^p(G.R^{ext})} - \sqrt[p]{\sum_* |q[i,j]|^p} = LB_{group}^{\triangle}(q,G)$.

We omit the proof for the second case as it is similar to the above argument. The proof for the third case (i.e., $LB_{group}^{\Delta}(q, G) = 0$) is trivial. □

We will discuss how to compute $LB_{group}(q, G)$ efficiently in the next subsection.

During our search procedure (cf. Fig. 5), we will apply $LB_{group}(q, G)$ on a group G. If we cannot filter G, then we partition its group region $G.R$ into four sub-groups G_1, G_2, G_3, G_4 accordingly, and apply $LB_{group}(q, G_i)$ on each sub-group G_i.

3.3 Supporting Group Filtering Efficiently

The lower bound $LB_{group}(q, G)$ involves the terms $\phi_{\min}(G.R^{ext})$, $\phi_{\max}(G.R^{ext})$, $\phi_{\min}^p(G.R^{ext})$, $\phi_{\max}^p(G.R^{ext})$, which require finding the smallest N_q and the largest N_q values in $G.R^{ext}$.

In this section, we design a data structure called *prefix histogram matrix* to support the above operations efficiently, namely in $O(\alpha)$ time. The parameter α allows trade-off between the time complexity and the bound tightness. A larger α tends to provide tighter bounds, but it incurs more computation time.

We proceed to elaborate on how to construct the prefix histogram matrix for a data matrix D. First, we partition the values in matrix D into α bins and convert each value $D[i, j]$ to the following bin number $D'[i, j]$:

$$D'[i, j] = \left\lfloor \alpha \cdot \frac{D[i, j] - D_{min}}{D_{max} - D_{min} + 1} \right\rfloor + 1$$

where D_{min} and D_{max} denote the minimum and maximum values in D, respectively.

We define the *prefix histogram matrix* PH_D as a matrix where each entry $PH_D[i, j]$ is a vector:

$$PH_D[i, j] = \langle P_1[i, j], P_2[i, j], \cdots, P_\alpha[i, j] \rangle$$

where

$$P_v[i, j] = \text{count}_{(x, y) \in [1..i, 1..j]}(D'[x, y] = v)$$

As a remark, the prefix histogram matrix occupies $O(\alpha N_D)$ space.

Figure 9a illustrates a histogram matrix PH_D in which each entry $PH_D[i, j]$ stores a count histogram for values in region $[1..i, 1..j]$ in the data matrix D.

Given an extended group region $G.R^{ext}$, we first retrieve count histograms at four corners of $G.R^{ext}$, and then combine them into the histogram as shown in Fig. 9b. With this histogram, we can derive bounds for the minimum/maximum N_q values of $G.R^{ext}$ in D by Definition 3.

Definition 3 (Sum of the smallest/largest N_q values in a count histogram). *Let CH be a count histogram for $G.R^{ext}$. We define $\phi_{\min}'(CH)$ as the sum of the smallest N_q values in CH, and $\phi_{\max}'(CH)$ as the sum of the largest N_q values in CH.*

While scanning the bins of CH from left to right, we examine the count and the minimum bound of each bin to derive $\phi'_{\min}(CH)$. A similar way can be used to derive $\phi'_{\max}(CH)$. The time complexity is $O(\alpha)$ as CH contains α bins.

As an example, consider the count histogram CH obtained in Fig. 9b. Assume that $\alpha = 6$ and $N_q = 4$. Thus, the width of each bin is $\frac{D_{max}-D_{min}+1}{\alpha} = \frac{12}{6} = 2$. Since the count of bin 9..10 is above N_q, we derive: $\phi'_{\min}(C\hat{H}) = 9 \cdot 4 = 36$. Note that $\phi'_{\min}(CH) = 36$ is looser than the actual value $\phi_{\min}(G.R^{ext}) = 38$ (obtained in Fig. 8).

Then we replace $LB^{\oplus}_{group}$ by the following function $LB'^{\oplus}_{group}$:

$$LB'^{\oplus}_{group}(q, G) = \begin{cases} \frac{\sqrt[p]{N_q}}{N_q}(\phi'_{\min}(CH) - \sum_* q) & \text{if } \phi'_{\min}(CH) > \sum_* q \\ \frac{\sqrt[p]{N_q}}{N_q}(\sum_* q - \phi'_{\max}(CH)) & \text{if } \phi'_{\max}(CH) < \sum_* q \\ 0 & \text{otherwise} \end{cases} \quad (11)$$

Since $\phi'_{\min}(CH) \le \phi_{\min}(G.R^{ext})$ and $\phi'_{\max}(CH) \ge \phi_{\max}(G.R^{ext})$, $LB'^{\oplus}_{group} \le LB^{\oplus}_{group}$.

Similarly, we can adapt the above technique to derive a lower bound of $LB^{\triangle}_{group}$ efficiently.

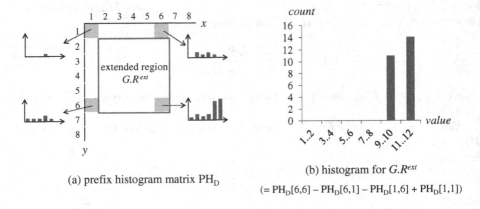

(a) prefix histogram matrix PH_D

(b) histogram for $G.R^{ext}$

($= PH_D[6,6] - PH_D[6,1] - PH_D[1,6] + PH_D[1,1]$)

Fig. 9. prefix histogram matrix, $\alpha = 6, D_{min} = 1, D_{max} = 12$

3.4   Algorithm for NN Search

In this section, we summarize our techniques in Algorithm 1. Like [16,23], we employ a min-heap H in order to process entries in ascending order of their lower bound distance. Also, we maintain the best distance found-so-far τ_{best} during the search. The main difference from [16,23] is that we apply multiple lower bound functions on candidates and also consider lower bound function for groups of candidates.

Initially, we create an entry e_{root} to represent the group of all candidates. In each iteration, we deheap an entry e and check whether it is a group entry

---

Algorithm 1. Progressive Search Algorithm for NN search

1: **procedure** PROGRESSIVE SEARCH(query matrix q, data matrix D)
2: $\quad \tau_{best} \leftarrow \infty$ $\qquad\qquad\qquad\qquad\qquad\qquad$ ▷ best NN distance found so far
3: $\quad$ create a min-heap H
4: $\quad$ create a heap entry e_{root}
5: $\quad e_{root}.G \leftarrow [0..L_D - 1, 0..W_D - 1]$ $\qquad\qquad$ ▷ the region covered by the group
6: $\quad e_{root}.bound \leftarrow LB_{group}(q, e.G)$
7: $\quad$ enheap e_{root} to H
8: $\quad$ **while** $H \neq \emptyset$ **do**
9: $\quad\quad e \leftarrow$ deheap an entry in H
10: $\quad\quad$ **if** $e.bound \geq \tau_{best}$ **then** $\qquad\qquad\qquad\qquad$ ▷ termination condition
11: $\quad\quad\quad$ **break**
12: $\quad\quad$ **if** $|e.G| \neq 1$ **then** $\qquad\qquad\qquad\qquad\qquad$ ▷ group entry
13: $\quad\quad\quad$ divide e into 4 entries e_1, e_2, e_3, e_4
14: $\quad\quad\quad$ **for** each e_i, $i \leftarrow 1$ to 4 **do**
15: $\quad\quad\quad\quad e_i.bound \leftarrow LB_{group}(q, e_i.G)$
16: $\quad\quad\quad\quad$ enheap e_i to H **if** $e_i.bound \leq \tau_{best}$
17: $\quad\quad$ **else** $\qquad\qquad\qquad\qquad\qquad\qquad\qquad\qquad$ ▷ candidate entry
18: $\quad\quad\quad$ **if** $e.\ell < \ell_{max}$ **then** $\qquad\qquad\qquad$ ▷ not at the deepest level
19: $\quad\quad\quad\quad e.bound \leftarrow LB_{level,\ell}(q, e)$
20: $\quad\quad\quad\quad$ increment $e.\ell$
21: $\quad\quad\quad\quad$ enheap e to H **if** $e.bound \leq \tau_{best}$
22: $\quad\quad\quad$ **else**
23: $\quad\quad\quad\quad$ compute $dist_p(q, c)$

---

or a candidate entry. When e is a group entry, we divide it into four group entries and enheap them into H. Otherwise, e is a candidate entry, and then we examine the level of e. If e has not reached the maximum level ℓ_{max}, we compute $LB_{level,\ell}(q, e)$, advance it to the next level, and enheap it into H. Otherwise, we compute the exact distance of e from q, and update τ_{best} if necessary. The loop terminates when H becomes empty or the lower bound of the current entry exceeds τ_{best}.

4   Experimental Evaluation

In this section, we compare the efficiency of our methods with the state-of-the-art method [22] called Dual-Bound (DB). Table 2 shows the bounding functions used in these methods. Our *progressive search* methods share the same prefix PS:

- PSL stands for progressive search with LB_{level} only, and
- PSLG stands for progressive search with both LB_{level} and LB_{group}.

The subscripts of our methods (e.g., $\oplus$ or Δ) indicate whether they use lower bound functions built on top of $LB_{\oplus}$ or LB_{Δ}. We implemented all algorithms in C/C++ and conducted experiments on an Intel i7 3.4 GHz PC running Ubuntu.

Table 2. The list of our methods and the competitor

Method	Bounding functions used in the method
DB	[22]
$PSL_\oplus$	$LB_\oplus, LB_{level}$
$PSL_\triangle$	$LB_\triangle, LB_{level}$
$PSLG_\oplus$	$LB_\oplus, LB_{level}, LB_{group}$

Table 3. Our datasets and queries

Dataset	Image size	Number of images	Number of queries per image
Photo	2560×1920	30	10
Weather	1800×1800	30	10

Note that each method (in Table 2) requires a preprocessing step — scan a data image D to compute its prefix-sum matrix. This step is done only once before queries arrive. It is negligible compared to the query response time.

Table 3 summarizes the details of our image data and queries. We collect image datasets from [1, 19]. *Photo* [19] contains 30 images of the size 2560×1920. *Weather* [1] contains 30 weather satellite images of the size 1800×1800; the timestamps of these images are from 00:00 on 1/4/2014 to 06:00 on 2/4/2014. For each image, we generate 10 random starting positions by the uniform distribution to extract queries from that image.

In each experiment, we execute the methods on 300 queries ($= 30$ images $\times$ 10 queries) and then report the average response time.

4.1 Experimental Results

First, we study the effect of the number of bins α on the response time of our method $PSLG_\oplus$. Figure 10 plots the running time as a function of α. When α increases, the group-based lower bound LB_{group} becomes tighter (i.e., higher pruning power) so the response time drops. Nevertheless, when α is too large, it incurs high overhead to compute LB_{group} so the response time rises slightly. In subsequent experiments, we set $\alpha = 16$ by default.

Next, we evaluate the scalability of methods with respect to the query size N_q. Figure 11 shows the response time of methods versus the query size N_q. Since DB [22] can only support query size of the form $2^r \times 2^r$, we use query sizes like $32^2, 64^2, \cdots$ in this experiment. Thanks to the group lower bound function, $PSLG_\oplus$ outperforms all other methods and scales better with respect to N_q. On the other hand, DB, $PSL_\triangle$ and $PSL_\oplus$ need to obtain candidates one-by-one and incur higher overhead on maintaining the min-heap.

Since $PSL_\oplus$ performs better than $PSL_\triangle$, we omit $PSL_\triangle$ in the next experiment.

Following [22], we then test the robustness of methods by adding noise to queries. As in [22], Gaussian noise with a standard deviation σ is added into each

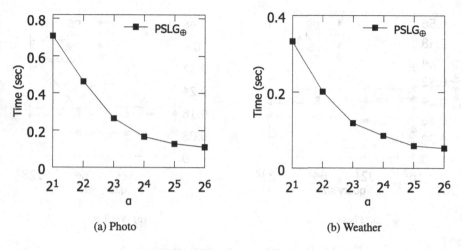

(a) Photo   (b) Weather

Fig. 10. Response time vs. the number of bins α

query image. The query size is fixed to 128×128 in this experiment. Figure 12 shows the response time of methods as a function of σ. The performance gap between our methods and DB widens as σ increases. At a high σ, the pruning power of all lower bound functions becomes weaker. For each worst-case candidate (that cannot be pruned), DB may invoke a long sequence of bounding functions on it, whereas our methods invoke only a logarithmic number of LB_{level} (in terms of N_q) on it. In summary, our methods are more robust against noise.

5 Related Work

5.1 Nearest Neighbor Search

The nearest neighbor (NN) search problem has been extensively studied in multimedia databases [14,15,21] and in time series databases [8,20,28].

Multimedia databases [14,15,21] usually conduct similarity search (i.e., NN search) on *feature vectors* of images (e.g., their color/texture histograms) rather than on raw pixel values in images. Various techniques on indexing [3,13,21], data compression [27], and hashing [12,24] have been developed to process NN search efficiently. Recall that those multimedia techniques require knowing feature vectors in advance. Those techniques are applicable to our problem context, when the query size N_q is fixed, as we can convert each candidate (sub-window) $c_{x,y}$ to a N_q-dimensional feature vector offline. However, those techniques become inapplicable if we need to support arbitrary query size (i.e., N_q only known at the query time). It is infeasible to do precomputation for every possible query size as it would blow up the storage space by a huge factor ($N_D{}^2$), where $N_D = L_D \times W_D$ is the size of the data image.

Generic NN search algorithms [16,23] are applicable to any types of objects and distance function $dist(q, c)$. Ref. [23] requires using a lower bound function

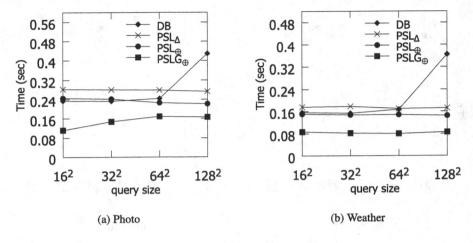

(a) Photo

(b) Weather

Fig. 11. Response time vs. vary query size N_q

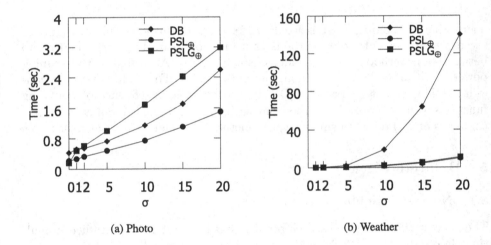

(a) Photo

(b) Weather

Fig. 12. Response time vs. the noise σ

$LB(q, c)$, where $LB(q, c) \leq dist(q, c)$ always holds. Its search strategy [23] is to examine candidates in ascending order of $LB(q, c)$ and then compute their exact distances to q, until the current $LB(q, c)$ exceeds the best NN distance found so far. Ref. [16] takes an additional upper bound function $UB(q, c)$ as input and utilizes it to further reduce the searching time. Observe that the lower bound functions for a specific problem (e.g., matrix similarity search problem) are not provided in [16,23]. In this paper, we focus on developing lower bound functions like LB_{level}, LB_{group} for matrix similarity search.

The NN search on a time series [8,20,28] can be considered as a special case of our problem, where both the data image D and the query q are modeled as vectors instead of matrices. While some simple lower bound functions originate from them, our proposed lower bound functions (LB_{level}, LB_{group}) are new and specific to matrix similarity search. Specifically, our LB_{level} is a generic function that can be built on top of any given LB_{basic}, and our LB_{group} can take a group of candidates as input.

5.2 Matrix Similarity Search Methods

Various lower bound functions [2,9,10,18,19,22,25] have been developed for the matrix similarity search problem, in order to prune unpromising candidates efficiently and thus avoid expensive exact distance computations. Most solutions focus on range search and a few study on NN search. Ouyang et al. [19] proposes a unified framework that covers range search solutions [2,9,10,18,25]. The state-of-the-art NN search method is [22]. It applies both lower and upper bound functions to accelerate NN search. Its lower/upper bound functions are based on a Fourier transform on matrix (called the Walsh-Hadamard transform), which can only support query of the size $2^r \times 2^r$. Thus, it cannot support arbitrary query size. Also, [22] has not explored our group-based lower bound function LB_{group}, which applies to a group of candidates instead of a single candidate.

6 Conclusion

We have developed a progressive NN search method for the matrix similarity search problem. It includes a generic lower bound function LB_{level} for candidates, and a group-based lower bound function LB_{group} for a group of candidates. Our proposed solution performs much better than the state-of-the-art method.

In the future, we plan to investigate approximate NN search methods for matrix similarity search. Sampling techniques may be adapted to address this problem.

References

1. Weather datasets. http://weather.is.kochi-u.ac.jp/sat/GAME/
2. Ben-Artzi, G., Hel-Or, H., Hel-Or, Y.: The gray-code filter kernels. IEEE Trans. Pattern Anal. Mach. Intell. **29**(3), 382–393 (2007)
3. Böhm, C., Berchtold, S., Keim, D.A.: Searching in high-dimensional spaces: index structures for improving the performance of multimedia databases. ACM Comput. Surv. **33**(3), 322–373 (2001)
4. Brad, R., Letia, I.A.: Extracting cloud motion from satellite image sequences. In: ICARCV, pp. 1303–1307 (2002)
5. Ciaccia, P., Patella, M., Zezula, P.: M-tree: An efficient access method for similarity search in metric spaces. In: VLDB, pp. 426–435 (1997)
6. Dufour, R.M., Miller, E.L., Galatsanos, N.P.: Template matching based object recognition with unknown geometric parameters. IEEE Trans. Image Process. **11**(12), 1385–1396 (2002)
7. Freeman, W.T., Jones, T.R., Pasztor, E.C.: Example-based super-resolution. IEEE Comput. Graph. Appl. **22**(2), 56–65 (2002)

8. Fu, A.W., Keogh, E.J., Lau, L.Y.H., Ratanamahatana, C.A., Wong, R.C.: Scaling and time warping in time series querying. VLDB J. **17**(4), 899–921 (2008)

9. Gharavi-Alkhansari, M.: A fast globally optimal algorithm for template matching using low-resolution pruning. IEEE Trans. Image Process. **10**(4), 526–533 (2001)

10. Hel-Or, Y., Hel-Or, H.: Real-time pattern matching using projection kernels. IEEE Trans. Pattern Anal. Mach. Intell. **27**(9), 1430–1445 (2005)

11. Ho, C., Agrawal, R., Megiddo, N., Srikant, R.: Range queries in OLAP data cubes. In: SIGMOD, pp. 73–88 (1997)

12. Indyk, P., Motwani, R.: Approximate nearest neighbors: towards removing the curse of dimensionality. In: STOC, pp. 604–613 (1998)

13. Jagadish, H.V., Ooi, B.C., Tan, K., Yu, C., Zhang, R.: idistance: an adaptive b^+-tree based indexing method for nearest neighbor search. ACM Trans. Database Syst. **30**(2), 364–397 (2005)

14. Keim, D.A., Bustos, B.: Similarity search in multimedia databases. In: ICDE, p. 873 (2004)

15. Korn, F., Sidiropoulos, N., Faloutsos, C., Siegel, E.L., Protopapas, Z.: Fast nearest neighbor search in medical image databases. In: VLDB, pp. 215–226 (1996)

16. Kriegel, H.-P., Kröger, P., Kunath, P., Renz, M.: Generalizing the optimality of multi-step k-nearest neighbor query processing. In: Papadias, D., Zhang, D., Kollios, G. (eds.) SSTD 2007. LNCS, vol. 4605, pp. 75–92. Springer, Heidelberg (2007)

17. Moshe, Y., Hel-Or, H.: Video block motion estimation based on gray-code kernels. IEEE Trans. Image Process. **18**(10), 2243–2254 (2009)

18. Ouyang, W., Cham, W.: Fast algorithm for walsh hadamard transform on sliding windows. IEEE Trans. Pattern Anal. Mach. Intell. **32**(1), 165–171 (2010)

19. Ouyang, W., Tombari, F., Mattoccia, S., di Stefano, L., Cham, W.: Performance evaluation of full search equivalent pattern matching algorithms. IEEE Trans. Pattern Anal. Mach. Intell. **34**(1), 127–143 (2012)

20. Rakthanmanon, T., Campana, B.J.L., Mueen, A., Batista, G.E.A.P.A., Westover, M.B., Zhu, Q., Zakaria, J., Keogh, E.J.: Searching and mining trillions of time series subsequences under dynamic time warping. In: KDD, pp. 262–270 (2012)

21. Samet, H.: Techniques for similarity searching in multimedia databases. PVLDB **3**(2), 1649–1650 (2010)

22. Schweitzer, H., Deng, R.A., Anderson, R.F.: A dual-bound algorithm for very fast and exact template matching. IEEE Trans. Pattern Anal. Mach. Intell. **33**(3), 459–470 (2011)

23. Seidl, T., Kriegel, H.: Optimal multi-step k-nearest neighbor search. In: SIGMOD, pp. 154–165 (1998)

24. Tao, Y., Yi, K., Sheng, C., Kalnis, P.: Quality and efficiency in high dimensional nearest neighbor search. In: SIGMOD, pp. 563–576 (2009)

25. Tombari, F., Mattoccia, S., di Stefano, L.: Full-search-equivalent pattern matching with incremental dissimilarity approximations. IEEE Trans. Pattern Anal. Mach. Intell. **31**(1), 129–141 (2009)

26. Viola, P.A., Jones, M.J.: Robust real-time face detection. Int. J. Comput. Vis. **57**(2), 137–154 (2004)

27. Weber, R., Schek, H., Blott, S.: A quantitative analysis and performance study for similarity-search methods in high-dimensional spaces. In: VLDB, pp. 194–205 (1998)

28. Yi, B., Faloutsos, C.: Fast time sequence indexing for arbitrary Lp norms. In: VLDB, pp. 385–394 (2000)

Discovering Non-compliant Window Co-Occurrence Patterns: A Summary of Results

Reem Y. Ali[1]([✉]), Venkata M.V. Gunturi[1], Andrew J. Kotz[2],
Shashi Shekhar[1], and William F. Northrop[2]

[1] Department of Computer Science and Engineering,
University of Minnesota, Minneapolis, MN 55455, USA
`alixx464@umn.edu`
[2] Department of Mechanical Engineering, University of Minnesota,
Minneapolis, MN 55455, USA

Abstract. Given a set of trajectories annotated with measurements of physical variables, the problem of Non-compliant Window Co-occurrence (NWC) pattern discovery aims to determine temporal signatures in the explanatory variables which are highly associated with windows of undesirable behavior in a target variable. NWC discovery is important for societal applications such as eco-friendly transportation (e.g. identifying engine signatures leading to high greenhouse gas emissions). Challenges of designing a scalable algorithm for NWC discovery include the non-monotonicity of popular spatio-temporal statistical interest measures of association such as the cross-K function. This challenge renders the anti-monotone pruning based algorithms (e.g. Apriori) inapplicable. To address this limitation, we propose two novel upper bounds for the cross-K function which help in filtering uninteresting candidate patterns. Using these bounds, we also propose a Multi-Parent Tracking approach (MTN-Miner) for mining NWC patterns. A case study with real world engine data demonstrates the ability of the proposed approach to discover patterns which are interesting to engine scientists. Experimental evaluation on real-world data show that MTNMiner results in substantial computational savings over the naive approach.

1 Introduction

Given a set of trajectories annotated with measurements of physical variables, the Non-compliant Window Co-occurrence (NWC) pattern discovery problem aims to determine temporal signatures in the explanatory variables which are highly associated with windows of undesirable behavior in a target variable (e.g. non-compliance with some standard). For instance, consider Fig. 1, which shows portions of trajectories of a metro transit bus in Minneapolis-St. Paul, MN, USA. In these trajectories, each point is annotated with physical measurements such as engine power, engine revolutions per minute (RPM), wheel speed, elevation and engine emissions. The red color marks temporal windows within the trajectories where a target variable (emissions of oxides of nitrogen (NO_x) in this example) shows a non-compliant behavior (i.e., the average emissions within the windows exceed US EPA regulations [1]). As shown in the

© Springer International Publishing Switzerland 2015
C. Claramunt et al. (Eds.): SSTD 2015, LNCS 9239, pp. 391–410, 2015.
DOI: 10.1007/978-3-319-22363-6_21

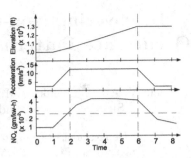

Fig. 1. Non-compliant emissions of oxides of nitrogen along a bus route in Minneapolis, MN (best viewed in color) (Color figure online).

Fig. 2. Example of candidate NWC patterns (best viewed in color) (Color figure online).

figure, while some journeys show this non-compliant behavior, others do not. NWC discovery aims to determine the underlying temporal signatures of the measured physical variables which are highly associated with those windows of elevated NO_x emissions. These signatures (aka "patterns") represent sequences defined on one or more variables that either coincide with or occur within a prespecified time lag from a non-compliant window. For instance, consider Fig. 2 which shows a non-compliant window $< 2, 6 >$, of length 5 s, in which the NO_x emissions within the window exceed the US EPA standard of 0.267 gm/kW-h. As shown in the figure, patterns of high acceleration and increase in elevation co-occur with this non-compliance behavior and thus would be considered as candidate patterns by the NWC discovery problem.

Importance: Discovering NWC patterns is important to several scientific and societal applications such as eco-friendly transportation (e.g. discovering engine behaviors leading to high greenhouse gas emissions), detection of engine signatures associated with engine malfunctions (e.g. patterns of sudden unintended acceleration [2]) which can help save people's lives, and industrial process control (e.g. understanding patterns of failure in an industrial process [3]). In this paper we use eco-friendly transportation as our illustrative application domain. Current efforts in the field of engine research are aimed at reducing harmful vehicle emissions such as NO_x and carbon dioxide (CO_2) due to their adverse effects on human health and the environment [4,5]. Despite recent advances in emissions reduction technologies and stricter standards imposed by regulatory government agencies, vehicles are emitting at rates higher than their certified limit [6,7]. These discrepancies are not a result of vehicles failing certification but rather the certification test not accurately reflecting real-world vehicle use. Therefore, identifying engine variable signatures co-occurring with elevated NO_x emissions in the real-world is key to understanding the cause of the excess emissions. Additionally, the time lag in co-occurrence patterns must be considered when analyzing vehicle systems due to the large timescale of some vehicle interactions. For instance, as engine load increases, engine temperature (a key factor

in NO_x production) may rise at a slower rate due to the heat capacity and iner-
tia of the engine and coolant and hence the non-compliant NO_x emissions might
only occur after a few seconds from the start of increase in engine load.

To measure the strength of an association between a pattern and non-
compliant windows in such a domain, a spatio-temporal statistical measure is
preferred in order to provide a statistical interpretation of the output patterns.
The cross-K function [8, 9] is a popular spatio-temporal statistical measure which
is often used to measure the interaction between pairs of events in space and
time. The cross-K function can express how much the association between a
given pattern and non-compliant windows deviates from the assumption of their
independence. Additionally, the cross-K function is not sensitive to the prevalence of the output pattern, unlike support measures, and hence can capture
rare signatures that are highly associated with non-compliant windows but not
prevalent in the input dataset.

Challenges: Designing an algorithm for NWC discovery that captures statistically meaningful patterns while maintaining the computational scalability is challenging for the following reasons: First, domain-preferred spatio-temporal statistical association measures (e.g. cross-K function) lack monotonicity: a pattern representing an engine signature over multiple variables may be interesting even though
its component single-variable signatures are not. For instance, the doubling of both
engine RPM and brake torque might be more strongly associated with a doubling
in NO_x than just the doubling of engine RPM. This property renders Apriori-based
pruning inapplicable. Second, there is a huge number of candidate patterns to consider. For each non-compliant window, the number of associated candidate patterns is exponential in the number of variables. This includes all combinations of
one, two, three, etc., variables. Third, the data volume is potentially huge due to
the large number of variables over a long time series.

Limitations of Related Work: Related
work for the NWC discovery problem
mainly consists of literature on mining multi-dimensional temporal association
rules [10–13]. In these rules, a consequent
occurs within T time points of an antecedent
(a single or multi-dimensional sequence).
However, these works have mainly focused
on finding the most frequent patterns using a
minimum support threshold and an Apriori-

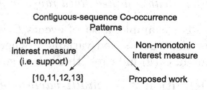

Fig. 3. Related work on mining contiguous sequence co-occurrence patterns

based pruning approach that relies on the anti-monotone property of the interest
measure. By contrast, in the NWC discovery problem, rare associations can still
be interesting since they can reveal rare patterns that are highly associated with
non-compliant windows but have low support. The statistical interest measure
used to capture these patterns does not have this anti-monotone property, rendering Apriori-based pruning inapplicable. Figure 3 shows a classification of the
related work.

Contributions: This paper makes the following contributions: (1) We formally define the Non-compliant Window Co-occurrence (NWC) pattern discovery problem (Sect. 2). (2) We provide two upper bounds for the cross-K function which are cheaper to compute than the computation of the exact cross-K function. We also show that one of the proposed upper bounds has a special monotone property that allows pruning of other candidate patterns without even calculating their upper bounds (Sect. 3). (3) We propose a Multi-parent Tracking approach (MTNMiner) that uses upper-bound filtering strategies to discover NWC patterns (Sect. 3.2). (4) We present a case study to evaluate the effectiveness of MTNMiner in finding statistically meaningful engine patterns that are associated with NO_x emissions in transit buses (Sect. 4). (5) We provide an experimental evaluation using real-world data and show that MTNMiner yields substantial computational savings compared to the naive approach (Sect. 5).

Scope: The process of mining association rules from time series data in continuous domains typically consists of two steps [10–12]: First, the time series values are discretized. Then, the data is used to extract interesting associations. In this paper, we focus on the problem of mining interesting co-occurrence patterns. We do not, however, address the problem of choosing the most suitable discretization technique. Instead, we assume that the discretization intervals for each variable are given as an input to this problem, possibly using representations suggested in [13–16].

2   Basic Concepts and Problem Statement

2.1   Basic Concepts

Definition 1. An event: *Given a variable v, an event $e_i(v)$ is a reading where v falls within a predefined range $[v_i, v_{i+1})$.*

For example, a set of events $E(v) = \{e_1(v), e_2(v), .., e_m(v)\}$ can be defined for the wheel speed variable v where $e_1(v)$ indicates that wheel speed $\in [0, 5)$ km/h, $e_2(v)$ indicates that wheel speed $\in [5, 10)$ km/h, and so on.

Definition 2. A multi-variate event trajectory (MET): *Given a set of explanatory variables V and a target variable y, a MET is a sequence of multi-variate points $p_t = (p_t^1, p_t^2, ..., p_t^{|V|}, y_t)$, $1 \le t \le \tau$, where t is a timestamp of p_t, τ is the trajectory length, p_t^k is an event defined for variable $v_k \in V, 1 \le k \le |V|$, and $y_t \in \mathbb{R}$.*

Figure 4 shows an example of a MET, of length $\tau = 8$, defined over two explanatory variables $V = \{v_1$:engine power, v_2:engine RPM$\}$, where $E(v_1) = \{a_1, a_2, a_3\}$ and $E(v_2) = \{b_1, b_2, b_3\}$ are their corresponding sets of events, and a target variable y of NO_x emissions.

Definition 3. An event-sequence S(v): *Given a variable v, an event-sequence S(v) is a sequence of events $e_i(v)$ that are temporally contiguous in a MET.*

For example, in Fig. 4, $a_2a_3a_2$ is an event sequence for engine power.

Definition 4. *A **non-compliant window** (W_N): Given a window length L and a MET m defined over a set of explanatory variables V and a target variable y, a non-compliant window $W_N = <t_i,t_j>$ is a temporal window in m, of length L, where y exceeds a user supplied standard. The length L is defined as the number of time instants within the window. i.e. $L = t_j - t_i + 1$.*

For example, given the MET in Fig. 4, a length $L = 3$ s, and a standard specifying that the percentage of increase in NO_x between the start and end of a window should not exceed 100 %, the window $< 1,3 >$ is identified as a non-compliant window since $\frac{0.023-0.011}{0.011} = 109\% > 100\%$.

Definition 5. *A **Non-compliant Window Co-occurrence (NWC) pattern**: Given a MET m defined over a set of explanatory variables V and a target variable y, and a time lag δ, an NWC pattern, C, is a set of equal-length event-sequences $\{S_i(u_i) \mid u_i \in U, U \subseteq V \text{ and } 1 \le i \le |U|\}$, that started at the same time point, and within a time lag δ preceding the start of a non-compliant window in m. Length(C) denotes the length of the event-sequences in C and is equal to the non-compliant window length L. Dim(C) denotes the dimensionality of pattern C (i.e. number of variables in C), where $Dim(C) = |U|$.*

For example, in Fig. 4, given a time lag $\delta = 1$ s, we can identify 6 NWC patterns as listed in Table 1. The first three patterns coincide with the non-compliant window, while the last three patterns precede the window by a lag of 1 s. Patterns with IDs =1, 2, 4 and 5 have a dimensionality of 1 since they are defined on only one variable, while that of patterns 3 and 6 is 2.

2.2 Interest Measure: Temporal Cross-K Function

In this work, the temporal cross-K function, a purely temporal form of the space-time cross-K function [8,9], is used as a statistical measure to express how much the association between a given pattern and non-compliant windows deviates from independence. A temporal cross-K function measuring the association between an NWC pattern, C, and the occurrence of non-compliant windows, W_N, at a time lag δ is calculated as follows: $K_{C,W_N}(\delta) = \lambda_{W_N}^{-1} E[$*number of non-compliant windows starting within time δ from the start of an instance of* $C]$, where λ_{W_N} is the expected number of non-compliant window events per unit time. Under the assumption of independence between the occurrences of pattern C and the non-compliant windows, $K_{C,W_N}(\delta)$ is equal to $(\delta + 1)$. Whenever $K_{C,W_N}(\delta)$ is greater than $\delta + 1$, this indicates an association between the pattern and the non-compliant behavior, with higher values indicating a stronger association. According to [17], $K_{C,W_N}(\delta)$ can be estimated by:

$$\hat{K}_{C,W_N}(\delta) = \lambda_{W_N}^{-1} \sum_i \sum_j \frac{I(0 \le d(C_i, W_{Nj}) \le \delta)}{|C|}$$

$$= \frac{T}{|W_N||C|} \sum_i \sum_j I(0 \le d(C_i, W_{Nj}) \le \delta) \tag{1}$$

where $d(C_i, W_{Nj})$ is the distance between the start of instance C_i of pattern C and the start of the non-compliant window W_{Nj}; I(.) is an indicator function that assumes a value of 1 if $0 \leq d(C_i, W_{Nj}) \leq \delta$, and a value of 0 otherwise; $T = \sum_{allMETs} \tau$ (referred to as the time series length in this paper), and $|W_N|$ and $|C|$ are the number of non-compliant windows and the number of instances of pattern C (i.e., cardinality of pattern C) across all METs (i.e., the time series) respectively. Hence, $\hat{K}_{C,W_N}(\delta)$ can be written as:

$$\hat{K}_{C,W_N}(\delta) = \frac{T \times |C \overset{\delta}{\bowtie} W_N|}{|W_N||C|} \qquad (2)$$

where $|C \overset{\delta}{\bowtie} W_N|$ denotes the cardinality of the temporal join set between the instances of pattern C and non-compliant windows, W_N, such that an instance C_i and W_{Nj} are only joined if C_i preceded W_{Nj} by time t where $0 \leq t \leq \delta$. For simplicity, in the rest of this paper, we will refer to $|C \overset{\delta}{\bowtie} W_N|$ as the cardinality of the join set of C.

2.3   Problem Statement

The problem of discovering Non-compliant Window Co-occurrence (NWC) patterns can be expressed as follows:

Given:
1. A set M of multivariate event trajectories (METs) defined on a set of explanatory variables V and a target variable y,
2. A window length L and a standard for defining non-compliant windows on y,
3. A time lag δ,
4. A temporal cross-K function threshold ϵ, and
5. A minimum support threshold $minsupp$ (optional, default value = 0)
Find: All NWC patterns C where $\hat{K}_{C,W_N}(\delta) > \epsilon$.
Objective: Reduce computational cost.
Constraints:
1. All output patterns should have a support at least equal to $minsupp$.
2. The input standard is cheaply computable (e.g. using an algebraic function).
3. All METs in M are sampled uniformly at the same sampling rate.
4. Correctness and completeness.

The $minsupp$ threshold is used to reduce the number of patterns output to the user (i.e. as a post-processing step), for example by discarding patterns that might have occurred only once in the dataset. $minsupp$ is not used as an interest measure to get frequent candidates, and it should generally be set to a very low value to allow the discovery of rare patterns.

Example: Figure 4 shows an example for the input to the NWC discovery problem. The input consists of a MET defined on two explanatory variables: v_1:engine power and v_2:engine RPM, and a target variable: NO_x emissions in gm/sec. $E(v_1) = \{a_1, a_2, a_3\}$ and $E(v_2) = \{b_1, b_2, b_3\}$ are the set of events defined for the engine power and engine RPM variables, respectively. Suppose that windows of non-compliant NO_x emission are defined as follows: windows of length $L = 3\,\mathrm{s}$ in which the percentage of increase in NO_x between the start and end of the window exceeds $100\,\%$. Based on that definition, only one non-compliant window $< 1,3 >$ can be identified and is marked by a red rectangle as shown in the figure. Hence, $|W_N| = 1$. The $\hat{K}_{C,W_N}(\delta)$ threshold ϵ is set to 5, δ is set to $1\,\mathrm{s}$, and $minsupp$ is set to $0.1\,\%$. The aim is to find all NWC patterns meeting the $\hat{K}_{C,W_N}(\delta)$ and $minsupp$ thresholds. Table 1 shows all the candidate NWC patterns. The first 3 patterns are those coinciding with the non-compliant window $< 1,3 >$, while the next 3 patterns are the patterns preceding the window by a lag of $1\,\mathrm{s}$. Columns 3, 4, 5 and 6 show the number of occurrences of each pattern in the time series (i.e. pattern cardinality), the cardinality of the join set between instances of this pattern and non-compliant windows, the pattern support, and the value of the interest measure, respectively. For example, the first pattern in the table occurred twice at time instants 1 and 5 (i.e. $|C|=2$). However, only one of those occurrences was associated with a non-compliant window: the pattern instance at t=1 coincided with the non-compliant window $< 1,3 >$ (i.e. $|C \overset{1}{\bowtie} W_N|=1$). Hence, for this pattern, $\hat{K}_{C,W_N}(\delta) = \hat{K}_{C,W_N}(1) = \frac{T \times |C \overset{1}{\bowtie} W_N|}{|W_N||C|} = \frac{8 \times 1}{1 \times 2} = 4 < 5$. However, the second pattern occurred only once at t=1, where it coincided with a non-compliant window. Hence for this pattern, $\hat{K}_{C,W_N}(1) = \frac{T \times |C \overset{1}{\bowtie} W_N|}{|W_N||C|} = \frac{8 \times 1}{1 \times 1} = 8 > 5$. As shown in Table 1, only patterns 2 and 3 have an interest measure exceeding ϵ, and thus these are the final output patterns as indicated in column 7.

Time	0	1	2	3	4	5	6	7
v_1: Engine Power	a_1	a_2	a_3	a_2	a_1	a_2	a_3	a_2
v_2: Engine RPM	b_1	b_1	b_2	b_3	b_1	b_1	b_2	b_2
NO_x (gm/sec)	0.011	0.011	0.015	0.023	0.023	0.021	0.019	0.019

Fig. 4. Input data with one non-compliant window identified (best in color) (Color figure online)

3   Proposed Approach

In this section, we first describe a naive approach for solving the NWC discovery problem. Then, we present the key ideas of the proposed approach including the two upper bounds developed for our interest measure and the properties of these bounds. Finally, we present our proposed **M**ulti-**P**arent **T**racking approach for mining **NWC** patterns (MTNMiner).

Naive Approach: The naive approach starts by finding all non-compliant windows in the given time series (i.e. the collection of input METs) using a sliding

Table 1. Candidate NWC patterns

ID	Candidate Pattern C	| C |	$\|C \overset{\delta}{\bowtie} W_N\|$	supp(C)	$\hat{K}_{C,W_N}(1)$	Is output?
1	$\{a_2a_3a_2\}$	2	1	1/4	4	NO (4 < 5)
2	$\{b_1b_2b_3\}$	1	1	1/8	8	YES (8 > 5)
3	$\{a_2a_3a_2, b_1b_2b_3\}$	1	1	1/8	8	YES (8 > 5)
4	$\{a_1a_2a_3\}$	2	1	1/4	4	NO (4 < 5)
5	$\{b_1b_1b_2\}$	2	1	1/4	4	NO (4 < 5)
6	$\{a_1a_2a_3, b_1b_1b_2\}$	2	1	1/4	4	NO (4 < 5)

window of the same length as the given non-compliant window length. Then, for each non-compliant window in a MET m, we enumerate all temporal windows in the MET that started within time t preceding this non-compliant window, where $0 \leq t \leq \delta$. Finally, for each of these temporal windows we enumerate all the candidate NWC patterns. Each pattern C is enumerated by calculating its cardinality $|C|$ using a single linear scan of the time series. Whenever an instance of the pattern is found, the algorithm examines the non-compliant windows table to count the number of windows that are within δ sec from this pattern. Hence, for a pattern C, both $|C|$ and $|C \overset{\delta}{\bowtie} W_N|$ are calculated using a single linear scan. Finally, if the pattern satisfies the *minsupp* threshold, its interest measure is calculated and the pattern is output if the measure exceeds the user-specified threshold ϵ.

Note that while enumerating the non-compliant windows and their corresponding candidate patterns, no non-compliant window or pattern is allowed to overlap two different METs. In addition, if the input METs belong to different moving objects (e.g. different vehicles), NWC patterns in a MET of one object should not be associated with a non-compliant window in a MET of another object. To achieve this, the non-compliant window table also stores the object ID to differentiate between the non-compliant windows of the different objects.

3.1   Key Proposed Ideas for Better Computational Performance

Our MTNMiner algorithm is founded on the following three key ideas:

Key Idea 1: Local Upper Bound of $\hat{K}_{C,W_N}(t)$: First, we define a subset/superset relation between NWC patterns.

Definition 6 (Subset/Superset Patterns). *Let $C=\{S_i(u_i) : u_i \in U, 1 \leq i \leq |U|\}$ and $C'=\{S_i(q_i) : q_i \in Q, 1 \leq i \leq |Q|\}$ be two NWC patterns. Then, C' is said to be a **subset** of C iff: (1) Length(C') = Length(C). (2) $Q \subseteq U$. (3) For every $S_i(q_i) \in C'$, we have $S_i(q_i) \in C$. Similarly, C is said to be a **superset** of C' (i.e. superset(C')).*

Definition 7 (Local Upper Bound). *Given an NWC pattern $C=\{S_i(u_i) \mid u_i \in U, U \subseteq V$ and $1 \leq i \leq |U|\}$ and a time lag δ, the **local upper bound** of*

$\hat{K}_{C,W_N}(\delta)$, denoted as $UB_{local}(\hat{K}_{C,W_N}(\delta))$, can be computed as follows:

$$UB_{local}(\hat{K}_{C,W_N}(\delta)) = \frac{T}{|W_N|} \times \frac{Upper_{Loc}(|C \overset{\delta}{\bowtie} W_N|)}{Lower(|C|)}$$

$$where: Upper_{Loc}(|C \overset{\delta}{\bowtie} W_N|) = \min_{\{S_i\} \in C, 1 \leq i \leq Dim(C)} (|\{S_i\} \overset{\delta}{\bowtie} W_N|)$$

$$and \ Lower(|C|) = |superset(C)| \tag{3}$$

Note that $Upper_{Loc}(|C \overset{\delta}{\bowtie} W_N|)$ is an upper bound of $|C \overset{\delta}{\bowtie} W_N|$ which exists in the numerator of $\hat{K}_{C,W_N}(\delta)$. It is computed using the minimum join set cardinality of all subset patterns of C that consist of only one event-sequence (i.e. one-variable subset patterns). $Lower(|C|)$ is a lower bound of $|C|$ which is in the denominator of $\hat{K}_{C,W_N}(\delta)$ and is equal to the cardinality of any superset pattern of C. Next, we prove that $UB_{local}(\hat{K}_{C,W_N}(\delta))$ is an upper bound of $\hat{K}_{C,W_N}(\delta)$.

Lemma 1. *Given an NWC pattern C and a time lag δ, $Upper_{Loc}(|C \overset{\delta}{\bowtie} W_N|)$ is an upper bound of $|C \overset{\delta}{\bowtie} W_N|$.*

Proof. For every NWC pattern $\{S_i\}$ consisting of a single event-sequence where $\{S_i\} \subseteq C$, $1 \leq i \leq Dim(C)$, we have $|\{S_i\}| \geq |C|$, where $|\{S_i\}|$ and $|C|$ are the cardinality of the patterns $\{S_i\}$ and C in the time series, respectively. Therefore, $|\{S_i\} \overset{\delta}{\bowtie} W_N| \geq |C \overset{\delta}{\bowtie} W_N|$, $\forall 1 \leq i \leq Dim(C)$. Then, $Upper_{Loc}(|C \overset{\delta}{\bowtie} W_N|) = \min_{\{S_i\} \in C, 1 \leq i \leq Dim(C)} (|\{S_i\} \overset{\delta}{\bowtie} W_N|) \geq |C \overset{\delta}{\bowtie} W_N|$. $\blacksquare$

Lemma 2. *Given an NWC pattern C and a δ time distance, $Lower(|C|)$ is a lower bound of $|C|$.*

Proof. Any superset pattern of C has a cardinality smaller than or equal to C. Therefore, $Lower(|C|) = |superset(C)| \leq |C|$. $\blacksquare$

Theorem 1. *Given an NWC pattern C and a time lag δ, $UB_{local}(\hat{K}_{C,W_N}(\delta))$ is an upper bound of $\hat{K}_{C,W_N}(\delta)$.*

Proof. Using Lemmas 1 and 2, we have $\hat{K}_{C,W_N}(\delta) = \frac{T}{|W_N|} \times \frac{|C \overset{\delta}{\bowtie} W_N|}{|C|} \leq \frac{T}{|W_N|} \times \frac{Upper_{Loc}(|C \overset{\delta}{\bowtie} W_N|)}{Lower|C|} = UB_{local}(\hat{K}_{C,W_N}(\delta))$

$\blacksquare$

Since $UB_{local}(\hat{K}_{C,W_N}(\delta))$ is an upper bound of $K_{C,\hat{W}_N}(\delta)$, then if this upper bound is less than the cross-K function threshold ϵ, this pattern will not be output and hence there is no need to compute the actual cardinality of the pattern or the cardinality of its join set.

Key Idea 2: Local Upper Bound of $\hat{K}_{C,W_N}(t)$: We also propose a second upper bound for the $K_{C,\hat{W}_N}(\delta)$ of a pattern C, namely, the lattice upper bound

$UB_{lattice}(\hat{K}_{C,W_N}(\delta))$. Although this bound is less tight than the local upper bound of a pattern, $UB_{lattice}(\hat{K}_{C,W_N}(\delta))$ has a conditional monotone property. Based on that property, if $UB_{lattice}(\hat{K}_{C,W_N}(\delta))$ is less than ϵ, then the lattice upper bound for all subset patterns of C is also less than ϵ and so they can be completely pruned without calculating their upper bounds.

Definition 8 (Lattice Upper Bound). *Given an NWC pattern $C=\{S_i(u_i) \mid u_i \in U, U \subseteq V \text{ and } 1 \leq i \leq |U|\}$ and a time lag δ, the **lattice upper bound** of $\hat{K}_{C,W_N}(lag)$, denoted as $UB_{lattice}(\hat{K}_{C,W_N}(\delta))$, can be computed as follows:*

$$UB_{lattice}(\hat{K}_{C,W_N}(\delta)) = \frac{T}{|W_N|} \times \frac{Upper_{Lat}(|C \overset{\delta}{\bowtie} W_N|)}{Lower(|C|)}$$

$$where: Upper_{Lat}(|C \overset{\delta}{\bowtie} W_N|) = \max_{\{S_i\} \in C, 1 \leq i \leq Dim(C)} (|\{S_i\} \overset{\delta}{\bowtie} W_N|)$$

$$and \; Lower(|C|) = |superset(C)| \tag{4}$$

Theorem 2. Given an NWC pattern C and a time lag δ, $UB_{lattice}(\hat{K}_{C,W_N}(\delta))$ is an upper bound of $\hat{K}_{C,W_N}(\delta)$.

Since Definition *8* differs from Definition *7* only in the min term being replaced by a max term, the proof of Theorem *2* is straightforward from Theorem *1*.

Conditional Monotone Property for the Lattice Upper Bound:

Lemma 3. *Given an NWC pattern C and a time lag δ, $UB_{lattice}(\hat{K}_{C,W_N}(\delta))$ is monotonically decreasing with decreasing Dim(C) if $Lower(|C|)$ is kept monotonically increasing. In other words, given two NWC patterns C and C' where $C' \subset C$, then if $Lower(|C'|) \geq Lower(|C|)$, then $UB_{lattice}(\hat{K}_{C',W_N}(\delta)) \leq UB_{lattice}(\hat{K}_{C,W_N}(\delta))$.*

Proof. Let $C' \subset C$ where C and C' are two NWC patterns. Then $\forall \; S_i(v_i) \in C'$, where $1 \leq i \leq Dim(C')$, we have $S_i(v_i) \in C$. Therefore, $Upper_{Lat}(|C' \overset{\delta}{\bowtie} W_N|) = \max_{\{S_i\} \in C', 1 \leq i \leq Dim(C')} (|S_i \overset{\delta}{\bowtie} W_N|) \leq \max_{\{S_i\} \in C, 1 \leq i \leq Dim(C)} (|S_i \overset{\delta}{\bowtie} W_N|) = Upper_{Lat}(|C \overset{\delta}{\bowtie} W_N|)$ (a). Also, since $Lower(|C|)$ is kept monotonically increasing as Dim(C) decreases, then $Lower(|C'|) \geq Lower(|C|)$ (b). From (a) and (b), we have $UB_{lattice}(\hat{K}_{C,W_N}(\delta)) = \frac{T}{|W_N|} \times \frac{Upper_{Lat}(|C \overset{\delta}{\bowtie} W_N|)}{Lower(|C|)} \geq \frac{T}{|W_N|} \times \frac{Upper_{Lat}(|C' \overset{\delta}{\bowtie} W_N|)}{Lower(|C'|)} = UB_{lattice}(\hat{K}_{C',W_N}(\delta))$. ∎

Key Idea 3: Efficiently Calculating the Pattern Cardinality: We propose a more efficient method to calculate the pattern cardinality by preprocessing the time series to create a *startingEdge* index. This index is a hash table where

the key is two events that occurred consecutively in time i.e., $s_1 \rightarrow s_2$. The value is a list of all the time instants where this edge appeared in the input time series. A separate index is kept for each of the input variables. To calculate the cardinality of a pattern, the first two consecutive events (i.e. first edge in the pattern) are used as the key, and the corresponding time instants where this edge occurred are retrieved from the hash table. Then, we only search the time series at these time instants to count the cardinality of the pattern.

3.2   MTNMiner: A Multi-parent Tracking Approach for Mining NWC Patterns

Our proposed MTNMiner algorithm starts by finding and then iterating through all non-compliant windows. For each window, it enumerates patterns starting at t time points preceding that window, where $0 \leq t \leq \delta$. In addition, MTNMiner uses the key ideas introduced in the previous subsection to efficiently traverse the candidate patterns enumeration space. For each value of t preceding a non-compliant window $< t_i, t_j >$, a lattice data structure is used to represent all the patterns starting at t, as shown in Fig. 6a. The lattice nodes represent all the possible patterns within the window $< t_i - t, t_j - t >$. Each node is labeled with the list of variables in the pattern it belongs to. For example, consider the input MET shown in Fig. 5. For the window $< 0, 2 >$, the lattice node labelled $\{a, b\}$ represents the pattern defined by the first two variables in that window, namely $\{a_1 a_2 a_3, b_1 b_1 b_2\}$.

Within the lattice of each window, MTNMiner starts by enumerating all leaf nodes representing one-variable patterns, and stores the join-set cardinality of these nodes in an array $LeafJoinSetCount$. Then, a top-down breadth first traversal/search (BFS) is performed to enumerate the rest of the lattice nodes while applying the proposed upper bounds. Since each node has multiple parents, a node can be pruned through the lattice upper bound of any of its parent nodes. Therefore, a node is inserted into the BFS queue for enumeration only if all its parents were already visited and none had a lattice upper bound $> \epsilon$. Hence, each node keeps track of the number of its unvisited parents (i.e. $unVisitedParents$). This also avoids adding duplicate copies of a node to the queue through the node's multiple parents. In addition, each node stores the following information: (1) $supersetCount$: the maximum cardinality found so far of a superset pattern of this node; and (2) $isPruned$: a flag to indicate if the node was already pruned through one of its ancestor nodes. Initially, for each node n, $isPruned$ is set to $False$, $unVisitedParents$ is set to the number of parent nodes of n, and $supersetCount$ is set to 1 since we are sure that there is at least one instance of the root node pattern in the current window, and this root node pattern is a superset of all the patterns in that window. An $enumeratedPatterns$ table is used to store the patterns already enumerated. This table also stores the cardinalities of the leaf nodes' patterns and their join sets. Finally, the algorithm uses a queue to perform a Breadth First Traversal for the lattice nodes.

Algorithm 1 shows the pseudo code of MTNMiner. First, the algorithm finds all the non-compliant windows and initializes the used data structures (lines 1–5).

Algorithm 1. MTNMiner

1: $H \leftarrow$ FINDALLNONCOMPLIANTWINDOWS
2: $enumeratedPatterns \leftarrow \{\}$
3: Queue $queue \leftarrow \{\}$
4: $startingEdgeIndex \leftarrow$ CREATESTARTINGEDGEINDEXFROMMETS
5: $lattice \leftarrow$ Create and initialize lattice
6: **for** each window $w=<t_i, t_j>$in H **do**   ▷ iterate through all non-compliant windows
7:     **for** t := δ **to** 0 **do**   ▷ iterate from 0 to the max lag δ preceding w
8:         $latticeCp \leftarrow$ CREATEDEEPCOPY($lattice$)
9:         $LeafJoinSetCount \leftarrow$ ENUMERATEONEVARIABLENODES($latticeCp$,$enumeratedPatterns$)
10:        $queue$.enqueue($latticeCp$.root)
11:        **while** $queue$ not empty **do**
12:            Node $node \leftarrow queue$.dequeue()
13:            ENUMERATEWITHUPPERBOUNDPRUNING($latticeCp$,$node$,$queue$,w,t,δ)

14: **function** ENUMERATEWITHUPPERBOUNDPRUNING($lattice$,n,$queue$,w,t,δ)
15:     **if** $UB_{lattice}(\hat{K}_{n.C,W_N}(\delta)) \leq \epsilon$ **then** PRUNEALLNODESUBSETS(n,$lattice$)
16:     **else if** $UB_{local}(\hat{K}_{n.C,W_N}(\delta)) \leq \epsilon$ **then**
17:         **for** each unpruned non-leaf child node ch of n **do**
18:             $ch.supersetCount \leftarrow$ max($ch.supersetcount$,$n.supersetcount$)
19:             Check if n is last visited parent of ch, then $queue$.enqueue(ch)
20:     **else**   ▷ no pruning occurred
21:         $C \leftarrow$ expandPattern(n)
22:         **if** C not in $enumeratedPatterns$ **then**
23:             $[|C|, |C \overset{\delta}{\bowtie} W_N|] \leftarrow$ Calculate cardinalities using $startingEdgeIndex$
24:             $enumeratedPatterns$.put(C)
25:             **if** $\frac{|C|}{T} \geq minsupp$ and $\hat{K}_{C,W_N}(\delta) > \epsilon$ **then** Output C.
26:             **for** each unpruned non-leaf child node ch of n **do**
27:                 $ch.supersetCount \leftarrow$ max($ch.supersetcount$,$|C|$)
28:                 Check if n is last visited parent of ch, then $queue$.enqueue(ch)
29:         **else**   ▷ C already enumerated
30:             PRUNEALLNODESUBSETS(n)

Next, the pattern enumeration step is performed (lines 6–13). The algorithm iterates through all temporal windows starting within a time lag t preceding a non-compliant window, where $0 \leq t \leq \delta$. For each temporal window, MTNMiner starts by creating a copy of the initial lattice to enumerate the patterns in that window (line 8). Within this window, pattern enumeration is performed in two phases: **In phase 1 (line 9)**: patterns represented by all the leaf nodes are enumerated. Each pattern is expanded by retrieving it from the input time series. If the pattern was already enumerated, its cardinality and join set cardinality are retrieved from the $enumeratedPatterns$ table and used to calculate its interest measure. The join set cardinality is also stored in the $LeafJoinSetCount$ array. Otherwise, the pattern cardinality and its join set cardinality are calculated and stored with the pattern in the $enumeratedPatterns$ table. Additionally, the join set cardinality is stored in the $LeafJoinSetCount$ array to be used in calculating the upper bounds for the rest of the lattice nodes. **In phase 2 (lines 10–13)**: the algorithm performs a top-down breadth first traversal starting from the root node of the lattice and continuing until the queue is empty. For each node in the queue, the function $EnumerateWithUpperBoundPruning(.)$ (lines 14–30) is called, as follows.

The $EnumerateWithUpperBoundPruning(.)$ function starts by calculating the lattice upper bound of the node using the maximum join set cardinality of all the one-variable subsets of the node (already stored in *LeafJoinSetCount*) and the *supersetCount* value of the node. If the lattice upper bound is $\leq \epsilon$, all subset patterns of this node (i.e. all descendant nodes) are marked as pruned (line 15). If not, the local upper bound is calculated. If this bound is $\leq \epsilon$ (line 16), then the cost of enumerating the pattern represented by this node (i.e. expanding it and calculating its cardinality and the cardinality of its join set) is saved. However, we still need to examine the children of this node (lines 17–19). For each child node not marked as pruned, we set its *supersetCount* variable to the maximum of its current *supersetCount* value and the *supersetCount* of its parent node. Then, we decrease the number of *unVisitedParents* for the child by one and if this was the last visited parent (i.e. unVisistedParents = 0), we insert the child node into the queue. Finally, if the local upper bound of the node was greater than ϵ, then we have to enumerate this node (lines 20–30). First the node is expanded by retrieving the actual pattern from the time series. If the pattern was already enumerated (lines 29–30), all its subset nodes are marked as pruned. Otherwise (lines 22–28), the cardinalities of the pattern and its join set are calculated using the *startingEdge* index, and the pattern is inserted into the *enumeratedPatterns* table. If the pattern satisfies the *minsupp* threshold, its interest measure is calculated and the pattern is output if its interest measure exceeded ϵ (line 25). Finally (lines 26–28), the child nodes are treated in the same way as described before; however, in this case, the *supersetCount* of each child is set to the maximum of its current value and the cardinality computed for the parent node (line 27). As the *supersetCount* value of each node increases, the lattice upper bound becomes tighter.

Table 2. MTNMiner Bottleneck Analysis

No. of variables	Cardinality counting time	Other tasks time	Total time
6	267.2 s	3.1 s	270.3 s
8	769.2 s	9.3 s	778.5 s
10	3772 s	39 s	3811 s

Time	0	1	2	3	4	5	6	7	8	9	10	11
var_a	a_1	a_2	a_3	a_2	a_1	a_2	a_3	a_2	a_1	a_2	a_3	a_2
var_b	b_1	b_1	b_2	b_3	b_1	b_1	b_2	b_2	b_1	b_2	b_3	b_2
var_c	c_1	c_2	c_1	c_1	c_1	c_2	c_1	c_1	c_2	c_1	c_2	c_2
var_d	d_2	d_2	d_3	d_3	d_2	d_2	d_3	d_2	d_1	d_2	d_2	d_3

Fig. 5. Input with 2 non-compliant windows

Bottleneck Analysis: Table 2 shows a break-down of the running time of MTN-Miner without pruning. A MET with T = 50,000 points is used, with L = 5 s, $\delta = 1$ s, $\epsilon = 15$, and no *minsupp* threshold is specified. As shown in the table, the main bottleneck is calculating the cardinality of the candidate patterns and

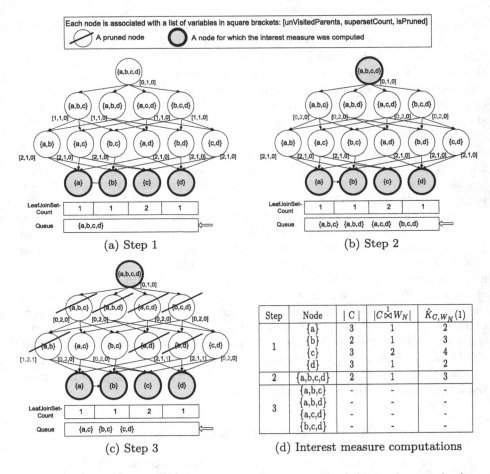

Fig. 6. MTNMiner Execution trace (Best viewed in color) (Color figure online)

their join set. The time required for copying and traversing the actual lattice (in addition to all other tasks) is negligible compared to the time required for the cardinality computation. Hence, our pruning strategies focus on avoiding this computation cost. It is also worth mentioning that although for each enumerated window, a lattice is created with nodes representing all candidate patterns within that window, only one lattice at a time is kept in memory.

Execution Trace: Figure 6 shows an example run of MTNMiner for the input data shown in Fig. 5. The MET is of length T=12 and we assume two non-compliant windows (with length L = 3 s) were identified (i.e., < 1, 3 > and < 7, 9 >). The cross-K function threshold $\epsilon = 3.5$, $\delta = 1$ s and no *minsupp* threshold is specified. For brevity, the execution trace shows only the enumeration of candidates within one window < 0, 2 >, which started 1 s before the non-compliant window < 1, 3 >. Similar enumerations will be done for the windows < 1, 3 >, < 6, 8 > and < 7, 9 >. Figure 6 shows the lattice for window < 0, 2 > after executing each step of the algorithm, where one whole level is enumerated at every step. Figure 6a shows the lattice created for window $< 0, 2 >$ after enumerating the leaf-nodes and calculating their interest measure. Their join set cardinalities are shown in the array $LeafJoinSetCount$. Figure 6d shows the computed values at step 1, in which only the pattern of node $\{c\}$ is output, namely $\{c_1 c_2 c_1\}$, since its interest measure equals $4 > \epsilon$. Next, the root node is inserted in the queue. Figure 6b shows the lattice after enumerating the root node by calculating its lattice upper bound ($= \frac{12}{2} \times \frac{max\{1,1,2,1\}}{1} = 12 > \epsilon$) and local upper bound ($= \frac{12}{2} \times \frac{min\{1,1,2,1\}}{1} = 6 > \epsilon$). Since both values exceed ϵ, the actual cardinalities of the root pattern and its join set are calculated. Then for its child nodes, the $unVisitedParents$ variable is decremented, their $supersetCount$ is set to the actual cardinality of the root node pattern just computed (changes are marked in red), and the child nodes are inserted into the queue. In Fig. 6c, the lattice upper bound is calculated for the first node in the queue, $\{a, b, c\}$. Although its value exceeds ϵ, the local upper bound value of 3 is less than ϵ and hence the node is pruned (from Theorem 1). Then, the $supersetCount$ of its child nodes is set to its $supersetCount$ value and their $unVisitedParents$ are decremented. Then, node $\{a, b, d\}$ which is next in the queue, is enumerated by calculating its lattice upper bound. Since the bound is equal to $3 < \epsilon$, the node is pruned and its three child nodes are also marked as pruned (from Theorem 2 and Lemma 3). The enumeration continues similarly for all the nodes in this level where nodes $\{a, c, d\}$ and $\{b, c, d\}$ will also be pruned through their local upper bounds ($=3 < \epsilon$). Finally, the next level of nodes will be enumerated similarly until the queue is empty.

4   Case Study

To evaluate the effectiveness of our proposed approach, we performed a case study on a real-world dataset collected from an on-board sensor on a transit bus in the Minneapolis-St. Paul area, USA. The dataset measured several engine and environmental variables at a rate of 1 Hz. Data points covered roughly 19 days (≈ 176 trips) on three different routes, ensuring the data was not biased by a specific route. The non-compliant windows of NO_x emissions were defined as windows of length $L = 5$ s in which the average NO_x in gm/kW-h exceeds the Environmental Protection Agency (EPA) test threshold ($NOxT$) of 0.267 [1] and the percentage increase in NO_x exceeds $PincT = 100\%$. We used a temporal cross-K function threshold $\epsilon = 15$, $\delta = 2$ s, and $minsupp = 0.01\%$. The variables used for this case study include engine RPM, engine torque, engine power, wheel speed, and acceleration as these parameters typically influence the increase in NO_x. Additionally, since the production of NO_x is heavily dependent on temperature [18], the engine intake temperature, coolant temperature, and selective catalytic reduction (SCR) system intake temperature were also added to the list of prescribed variables. Most of the variables had equal length intervals, however for the engine RPM, additional modified windows were created to account for the narrow windows of engine idling.

Table 3. Interesting case study patterns

ID	NWC Pattern	C	$\hat{K}_{C,W_N}(2)$
1	Wheel speed:	$\{w_0\ w_0\ w_0\ w_1\ W_2\}$	21.57
2	Engine RPM:	$\{s_1\ s_2\ S_3\ S_3\ S_3\}$	16.28
	Engine power:	$\{r_5r_5r_5r_5r_5\}$	
	Wheel speed:	$\{w_0w_0w_0w_0w_0\}$	
	Acceleration:	$\{a_{16}\ a_{16}\ a_{17}\ a_{17}\ a_{17}\}$	
3	Engine RPM:	$\{s_1\ s_1\ s_2\ S_3\ S_3\}$	17.15
	Engine power:	$\{r_5r_5r_5r_5r_5\}$	
	Wheel speed:	$\{\ w_1\ w_0w_0w_0w_0\}$	

The number of identified non-compliant windows was 98,290, generating 1,159 NWC patterns. Analysis of the output shows that MTNMiner was able to correctly identify the high NO_x association with low engine load and slow speed driving that was previously found by Misra et al. [6]. The NWC pattern showing the slow speed association with high NO_x is shown in the first row of Table 3. This pattern illustrates that accelerating with speeds between 0 and 15 km/h is highly associated with elevated NO_x conditions. The output pattern shown in the second row of Table 3 illustrates the association between high NO_x output and low engine load. In this instance, the wheel speed was between 0 and 5 km/h, and the engine load is around 10 % of the rated load which would constitute a low load condition. These findings confirm that NWC pattern discovery can correctly identify patterns associated with high NO_x. An interesting NWC pattern finding is also shown in the last row of Table 3. In this instance the wheel speed appears to decrease from the start of the window, but the engine RPM appears to increase substantially, resulting in counter intuitive vehicle operation. A potential explanation of this case could be the effect of some factors such as a down-shift in the transmission. Further investigation is required to understand the true cause of this finding.

5   Experimental Evaluation

The goal of our experiments was to evaluate the performance of the pruning filters proposed in MTNMiner as compared to the naive approach. The evaluation was performed on real-world data by varying and observing the effect of the following workload parameters: time series length T, number of variables $|V|$, temporal cross-K function threshold ϵ, time lag value δ, NWC pattern length L (i.e. non-compliant window length), and the non-compliant window definition.

Experimental Setup: Experiments were performed using the real-world dataset used in the case study with a time series of length T=100,000 points. The non-compliant windows were defined as windows of length 5 s in which the average of NO_x emissions in gm/kW-h exceeded the EPA standard threshold ($NOxT$) of 0.267, and the percentage of increase in NO_x exceeded $PincT =$

100 %. The default parameter values were: T = 50,000 points, $|V| = 8$, $\epsilon = 15$, $\delta = 2$ s, $L = 5$ s, $minsupp = 0.001$, $NOxT = 0.267$ and $PincT = 100$ %, unless stated otherwise. Algorithms were implemented using the Java programming language. All experiments were run on a machine with an Intel Xeon Quad Core 3.00 GHz processor with 64 GB RAM.

5.1 Experimental Results

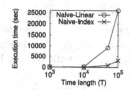

Fig. 7. Execution time with varying T

In this subsection, we focus on evaluating the performance of our pruning filters which are the main contribution of this work. However, we also evaluated the effect of the *startingEdge* index by running two versions of the naive approach: one using a linear scan of the data to calculate the cardinality of each pattern and its join set, and another version using the *startingEdge* index. Figure 7 shows the execution times of both versions. As can be seen, the *startingEdge* index leads to substantial computational savings by reducing the time required for cardinality counting. At $T=10^5$ points, the naive approach using the *startingeEdge* index was 8.5 times faster than the linear scan version. Therefore, for the rest of our experiments we used the naive approach with the *startingEdge* index as the baseline method to evaluate the performance of MTNMiner.

Effect of Time Series Length (T): We ran the naive approach and MTN-Miner on subsets of the dataset with 1000, 10,000, 50,000 and 100,000 points where each subset was a contiguous set of trips. Figure 8a shows the execution times for both algorithms and Fig. 8b shows the corresponding speedup ($=\frac{Naive\,execution\,time}{MTNMiner\,execution\,time}$). As can be seen, MTNMiner reduces the computation cost of the naive approach as a result of the pruning filters, where computational savings increase as the length of the time series increases. At 100,000 points, MTNMiner is 2.4 times faster than the naive approach.

Effect of the Number of Variables ($|V|$): Figure 8c shows the execution times for both algorithms as the number of variables increases and Fig. 8d shows the corresponding speedup. Although the effect of the increase in the number of variables is exponential on both algorithms, the results show a significant separation between the two approaches where the computational savings of MTNMiner increase as the number of variables increases. This is because as the number of

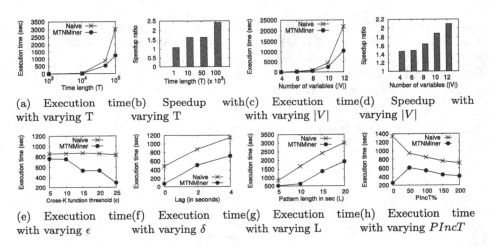

(a) Execution time with varying T
(b) Speedup with varying T
(c) Execution time with varying $|V|$
(d) Speedup with varying $|V|$

(e) Execution time with varying ϵ
(f) Execution time with varying δ
(g) Execution time with varying L
(h) Execution time with varying $PIncT$

Fig. 8. Execution time for MTNMiner vs. the naive approach.

variables increases, a larger number of nodes occurs in the middle layer of the lattice, which allows MTNMiner to prune more patterns.

Effect of Temporal Cross-K Function Threshold (ϵ): Figure 8e shows the execution times for both the naive and MTNMiner algorithms as the temporal cross-K function threshold ϵ increases. As can be seen, the naive approach execution time remains constant since no pruning takes place. However, the execution time for MTNMiner dramatically decreases since most of the candidate patterns can be pruned at higher thresholds for the interest measure.

Effect of Time Lag (δ) and Pattern Length (L): To observe the effect of the maximum time lag between a pattern and a non-compliant window, we measured the execution times at $\delta = 0$, 2 and 4 s. Figure 8f shows that a larger δ value increases the execution time for both algorithms. The reason is that more time is needed to enumerate the larger number of temporal windows preceding each non-compliant window. Nevertheless, MTNMiner consistently outperforms the naive approach because of its pruning filters. Figure 8g shows a similar trend for the effect of pattern length due to the increase in the cost of calculating the pattern cardinality. Still, MTNMiner always outperforms the naive approach.

Effect of the Non-compliant Window Definition: Figure 8h shows the execution times as $PIncT$ increases from 0 % to 200 %. The computational cost of the naive approach decreases as $PIncT$ increases. This is due to the decrease in the number of non-compliant windows, which reduces the total number of patterns enumerated. Similarly, for MTNMiner, the computational cost decreases as $PIncT$ increases from 50 % to 200 %. However, at $PIncT = 0$ %, the number of non-compliant windows was very high, resulting in a large decrease in the interest measure values for all the candidate patterns. This occurred because the number of non-compliant windows $|W_N|$ exists in the interest measure's denominator and no output patterns were produced. As a result, most patterns were pruned by MTNMiner, leading to a large reduction in execution time. However, as $PIncT$ increased from 0% to 50%, the number of non-compliant windows decreased substantially (from 12,931 to 6,026 windows), leading to higher interest measures and less pruning. Then, as $PIncT$ increased to 100%, the number of non-compliant windows exhibited a smaller decrease (from 6,026 to 4,845 windows). At this smaller decrease, the reduction in the overall computation pattern enumeration time was higher than the pruning lost by the increase of the interest measure values, leading to an overall reduction in execution time.

6   Discussion

Some other studies in the literature have addressed the problem of mining rare (i.e. low support) co-occurrence/association patterns with a high confidence threshold [19–22]. Although these methods can capture rare associations, they only focus on associations between single events and do not model associations between contiguous sequences (e.g. a temporal signature of engine variables co-occurring with non-compliant windows). For instance, they would not be able to capture the association of a continuous acceleration or braking pattern with a window of elevated emissions as shown in our case study. In addition, none of these methods except [22] guarantee completeness.

7   Conclusion and Future Work

This work explored the problem of Non-compliant Window Co-occurrence (NWC) pattern discovery in relation to an important real-world application: eco-friendly transportation. The NWC discovery problem is challenging due to the large number of candidate patterns, large data volume and the lack of monotonicity in the temporal cross-K function used to measure the interestingness of a pattern. We proposed two upper bounds on the interest measure which are much less expensive to compute, and showed that one upper bound exhibits a conditional monotone property that allows other patterns to be pruned without calculating their upper bound. Using these bounds, the proposed MTNMiner algorithm showed substantial computational savings over the naive approach. We also presented a case study using engine measurement data that also validated the effectiveness of MTN-Miner. In the future, we plan to explore spatial aspects of the NWC discovery (e.g. the effect of left/right turns on non-compliant engine emissions). Moreover, we will investigate the discovery of statistically significant NWC patterns.

Acknowledgement. This material is based upon work supported by the National Science Foundation under Grant No. 1029711, IIS-1320580, 0940818 and IIS-1218168, the USDOD under Grant No. HM1582-08-1-0017 and HM0210-13-1-0005, and the University of Minnesota under the OVPR U-Spatial. We are particularly grateful to Kim Koffolt and the members of the University of Minnesota Spatial Computing Research Group for their valuable comments.

References

1. Heavy-Duty Onroad Engines. https://www.dieselnet.com/standards/us/hd.php (2015)
2. Sudden unintended acceleration. http://en.wikipedia.org/wiki/Sudden_unintended_acceleration (2015)
3. Wang, J., He, Q.P.: Multivariate statistical process monitoring based on statistics pattern analysis. Ind. Eng. Chem. Res. **49**(17), 7858–7869 (2010)
4. Vijayaraghavan, K., et al.: Effects of light duty gasoline vehicle emission standards in the united states on ozone and particulate matter. Atmos. Environ. **60**, 109–120 (2012)
5. US EPA: Ground level ozone health effects (2014)
6. Misra, C., et al.: In-use nox emissions from model year 2010 and 2011 heavy-duty diesel engines equipped with aftertreatment devices. Env. Sci. tech. **47**(14), 7892–7898 (2013)
7. Office of Transportation & Air Quality: Mpg: Label values vs. corporate average fuel economy (cafe) values label mpg (2014)
8. Diggle, P.J., Chetwynd, A.G., Häggkvist, R., Morris, S.E.: Second-order analysis of space-time clustering. Stat. Methods Med. Res. **4**(2), 124–136 (1995)
9. Gabriel, E., Diggle, P.J.: Second-order analysis of inhomogeneous spatio-temporal point process data. Stat. Neerl. **63**(1), 43–51 (2009)
10. Schluter, T., Conrad, S.: About the analysis of time series with temporal association rule mining. In: IEEE Symposium on Computational Intelligence and Data Mining, pp. 325–332 (2011)
11. Harms, S.K., Deogun, J.S.: Sequential association rule mining with time lags. J. Intell. Inf. Syst. **22**(1), 7–22 (2004)
12. Sacchi, L., Larizza, C., Combi, C., Bellazzi, R.: Data mining with temporal abstractions: learning rules from time series. Data Min. Knowl. Disc. **15**(2), 217–247 (2007)
13. Das, G., et al.: Rule discovery from time series. In: Proceedings of the ACM International Conference on Knowledge and Data Discovery, pp. 16–22 (1998)
14. Daw, C.S., Finney, C.E.A., Tracy, E.R.: A review of symbolic analysis of experimental data. Rev. Sci. Instrum. **74**(2), 915–930 (2003)
15. Kotsiantis, S., Kanellopoulos, D.: Discretization techniques: a recent survey. GESTS Intl. Trans. Computer Sci. Eng. **32**(1), 47–58 (2006)
16. Lin, J., Keogh, E., Wei, L., Lonardi, S.: Experiencing sax: a novel symbolic representation of time series. Data Min. Knowl. Disc. **15**(2), 107–144 (2007)
17. Dixon, P.M.: Ripley's k function. Encyclopedia of environmetrics. Wiley, New York (2002)
18. Turns, S.R.: An Introduction to Combustion: Concepts and Applications, vol. 287, 3rd edn. McGraw-hill, New York (2012)
19. Cohen, E., et al.: Finding interesting associations without support pruning. IEEE Trans. Knowl. Data Eng. **13**(1), 64–78 (2001)
20. McIntosh, T., Chawla, S.: High confidence rule mining for microarray analysis. IEEE/ACM Trans. Comput. Biol. Bioinf. **4**(4), 611–623 (2007)
21. Zakaria, W., Kotb, Y., Ghaleb, F.: Mcr-miner: maximal confident association rules miner algorithm for up/down-expressed genes. Appl. Math **8**(2), 799–809 (2014)
22. Huang, Y., et al.: Mining confident co-location rules without a support threshold. In: Proceedings of the 2003 ACM symposium on Applied computing, pp. 497–501 (2003)

Keyword and Pattern

Maximizing Influence of Spatio-Textual Objects Based on Keyword Selection

Orestis Gkorgkas[1]([✉]), Akrivi Vlachou[1,2], Christos Doulkeridis[3],
and Kjetil Nørvåg[1]

[1] Norwegian University of Science and Technology (NTNU), Trondheim, Norway
{orestis,vlachou,noervaag}@idi.ntnu.no
[2] Institute for the Management of Information Systems,
R.C. "Athena", Marousi, Greece
[3] Department of Digital Systems, University of Piraeus, Piraeus, Greece
cdoulk@unipi.gr

Abstract. In modern applications, spatial objects are often annotated
with textual descriptions, and users are offered the opportunity to formulate *spatio-textual queries*. The result set of such a query consists of
spatio-textual objects ranked according to their distance from a desired
location and to their textual relevance to the query. In this context,
a challenging problem is how to select a set of at most b keywords to
enhance the description of the facilities of a spatial object, in order to
make the object appear in the top-k results of as many users as possible.
In this paper, we formulate this problem, called *Best-terms* and we show
that it is NP-hard. Hence, we present a baseline algorithm that provides
an approximate solution to the problem. Then, we introduce a novel
algorithm for keyword selection that greatly improves the efficiency of
query processing. By means of a thorough experimental evaluation, we
demonstrate the performance gains attained by our approach.

1 Introduction

Spatio-textual search has attracted increased attention recently, due to the
numerous applications that provide value-added services to the users by combining spatial location with textual relevance. Given a database of geographical
points of interest that are annotated with textual information (also called *spatio-
textual objects*), the objective of a spatio-textual query is to retrieve a ranked
set of top-k spatio-textual objects that are close to the query point and have
high textual similarity to the query keywords. As a notable example, consider
hotels that are annotated with their facilities (e.g., in the form of keywords),
and tourists that search for hotels close to some location of interest and a set of
query keywords indicating desired facilities (for example "pool" or "Wi-Fi").

An interesting problem encountered in real-life applications that rely on
spatio-textual retrieval is how to improve the ranking of a spatio-textual object
for as many users as possible. For instance, for a newly established hotel at some
location, the question is how to enrich its textual annotation in order to maximize its rank for many different users. To address this challenging problem, we

© Springer International Publishing Switzerland 2015
C. Claramunt et al. (Eds.): SSTD 2015, LNCS 9239, pp. 413–430, 2015.
DOI: 10.1007/978-3-319-22363-6_22

capitalize on reverse top-k queries [19], which retrieve the set of users that have a given object in their top-k results. We model the problem as a maximization of the cardinality of the reverse top-k result set, and we explore the different combinations of keywords that will increase the query object's rank for many users, when added to its textual annotation. We call this problem as *Best-terms*, we show that it is NP-hard, and we present a greedy solution that serves as baseline. Then, we propose a novel algorithm that boosts the performance of query processing, by deliberately selecting keywords that increase the score of the query object for many users simultaneously. Finally, we present the results of our experimental evaluation that verifies the performance gains of our algorithm.

In summary, our main contributions are outlined below:

- We formulate the novel problem, called Best-terms, of increasing the rank of a spatio-textual object for many different users, by enriching its textual description.
- We show that the Best-terms problem is NP-hard and we provide a baseline solution.
- We propose an efficient query processing algorithm that significantly outperforms the baseline consistently.
- We provide an experimental evaluation that demonstrates the merits of our approach.

The rest of this paper is structured as follows: Sect. 2 provides an overview of the related work. Section 3 presents the necessary background and preliminary concepts. Then, in Sect. 4, we formally describe the problem statement. Section 5 presents the baseline algorithm, while Sect. 6 describes our efficient query processing algorithm. Section 7 presents the experimental evaluation, and Sect. 8 concludes the paper.

2   Related Work

In this section, we provide an overview of the related research literature.

Keyword Recommendation. Zhang et al. [23] present a method for recommending keywords for advertisements in keyword search results using Wikipedia. They focus mostly in cases where the advertisement (target) consists of short-text web pages that contain inadequate textual content to describe the advertised entity. Based on the fact that a large number of entities are described in Wikipedia, they use Wikipedia articles relevant to the advertised entity in order to recommend keywords to connect to the target. Fuxman et al. [9] follow a different approach. They suggest keyword queries to advertisers using logs that store the queries posed by the users and the URLs of the result set that were selected by the users. Some of the URLs are also connected to a set of concepts. The target of the authors is to connect the set of concepts to the queries using the Markov Random Field model and suggest the most relevant queries for each concept to the advertisers. Ravi et al. [16] propose a variety of methods for automatic generation of bid phrases. Among others they introduce the usage of a translation model that extends a predefined mapping between bidding phrases and target web pages. Papadimitriou et al. [15] study the problem of mapping an advertisement in a set of URLs based on a set of keyword queries. In particular, they assume that each advertisement is mapped to a set of keyword queries and their aim is to map each advertisement in a set of URLs which will be representative of the results produced by the attached keyword queries. Choi et al. [4] create a representative summary of the advertisement based on the context of the advertised material. Their method makes use of co-occurrence and semantic vectors in order to enrich the ad context and create a representative set of terms. Cholette et al. [5] study the problem of finding optimal bids in search-based algorithms. Agrawal et al. [1] introduce an approach for recommending bid phrases from a given ad landing page by classifying a set of labels generated by click logs. Their classifier has logarithmic complexity and can efficiently make predictions on large sets of labels.

The aim of the aforementioned approaches is to identify potentially relevant queries to the advertised products and form bid phrases based on the identified queries. Our approach is inherently different, because the above techniques try to predict relevant queries and do not consider the relevance of the advertised product in relation to similar products. In addition, they do not consider top-k search criteria as the appearance of a product in a search result is decided mainly on the bidding strategy. On the contrary, our aim is to enhance the description of a spatio-textual object and to increase the number of queries for which the target product appears in the top-k list of the search results. In this effort, we take into consideration not only the user preferences, but also the rest of the spatio-textual objects that are relevant to those queries.

Spatial Keyword Search. Spatial keyword search has been well studied during the recent years and several index structures have been introduced for efficient search. A detailed evaluation of existing spatio-textual indexes can be found in [3]. Cong et al. [6] introduced the IR-tree and its variants. The IR-tree is based on the R-tree structure. Each node of the tree is also associated with inverted index containing the textual information of the children of the node. Rocha et al. [17] proposed the S2I index which uses different strategies for frequent and infrequent terms and outperforms the IR-tree. Zhang et al. proposed the I^3 index [22], which is based on the quadtree, and the RCA approach [21], which is based on Fagin's CA algorithm [8]. Both approaches outperform the S2I and IR-tree index structures. Nonetheless, the IR-tree is able to perform a spatial only search, retrieving objects that are not textually relevant to a spatio-textual query. This possibility is not offered by any of the S2I, I^3 and RCA approaches. Cao et al. [2] introduce the concept of *prestige* where a spatio-textual object has a higher prestige if it is collocated with other textually similar objects. They calculate the prestige of a spatio-textual object based on a graph where each node corresponds to an object and two nodes are connected if and only if their textual similarity and spatial proximity exceed certain thresholds. Deng et al. [7] suggested an approach of finding a set of spatio-textual objects that are relevant to a spatio-textual query and at the same time they fulfill a desired spatial property. In particular, their aim is to identify a keyword-cover of optimal score, where as keyword cover is defined a set of objects where each object is associated with exactly one term of the spatio-textual query.

Lu et al. [14] and Lu et al. [13] studied the problem of reverse spatial and textual k nearest neighbor search where, given a query point q, the objective is to locate the set of spatio-textual objects for which q is among the k nearest neighbors. The distance between the objects is a linear combination of the textual and the Euclidean distance of the objects. The authors introduce the IUR-tree which is an adaptation of the IR-tree. Each node of the IUR-tree contains the union and the intersection of the terms contained in the objects in the subtree rooted at the node. Our approach is different, as we do not evaluate the similarity between elements of a set of spatio-textual objects, but our aim is to increase the relevance and therefore the visibility of an object against a set of user preferences which constitutes a different set from that of the spatio-textual objects that our query object belongs.

Wu et al. [20] propose the W-IR-tree which is similar to the IR-tree but it is constructed based primarily on textual distance. The W-IR-tree shows improved performance for batch queries where objects are considered relevant to the query only if they contain all terms of the query. The W-IR-tree cannot be applied in our case as we consider it possible for a spatio-textual object to be relevant to a user preference even if it does not contain all terms of the user preference. Gao et al. [10] propose a filter-and-refinement framework for processing reverse boolean top-k spatial keyword queries. They focus on queries where a spatio-textual object must contain all terms of a query to be considered a valid result. Lin et al. [12] study the problem of identifying important terms in the textual description of a spatio-textual object that cause the object to be highly ranked for a specific query or a specific region. Our approach is different, as we focus on enriching the textual description of a an object with new terms.

3   Preliminaries

Let D be a set of objects, where each object o is represented by a tuple of the form $o = \langle o.T, o.L \rangle$ where $o.T$ is a set of keywords describing the features of o and L is a point in $\mathbb{R}^2$ describing the location of o. We denote as $\mathcal{A} = \bigcup_{o \in D} o.T$ to be the set of all keywords in D. In the scope of this paper, we call these objects *spatio-textual objects*. For a given object o, we consider the *size* of o to be equal to $|o.T|$, namely the size of an object is the number of terms it contains.

3.1   Top-k Spatial Keyword Queries

Let u be a user preference query on D, where u is represented by the a tuple $u = \langle u.T, u.L, \alpha \rangle$, $u.T \subseteq \mathcal{A}$ is the text describing the user's desired features, $u.L \in \mathbb{R}^2$ denotes the desired location and $\alpha \in [0, 1]$ denotes the importance of location over matching the desired features. Given a preference u, we can assign a score to each object using the following equation:

$$f(o, u) = \alpha \times \delta(o.L, u.L) + (1 - \alpha) \times \theta(o.T, u.T) \qquad (1)$$

where $\delta(o.L, u.L)$ is the spatial distance, and $\theta(o.T, u.T)$ is the textual distance between the object o and the user preference u. Given an integer k, we can return the top-k spatio-textual objects according to their score. In the scope of this paper, we assume that lower scores are better, both spatial and textual distances are normalized in the interval [0,1] and $f(o, u) = 1.0$ if $\theta(o.T, u.T) = 1$. The latter assumption implies that objects that are not textually relevant to the query cannot be considered as a valid result.

The textual relevance we employ is the normalized intersection of terms between the description of a spatio-textual object $o.T$ and a user preference keyword set $u.T$, i.e., $\theta(o.T, u.T) = 1 - |o.T \bigcap u.T||u.T|^{-1}$. Although in large documents different textual similarity functions are more appropriate, the intersection is more representative in cases of feature selection. For instance if a user is looking for a hotel with a restaurant and a pool, any hotel offering more features (e.g. restaurant, pool, bar) than the ones specified by the user should not be less textually relevant than a hotel which offers only the features specified by the user preference (restaurant, pool).

Definition 1 Top-k query. *Given a set D of spatio-textual objects, a set of terms $\mathcal{A}$, a scoring function f, an integer k, and a query u, the result set $TOP_k(u)$ of a top-k query is a set of spatio-textual objects such that $TOP_k(u) \subseteq D$, $|TOP_k(u)| = k$ and $\forall o_1, o_2 : o_1 \in TOP_k(u), o_2 \in D - TOP_k(u)$ it holds that $o_1.T \bigcap u.T \neq \emptyset$ and $f(o_1, u) \leq f(o_2, u)$.*

If an object o belongs to the $TOP_k(u)$ set of a user preference u, we say that o is *visible* to u or that u *sees* o. For a specific set of objects D and a set of user preferences U, it is possible to identify for a query object q the set of users who can see q. This is the reverse procedure of a top-k query and therefore it is called *reverse top-k ($RTOP_k$) query* [18].

Definition 2 $RTOP_k$ query. *Given a set D of spatio-textual objects, a set of user queries U, a scoring function f, integer k, and a spatio-textual object q, the result set $RTOP_k(q)$ of a reverse top-k query is set such that $RTOP_k(q) \subseteq U$ and $u \in RTOP_k(q)$ if and only if $\exists o \in TOP_k(u)$ such that $f(q, u) \leq f(o, u)$.*

The cardinality of the $RTOP_k$ set of a query-object q is called *influence score* of the object and we denote it as $I(q)$. The influence score indicates the number of users to whom q is visible.

3.2   IR-tree

We employ a state-of-the-art index structure to process spatial keyword queries, namely the IR-tree [6]. The IR-tree is an R-tree where each node is associated with an inverted index of the objects contained in the respective sub-tree rooted at the node. The IR-tree offers the possibility of retrieving objects that are near a query point but not textually relevant to it, a property that is essential in identifying possibly interesting terms. In more detail, each leaf node contains an inverted index of the spatio-textual objects contained in the node. The leaf node is characterized by a spatio-textual pseudo-object which consists of a *minimum bounding rectangle* (MBR) that encloses all objects of the node and a pseudo-document that consists of the union of all the terms contained in the children of the node. Each non-leaf node contains an inverted index of the spatio-textual pseudo-objects of the children nodes it contains. Non-leaf nodes are also characterized by spatio-textual pseudo-objects which are constructed similarly to the pseudo-objects of the leaf nodes.

4   Problem Definition

Given a set of spatio-textual objects D and a set of spatio-textual preferences U, the influence score of an object q is the number of preferences to which q is visible. Assuming that the location of a spatio-textual object cannot change, the only the way to improve the influence score of q is to enhance its textual description, in order to increase the textual relevance between q and the user preferences in U. In this paper, we study the problem of finding a set of b terms, which when added to the textual description of q, they maximize the influence score of q. We refer to this problem as *Best-terms*.

Definition 3 *Best-terms query*. *Given a set D of spatio-textual objects, a set of terms $\mathcal{A} = \bigcup_{o \in D} o.T$, a set of queries U, a scoring function f, an integer k, a spatio-textual object $q = \langle q.T, q.L \rangle$, and an integer b, the set BT is a set of terms such that $\text{BT} \subseteq \mathcal{A}$, $\text{BT} \cap q.T = \emptyset$, $|\text{BT}| \leq b$ and $\forall T \subseteq \mathcal{A} - \text{BT}, |T| \leq b$ it holds that $I(q_1) \geq I(q_2)$ where $q_1 = \langle q.T \bigcup \text{BT}, q.L \rangle$ and $q_2 = \langle q.T \bigcup T, q.L \rangle$.*

The Best-terms problem is NP-hard. We show that by studying a special case of a Best-terms query, namely the respective decision problem of finding whether there exists a set of terms T with $|T| \leq b$ such that $I(\langle q.T \bigcup T, q.L \rangle) = |U|$.

Problem 1 *Best-terms (decision problem)*. *Given a set D of spatio-textual objects, a set of terms $\mathcal{A} = \bigcup_{o \in D} o.T$, a set of queries U, a scoring function f, an integer k, and a spatio-textual object $q = \langle q.T, q.L \rangle \in D$, decide if there is a set BT such that $\text{BT} \subseteq \mathcal{A}$, $\text{BT} \cap q.T = \emptyset$, $|\text{BT}| \leq b$ for which it holds that $I(q_1) = U$ where $q_1 = \langle q.T \bigcup \text{BT}, q.L \rangle$*

We will show that Problem 1 is NP-complete by reducing the set cover problem in Problem 1 using the restriction technique [11].

Definition 4 *Set cover problem*. *Let U be a set of elements (universe) and $\mathcal{T} = \{T_1, \ldots, T_n\}$ be a collection of sets where $\bigcup_{i=1}^{n} T_i = U$. The   set cover problem decides if there is a subset of $\mathcal{T}$, $\mathcal{T}' \subseteq \mathcal{T}$ of size $|T| \leq b$ such that $\mathcal{T}'$ is a cover of U.*

Theorem 1. *The decision problem of Best-terms is NP-complete.*

Proof. Let an oracle machine select the BT set for a query object q. We set $p = \langle q.T \bigcup BT, q.L \rangle$ and by performing a TOP_k query for each user preference we can calculate the $RTOP_k(p)$ set and the influence score $I(p)$ of object p in polynomial time. Therefore the solution can be verified in polynomial time and our problem belongs to the NP class.

We set U a to be a set of users and $D = \{q\}$, where $q.T = \emptyset$. We define a collection $\mathcal{T} = \{T_1, \ldots, T_{|A|}\}$ of sets, one for each term t_i in A where a user u belongs in T_i only if $t_i \in u.T$. If we consider $k = 1$, then, for all users that $q.T \bigcap u.T = \emptyset$ it holds that $q \notin TOP_k(u)$ since q is not relevant to $u.T$. If $q.T \bigcap u.T \neq \emptyset$ then $q \in TOP_k(u)$ as it is the only object. Therefore any selection of a term t_i is equivalent of selecting a subset of T_i of U. The set cover problem is consequently reduced to Problem 1, as it can be seen as a special case of Problem 1. Problem 1 is therefore NP-complete. Best-terms is at least as hard as Problem 1, which leads us to the conclusion that the Best-terms problem is NP-hard.

5   The Best Term First (BTF) Algorithm

Since the Best-terms problem is NP-hard, an exact solution is infeasible, even for medium-sized datasets. Motivated by this observation, in this section we describe a greedy algorithm, termed *Best Term First* (BTF), that provides an approximate solution to the Best-terms problem. BTF operates in an iterative way consisting of b steps, and in each step it adds to the query object the term that induces the highest increase in influence score.

5.1   Algorithmic Description

Algorithm 1 describes the BTF approach in more detail. BTF takes as input an IR-tree index containing the set of spatio-textual objects D, and an IR-tree index containing the set of user preferences U. BTF works in b iterations, and in each iteration the best term (i.e., the term that induces the maximum increase in the influence of q) is selected and added to the terms of the query object.

Initially, BTF creates a pseudo-preference q' defined by q and using $\alpha = 1$, which indicates that q' uses only distance to data objects, not textual similarity, for ranking. The role of q' is to enable traversing the preference dataset solely based on distance to the query object q. This imitates a sorted access to the preferences, yet this is achieved by means of the IR-tree index on U, without having to sort U.

In each iteration, BTF first creates a set C of candidate spatio-textual objects, one for each term that can be added to q. The size of C is equal to $|\mathcal{A} - q.T|$. In lines 9 and 10 the algorithm exploits the sorted access to the preference dataset, in order to avoid processing some top-k queries. More accurately, given the current user preference u, the score of the last retrieved spatio-textual

---

Algorithm 1. Best Term First (BTF) Algorithm

Input: U:set of users, D: set of objects,
         q:query point, b : number of new terms
Output: BT: set of new terms

1 $C \leftarrow \emptyset$, buffer $\leftarrow \emptyset$
2 $q' \leftarrow \langle q.T, q.L, 1 \rangle$
3 bestCandidate$\leftarrow q$
4 **for** $i = 0; i < b; i + +$ **do**   // repeat until b new terms have been found
5   **forall the** $t \in \mathcal{A} - q.T$ **do**
6         $C \leftarrow C \bigcup \{\langle \text{bestCandidate}.T \bigcup \{t\}, \text{bestCandidate}.L \rangle\}$
7      $u \leftarrow$next(U,q')
8   **while** $u \neq null$ **do**
9         $\tau \leftarrow \max\limits_{p \in \text{buffer}} (f(p, u))$ // empty buffer in first iter., so we set $\tau \leftarrow \infty$
10   **if** $\exists c \in C : f(c, u) \leq \tau$ **then**
11   buffer$\leftarrow$ TOP$_k(u)$
12           $\tau \leftarrow \max\limits_{p \in \text{buffer}} (f(p, u))$
13   **forall the** $c \in C$ **do**
14   **if** $f(c, u) \leq \tau$ **then**
15                 $I(c) \leftarrow I(c) + 1$
16        $u \leftarrow$next(U,q')
17   bestCandidate$\leftarrow$ $\underset{c}{\text{argmax}}(I(c))$
18 BT$\leftarrow$ bestCandidate.T-q.T
19 **return** BT

---

objects is compared with the scores of the candidate objects C, and if no candidate object has a better score than the k-th ranked spatio-textual object, the user preference is ignored (*pruning condition*) as no candidate object can be in its TOP$_k$ set. Otherwise, the top-k query needs to be executed and its TOP$_k$ result set is stored in the buffer. All candidate objects that are no worse than the k-best element of the calculated TOP$_k$ set belong also to the TOP$_k$ set of u and therefore their influence score is increased. When all user preferences have been examined, the object with the highest influence score is selected and a new set of candidate objects is created based on that object. The procedure is repeated b times until an object with b new terms is created. The b terms that were selected constitute the resulting BT set.

Although BTF adopts a greedy technique to select the b terms, the use of sorted access to dataset U together with the pruning condition reduce the number of processed top-k queries, thereby saving computational costs.

5.2   Complexity Analysis

The cost of the BTF algorithm is determined by the cost of selection of each of the b terms. The main factors that affect the cost of term selection are the construction of set C with cost $O(|\mathcal{A}|)$, and the cost C_{topk} of processing a top-k query which in worst case will be processed $|U|$ times. Thus, the overall complexity of BTF is equal to $C_{BTF} = O(b(|\mathcal{A}| + |U|C_{topk}))$. However, in practice the number of processed top-k queries is much smaller than $|U|$.

6   Graph-Based Term Selection

BTF extends the textual description of a spatio-textual object iteratively, which forces the algorithm to scan the preferences set U multiple times. In this section we present a novel algorithm, named Graph-Based Term Selection (GBTS), which examines the set of preferences only once and creates a graph of terms that provides an estimation of the influence gain any combination of terms may provide.

Essentially, GBTS consists of two separate algorithms. The first algorithm, named Graph Construction (GC), creates a graph connecting the terms which when added to the spatio-textual query object q, they can induce an increase in its influence score. The second algorithm, named Best Subgraph Selection (BSS), traverses the graph in a deliberate manner, in order to identify the sets of terms that will induce the highest increase in the influence score of q.

6.1   Graph Construction Algorithm

Given a set of objects D, a set of user preferences U and a spatio-textual object q, we denote as $\widehat{U}(q)$ the subset of all preferences $(\widehat{U}(q) \subseteq U)$ for which q is not visible and at most b terms are needed for q to become visible. The Graph Construction algorithm builds a weighted graph $G = (V, E)$ where each node of the graph represents a candidate term, and the weights on edges indicate the maximum increase in the influence score of q that can be induced, if the respective set of terms is added to q.

In more detail, for each examined user preference u, the algorithm adds to graph G a node for each previously unseen term. The edges connecting the nodes and the weights of the edges are determined by the number of terms λ that need to be added to u for it to be included in RTOP$_k(q)$. The value of λ is calculated based on Eq. 2, where τ is the worst score that q is required to have in order to be in the $TOP_k(u)$ set and derives directly from Eq. 1.

$$\tau = \alpha \times \delta(q.L, u.L) + (1 - \alpha) \times \frac{|q.T \bigcap u.T| + \lambda}{|u.T|} \qquad (2)$$

- If $\lambda \leq 1$, the algorithm adds a loop edge with weight equal to 1 to each term t that is not contained in q. If the edge already exists, the weight is simply added to the weight of the existing edge.
- In the case that $\lambda > 1$, i.e., more than one terms are necessary for q to be included to $TOP_k(u)$, the procedure is slightly different. Let $T = u.T - q.T = \{t_1, \ldots, t_n\}$ be the terms that are included in u but not in q. For each pair of terms in $u.T - q.T$, the algorithm adds an edge with weight w_e. As before, if an edge already exists, the weight is added to the existing edge.

Algorithm 2. Graph Construction (GC) Algorithm

Input: U:set of users, D: set of objects, q:query point, b : number of new terms
Output: $G = (V, E)$: resulting graph

1 $V \leftarrow \emptyset, E \leftarrow \emptyset$, buffer$\leftarrow \emptyset, G \leftarrow (V, E)$ // graph initialization
2 $q' \leftarrow \langle q.T, q.L, 1 \rangle$
3 $u \leftarrow next(U, q')$
4 **while** $u \neq null$ **do**
5 $\quad$ buffer$\leftarrow TOP_k(u)$
6 $\quad$ $\tau \leftarrow \max\limits_{p \in \text{buffer}} (f(p, u))$
7 $\quad$ **if** $f(q, u) > \tau$ **then**   // if $q \notin TOP_k(u)$
8 $\quad\quad$ $T \leftarrow u.T - q.T$
9 $\quad\quad$ $V \leftarrow V \bigcup T$
10 $\quad\quad$ $\lambda \leftarrow \max \left(1, \left\lceil \left(1 - \dfrac{\tau - a\delta(q, u)}{1-a}\right) |u.T| - |q.T \bigcap u.T| \right\rceil \right)$ // from Eq. 2
11 $\quad\quad$ **if** $\lambda = 1$ **then**
12 $\quad\quad\quad$ $E \leftarrow E \bigcup \{e = (t_i, t_i, 1) : t_i \in T\}$
13 $\quad\quad$ **else if** $1 < \lambda \leq b$ **then**
14 $\quad\quad\quad$ $E \leftarrow E \bigcup \left\{ e = \left(t_i, t_j, \dfrac{2}{\lambda(\lambda - 1)} \right) : \forall t_i, t_j \in T \text{ and } t_i \neq t_j \right\}$
15 $\quad$ $u \leftarrow next(U, q')$
16 **return** G

Since we add λ terms that correspond to $\lambda(\lambda-1)/2$ pairs of terms, the weight of each edge w_e is set to $2\left(\lambda(\lambda - 1)\right)^{-1}$, which is a normalization that makes the sum of weights added equal to 1. Intuitively, we add a total weight of 1 to each subgraph $G' = (V', E')$ where $V' \subseteq T$ and $|V'| = \lambda$, indicating the potential increase in the influence score of q if the terms contained in G' were added to q.

Algorithm 2 describes the construction of the term graph G. Similarly to Algorithm 1, GC traverses the preferences based on their distance to q. For each user preference u, if q is not in the $TOP_k(u)$ set, GC updates the node set of G and calculates λ (line 10), the number of terms that need to be added in q for it to be included in the $TOP_k(u)$ set. A non-positive value of λ indicates that u is located near q but $q.T \bigcap u.T = \emptyset$ and therefore q is not included in the $TOP_k(u)$ set. The addition of any term will allow q to be added to $TOP_k(u)$ set and therefore one loop edge is added to each term t for which it holds $t \in u.T - q.T$. If more than one terms are necessary to be added in q ($\lambda > 1$), GC adds all necessary edges in the graph. The algorithm continues until all user preferences have been examined.

The size of the graph depends on the number of distinct terms contained in $\widehat{U}(q)$. The terms correspond to the features extracted from the textual descriptions of spatio-textual objects that describe the offered facilities. In practice, we have noticed that the vocabulary for the targeted applications is limited and therefore the graph is expected to fit in main memory.

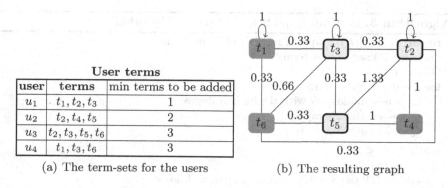

User terms		
user	terms	min terms to be added
u_1	t_1, t_2, t_3	1
u_2	t_2, t_4, t_5	2
u_3	t_2, t_3, t_5, t_6	3
u_4	t_1, t_3, t_6	3

(a) The term-sets for the users   (b) The resulting graph

Fig. 1. Example graph: The nodes of the suggested solution are colored with light gray

Example 1. *As an example, let the user preferences in Fig. 1(a) be the $\widehat{U}(q)$ set for $b = 3$, i.e., the set of user preferences that can be added to the $RTOP_k(q)$ set if 3 more terms are added to the spatio-textual object q. We also assume that the shown terms for each user preference are not included in q. The first step of the algorithm is the evaluation of the user preference u_1. The algorithm adds to the graph the nodes t_1, t_2, t_3 and since only one term needs to be added to q for u_1 to be added to $RTOP_k(q)$, it adds one loop edge with weight 1 to all terms. On the next step u_2 is processed and two more nodes (t_4, t_5) are added to the graph. For each pair of the terms contained in u_2 we add an edge to the graph with weight equal to $2(\lambda(\lambda - 1))^{-1}$ where λ is equal to 2 which the number of terms needed to be added to q, for u to be in $RTOP_k(q)$. When u_3 is processed, t_6 is added to the graph and for each pair of terms in u_3 an edge with weight $1/3$ is added to graph. Finally, u_4 is processed and the graph is updated accordingly.*

6.2   Best Subgraph Selection Algorithm

When the graph has been created, the Best Subgraph Selection algorithm (BSS) chooses as seed nodes the b nodes (terms) of the graph with the highest degree and creates a set of b subgraphs with initially one node each. Next, each subgraph is expanded by adding at each step the node with highest degree that is adjacent to a node of the subgraph. The expansion of each subgraph is continued until each subgraph has b nodes or the subgraph cannot be expanded. Finally, the subgraph with the highest sum of edge weights is selected as solution and the set of terms included in the subgraph are the ones that constitute the BT set.

Algorithm 3 describes the algorithm of term selection. Initially an empty priority queue (Q) is constructed. Subsequently, at line 3 the algorithm chooses as seed the highest degree node t_i that has not yet been selected and constructs the subgraph G_{t_i} (line 4). The subgraph is constructed by repeatedly selecting the highest degree node adjacent to the G_{t_i} until $|G_{t_i}| = b$ or until no nodes can be added to G_{t_i}. When each subgraph is constructed, it is pushed to Q. The sorting key of Q is the sum of weights of the edges in the subgraph. The BT set is constructed by selecting the subgraph with the highest sum of edges and

---

Algorithm 3. Best Subgraph Selection (BSS) Algorithm

    Input: $G = (V, E)$: graph, b: number of desired terms
    Output: BT:set of new terms
1 $Q \leftarrow \emptyset$, BT $\leftarrow \emptyset$
2 **for** $i = 0; i < b; i + +$ **do**
3      $t_i \leftarrow$ next node of G with the highest degree
4      $G_{t_i} \leftarrow$ createSubgraph(t_i)
5   Q.add(sumOfWeights(G_{t_i}),G_{t_i})
6 **while** $|\text{BT}| \leq b$ **do**
7      $G_S \leftarrow$ Q.pop()
8   add to BT the $b - |\text{BT}|$ highest degree nodes from G_S
9 **return** BT

---

adding the terms of the subgraph to BT. If the subgraphs contain less than b terms, more subgraphs are pulled from the priority queue until BT contains b terms. In such cases we add from each subsequent subgraph to BT the $b - |\text{BT}|$ highest degree nodes of the subgraph.

Example 2. *Continuing the previous example, during the execution of BSS, 3 subgraphs are created with seed nodes the terms t_2, t_5 and t_3. We denote the respective subgraphs as G_{t_i} where t_i is the seed node of the subgraph. Each subgraph G_{t_i} is extended to the highest degree node adjacent to G_{t_i}. In the case of G_{t_2}, the subgraph is expanded by adding first node t_5, which is the node with the highest degree adjacent to t_2 and subsequently with node t_3 which is the highest degree node adjacent to either t_2 or t_5. After the addition of t_3, the size of G_{t_2} becomes equal to 3 and therefore the expansion stops and the next subgraph is processed. In the case of the example all subgraphs produce the same result which includes the light gray nodes in Fig. 1(b). The nodes contained in the result are the ones to be added to q.*

6.3 Complexity Analysis

The overall complexity of Graph-Based Term Selection is determined by the two algorithms that comprise it.

$$C_{GBTS} = C_{GC} + C_{BSS}$$

    GC consists of two parts: the processing of $|U|$ top-k queries and the addition of edges $\widehat{U}(q)$ times. The addition of an edge is done in constant time and therefore the cost of GC is equal to: $C_{GC} = O(|U| \cdot C_{topk} + \widehat{U}(q))$.

    BSS also consists of two parts: the construction of b subgraphs, and the selection of nodes (terms) from the best of these subgraphs. The main cost of BSS is the construction of the b subgraphs, which is $O(b \cdot (b^3 + \log b))$. The cost of expanding a single-node subgraph b times and finding the highest degree node

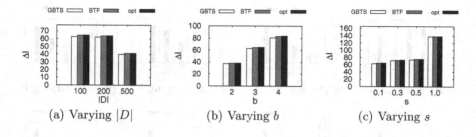

Fig. 2. Evaluating the quality of results

is equal to $O(b^3)$, while the cost of insertion to the priority queue is equal to $O(logb)$. The node selection is in worst case $O(b \cdot logb)$, though in practice it is $logb$. Hence, we derive: $C_{BSS} = O(b \cdot (b^3 + logb))$.

Consequently, the overall complexity of Graph-Based Term Selection is equal to: $C_{GBTS} = O(|U| \cdot C_{topk} + \widehat{U}(q) + b \cdot (b^3 + logb))$.

7   Experimental Evaluation

In this section, we present the results of the experimental evaluation. All algorithms were implemented in Java and the experiments were executed on an AMD Opteron 4130 Processor (2.60 GHz), with 32 GB of RAM and 2 TB of disk.

Datasets and Metrics. For the data set D of spatio-textual objects, we used a set of 200000 descriptions of hotels from the site of Booking.com[1]. The dataset contains 188 distinct features. The set of preferences U was generated using a uniform distribution for creating the location and the α parameter of each preference, while the terms were randomly chosen from the vocabulary generated by processing the set of hotels. The location of the user preferences was bounded in the MBR defined by set of hotels. We also tested our algorithm against a Zipfian distribution of terms. We used the Zipfian distribution generator provided by the Apache Commons project[2]. The metrics under which we evaluated the implemented algorithms were: (a) increase in the influence score ΔI, (b) number of I/O's performed by each algorithm, and (c) processing time.

Experimental Procedure. Both datasets D and U were indexed using an IR-tree where the maximum capacity of each node was 100 entries. We employed a buffer which was fixed at the size of 4MB, for both the tree index and the inverted files. The performance of the proposed algorithms was evaluated through a series of experiments varying the parameters of (a) the cardinality of D in the interval $[10\,K, 200\,K]$, (b) the cardinality of U, $[10\,K, 200\,K]$, (c) the number of returned results per user preference k, $[5, 50]$, (d) the maximum size of user preferences, $[1, 5]$, and (d) the number of returned terms for a query object b, $[2\text{–}5]$.

---

[1] http://www.booking.com.
[2] http://commons.apache.org/proper/commons-math/.

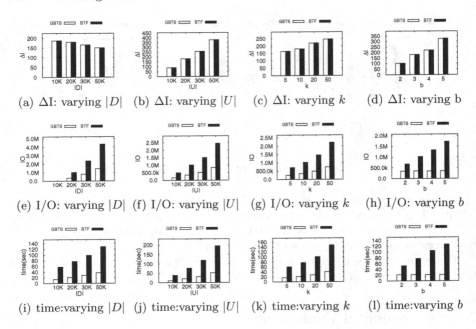

(a) ΔI: varying $|D|$ (b) ΔI: varying $|U|$ (c) ΔI: varying k (d) ΔI: varying b

(e) I/O: varying $|D|$ (f) I/O: varying $|U|$ (g) I/O: varying k (h) I/O: varying b

(i) time:varying $|D|$ (j) time:varying $|U|$ (k) time:varying k (l) time:varying b

Fig. 3. Analysis for varying $|D|, |U|, k$, and b

For the Zipfian distribution we varied the value of the characteristic exponent s in the interval $[0.1–1.0]$. The default setup for the experiments was: $|D| = 20\,K$, $|U| = 20\,K$, $k = 10$, $b = 3$ and each the maximum preference size was set to 5. For each experiment a random set of 20 query objects was selected from D.

Quality Evaluation. We compared the proposed algorithms against an exhaustive algorithm, which examines all $\binom{|A-q.T|}{b}$ term combinations[3] and calculates the optimal set of terms BT. Due to the high processing cost of the exhaustive algorithm even for small values of b, we employed datasets of limited size. The default setting for this series of experiments was $|D| = 100, |U| = 1000$, and $b = 3$. The set of objects D, consisted of a random set of hotels from the area of Catalonia in Spain, and they were selected from Booking.com. The set of preferences U, follows a uniform distribution in Figs. 2(a) and (b), and a Zipfian distribution in Fig. 2(c). Figure 2 indicates that both algorithms achieve an increase to the influence score which is very close to the optimal value. The execution time of the exhaustive algorithm was in all cases orders of magnitude larger than the execution time of BTF and GBTS.

Varying $|D|$. Figures 3(a), (e), and (i) illustrate the performance of the algorithms as we vary the number of spatio-textual objects. Figure 3(a) indicates

---

[3] Based on the adopted similarity function, the addition of a term does not have a negative effect on the influence score. In the general case, an exact algorithm should examine $2^{|A|}$ term combinations.

that both algorithms perform similarly with respect to the increase of the influence score. As the number of objects increase the gain in influence score drops as more spatio-textual objects compete for the same number of user-preferences and therefore it becomes harder for a query object to increase its influence score. Figures 3(e) and (i) indicate that the I/O accesses and the processing time for both algorithms increase when the dataset size increases. As the dataset size increases the cost of a single TOP_k query increases as well and therefore both algorithms are affected by the dataset size. The effect on BTF is larger than in GBTS as BTF accesses the data multiple times in order to create the set of new terms.

Varying $|U|$. Figures 3(b), (f), and (j) depict the performance of both algorithms as more preferences are processed. When the number of preferences increases there are more user preferences that can be added to the $RTOP_k$ set of an object with an addition of a new set of terms and therefore the gain in influence score increases as well. The processing cost for both algorithms is expected to raise for a larger number of user preferences, as more preferences have be to examined. Both processing time and I/O cost raise faster for BTF than for GBTS. In particular the processing cost for BTF grows almost by a factor of b faster than GBTS as BTF has to process the set of preferences b times in order to identify the set of new terms.

Varying k. As the size of the TOP_k set of each preference increases, the cost of a single TOP_k query increases as well. Figures 3(c), (g), and (k) indicate that the increased I/O and processing cost of a TOP_k query affects both algorithms. Similarly to the increase on the size of datasets, the effect on BTF is magnified by a factor of b. The influence score gain raises as well, since with the increase in k more objects can be included in the TOP_k set of a user preference and the necessary increase in the text similarity for a query object q to be added to a TOP_k set of a user preference u becomes smaller.

Varying b. Figures 3(d), (h), and (l) illustrate the performance of the algorithms as we vary the number of new terms added to each query object. It is noteworthy that both algorithms behave similarly with respect to the increase of the influence score. The cost of BTF raises linearly with respect to b, which is expected as it has to process the data b times before returning the resulting BT set. On the other hand, GBTS remains unaffected by the increase of the b parameter, as it has to access the preferences set only once.

Varying the Query Size. Figure 4 indicates that as the maximum preference size increases, the possible gain of influence score for a spatio-textual object drops. The reason lies in the fact that for a large user preference u, more terms are required to be added to a spatio-textual object q, for q to enter the $TOP_k(u)$ set. Larger queries require more complex TOP_k queries on the indexes and consequently the performance of both algorithms is affected. As expected BTF is affected in a larger degree than GBTS by the increased cost of the TOP_k queries.

Zipfian Distribution. It is quite common that the terms of user-preferences follow a Zipfian distribution. We tested our algorithms against a set of user

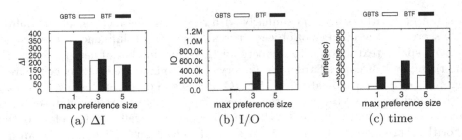

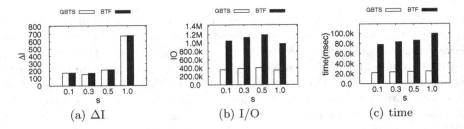

Fig. 4. Varying max preference size

Fig. 5. Varying zipf distribution

preferences where the occurrences of terms follow a Zipfian distribution. Figure 5 illustrates the experimental results. Similarly to the uniform distribution, GBTS outperforms BTF in terms of I/O accesses and processing time while producing the same gain in influence score. In cases where the exponent of the Zipfian distribution takes high values the gain in influence score raises significantly. Such behavior is expected as when a small number of distinct terms appear in a large number of user preferences, adding those terms to a spatio-textual object will result in a significant increase of its influence score since the addition of those terms will allow it to enter the TOP_k set of many user preferences.

Scalability Analysis. We evaluated the performance of GBTS against larger datasets to evaluate the scalability of our approach. BTF is not included in the results as it needed excessive time to produce results. The experimental results shown in Fig. 6 indicate that the processing time of GBTS grows logarithmically

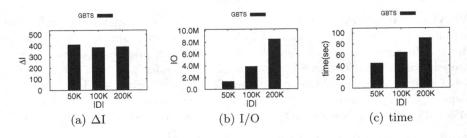

Fig. 6. Varying data cardinality

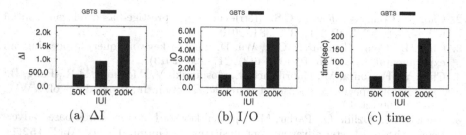

Fig. 7. Varying preferences cardinality

with respect to the size of the D, while the I/O cost increases linearly. The performance difference between processing time and I/O accesses lies in the fact that in the first TOP_k queries we have an increased number of I/Os, however after a certain number of queries, several nodes of the IR-tree are buffered and as a result the subsequent TOP_k queries induce a limited number of I/O accesses. Figure 7 illustrates the performance of GBTS with respect to the cardinality of user preferences set. Both the processing time and the I/O increase linearly with respect to time.

8 Conclusions

In this paper, we address the challenging problem of increasing the influence of a spatio-textual object, by enriching its textual description with at most b carefully selected keywords. In this way, the spatio-textual object's textual relevance to user queries is increased, with the ultimate objective being for the object to become part of the top-k result for many different users. We provide a formal problem statement that is novel and relies on concepts related to top-k and reverse top-k queries. We show that the problem is NP-hard, and we present a greedy solution to the problem. Then, we propose a more efficient algorithm that achieves results of comparable quality, but with significantly lower processing cost. We demonstrate the performance gains of the proposed approach by means of a thorough experimental evaluation that includes real data.

Acknowledgments. A. Vlachou was supported by the Action "Supporting Postdoctoral Researchers" of the Operational Program "Education and Lifelong Learning" (Action's Beneficiary: General Secretariat for Research and Technology), and is co-financed by the European Social Fund (ESF) and the Greek State. C. Doulkeridis has been co-financed by ESF and Greek national funds through the Operational Program "Education and Lifelong Learning" of the National Strategic Reference Framework (NSRF) - Research Funding Program: Aristeia II, Project: ROADRUNNER.

References

1. Agrawal, R., Gupta, A., Prabhu, Y., Varma, M.: Multi-label learning with millions of labels: recommending advertiser bid phrases for web pages. In: Proceedings of WWW, pp. 13–24 (2013)

2. Cao, X., Cong, G., Jensen, C.S.: Retrieving top-k prestige-based relevant spatial web objects. PVLDB **3**(1), 373–384 (2010)
3. Chen, L., Cong, G., Jensen, C.S., Wu, D.: Spatial keyword query processing: an experimental evaluation. PVLDB **6**(3), 217–228 (2013)
4. Choi, Y., Fontoura, M., Gabrilovich, E., Josifovski, V., Mediano, M.R., Pang, B.: Using landing pages for sponsored search ad selection. In: Proceedings of WWW, pp. 251–260 (2010)
5. Cholette, S., Özlük, Ö., Parlar, M.: Optimal keyword bids in search-based advertising with stochastic advertisement positions. J. Optim. Theory Appl. **152**(1), 225–244 (2012)
6. Cong, G., Jensen, C.S., Wu, D.: Efficient retrieval of the top-k most relevant spatial web objects. PVLDB **2**(1), 337–348 (2009)
7. Deng, K., Li, X., Lu, J., Zhou, X.: Best keyword cover search. IEEE Trans. Knowl. Data Eng. **27**(1), 61–73 (2015)
8. Fagin, R., Lotem, A., Naor, M.: Optimal aggregation algorithms for middleware. J. Comput. Syst. Sci. **66**(4), 614–656 (2003)
9. Fuxman, A., Tsaparas, P., Achan, K., Agrawal, R.: Using the wisdom of the crowds for keyword generation. In: Proceedings of WWW, pp. 61–70 (2008)
10. Gao, Y., Qin, X., Zheng, B., Chen, G.: Efficient reverse top-k boolean spatial keyword queries on road networks. IEEE Trans. Knowl. Data Eng. **27**(5), 1205–1218 (2015)
11. Garey, M.R., Johnson, D.S.: Computers and Intractability: A Guide to the Theory of NP-Completeness. Macmillan Higher Education, Basingstoke (1979)
12. Lin, X., Xu, J., Hu, H.: Reverse keyword search for spatio-textual top-k queries in location-based services. To appear in IEEE Trans. Knowl. Data Eng. (2015)
13. Lu, J., Lu, Y., Cong, G.: Reverse spatial and textual k nearest neighbor search. In: Proceedings of SIGMOD, pp. 349–360 (2011)
14. Lu, Y., Lu, J., Cong, G., Wu, W., Shahabi, C.: Efficient algorithms and cost models for reverse spatial-keyword k-nearest neighbor search. ACM Trans. Database Syst. **39**(2), 13 (2014)
15. Papadimitriou, P., Garcia-Molina, H., Dasdan, A., Kolay, S.: Output URL bidding. PVLDB **4**(3), 161–172 (2010)
16. Ravi, S., Broder, A.Z., Gabrilovich, E., Josifovski, V., Pandey, S., Pang, B.: Automatic generation of bid phrases for online advertising. In: Proceedings of WSDM, pp. 341–350 (2010)
17. Rocha-Junior, J.B., Gkorgkas, O., Jonassen, S., Nørvåg, K.: Efficient processing of top-k spatial keyword queries. In: Pfoser, D., Tao, Y., Mouratidis, K., Nascimento, M.A., Mokbel, M., Shekhar, S., Huang, Y. (eds.) SSTD 2011. LNCS, vol. 6849, pp. 205–222. Springer, Heidelberg (2011)
18. Vlachou, A., Doulkeridis, C., Kotidis, Y., Nørvåg, K.: Reverse top-k queries. In: Proceedings of ICDE, pp. 365–376 (2010)
19. Vlachou, A., Doulkeridis, C., Nørvåg, K., Kotidis, Y.: Branch-and-bound algorithm for reverse top-k queries. In: Proceedings of SIGMOD, pp. 481–492 (2013)
20. Wu, D., Yiu, M.L., Cong, G., Jensen, C.S.: Joint top-k spatial keyword query processing. IEEE Trans. Knowl. Data Eng. **24**(10), 1889–1903 (2012)
21. Zhang, D., Chan, C., Tan, K.: Processing spatial keyword query as a top-k aggregation query. In: Proceedings of SIGIR, pp. 355–364 (2014)
22. Zhang, D., Tan, K., Tung, A.K.H.: Scalable top-k spatial keyword search. In: Proceedings of EDBT, pp. 359–370 (2013)
23. Zhang, W., Wang, D., Xue, G.-R., Zha, H.: Advertising keywords recommendation for short-text web pages using Wikipedia. ACM TIST **3**(2), 36 (2012)

Geo-Social Keyword Search

Ritesh Ahuja, Nikos Armenatzoglou, Dimitris Papadias$^{(\boxtimes)}$,
and George J. Fakas

Department of Computer Science and Engineering,
Hong Kong University of Science and Technology, Hong Kong, China
{rahuja,nikos,dimitris,gfakas}@cs.ust.hk

Abstract. In this paper, we propose Geo-Social Keyword (GSK) search, which enables the retrieval of users, points of interest (POIs), or keywords that satisfy geographic, social, and/or textual criteria. We first introduce a general GSK framework that covers a wide range of real-world tasks, including advertisement, context-based search, and market analysis. Then, we present three concrete GSK queries: (i) NPRU that returns the top-k users based on their spatial proximity to a given query location, their popularity, and their similarity to an input set of terms; (ii) NSTP that outputs the top-k POIs based on their proximity to a user v, the number of check-ins by friends of v, and their similarity to a set of terms; (iii) FSKR that discovers the top-k keywords based on their frequency in pairs of friends located within a spatial area. For each query, we develop a processing algorithm that utilizes a novel hybrid index. Finally, we evaluate our framework with thorough experiments using real datasets.

1 Introduction

The rising popularity of social networks and smart-phones has led to the development of techniques for personalized search and targeted advertisement that combine social, geographic and textual criteria. As an instance of social and textual fusion, social networks, such as Facebook, permit the promotion of products to connected users that share common interests, e.g. the advertisement of a rock festival to a group of friends that like rock music [1]. As an example of geographic and textual integration, Web search engines, such as Google, allow search for Points Of Interest (POIs) that match some description and are near the query location , e.g., "Chinese restaurants nearby" [2]. Finally, Geo-Social Networks (GeoSNs), such as Foursquare, combine geographic and social aspects by enabling users to check-in at POIs, i.e., publish their current location to friends. Moreover, advertisers can send GroupON-like offers to users in their vicinity to attract them, as well as their friends [3].

Similar combinations of social, geographic and textual criteria have been investigated in the research literature. (i) *Keyword search in social networks* focuses of queries that seek groups of users forming a particular social structure

R. Ahuja—Supported by GRF grant 617412 from Hong Kong RGC.

C. Claramunt et al. (Eds.): SSTD 2015, LNCS 9239, pp. 431–450, 2015.
DOI: 10.1007/978-3-319-22363-6_23

(e.g. clique), and their members' profiles cover a set of input terms [13,14,16]. (ii) *Spatial keyword search* queries return POIs that satisfy various spatial (e.g., range, nearest neighbor) and textual (e.g., text similarity) constraints [7,10,11, 18,20,24]. (iii) *GeoSN queries* output individual users, or groups of friends, that exhibit some spatial and social properties, e.g., the closest clique of m friends to a query point [5,17,19,22].

All the above cases consider only two out of the three criteria, focusing on a single output type (e.g., users or POIs, but not both). On the other hand, we introduce *Geo-Social Keyword* (GSK) search, a class of top-k queries that combine all spatial, social, and textual attributes, and may return users, POIs or keywords. We present three concrete GSK queries: (i) *Top-k Nearest, Popular and Relevant Users* (NPRU) that, given a query location q and a set of terms T_q, outputs the top-k users based on their proximity to q, their social connectivity, and the similarity of their profiles to T_q; (ii) *Top-k Nearest Socially and Textually Relevant POIs* (NSTP), which, given a user v and a set of terms T_q, returns the top-k POIs based on their proximity to v, the number of check-ins by friends of v, and their similarity to T_q; and (iii) *Top-k Frequent Social Keywords in Range* (FSKR) that discovers the top-k keywords based on their frequency in pairs of friends located within a geographic area.

Each query is suitable for a different type of task, including advertisement, context-based search, and market analysis. For instance, NPRU could be used by a restaurant to send promotions to nearby users, who are well-connected and have expressed interest in its cuisine type. Conversely, a user could issue an NSTP query to locate nearby restaurants of a specific type that are 'liked' by his friends. Finally, FSKR could identify trends or word-of-mouth effects in a geographic area, using the frequency of keywords shared by friends.

For each query, we provide a query processing algorithm that utilizes the *GSK Index* (GSKI), a novel hybrid structure that stores users and POIs, based on spatial, social, and textual attributes. GSKI is a lightweight multi-level grid that supports efficient updates. Summarizing, our contributions are:

- We define GSK search as a general framework for retrieval of the top-k users, POIs or keywords using various types of criteria.
- We present the GSKI, a hybrid structure for indexing users and POIs.
- We propose three GSK queries and the respective processing algorithms that utilize the GSKI.
- We conduct a thorough experimental evaluation on real datasets.

The rest of the paper is organized as follows. Section 2 overviews related work. Section 3 formalizes the GSK problem and introduces the general framework. Section 4 presents the GSK Index. Sections 5, 6 and 7 propose the GSK queries and the corresponding query processing methods. Section 8 contains the experimental evaluation. Finally, Sect. 9 concludes the paper with directions for future work.

2   Related Work

We overview (i) keyword search in social networks, (ii) spatial keyword search, and (iii) GeoSN queries.

Keyword Search in Social Networks. Although, there has been extensive work on keyword search for general graphs, here we focus on social networks. Lappas et al. [16] propose the *Team Formation* (*TF*) query: given a weighted social graph and a set of terms T_q, *TF* returns a subgraph of users, whose textual descriptions cover T_q and their diameter (i.e., maximum shortest-path distance between any two nodes) is minimized. The authors also devise a variant, where the subgraph must be a minimum spanning tree, and show that both problems are NP-Complete. [13] extends *TF* by additionally seeking a team leader, i.e., the member of the resulting group with the minimum total social shortest-path distances from all members. Finally, [14] proposes the r-cliques query: given a weighted social graph and a set of terms T_q, return a sugbraph of users that covers T_q, and has diameter no larger than r. In the above methods, textual information is stored in inverted files and the graph is kept in adjacency lists.

Spatial Keyword Search. Four types of spatial-keyword queries have received particular attention in the literature [8] namely, the *Boolean Range* (*BR*), the *Boolean k-NN* (*BkNN*), the *Spatial Aware Top-k text retrieval* (*SATopk*), and the *Spatial Group Keyword* (*SGK*) query. Given a spatial region R and a set of terms T_q, *BR* returns all POIs in R, whose textual description contains all terms in T_q [20,24]. *BkNN* outputs the k nearest POIs to a query point q each of which covers all the query terms [11]. Given q, T_q and a positive integer k, *SATopk* returns a list of k POIs ranked based on their spatial proximity to q and textual similarity to T_q [10]. Finally, *SGK* discovers a set of POIs that collectively cover the query terms and either the sum of their distances to the query location is minimized [7], or the maximum distance between any two POIs in the group is minimized [18]. A recent work [21] introduces the *Social-aware top-k Spatial Keyword* (*SkSK*) query, which enhances personalized spatial-keyword search by additionally taking into consideration the social connectivity of the query issuer to all users, who have liked or recommended the POIs.

Spatial-keyword indices can be broadly classified according to the spatial and textual structures employed. They are usually based on the R-Tree and its variants, where each minimum bounding rectangle (MBR) keeps the textual information of the POIs located within its bounds. Specifically, MBRs in [7,10] utilize inverted files, while in [11,23] use bitmaps. Grid-based spatial-keyword structures decompose the space into cells; each cell has a unique id according to a global order (e.g., Hilbert curves [9]). Then, inverted files are primarily used for indexing the cells based on the textual description of the POIs located within their bounds [15,20]. Indices based on trees are in general more efficient than grid-based structures [24], but the latter are easier to maintain. The *Social Network-aware IR-Tree* [21] is an R-Tree, where each node also contains a set of users relevant to the POIs indexed by the subtree rooted at the node; contrary to its name, it does not index social information (i.e., user connections).

Geo-Social Networks. GeoSN queries return users, or groups of users, that satisfy spatial and social criteria. Given a location q and two positive integers k and m $(k < m)$, the *Socio-Spatial Group* query outputs a group of m users, such that the total distance of the users to q is minimized, and each user is connected to at least $m - k$ other group members [22]. Given a location q and two positive integers m, k, the *Nearest Star Group* query [5] returns the k nearest subgraphs of m users, such that each subgraph (i.e., star) has a user, who is socially connected to all users. Given a user v, the *k-Geo-Social Circle of Friends* query [17] finds a group of $k + 1$ users that contains v and k friends with small pairwise social distances, so that the diameter of the group is minimized. Finally, [19] introduces the *Social and Spatial Ranking* query, which given a user v, reports the top-k users based on their spatial proximity and social connectivity to v.

Most GeoSN approaches maintain separate structures for the spatial and social attributes. For instance, Liu et al. [17] store the social graph in an adjacency matrix, and employ the R*-Tree for spatial indexing. Similarly, [5] uses adjacency lists and a regular spatial grid, respectively. On the other hand, Yang et al. [22] propose a hybrid index that constructs an R-tree while ensuring a specified degree of connectivity among the users within the same node.

3   GSK Query Framework

Our setting consists of a social graph network and a set of POIs. The social network is modeled as an unweighted, undirected graph $G = (V, E)$, where a node $v \in V$ represents a user and an edge $(v, u) \in E$ indicates the friendship between v and $u \in V$. Each user $v \in V$ may be associated with textual and spatial information that represent his preferences and his most recent location, respectively. Each POI $p \in P$ has a spatial location, a textual description and a set of users V_p that have checked-in at p in the past. T denotes a set of terms/keywords; specifically, T_v (resp. T_p) is the set that appears in the preference of user v (resp. the description of POI p).

Figure 1 depicts a running example of a social network with the locations of 10 users as grey points, and the incident edges as their social relations. The black squares represent the location of 4 POIs. Next to each user v and POI p is the corresponding set of terms T_v and T_p, e.g., $\{c, f\}$ for v_4 and $\{c, e\}$ for p_1. Moreover, the list below each POI (e.g., $[v_2, v_4, v_5, v_6]$ for p_1) represents the users that have checked-in there. Depending on the application, the setting may vary; e.g., the textual information of users may correspond to their query history or profile data (instead of preferences), V_p may denote the current (instead of all) check-ins at p, etc.

Geo-Social Keyword (GSK) search constitutes a family of top-k queries that return results of type $RT = (C, l)$, where C denotes the object class (i.e., V, P or T) and l represents the cardinality. For example, $RT = (V, 3)$ denotes that the output contains k groups of 3 users each, whereas $RT = (P, 1)$ signifies that the output consists of k individual POIs. Given a GSK query q, each object o of type RT (e.g., a group of 3 users, or a single POI) is assigned a geographic $f_g(o)$,

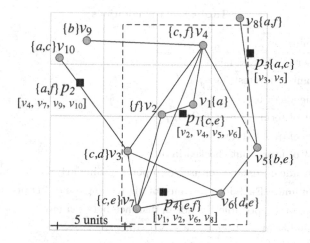

Fig. 1. Running example

social $f_s(o)$ and a textual $f_t(o)$ score. In general, $f_g(o)$ depends on the proximity of o to q, $f_s(o)$ on the social connectivity of o, and $f_t(o)$ on the similarity between the terms of o and q.

The total score of an object is obtained by combining the partial ones using a ranking function F. We implement F as a weighted combination of the partial scores, i.e., $F(o) = \alpha_g \cdot f_g(o) + \alpha_s \cdot f_s(o) + \alpha_t \cdot f_t(o)$, where $\alpha_g, \alpha_s, \alpha_t$ are non-negative real numbers such that $\alpha_g + \alpha_s + \alpha_t = 1$, but any monotone[1] function can be used. A criterion (e.g., textual) can be omitted by setting the corresponding weight (e.g., α_t) to zero. Moreover, in some cases we may only be interested in objects that satisfy a set of constraints CN, i.e., POIs in a geographic area, or users who have certain characteristics (e.g., males above 30 years old). Finally, we define a GSK query as follows:

Definition 1. *Given a positive integer k, a result type RT, functions f_g, f_s, f_t, F, and a set of constraints CN, a GSK query returns the k objects of type RT that have the highest scores according to F and satisfy all constraints in CN.*

By employing different combinations of result types, ranking functions and constraints, we can devise a wide range of GSK queries. In this paper, we will present three diverse queries that retrieve individual users, POIs and keywords. All the queries utilize the index of the next section. Table 1 contains the frequent symbols.

---

[1] F should satisfy the condition $\forall o, o' : f_g(o) \geq f_g(o') \land f_s(o) \geq f_s(o') \land f_t(o) \geq f_t(o') \Rightarrow F(o) \geq F(o')$.

Table 1. Basic notations

Notation	Definition
v	User in GeoSN, i.e., $v \in V$
p	Point of interest, i.e., $p \in P$
N_v	Friends of user v
T_v (T_p)	Set of terms of user v (POI p)
T_q	Set of query terms
V_p	Set of users that checked-in at p
$\|v, q\|$	Euclidean distance of user v to point q. Similarly for p, i.e., $\|p, q\|$
$\|c, q\|_{min}$	Minimum Euclidean distance of 2D rectangle c to 2D point q
max_{dist}	Maximum possible Euclidean distance between any two points
$deg(v)$	Number of v's friends, i.e., $|N_v| = deg(v)$
max_{deg}	Maximum number of friends in the graph
$TS(T_1, T_2)$	Normalized textual similarity between term sets T_1 and T_2

4   Geo-Social Keyword Index

The *Geo-Social Keyword Index* (GSKI) stores users and POIs based on their geographical, social and textual attributes. Given a granularity factor g and a height parameter h, GSKI partitions the geographical space into $g^h \times g^h$ equally sized leaf cells. Each leaf cell lc contains:

- a rectangle R_{lc} that represents the area covered by lc,
- a list of users V_{lc} and a list of POIs P_{lc} that lie in R_{lc},
- the maximal degree D_{lc} of any user in R_{lc},
- inverted files IV_{lc} and IP_{lc}, consisting of lists of keywords appearing in the preferences of users and in the descriptions of POIs in R_{lc}, respectively. Lists are sorted by the *impact* of keywords based on the *cosine-normalized tf-idf* [25], and
- a bloom filter[2] B_{lc} of the union of all users checked-in at POIs in R_{lc}, i.e. B_{lc} = bloom filter of $\bigcup_{p \in P_{lc}} V_p$.

Next, a hierarchical grid of height h is constructed in a bottom-up fashion, where each intermediate cell points to g^2 cells at the lower level that lie inside its spatial extent. Every intermediate cell ic keeps only a small amount of information summarizing its children cells. Specifically, ic is associated with a rectangle R_{ic}, maximum degree of users in R_{ic}, namely D_{ic}, and bloom filter B_{ic}. Additionally, for each term that appears in users or POIs located within the bounds of R_{ic}, ic keeps the term's maximum textual impact in sets SV_{ic} and SP_{ic}, respectively.

---

[2] A bloom filter is a space-efficient probabilistic data structure that is used to test whether an element is a member of a set [6].

Figure 2 illustrates the GSKI and Table 2 shows the corresponding cell contents for our running example, assuming $g = 2$ and $h = 2$. Leaf cell $C0_{2,2}$ is a child of $C1_{1,1}$, which in turn is a child of $C2_{0,0}$. $C0_{2,2}$ contains users v_1, v_2 and POI p_1 in its spatial extent. Consequently, as elaborated in the fourth to last row of Table 2, $D_{C0_{2,2}} = 2 = deg(v_1) = deg(v_2)$, $IV_{C0_{2,2}}$ stores terms a, f, since they appear in v_1's and v_2's preferences, and $IP_{C0_{2,2}}$ keeps terms c, e occurring in p_1's description. Each term is associated with an *impact* value [25] in the range $[0,1]$. $B_{C0_{2,2}}$ contains users v_2, v_4, v_5, v_6 who checked-in at p_1. Intermediate cell $C1_{1,1}$ aggregates the information of its children $C0_{2,2}$, $C0_{2,3}$, $C0_{3,2}$, and $C0_{3,3}$. $D_{C1_{1,1}} = 5 = deg(v_4)$, since v_4 is located in $C0_{2,3}$. $C1_{1,1}$ keeps $SV_{C1_{1,1}}$ and $SP_{C1_{1,1}}$ with the terms that appear in the children, namely $\{a, c, f\}$ and $\{a, c, e\}$, respectively. Finally, $B_{C1_{1,1}}$ contains the union of $B_{C0_{2,2}}$ and $B_{C0_{3,3}}$ ($C0_{2,3}$ and $C0_{3,2}$ do not contain POIs).

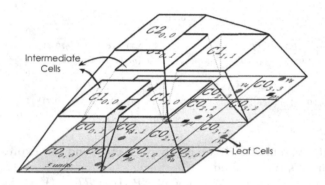

Fig. 2. Hierarchical grid

To enable effective pruning during query processing, the GSKI preserves monotonicity across the height of the hierarchical grid, i.e., assuming a monotone function F, the overall score of an intermediate cell ic constitutes an upper bound for the score of any user or a POI within R_{ic}. Moreover, since the GSKI only keeps concise aggregated data at the intermediate levels, the size of the inverted file at a non-leaf cells is smaller than that of the original inverted file. Finally, we chose a grid-based structure because grids in general are usually significantly faster that R-trees for highly dynamic settings [12] such as ours, where there are numerous location updates from users.

5 Top-K Nearest, Popular and Relevant Users

A *Top-k Nearest, Popular and Relevant Users* (NPRU) query returns the top-k users based on their spatial proximity to a location q, their social connectivity, and their textual similarity to an input set of terms T_q. NPRU is useful for advertisement and promotion purposes. For instance, consider a restaurant owner who wishes to send lunch coupons. Promising targets are users that

Table 2. GSKI contents

Cell c	IV_c / SV_c	IP_c / SP_c	D_c	B_c
$C2_{0,0}$				
$C1_{0,0}$	$\langle c, 0.71 \rangle, \langle d, 0.71 \rangle, \langle e, 0.71 \rangle$		4	
$C0_{0,0}$			0	
$C0_{1,0}$	$c:\ \langle v_7, 0.71 \rangle, \langle e: \rangle\ \langle v_7, 0.71 \rangle$		4	
$C0_{0,1}$			0	
$C0_{1,1}$	$c:\ \langle v_3, 0.71 \rangle, \langle d: \rangle\ \langle v_3, 0.71 \rangle$		4	
$C1_{1,0}$	$\langle b, 0.71 \rangle, \langle d, 0.71 \rangle, \langle e, 0.71 \rangle$	$\langle e, 0.71 \rangle, \langle f, 0.71 \rangle$	3	$\langle v_1, v_6, v_8, v_9 \rangle$
$C0_{2,0}$		$e:\ \langle p_4, 0.71 \rangle, f:\ \langle p_4, 0.71 \rangle$	0	$\langle v_1, v_2, v_6, v_8 \rangle$
$C0_{3,0}$	$d:\ \langle v_6, 0.71 \rangle, e:\ \langle v_6, 0.71 \rangle$		3	
$C0_{2,1}$			0	
$C0_{3,1}$	$b:\ \langle v_5, 0.71 \rangle, e:\ \langle v_5, 0.71 \rangle$		3	
$C1_{0,1}$	$\langle a, 0.71 \rangle, \langle b, 1.0 \rangle, \langle c, 0.71 \rangle$	$\langle a, 0.71 \rangle, \langle f, 0.71 \rangle$	1	$\langle v_4, v_7, v_9, v_{10} \rangle$
$C0_{0,2}$		$a:\ \langle p_2, 0.71 \rangle, f:\ \langle p_2, 0.71 \rangle$	0	$\langle v_4, v_7, v_9, v_{10} \rangle$
$C0_{1,2}$			0	
$C0_{0,3}$	$a:\ \langle v_{10}, 0.71 \rangle, b:\ \langle v_9, 1.0 \rangle, c:\ \langle v_{10}, 0.71 \rangle$		1	
$C0_{1,3}$			0	
$C1_{1,1}$	$\langle a, 1.0 \rangle, \langle c, 0.71V \rangle, \langle f, 1.0 \rangle$	$\langle a, 0.71 \rangle, \langle c, 0.71 \rangle, \langle e, 0.71 \rangle$	5	$\langle v_2, v_3, v_4, v_5, v_6 \rangle$
$C0_{2,2}$	$a:\ \langle v_1, 1.0 \rangle, f:\ \langle v_2, 1.0 \rangle$	$c:\ \langle p_1, 0.71 \rangle, e:\ \langle p_1, 0.71 \rangle$	2	$\langle v_2, v_4, v_5, v_6 \rangle$
$C0_{3,2}$			0	
$C0_{2,3}$	$c:\ \langle v_4, 0.71 \rangle, f:\ \langle v_4, 0.71 \rangle$		5	
$C0_{3,3}$	$a:\ \langle v_8, 0.71 \rangle, f:\ \langle v_8, 0.71 \rangle$	$a:\ \langle p_3, 0.71 \rangle, c:\ \langle p_3, 0.71 \rangle$	1	$\langle v_3, v_5 \rangle$

(i) are near the restaurant, (ii) are highly connected, and (iii) express preference to the restaurant's type of food.

In our framework, the output type of NPRU is $RT = (V, 1)$, i.e., the result consists of individual users, and $CN = \emptyset$, i.e., there are no constraints on the users to be retrieved. Regarding the geographic $f_g(v)$, social $f_s(v)$ and textual $f_t(v)$ scores of each user $v \in V$, there are several alternatives. In our implementation, we set $f_g(v) = 1 - \frac{\|v,q\|}{max_{dist}}$, where max_{dist} denotes the maximum Euclidean distance in the data space. Intuitively, the spatial score of a user v decreases as his Euclidean distance $\|v, q\|$ from q increases. The social score of v is defined as $f_s(v) = \frac{deg(v)}{max_{deg}}$, where $deg(v)$ is the number of v's friends, and max_{deg} is the maximum degree of any user in the network. The textual score $f_t(v)$ is the *cosine-normalized tf-idf* similarity $TS(T_v, T_q)$ [25] between the terms T_v of v and those in T_q. All partial scores are in the range [0,1]. The total score of v is $F(v) = \alpha_g \cdot f_g(v) + \alpha_s \cdot f_s(v) + \alpha_t \cdot f_t(v)$, as discussed in Sect. 3.

Consider, for instance an NPRU query with $k = 2$, $q = p_1$, $T_q = \{c, e\}$ and $\alpha_g = \alpha_s = \alpha_t = \frac{1}{3}$ in the running example of Fig. 1, e.g., a Chinese restaurant p_1 wishes to discover the top-2 users in its vicinity, that have many friends and at the same time have matching keywords c, e (Chinese, Restaurant). The best user is u_7 because both keywords c and e are in his preferences. The top-2 user is v_4 with keyword c. Note that v_4 out-ranks v_3, which is slightly closer to p_1 and contains c, because he has higher degree (5 as opposed to 4 for v_3). Although users v_1 and v_2 are the nearest to p_1, they are not in the result because neither contains keyword c or e; accordingly, their f_t score is zero.

Processing NPRU queries is based on the branch-and-bound paradigm using the GSKI. Specifically, a priority heap H maintains visited cells and users along with their score according to F. The score of a cell c takes into consideration (i) the minimum Euclidean distance of the cell to q, (ii) the maximum degree of any user in c, and (iii) the maximum textual similarity of the queried terms amongst the preferences of the users in c. This guarantees that the score of c is an upper bound for the score of child cells and users within its extent. Consequently, if the score of c does not exceed that of the top-kth user, then c can be safely pruned.

Figure 3 illustrates the pseudo-code of NPRU processing. Initially, the algorithm adds GSKI's root cell to H (Line 2). Then, in an iterative manner, it removes the entity with the highest score from H, namely e, and (i) if e is an intermediate cell, then it adds all its children cells to H (Lines 5–7), or (ii) if e is a leaf cell, then it adds all users within e's spatial extent to H (Lines 8–10), or (iii) if e is a user, it adds him to the result set (Lines 11–12). The algorithm terminates when the result set contains k users (Lines 13–14). The cells and users remaining in H have score at most as high as that of the k-th result and, hence, can be ignored.

Input: Social Graph $G = (V, E)$, integer k, location q, set of terms T_q, weights α_g, α_s, α_t
Output: Top-k users according to F

1. Define H as an empty heap of $GSKI$ cells sorted according to their scores in decr. order
2. Add the root cell of $GSKI$ to H
3. **While** H is not empty
4. e = top entity of H // it also removes e from H
5.   **If** e is an intermediate cell of $GSKI$
6.   **For** each child c of e
7.   Add to H cell c with score $\alpha_g \cdot (1 - \frac{\|c,q\|_{min}}{max_{dist}}) + \alpha_s \cdot \frac{D_c}{max_{deg}} + \alpha_t \cdot TS(T_c, T_q)$
8.   **Else If** e is a leaf cell of $GSKI$
9.   **For** each user $v \in V_e$
10.   Add to H user v with score $\alpha_g \cdot (1 - \frac{\|v,q\|}{max_{dist}}) + \alpha_s \cdot \frac{deg(v)}{max_{deg}} + \alpha_t \cdot TS(T_v, T_q)$
11.   **Else** // e is a user
12.   Add e to R
13.   **If** $|R| = k$ then stop the execution
14. **Return** R

Fig. 3. NPRU Algorithm

Table 3 shows the heap state during the execution of the example query: $k = 2$, $q = p_1$, $T_q = \{c, e\}$ and $\alpha_g = \alpha_s = \alpha_t = \frac{1}{3}$, using the GSKI contents of Table 2. Heap entries consist of a cell or a user, and the corresponding score according to F. Cells and users added to H are shown in bold. First, the algorithm inserts the root of GSKI in H. At iteration 1, it removes the root cell and adds its children along with their scores to H. Next, the intermediate cell with the highest score, $C1_{0,0}$, is removed and its child leaf cells $\{C0_{0,0}, C0_{1,0}, C0_{0,1}, C0_{1,1}\}$ are added to H. Similarly, $C0_{1,0}$ is removed at the next iteration and user v_7 is added to H. Next, intermediate cell $C1_{1,1}$ is de-heaped and its child leaf nodes are en-heaped. Then, user v_7 is removed and becomes the top-1 result. The algorithm continues in the same manner and terminates after the 6th iteration, when the top-2 user v_4 is de-heaped.

Table 3. Heap of NPRU

Interation #	Heap Contents
0	$\langle C2_{0,0}, \infty \rangle$
1	$\langle \mathbf{C1_{0,0}, 0.90} \rangle$, $\langle \mathbf{C1_{1,1}, 0.81} \rangle$, $\langle \mathbf{C1_{1,0}, 0.71} \rangle$, $\langle \mathbf{C1_{0,1}, 0.51} \rangle$
2	$\langle \mathbf{C0_{1,0}, 0.85} \rangle$, $\langle C1_{1,1}, 0.81 \rangle$, $\langle C1_{1,0}, 0.71 \rangle$, $\langle \mathbf{C0_{1,1}, 0.71} \rangle$, $\langle C1_{0,1}, 0.51 \rangle$, $\langle \mathbf{C0_{0,1}, 0.24} \rangle$, $\langle \mathbf{C0_{0,0}, 0.21} \rangle$
3	$\langle C1_{1,1}, 0.82 \rangle$, $\langle \mathbf{v_7, 0.80} \rangle$, $\langle C1_{1,0}, 0.71 \rangle$, $\langle C0_{1,1}, 0.71 \rangle$, $\langle C1_{0,1}, 0.51 \rangle$, $\langle C0_{0,1}, 0.24 \rangle$, $\langle C0_{0,0}, 0.21 \rangle$
4	$\langle v_7, 0.80 \rangle$, $\langle \mathbf{C0_{2,3}, 0.75} \rangle$, $\langle C1_{1,0}, 0.71 \rangle$, $\langle C0_{1,1}, 0.71 \rangle$, $\langle C1_{0,1}, 0.51 \rangle$, $\langle \mathbf{C0_{2,2}, 0.46} \rangle$, $\langle \mathbf{C0_{3,3}, 0.33} \rangle$, $\langle \mathbf{C0_{3,2}, 0.30} \rangle$, $\langle C0_{0,1}, 0.24 \rangle$, $\langle C0_{0,0}, 0.21 \rangle$
5	$\langle C0_{2,3}, 0.75 \rangle$, $\langle C1_{1,0}, 0.71 \rangle$, $\langle C0_{1,1}, 0.71 \rangle$, $\langle C1_{0,1}, 0.51 \rangle$, $\langle C0_{2,2}, 0.46 \rangle$, $\langle C0_{3,3}, 0.33 \rangle$, $\langle C0_{3,2}, 0.30 \rangle$, $\langle C0_{0,1}, 0.24 \rangle$, $\langle C0_{0,0}, 0.21 \rangle$
6	$\langle \mathbf{v_4, 0.72} \rangle$, $\langle C1_{1,0}, 0.71 \rangle$, $\langle C0_{1,1}, 0.71 \rangle$, $\langle C1_{0,1}, 0.51 \rangle$, $\langle C0_{2,2}, 0.46 \rangle$, $\langle C0_{3,3}, 0.33 \rangle$, $\langle C0_{3,2}, 0.30 \rangle$, $\langle C0_{0,1}, 0.24 \rangle$, $\langle C0_{0,0}, 0.21 \rangle$

6   Top-K Nearest Socially and Textually Relevant POIs

Given a user v and a set of terms T_q, a *Top-k Nearest Socially and Textually Relevant POIs* (NSTP) query returns the top-k POIs based on their proximity to v, the textual similarity of their descriptions to T_q, and the number of v's friends that checked-in. NSTP enables location-aware, socially-aware, and/or context-aware search. For instance, consider a user who wants to visit a restaurant. NSTP could locate nearby restaurants offering cuisine similar to the user's preferences that are also visited (or 'liked') by his friends.

The output type of NSTP query is $RT = (P, 1)$, i.e., the result consists of individual POIs, and $CN = \emptyset$, i.e., there are no constraints on the POIs to be retrieved[3]. The geographic and textual score definitions are similar to NPRU, i.e., $f_g(p) = 1 - \frac{\|v,p\|}{max_{dist}}$ and $f_t(p)$ is based on *cosine-normalized tf-idf* between T_p and T_q. The social score is defined as $f_s(p) = \frac{|N_v \cap V_p|}{|N_v|}$, where set N_v consists

---

[3] Additional constraints in this case could restrict the top-k POIs to be in a certain area, or enforce certain properties (e.g., restaurant must be open after 10 pm).

of v's friends (i.e., $|N_v| = deg(v)$), and V_p contains the ids of the users who checked-in at p. The partial scores are combined by the linear function F also used in NPRU.

For example, consider an NSTP query with $v = v_7$, $k = 2$, $T_q = \{c, e\}$, and $\alpha_g = \alpha_s = \alpha_t = \frac{1}{3}$ using the running example, e.g., user v_7 searches for two nearby Chinese restaurants (c, e) that have been visited by many of his friends. The best POI is p_1 since it is relatively close to v_7, contains both queried terms, and it has been visited by 3 of his 4 friends (v_2, v_4, v_6). The top-2 POI is p_4 because it is the closest POI to v_7, contains term e, and was visited by two of v_7's friends (v_2, v_6). POIs p_2 and p_3 are not in the result set since they are far from v_7, are not relevant to T (only p_3 contains one of the queried terms), and are not popular among v_7's friends (each is visited by only one friend).

NSTP query processing is similar to NPRU. Specifically, the algorithm uses a max-heap to store cells and POIs sorted in decreasing order of their scores. The score of a cell c is based on: (i) the minimum distance of c to v, (ii) an upper bound for the number of v's friends that checked-in at any POI within c, and (iii) the maximum textual similarity of T to the descriptions of the POIs in c. For the computation of (ii), the algorithm examines if each friend of v is in the bloom filter of c. Bloom filters may falsely indicate the presence of a user. However, although false positives increase the score of c, they do not affect correctness because the score of c is always an upper bound (albeit, in some cases, loose) for that of any child cell or POI in c. The algorithm terminates after it retrieves k POIs from the priority heap.

Consider again the example query with input: $v = v_7$, $k = 2$, $T_q = \{c, e\}$, and $\alpha_g = \alpha_s = \alpha_t = \frac{1}{3}$, using the GSKI contents of Table 2. Table 4 shows the state of the heap at each iteration. Starting from the root cell, the algorithm retrieves the top-1 POI p_1 at iteration 3. Then, it continues until iteration 6, when it discovers p_4 and terminates.

7   Frequent Social Keywords in Range

A *Frequent Social Keywords in Range* (FSKR) query returns the top-k terms based on their frequency in pairs of friends located within a spatial area SR. FSKR allows the discovery of trends or word-of-mouth effects. For instance, FSKR on textual content derived from Twitter/Facebook posts can reveal topics that are trending among friends in a geographic area. This information can be then utilized by businesses towards social media marketing.

The output of FSKR query is $RT = (T, 1)$, i.e., the result consists of individual terms. In addition, CN contains the constraint that valid terms must appear jointly in the preferences of friends in SR. FSKR does not apply geographic or social scores; instead, the total score of a term t is based solely on its frequency among friends, i.e., $F(t) = f_t(t) = |\{(v, u) \in E / t \in T_v \wedge t \in T_u \wedge v, u \; inside \; SR\}|$, where T_v (resp. T_u) denotes the terms associated with v (resp. u). Note that an edge (v, u) contributes 2 to the score of t; once per incident user v and u. This does not affect the ranking of the top-k results.

Table 4. Heap of NSTP

Interation #	Heap H Contents
0	$\langle C2_{0,0}, \infty \rangle$
1	$\langle C1_{1,1}, \mathbf{0.75} \rangle$, $\langle C1_{1,0}, \mathbf{0.55} \rangle$, $\langle C1_{0,0}, \mathbf{0.33} \rangle$, $\langle C1_{0,1}, \mathbf{0.28} \rangle$
2	$\langle C0_{2,2}, \mathbf{0.75} \rangle$, $\langle C1_{1,0}, 0.55 \rangle$, $\langle C0_{3,3}, \mathbf{0.44} \rangle$, $\langle C1_{0,0}, 0.33 \rangle$, $\langle C1_{0,1}, 0.28 \rangle$, $\langle C0_{3,2}, \mathbf{0.19} \rangle$, $\langle C0_{2,3}, \mathbf{0.15} \rangle$
3	$\langle p_1, \mathbf{0.75} \rangle$, $\langle C1_{1,0}, 0.55 \rangle$, $\langle C0_{3,3}, 0.44 \rangle$, $\langle C1_{0,0}, 0.33 \rangle$, $\langle C1_{0,1}, 0.28 \rangle$, $\langle C0_{3,2}, 0.19 \rangle$, $\langle C0_{2,3}, 0.15 \rangle$
4	$\langle C1_{1,0}, 0.55 \rangle$, $\langle C0_{3,3}, 0.44 \rangle$, $\langle C1_{0,0}, 0.33 \rangle$, $\langle C1_{0,1}, 0.28 \rangle$, $\langle C0_{3,2}, 0.19 \rangle$, $\langle C0_{2,3}, 0.15 \rangle$
5	$\langle C0_{2,0}, \mathbf{0.55} \rangle$, $\langle C0_{3,3}, 0.44 \rangle$, $\langle C1_{0,0}, 0.33 \rangle$, $\langle C1_{0,1}, 0.28 \rangle$, $\langle C0_{2,1}, \mathbf{0.28} \rangle$, $\langle C0_{3,0}, \mathbf{0.25} \rangle$, $\langle C0_{3,1}, \mathbf{0.23} \rangle$, $\langle C0_{3,2}, 0.19 \rangle$, $\langle C0_{2,3}, 0.15 \rangle$
6	$\langle p_4, \mathbf{0.59} \rangle$, $\langle C0_{3,3}, 0.44 \rangle$, $\langle C1_{0,0}, 0.33 \rangle$, $\langle C1_{0,1}, 0.28 \rangle$, $\langle C0_{2,1}, 0.28 \rangle$, $\langle C0_{3,0}, 0.24 \rangle$, $\langle C0_{3,1}, 0.23 \rangle$, $\langle C0_{3,2}, 0.19 \rangle$, $\langle C0_{2,3}, 0.15 \rangle$

Consider, for instance, the FSKR query with $k = 2$ and an area SR represented by the dashed-line rectangle in Fig. 1. The top-1 term is c, with score $F(c) = 6$, since it appears in 3 pairs of friends within the range, i.e., (v_3, v_4), (v_3, v_7), and (v_4, v_7). The top-2 term can be either e (v_6, v_7), or d (v_3, v_6), both with score 2. The remaining terms in SR (a, d, f) are not shared by any pair of friends.

FSKR query processing is performed in two steps: first, for every term t in SR, a list $PL[t]$ is created with the users (in SR) containing t; then, the score $F(t)$ of each term t is computed by examining the connections of users appearing in $PL[t]$. Specifically, the contribution of each $v \in PL[t]$ to $F(t)$ is $|N_v \cap PL[t]|$, where N_v is the set of v's friends. Let $best_{score}$ be the score of the current top-kth term. The upper bound score of any (not-yet-examined) term t is $|PL[t]| \cdot (|PL[t]| - 1)$, when all users containing t form a clique. Consequently, if $|PL[t]| \cdot (|PL[t]| - 1) \leq best_{score}$, then t can be safely pruned. Based on this observation, FSKR examines terms in decreasing order of their list sizes, until the first term that can be eliminated by its upper bound score.

Figure 4 elaborates the procedure. The algorithm first retrieves the nonempty leaf cells of GSKI that intersect with the spatial range SR. For each keyword t in the inverted lists of these cells, Lines 3–13 generate $PL[t]$. Next, the terms are sorted in decreasing order of $|PL[t]|$ size. For each term t, Lines 18–20, compute the score of t, and update $best_{score}$ accordingly. The algorithm terminates at the first term for which $|PL[t]| \cdot (|PL[t]| - 1) \leq best_{score}$ (Lines 16–17), and returns the top-k set (Line 21). Unexamined terms cannot be in the result set, and are pruned.

We describe the algorithm using our running example of Fig. 1, where $k = 1$ and the spatial range SR is depicted as a dashed rectangle. Initially, $best_{score} = 0$. The terms associated with users in SR are a, c, d, e, f with lists

---

Input: Social Graph $G = (V, E)$, integer k, spatial range SR
Output: Top-k terms according to F

1. Initialize list PL as an empty list of sets, $best_{score} = 0$
2. Set $C = $ all non-empty leaf cells in GSKI that intersect with SR
3. **For** each cell $c \in C$
4.   **For** each term $t \in IV_c$
5. $Occur_t = $ posting list of t in IV_c
6.   **If** t appears for first time
7. $PL[t] = \{\emptyset\}$
8.   **Else**
9.   **If** R covers c
10. $PL[t] = PL[t] \cup Occur_t$
11.   **Else**
12. $Occur_{t,valid} = $ Exclude from $Occur_t$ all users not in SR
13. $PL[t] = PL[t] \cup Occur_{t,valid}$
14. Sort PL according to sets' sizes in decreasing order
15. **For** each term $t \in PL$
16.   **If** $|PL[t]| \cdot (|PL[t]| - 1) \leq best_{score}$
17.   **Exit** For Loop
18.   **For** each user $v \in PL[t]$
19. $Score_t = Score_t + |N_v \cap PL[t]|$
20. $best_{score} = k^{th}$ highest score
21. **Return** the terms with the k highest scores

Fig. 4. FSKR algorithm

$PL[a] = \{v_1\}$, $PL[c] = \{v_3, v_4, v_7\}$, $PL[d] = \{v_3, v_6\}$, $PL[e] = \{v_6, v_7\}$ and $PL[f] = \{v_2, v_4\}$. FSKR iterates over the lists in sorted order, starting from c. It computes $|PL[c] \cap N_{v_3}| = 2$, $|PL[c] \cap N_{v_4}| = 2$, $|PL[c] \cap N_{v_7}| = 2$, and $F(c) = 6$. Since $k = 1$, it sets $best_{score} = 6$ and retrieves the second most frequent keyword e. The upper bound score for e is 2, which is below $best_{score}$. Consequently, the algorithm stops and outputs c as the top-1 result.

8 Experimental Evaluation

Section 8.1 presents the real datasets, Sect. 8.2 contains a qualitative evaluation of the proposed queries, and Sect. 8.3 evaluates their performance experimentally.

8.1 Datasets

We use two real datasets obtained from *Yelp* [4] that consist of users and POIs located in Las Vegas (LV) and Phoenix (PX). In particular, each dataset includes: (i) a social graph, (ii) latest and past user check-ins, (iii) user preferences, (iv) POI locations, and (v) POI descriptions. Table 5 summarizes the

Table 5. Datasets

Statistic	LV	PX
$|V|$	40,297	30,056
Avg. Degree	9.66	5.41
Max. Degree	2451	1246
Avg. $|T_v|$	161	166
$|P|$	12,773	16,154
Avg. $|V_p|$	14.98	8.89
Avg. $|T_p|$	5.35	9.7
Area	$37\,km \times 46\,km$	$71\,km \times 87\,km$
Max. Dist	$60\,km$	$112\,km$

characteristics of LV and PX. Note that LV contains more users in a smaller geographic area, whose distribution is skewed. Users and POIs in PX are distributed more uniformly.

8.2   Visualization

We qualitatively evaluate the proposed queries using LV. In the following visualizations, users and POIs are depicted as grey points and rectangles, respectively. Query points and top-k results are colored black, and each points to an information table that presents their parameters and partial scores.

Top-k Nearest, Popular and Relevant Users. Figure 5 illustrates the results of an NPRU query issued by a Mexican bar, where $T_q = \{mexican, alcohol, bar\}$, $k = 3$ and $\alpha_g = \alpha_s = \alpha_t = \frac{1}{3}$. The top-1 user is the closest to the query point, the most popular and the most relevant to T_q. Although the top-2 user is farther than top-3, he receives a better score because he has a higher degree and his preferences are more similar to T_q.

Top-k Nearest Socially and Textually Relevant POIs. Figure 6 depicts the results of an NSTP query issued by a user v, who searches for 3 nearby POIs that contain terms *"mexican, alcohol, bar"* and have been visited by his friends ($\alpha_g = \alpha_t = 0.25$ and $\alpha_s = 0.5$). The top-1 bar is 400 meters away from v, and has been visited by one of v's friends. The top-2 bar is $1.53\,km$ far from v, and has also been visited by one friend. Note that the top-3 bar has the highest textual similarity, but it is relatively far, and has not been visited by any of v's friends.

Frequent Social Keywords in Range. Figure 7 visualizes the results of an FSKR query, where a dashed-lined rectangle represents SR and $k = 1$. The top-1 keyword *"food"* is shared among 9 pairs of friends, connected by the bold edges. The remaining edges denote social connections of users in SR.

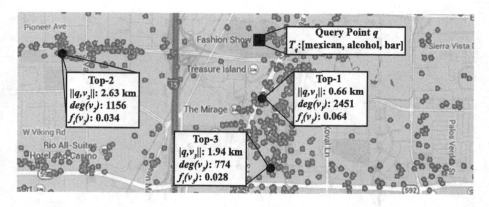

Fig. 5. Top-3 users in NPRU

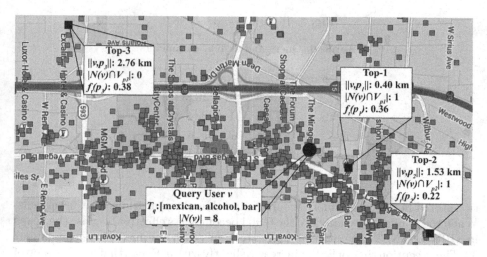

Fig. 6. Top-3 POIs in NSTP

8.3 Performance

The query processing algorithms were implemented in C++ under Linux Ubuntu, and executed on an Intel Xeon E5-2660 2.20 GHz with 8 GB RAM. All data and indices are stored in the main memory. The social graph is kept as a collection of adjacency lists, one per user. The reported times are the average of 20 query executions for each of LV and PX. Table 6 includes the tested value ranges for the query and system parameters in our setup; r corresponds to the radius of the circular spatial range SR of FSKR.

Geo-Social Keyword Index. Figure 8 studies the effect of GSKI granularity g on the running time of NPRU, NSTP, and FSKR using LV, for $h = 4$, $k = 16$, $|T_q| = 3$, and $r = 3\,km$. For granularity up to 5, the running time of NPRU and NSTP decreases with g. Since the cells cover smaller areas, the aggregate informa-

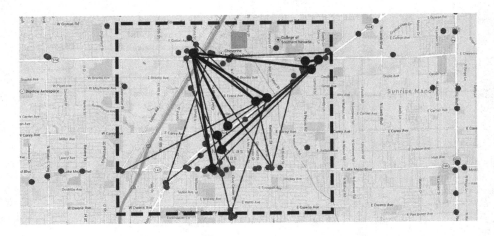

Fig. 7. Top-1 keyword in FSKR

Table 6. Query and system parameters

Parameter	Default	Range
k	16	4, 8, 16, 32, 64
$|T_q|$	3	1, 2, 3, 4, 5
g	5	3, 4, 5, 6
r (km)	3	1, 2, 3, 4, 5

tion stored in the cells is more accurate, and thus the algorithms visit fewer cells. When the granularity exceeds 5, the GSKI becomes less effective because the heaps in NPRU and NSTP maintain numerous cells, i.e., each intermediate cell has fanout 36. The execution time of FSKR increases slightly with g. Recall that the first step of FSKR creates the occurrence lists of terms in SR by merging the inverted files of the cells that intersect with SR. Consequently, the CPU time grows as the algorithm merges more inverted lists, but the impact is negligible. In the remaining experiments, we set $g = 5$ because it minimizes the execution time of NPRU, NSTP, and it marginally affects FSKR.

Table 7 assesses the total construction time of GSKI indices under the setup of Fig. 8 in both datasets. In the most challenging setting, i.e., $g = 6$ and $h = 4$ (1.6M leaf cells), GSKI needs only 45 s for both datasets since it only keeps concise aggregated data at the intermediate levels.

Top-k Nearest, Popular and Relevant Users. Figure 9(a) presents the query time of NPRU as a function of the result size k in LV and PX, for $|T_q| = 3$. In both datasets, the cost increases with k because the algorithm retrieves more users from the priority heap, and thus performs more iterations. NPRU is faster in PX because it contains relatively few users, who are rather uniformly

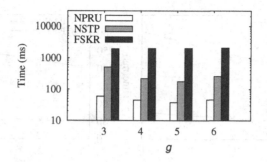

Fig. 8. Effect of GSKI granularity (LV Dataset, $h = 4$)

Table 7. GSKI construction time

Granularity g	Height h	# Leaf cells	LV Time (sec)	PX Time (sec)
3	4	6561	10.2	8.7
4	4	65536	13.3	11.6
5	4	390625	16.6	14.7
6	4	1679616	23.7	21.3

distributed. Therefore, the cells contain more accurate information that leads to better pruning.

Figure 9(b) plots the running time versus the number of queried terms, i.e., $|T_q|$, for $k = 16$. In both datasets, the cost increases with $|T_q|$ as the algorithm requires more computations to calculate the textual similarity of each visited cell or user. In addition, when $|T_q|$ increases, more cells become textually relevant to the query, reducing the pruning power of the algorithm.

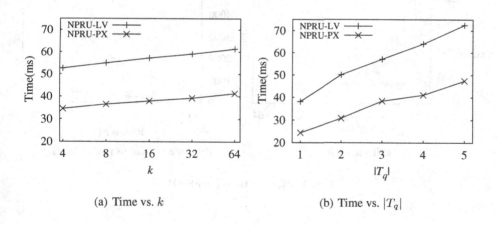

(a) Time vs. k

(b) Time vs. $|T_q|$

Fig. 9. Query time for NPRU

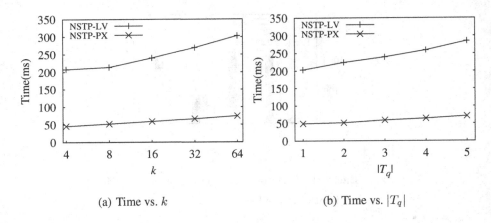

(a) Time vs. k

(b) Time vs. $|T_q|$

Fig. 10. Query time for NSTP

Top-k Nearest Socially and Textually Relevant POIs. Figure 10(a) shows the execution time of NSTP versus the result size k in LV and PX, for $|T_q| = 3$. Similar to NPRU, the running time increases with k since the algorithm executes more iterations. Compared to PX, the cost in LV increases faster because the distribution of POIs is highly skewed. This leads to inaccurate aggregate information at cells covering dense areas, burdening the reported average time. Figure 10(b) measures the running time as a function of $|T_q|$, for $k = 16$. The diagrams and the explanations are similar to those of Fig. 9(b).

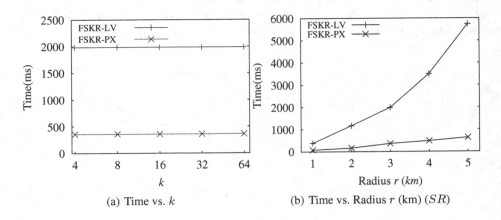

(a) Time vs. k

(b) Time vs. Radius r (km) (SR)

Fig. 11. Query time for FSKR

Frequent Social Keywords in Range. Figure 11(a) plots the running time of FSKR versus k, for $r = 3\,km$. Recall that FSKR initially creates the occurrence lists of the terms in SR by merging the inverted lists of the leaf cells that overlap SR. Then, the terms are sorted in decreasing order of list size. These steps dominate the total cost. Consequently, the value of k does not affect the execution time. FSKR is slower in LV since the average number of users in SR is greater, i.e., 2105 in LV and 464 in PX.

Figure 11(b) shows the execution time as a function of the radius r of SR, for $k = 16$. In both datasets the running time grows with r. In LV, the cost exhibits a steep increase because many new users are covered by the expanded SR. For instance, for $r = 4\,km$, SR includes on average 3662 users in LV and 627 in PX, whereas for $r = 5\,km$, it covers 6092 and 776 users, respectively.

Summarizing the experimental evaluation, all algorithms are very fast (at most, a few seconds) under all settings. In addition, the construction of GSKI only takes up to 23 s for the selected g, h, and the largest dataset. Finally, the GSKI supports efficient location updates as it is based on a grid structure.

9 Conclusion

This paper introduces a class of top-k queries that enable retrieval of users, POIs or keywords based on geographic, social and textual criteria. We propose three concrete queries that can be used in various tasks involving context-based search, profile-based advertisement and market analysis. For each query we provide a processing algorithm that exploits a specialized index. Our experiments with real datasets confirm the effectiveness and efficiency of the proposed methods.

An interesting direction for future work concerns additional GSK queries, applicable to different tasks. Even the same queries can be altered to support alternative partial scores. For instance, instead of the Euclidean, we could apply the road network distance to the definition of geographic score in NPRU and NSTP. Similarly, FSKR could be based on co-occurrences of terms in triangles (instead of pairs) of friends.

References

1. Facebook ads, audience targeting. https://www.facebook.com/help/229438340 403916
2. Google mobile maps. http://www.google.com/mobile/maps/
3. GroupOn Now! deals available on Foursquare. https://blog.groupon.com/cities/ groupon-now-deals-available-in-foursquare/
4. Yelp academic dataset. http://www.yelp.com/dataset_challenge/
5. Armenatzoglou, N., Papadopoulos, S., Papadias, D.: A general framework for geo-social query processing. Proc. VLDB Endow. **6**(10), 913–924 (2013)
6. Bloom, B.H.: Space/time trade-offs in hash coding with allowable errors. Commun. of ACM **13**(7), 422–426 (1970)
7. Cao, X., Cong, G., Jensen, C.S., Ooi, B.C.: Collective spatial keyword querying. In: SIGMOD (2011)
8. Chen, L., Cong, G., Jensen, C.S., Wu, D.: Spatial keyword query processing: an experimental evaluation. Proc. VLDB Endow. **6**(3), 217–228 (2013)

9. Chen, Y.-Y., Suel, T., Markowetz, A.: Efficient query processing in geographic web search engines. In: SIGMOD (2006)
10. Cong, G., Jensen, C.S., Wu, D.: Efficient retrieval of the top-k most relevant spatial web objects. Proc. VLDB Endow. 2(1), 337–348 (2009)
11. De Felipe, I., Hristidis, V., Rishe, N.: Keyword search on spatial databases. In: ICDE (2008)
12. Kalashnikov, D.V., Prabhakar, S., Hambrusch, S.E.: Main memory evaluation of monitoring queries over moving objects. Distrib. Parallel Databases 15(2), 117–135 (2004)
13. Kargar, M., An, A.: Discovering top-k teams of experts with/without a leader in social networks. In: CIKM (2011)
14. Kargar, M., An, A.: Keyword search in graphs: finding r-cliques. Proc. VLDB Endow. 4(10), 681–692 (2011)
15. Khodaei, A., Shahabi, C., Li, C.: Hybrid indexing and seamless ranking of spatial and textual features of web documents. In: Bringas, P.G., Hameurlain, A., Quirchmayr, G. (eds.) DEXA 2010, Part I. LNCS, vol. 6261, pp. 450–466. Springer, Heidelberg (2010)
16. Lappas, T., Liu, K., Terzi, E.: Finding a team of experts in social networks. In: SIGKDD (2009)
17. Liu, W., Sun, W., Chen, C., Huang, Y., Jing, Y., Chen, K.: Circle of friend query in geo-social networks. In: Lee, S., Peng, Z., Zhou, X., Moon, Y.-S., Unland, R., Yoo, J. (eds.) DASFAA 2012, Part II. LNCS, vol. 7239, pp. 126–137. Springer, Heidelberg (2012)
18. Long, C., Wong, R.C.-W., Wang, K., Fu, A.W.-C.: Collective spatial keyword queries: a distance owner-driven approach. In: SIGMOD (2013)
19. Mouratidis, K., Li, J., Tang, Y., Mamoulis, N.: Joint search by social and spatial proximity. IEEE Trans. Knowl. Data Eng. 10(16), 1169–1184 (2015)
20. Vaid, S., Jones, C.B., Joho, H., Sanderson, M.: Spatio-textual Indexing for geographical search on the web. In: Medeiros, C.B., Egenhofer, M., Bertino, E. (eds.) SSTD 2005. LNCS, vol. 3633, pp. 218–235. Springer, Heidelberg (2005)
21. Otto, A., Kaulfersch, E., Brinkfeldt, K., Neumaier, K., Zschieschang, O., Andersson, D., Rzepka, S.: Reliability of new SiC BJT power modules for fully electric vehicles. In: Fischer-Wolfarth, J., Meyer, G. (eds.) Advanced Microsystems for Automotive Applications 2014. LNMOB, vol. 1, pp. 235–244. Springer, Heidelberg (2014)
22. Yang, D.-N., Shen, C.-Y. Lee, W.-C., Chen, M.-S.: On socio-spatial group query for location-based social networks. In: KDD (2012)
23. Zhang, D., Chee, Y.M., Mondal, A., Tung, A., Kitsuregawa, M.: Keyword search in spatial databases: towards searching by document. In: ICDE (2009)
24. Zhou, Y., Xie, X., Wang, C., Gong, Y., Ma. W.-Y.: Hybrid index structures for location-based web search. In: CIKM (2005)
25. Zobel, J., Moffat, A.: Inverted files for text search engines. ACM Comput. Surv. 38(2), 6.1–6.56 (2006)

RCP Mining: Towards the Summarization of Spatial Co-location Patterns

Bozhong Liu[1,2(✉)], Ling Chen[2], Chunyang Liu[2], Chengqi Zhang[2], and Weidong Qiu[1]

[1] School of Information Security and Engineering,
Shanghai Jiao Tong University, Shanghai, China
bozhong.liu@student.uts.edu.au
[2] Centre for Quantum Computation and Intelligent Systems,
University of Technology, Sydney, Australia

Abstract. Co-location pattern mining is an important task in spatial data mining. However, the traditional framework of co-location pattern mining produces an exponential number of patterns because of the downward closure property, which makes it hard for users to understand, or apply. To address this issue, in this paper, we study the problem of mining *representative co-location patterns* (RCP). We first define a covering relationship between two co-location patterns by finding a new measure to appropriately quantify the distance between patterns in terms of their prevalence, based on which the problem of RCP mining is formally formulated. To solve the problem of RCP mining, we first propose an algorithm called *RCPFast*, adopting the *post-mining* framework that is commonly used by existing distance-based pattern summarization techniques. To address the peculiar challenge in spatial data mining, we further propose another algorithm, *RCPMS*, which employs the *mine-and-summarize* framework that pushes pattern summarization into the co-location mining process. Optimization strategies are also designed to further improve the performance of *RCPMS*. Our experimental results on both synthetic and real-world data sets demonstrate that RCP mining effectively summarizes spatial co-location patterns, and *RCPMS* is more efficient than *RCPFast*, especially on dense data sets.

1 Introduction

As one of the most fundamental tasks in spatial data mining, *co-location mining* aims to discover co-location patterns where each is a group of spatial features whose instances are frequently located close to each other [14]. Spatial co-location patterns yield important insights for various applications such as epidemiology [21], ecology [4] and e-commerce [22]. A common framework of co-location pattern mining uses the frequencies of a set of spatial features participating in a co-location to measure the prevalence (known as *participation index* [14], or *PI* for short) and requires a user-specified minimum threshold to find interesting patterns. Typically, if the threshold is high, the framework may generate commonsense patterns. However, with a low threshold, a great number of patterns

© Springer International Publishing Switzerland 2015
C. Claramunt et al. (Eds.): SSTD 2015, LNCS 9239, pp. 451–469, 2015.
DOI: 10.1007/978-3-319-22363-6_24

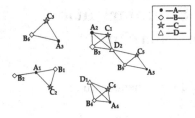

Table 1. Prevalent patterns.

ID	Feature Sets	Events	PI
F_1	$\{A, B\}$	A_1B_1, A_1B_2, A_2B_3	1
		A_4B_4, A_5B_5, A_3B_6	
F_2	$\{A, B, C\}$	$A_1B_1C_2, A_2B_3C_1$	5/6
		$A_3B_6C_3, A_4B_4C_4, A_5B_5C_5$	
F_3	$\{A, B, C, D\}$	$A_2B_3C_1D_2, A_4B_4C_4D_1$	1/3
F_4	$\{B, C, D\}$	$B_3C_1D_2, B_4C_4D_1, B_5C_5D_2$	1/2
F_5	$\{C, D\}$	C_1D_2, C_4D_1, C_5D_2	3/5

Fig. 1. A motivating example.

will be found. This is further exacerbated by the downward closure property that holds for the PI measure. That is, if a set of features is prevalent with respect to a threshold of PI, then all of its subsets will be discovered as prevalent co-location patterns. A huge pattern number will jeopardize the usability of resulted patterns, as it demands great efforts to understand or examine the discovered knowledge.

The key idea of solving this problem is to find an effective way to summarize the co-location patterns, e.g., to find a high-quality representation that describes the complete set of resulted patterns precisely and concisely. Two types of compressed co-location patterns have been explored in the literature: *maximal co-location patterns* (MCP) [15] and *closed co-location patterns* (CCP) [19]. A co-location pattern is a MCP if it is prevalent itself and none of its super-patterns are prevalent. MCP mining may significantly reduce the number of co-location patterns, but it fails to preserve the prevalence information. It is therefore a lossy approximation. As for the second type, a co-location pattern is a CCP if it is prevalent itself and none of its super-patterns have the same PI as it does. CCP mining not only diminishes the number of co-location patterns but also preserves the complete PI information. However, by emphasizing too much on the PI information, the compression power of CCP mining is limited.

For example, given a spatial data set shown in Fig. 1, where instances/events of four spatial features, A, B, C and D, are represented by different symbols and edges connecting events denote spatial neighborhood relationships, Table 1 lists a set of five prevalent co-location patterns and their corresponding PI in the data set (the definition of PI is provided in Sect. 2). If MCP mining is adopted, only F_3 will be output as the others are all sub-patterns of F_3. However, F_3 is significantly different from others in terms of their PIs. In contrast, if CCP mining is used, then all patterns will be returned since each of them is a closed pattern. That is, CCP mining provides no compression on this set of patterns. Therefore, to address the limitations of MCP and CCP mining, a method that not only provides optimal compression rate but also preserves reasonable prevalence information will be favored.

Similar idea has been explored in the studies of summarizing frequent itemsets [10,17,18]. Xin et al. [17] proposed the notion of a ϵ-*cover relationship* between itemsets. An itemset X_1 is ϵ-covered by another itemset X_2 if X_1 is a subset of X_2 and $1 - \frac{|T(X_1) \cap T(X_2)|}{|T(X_1) \cup T(X_2)|} \leq \epsilon$, where $T(X_i)$ is the set of supporting transactions of pattern X_i. The goal is then to find a minimum set of representative itemsets that can ϵ-cover all frequent itemsets. In this paper, we follow their idea and propose to summarize co-location patterns using a set of *representative co-location patterns* (RCPs), which strikes a fine balance between improving compression rate and preserving prevalence information.

However, existing methods for representative itemsets mining cannot be applied directly to representative co-location pattern mining, neither the framework of problem definition nor the mining process. This is mainly because there is no natural notion of *transactions* in co-location mining [14]. Consequently, the original definition of the ϵ-cover relationship cannot be adopted straightforwardly because it is defined on a supporting transaction-based distance measure. Moreover, the mining process will be more complicated as it is more expensive to examine whether a set of feature instances participate in a co-location than checking whether a set of items appear in one transaction.

To formulate the problem of representative co-location pattern mining, we first define a new measure to appropriately quantify the distance between two co-location patterns in terms of their prevalence, based on which the ϵ-cover relationship can be stated on a pair of co-location patterns. To solve the problem of RCP mining, we first propose an algorithm, *RCPFast*, which follows existing distance-based pattern summarization techniques to adopt the *post-mining* framework that finds RCPs from the set of discovered co-location patterns. Observing a peculiar challenge in spatial data mining, we then develop another algorithm, called *RCPMS*, which employs a *mine-and-summarize* framework to discover RCPs directly from the spatial data. To our knowledge, *RCPMS* is the first work among existing distance-based pattern summarization that pushes summarization into the pattern mining process. Optimization strategies are also devised to further improve the efficiency of *RCPMS*. We evaluate the performance of the developed algorithms on both synthetic and real-world data sets. Our experimental results demonstrate the effectiveness of RCP mining, and the efficiency of *RCPMS* compared with *RCPFast*, especially on dense data sets.

The remainder of this paper is organized as follows. In Sect. 2, we define relevant concepts and formally formulate the problem. Section 3 introduces the *RCPFast* algorithm. Section 4 describes the *RCPMS* algorithm and optimization strategies. In Sect. 5, we evaluate the performance of the developed algorithms. Existing works related to our research are reviewed in Sect. 6. Section 7 closes this paper with some conclusive remarks.

2   Preliminary

In this section we first review definitions related to traditional co-location patterns. Then, we introduce a distance metric to measure the prevalence difference between two patterns. Finally, we formally define the problem of representative co-location pattern mining.

2.1 Co-Location Patterns

Given a set of spatial features $\mathcal{F} = \{f_1, f_2, \ldots, f_K\}$, a spatial data set is a collection of instances/events $\mathcal{E} = \{e_1, e_2, \ldots, e_N\}$, where each $e_i \in \mathcal{E}$ is represented by a vector $\langle event\_id, \ spatial\_feature\_type, \ location \rangle$. We review the measures used to characterize the interestingness of a subset of features $F \subseteq \mathcal{F}$ as follows. Please refer to [14] for the details.

Definition 1. *Given* $F = \{f_1, \ldots, f_k\} \subseteq \mathcal{F}$, $E = \{e_1, \ldots, e_k\} \subseteq \mathcal{E}$ *is a **Row Instance (RI)** of* F, *denoted as* $RI(F)$, *if* $\forall i \in [1, k]$, e_i *is an instance of* f_i *and* $\forall i, j \in [1, k], \|e_i - e_j\| \leq \tau$, *where* $\|e_i - e_j\|$ *refers to the spatial distance between two events and* τ *is a user-specified spatial distance threshold.*

Definition 2. *Given a spatial data set* $\mathcal{E}$ *of a set of spatial features* $\mathcal{F}$, *the **Table Instance (TI)** of a subset of features* $F \subseteq \mathcal{F}$, *denoted as* $TI(F)$, *is the collection of all its row instances in* $\mathcal{E}$. *That is,* $TI(F) = \{RI_1(F), \ldots, RI_m(F)\}$.

For example, consider the spatial data set in Fig. 1 and $F_5 = \{C, D\}$ in Table 1. $\{C_1 D_2\}$ is a RI of F_5. $TI(F_5) = \{C_1 D_2, C_4 D_1, C_5 D_2\}$.

Definition 3. *Given* $F = \{f_1, \ldots, f_k\}$, *the **Participation Ratio** of a feature* $f_i \in F$, *denoted as* $PR(f_i, F)$, *is the fraction of events of feature* f_i *that participate in the table instance of* F. *That is,*

$$PR(f_i, F) = \frac{|\{e_j | e_j \in TI(\{f_i\}), e_j \in \widehat{TI}(F)\}|}{|TI(\{f_i\})\}|}, \tag{1}$$

where $\widehat{TI}(\cdot)$ *is the union of elements in TI set.*

Hence, the denominator refers to the total number of events of feature f_i and the numerator refers to the number of distinct events of feature f_i that appear in the table instance of F.

Definition 4. *The* ***Participation Index*** *of a subset of features* $F = \{f_1, \ldots, f_k\}$, *denoted as* $PI(F)$, *is defined as* $PI(F) = \min_{i \in [1,k]} PR(f_i, F)$.

For example, consider the spatial data set in Fig. 1 and $F_2 = \{A, B, C\}$ in Table 1. Since $PR(A, F_2) = 5/5$, $PR(B, F_2) = 5/6$, $PR(C, F_2) = 5/5$, we have $PI(F_2) = \min(5/5, 5/6, 5/5) = 5/6$.

Definition 5. *Given a user-specified threshold minpi, a subset of features* $F \subseteq \mathcal{F}$ *is a **Prevalent Co-location Pattern (PCP)** if* $PI(F) \geq minpi$.

2.2 Co-Location Distance Measure

A distance measure between traditional frequent itemsets has been proposed in [17]. It compares the supporting transactions of two itemsets and deduces a numerical value as $D(I_1, I_2) = 1 - \frac{|T(I_1) \cap T(I_2)|}{|T(I_1) \cup T(I_2)|}$, where $T(I_i)$ denotes the set of transactions supporting the itemset I_i. However, it is difficult to apply this measure to co-location patterns because there is no natural notion of transactions in co-location mining [14]. We explore a new distance measure that appropriately quantifies the prevalence difference between two co-location patterns which can be computed efficiently without manipulating the spatial data set.

For simplicity, we denote the set in the numerator of Eq. (1) as $E_F(f_i)$ (i.e., $E_F(f_i) = \{e_j | e_j \in TI(\{f_i\}), e_j \in \widehat{TI}(F)\}$). It refers to the set of events of feature f_i that participate in the table instance of F.

Definition 6. *Let F_1 and F_2 be two co-location patterns and f be a feature shared by them, namely, $f \in F_1 \cap F_2$, the **Feature Distance** between F_1 and F_2 w.r.t. f is defined as*

$$FD_f(F_1, F_2) = 1 - \frac{|E_{F_1}(f) \cap E_{F_2}(f)|}{|E_{F_1}(f) \cup E_{F_2}(f)|} \tag{2}$$

Definition 7. *Given two co-location patterns F_1 and F_2, the **Co-location Distance** between them is defined as*

$$D(F_1, F2) = \begin{cases} \max\limits_{\forall f \in F_1 \cap F_2} FD_f(F_1, F_2), & \text{if } F_1 \cap F_2 \neq \emptyset \\ 1, & \text{otherwise} \end{cases} \tag{3}$$

Let us apply this new distance measure to co-location patterns in Table 1 to see if it reasonably reflects the distance/proximity between patterns in terms of their prevalence. Firstly, we consider $F_1 = \{A, B\}$ and $F_2 = \{A, B, C\}$. According to the above definitions, $E_{F_1}(A) = E_{F_2}(A) = \{A_1, A_2, \cdots, A_5\}$, $E_{F_1}(B) = \{B_1, B_2, \cdots, B_6\}$, $E_{F_2}(B) = \{B_1, B_3, B_4, B_5, B_6\}$, then $FD_A(F_1, F_2) = 1 - \frac{|E_{F_2}(A)|}{|E_{F_1}(A)|} = 1 - \frac{5}{5} = 0$, $FD_B(F_1, F_2) = 1 - \frac{5}{6} = \frac{1}{6}$. Hence, $D(F_1, F_2) = \max(0, \frac{1}{6}) = \frac{1}{6}$. This small distance value suggests that F_1 and F_2 are quite similar in terms of prevalence. Similarly, let us consider $F_2 = \{A, B, C\}$ and $F_3 = \{A, B, C, D\}$. We can have $D(F_2, F_3) = \max(1 - \frac{2}{5}, 1 - \frac{2}{5}, 1 - \frac{2}{5}) = \frac{3}{5}$, which indicates that the two patterns (F_2, F_3) are quite different. We observe that the new distance measure captures the prevalence distance between co-location patterns appropriately.

2.3   Problem Statement

Based on the proposed distance measure, we define the ϵ-cover relationship between two co-location patterns as follows.

Definition 8. *Given two co-location patterns F_1 and F_2, and a real number $\epsilon \in [0, 1]$, we say F_2 ϵ-covers F_1 if (1) $F_1 \subseteq F_2$ and (2) $D(F_1, F_2) \leq \epsilon$.*

Then, given a set of prevalent co-location patterns, we can group them into ϵ-clusters, where each ϵ-cluster consists of a centroid pattern F_r that ϵ-covers all patterns in the cluster. It seems that we may return centroid patterns of ϵ-clusters as representative patterns. However, by doing so, we restrict the representative patterns to be prevalent themselves (i.e. $PI(F_r) \geq minpi$). The minimum number of representative co-location patterns that can be achieved using this method is the number of MCPs.

In [17], it shows that an itemset only needs to satisfy a relaxed condition (i.e., $Supp(X) \geq (1 - \epsilon) * minsupp$) to ϵ-cover a frequent itemset. We find that this property holds as well for our newly defined distance measure and the induced ϵ-cover relationship. The details are given in the long version of this paper [8].

Thus, to ϵ-cover a prevalent co-location pattern F_1, F_2 only needs to be prevalent with respect to a lower threshold $minpi^* = (1-\epsilon) * minpi$. Our experimental results in Sect. 5 show that this relaxation contributes to an improved compression rate.

Definition 9 (Problem Statement). *Given a set of spatial features $\mathcal{F}$, a spatial data set $\mathcal{E}$ on $\mathcal{F}$, a spatial distance threshold τ, a co-location distance threshold ϵ, and a prevalence threshold $minpi$, the problem of representative co-location pattern mining is to discover a minimal set of co-location patterns $\mathcal{R}$ such that: (1) For all $F_r \in \mathcal{R}, PI(F_r) \geq (1 - \epsilon) * minpi$; (2) For any prevalent co-location patterns F, i.e., $PI(F) \geq minpi$, there exits a $F_r \in \mathcal{R}$ s.t. F_r ϵ-covers F.*

3   The *RCPFast* Algorithm

In this section, we first introduce an algorithm, *RCPFast*, which follows existing distance-based pattern summarization approaches to mine RCPs by adopting a *post-mining* framework.

Similar to [17], the mining framework of *RCPFast* consists of three stages. Stage 1 discovers two sets of prevalent co-location patterns, PCP and PCP^*, with respect to $minpi$ and $(1 - \epsilon) * minpi$, respectively. The objective is then to select minimal number of patterns from PCP^* to cover all patterns in PCP. Stage 2 generates the complete coverage information by finding all prevalent co-location patterns $F \in PCP$ that can be ϵ-covered by each pattern $F_r \in PCP^*$. All prevalent co-location patterns ϵ-covered by F_r is stored in $set(F_r)$. Stage 3 finds the set of desired RCPs based on the coverage information. As discussed in [17], it can be solved by a greedy strategy that always selects the representative pattern that covers the most number of prevalent co-location patterns. According to [3], this is a set cover problem which is NP-hard and the time complexity of the greedy algorithm is $O(\sum_{F_r \in PCP^*} |set(F_r)|)$. Hence the computational cost of *RCPFast* mainly comes from the first two stages.

Since the time complexity of the greedy algorithm is $O(\sum_{F_r \in PCP^*} |set(F_r)|)$, the computational cost of *RCPFast* mainly comes from the first two stages.

For the first stage, mining prevalent co-locations is a well-studied topic. Many efficient algorithms have been proposed, e.g., the spatial-join method [5] and the join-less method [21]. Note that it is unnecessary to run the mining process twice to discover PCP and PCP^*. We can find prevalent patterns w.r.t. $(1-\epsilon) * minpi$ first, and then filter the results to obtain those prevalent w.r.t. $minpi$.

For the second stage, the bottleneck lies in the computations of co-location distance between two patterns to verify the ϵ-cover relationship. The complexity

Algorithm 1. *RCPFast*

Input: $(1)\mathcal{E}$, $(2)\tau$, $(3)minpi$, $(4)\epsilon$.
Output: The set of RCPs $\mathcal{R}$
1: $PCP = \text{MinePCP}(\mathcal{E}, \tau, minpi)$
2: $PCP^* = \text{MinePCP}(\mathcal{E}, \tau, (1-\epsilon)*minpi)$
3: Sort PCP^* in decreasing order by pattern length, and PCP in increasing order.
4: **for all** $F_r \in PCP^*$ **do**
5:     **for all** $F \in PCP$ **do**
6:         **if** $F \subseteq F_r$ **then**
7:             Insert F into $CandList(F_r)$
8: **for all** $F_r \in PCP^*$ **do**
9:     **for all** $F \in CandList(F_r)$ **do**
10:         **if** F_r ϵ-covers F **then**
11:             Insert F into $set(F_r)$
12:             Find a set of patterns $\mathcal{Q} \subseteq PCP^*$ s.t. $\forall Q \in \mathcal{Q}$, $F \subseteq Q \subseteq F_r$
13:             **for all** $Q \in \mathcal{Q}$ **do**
14:                 Remove F from $CandList(Q)$ to $set(Q)$
15: **while** $PCP \neq \emptyset$ **do**
16:     Find a F_r that maximizes $|set(\mathcal{R})|$
17:     **for all** $F \in set(F_r)$ **do**
18:         Delete F from PCP
19:     $\mathcal{R} = \mathcal{R} \cup \{F_r\}$
20: Return $\mathcal{R}$

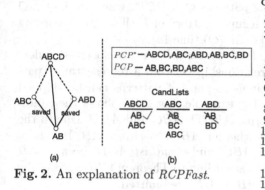

$PCP^* -$ ABCD,ABC,ABD,AB,BC,BD
$PCP -$ AB,BC,BD,ABC

CandLists

ABCD	ABC	ABD
AB✓	AB	AB
ABC	BC	BD
	ABC	

(a)   (b)

Fig. 2. An explanation of *RCPFast*.

of generating the complete coverage information is $O(|PCP| * |PCP^*|)$, which will become a performance issue when there are many prevalent patterns. Therefore, we aim to exploit strategies to skip verifying the ϵ-cover relationship for as many pairs of patterns as possible.

Theorem 1. *Given three co-location patterns F_1, F_2, and F_3 s.t. $F_1 \subseteq F_2 \subseteq F_3$, if F_3 ϵ-covers F_1, then F_2 ϵ-covers F_1.*

Proof. From $D(F_1, F_3) \leq \epsilon$, we have $\forall f \in F_1$, $1 - \frac{|E_{F_3}(f)|}{|E_{F_1}(f)|} \leq \epsilon$. Because $\forall f \in F_1$, $|E_{F_2}(f)| \geq |E_{F_3}(f)|$. Thus, we have $1 - \frac{|E_{F_2}(f)|}{|E_{F_1}(f)|} \leq 1 - \frac{|E_{F_3}(f)|}{|E_{F_1}(f)|} \leq \epsilon$, or $D(F_1, F_2) \leq \epsilon$, which proves the result. □

According to the theorem, we are allowed to skip computing co-location distance for certain pairs of co-location patterns. For example, as shown in Fig. 2 (a), if we have found that $\{A, B, C, D\}$ ϵ-covers $\{A, B\}$, then we conclude immediately that $\{A, B, C\}$ ϵ-covers $\{A, B\}$ and $\{A, B, D\}$ ϵ-covers $\{A, B\}$ without computing their corresponding co-location distances. To maximize the benefit introduced by Theorem 1, we order the co-location patterns according to pattern lengths. Then, the procedure of *RCPFast* is illustrated in Algorithm 1.

Algorithm 1 follows the three-stage framework. The first stage (lines 1–2) mines two sets of prevalent co-location patterns and the third stage (lines 15–19) discovers the RCPs using a greedy strategy. The second stage starts with sorting the patterns in PCP^* in decreasing order of pattern length, and sorting patterns in PCP in the reverse order (line 3). Then, a candidate list (*CandList*) is constructed for each representative pattern F_r in PCP^*, which stores all prevalent patterns that may be ϵ-covered by F_r (lines 4–7). Lines 8–14 find the complete coverage information for each pattern F_r in PCP^* by implementing the optimization enabled by Theorem 1. In particular, once it is confirmed that F_r ϵ-covers a prevalent pattern F (line 10), we find a set of patterns $\mathcal{Q} \subseteq PCP^*$ where each $Q \in \mathcal{Q}$ is a sub-pattern of the current F_r and a super-pattern of F (line 12). According to Theorem 1, F can be directly added to $set(Q)$ (line 14).

Note that, the purpose of sorting PCP and PCP^* in the specified orders is to allow early discovery of ϵ-cover relationship between a representative pattern and its short sub-patterns so that more pairs of patterns can be skipped for co-location distance computation. For example, Fig. 2 (b) shows the candidate lists of three representative patterns, $ABCD$, ABC and ABD. Due to the ordering of patterns, we examine first whether $ABCD$ ϵ-covers AB. If it happens, we can delete AB from $CandList(ABC)$ and $CandList(ABD)$ because AB should be covered by these two patterns according to Theorem 1. Therefore, the computations of $D(ABC, AB)$ and $D(ABD, AB)$ are omitted.

4   The *RCPMS* Algorithm

Recall that, to verify the ϵ-cover relationship in the second stage of *RCPFast*, we need to compute the co-location distance between two patterns, which requires the *table instance* information of the corresponding patterns. However, the output of prevalent co-location pattern mining in the first stage contains only the prevalent patterns as well as their *PI* information. It may not be an issue for frequent itemset summarization as the supporting transactions of an itemset can be retrieved easily. However, for spatial data mining, it is expensive to re-scan the data to obtain the table instance of a co-location pattern whenever it is required. One possible solution is to output the information of table instances as additional results. However, if the information is stored in disk, extra I/O cost will be incurred. If the information is stored in memory, it will become problematic when the number patterns is huge. Therefore, we are motivated to push coverage validation into the co-location mining process, thereby integrating the first two stages in order to address the table instance acquisition problem.

Based on the idea, we devise the *RCPMS* algorithm that employs a novel *mine-and-summarize* framework, while all existing distance-based pattern summarization techniques adopt the *post-mining* paradigm. More specifically, whenever a representative pattern, prevalent w.r.t. $(1 - \epsilon) * minpi$, is discovered, all prevalent patterns, w.r.t. *minpi*, which can be ϵ-covered by it will be found. The feasibility of this idea is supported by the following two facts.

(1) Traditional prevalent co-location pattern mining algorithms usually use an Apriori-based level-wise scheme to generate patterns [5, 21]. When a representative pattern is mined, all its prevalent sub-patterns have already been found. Hence, it is sufficient to find the coverage information for the current representative pattern.

Algorithm 2. RCPMS

Input: Same as *RCPFast*.
Output: Same as *RCPFast*.
1: $P_1 = \mathcal{F}, k = 2$
2: **while** $P_{k-1} \neq \emptyset$ **do**
3:    $C_k = $ gen_candidate_colo(P_{k-1})
4:    **for all** $C \in C_k$ **do**
5:       $pi = $ calculate_PI(C)
6:       **if** $pi \geq (1 - \epsilon) * minpi$ **then**
7:          $D\_Table \leftarrow$ cal_preval_child_dis(C)
8:          $set(C) = $ gen_cover_set$(C, C, 0)$
9:          **if** $pi \geq minpi$ **then**
10:             Insert C into P_k and $set(C)$
11:    $k = k + 1$
12: Obtain RCPs using the greedy algorithm

Algorithm 3. gen_cover_set(F_r, F, dis)

Input: F_r: current RCP; F: a sub-pattern;
   dis: accumulated distance
Output: S: all prevalent patterns ϵ-covered
   by F_r
1: **for all** $P \subset F$ s.t. $|F| - |P| = 1$ &
   $PI(P) \geq minpi$ **do**
2:    $dis = dis + $ TableLookup(P, F)
3:    **if** $dis \leq \epsilon$ **then**
4:       Insert P to S
5:       gen_cover_set(F_r, P, dis)
6:    **else**
7:       $dis = D(F_r, P)$
8:       **if** $dis \leq \epsilon$ **then**
9:          Insert P to S
10:          gen_cover_set(F_r, P, dis)
11: **return** S

(2) When a representative pattern is output in the mining process, its information of table instance is available, which can be used to compute its co-location distances with its sub-patterns. For its sub-patterns, we store their table instance information in memory if they are child/immediate sub-patterns of the current representative pattern (e.g., $F \subset F_r$ and $|F_r| - |F| = 1$). Otherwise, we will retrieve the table instance information of a sub-pattern F (e.g. $F \subset F_r$ and $|F_r| - |F| > 1$) only if the ϵ-cover relationship between F_r and F cannot be inferred using our devised optimization and approximation strategies, which will be discussed in Subsects. 4.1 and 4.2.

The general idea of *RCPMS* is summarized in Algorithm 2. In the beginning, it assigns all unique spatial features to P_1 (line 1). From line 2 to line 11, an iterative process is used to generate patterns of length k from patterns of length $k - 1$. In particular, line 3 calls the function *gen_candidate_colo* to generate candidate co-location patterns (e.g., using an Apriori-like strategy). For each candidate co-location pattern, we first calculate its PI (line 5). If the candidate pattern is prevalent w.r.t. $(1 - \epsilon) * minpi$ (line 6), we compute its co-location distances with its child sub-patterns which are prevalent w.r.t. $minpi$ and store it in a distance table (line 7). Note that, only the co-location distances between the current pattern and its prevalent child sub-patterns need to be computed at this stage. As discussed later, its co-location distances with other descendent prevalent sub-patterns will be computed only if they can't be inferred using our proposed optimization and approximation strategies. In line 8, we call the method *gen_cover_set* to find all prevalent sub-patterns that can be covered by the current representative pattern. Finally, if the current pattern is prevalent w.r.t. *minpi*, it should be used to generate candidate patterns in the next round and should be included into its own cover set (lines 9 and 10). Line 12 is the same as the third stage of *RCPFast* which finds the minimal RCPs.

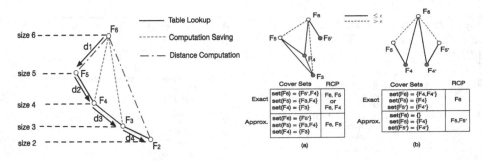

Fig. 3. Optimization strategy   **Fig. 4.** Approximation strategy.

In the following, we describe the details of the function *gen_cover_set* which finds all prevalent sub-patterns that can be ϵ-covered by the current representative pattern. Before presenting the function, we first introduce an optimization strategy and an approximation strategy that are used by the function.

4.1   Optimization Strategy

Note that, the optimization strategy used by *RCPFast* (i.e., *Theorem* 1) is not applicable here. This is because when we output a representative pattern of length k in *RCPMS*, the coverage information of representative patterns of length $(k-1)$ has already been found. Therefore, we exploit a new optimization strategy based on the following theorem:

Theorem 2. *Given three co-location patterns F_1, F_2 and F_3 s.t. $F_1 \subseteq F_2 \subseteq F_3$, $D(F_1, F_2) + D(F_2, F_3) \geq D(F_1, F_3)$.*

Due to the space constrain, the proof of Theorem 2 is given in [8].

Figure 3 illustrates how to use Theorem 2 to skip computing co-location distance between a representative pattern and its non-child prevalent sub-patterns. Suppose F_6 is the current representative pattern and we have computed its co-location distance with its child sub-pattern F_5, $D(F_6, F_5) = d_1$, stored in the *D_Table* (e.g., line 7 in Algorithm 2). Next, we need to examine whether F_6 ϵ-covers F_5's child, e.g., F_4. Note that $D(F_5, F_4) = d_2$ should have been computed and stored in the *D_Table* when outputting F_5 in the previous round. According to Theorem 2, we infer that $D(F_6, F_4) < D(F_6, F_5) + D(F_5, F_4) = d_1 + d_2$. Therefore, if $d_1 + d_2 \leq \epsilon$, we can conclude that F_6 ϵ-covers F_4 without computing $D(F_6, F_4)$. Similarly, when examining whether F_6 ϵ-covers F_3, which is a child sub-pattern of F_4, we have $D(F_6, F_3) < D(F_6, F_5) + D(F_5, F_4) + D(F_4, F_3) = d_1 + d_2 + d_3$. As indicated in the figure, $d_1 + d_2 + d_3 \leq \epsilon$, we conclude that F_6 ϵ-covers F_3 and skip computing the distance $D(F_6, F_3)$. When it comes to F_2, since $d_1 + d_2 + d_3 + d_4 > \epsilon$, we have to compute the exact value of $D(F_6, F_2)$ (we will have to re-gain the table instance of F_2 in this case). Therefore, in this particular example, Theorem 2 enables us to skip two of the three co-location distance computations (i.e., $D(F_6, F_4)$, $D(F_6, F_3)$ and $D(F_6, F_2)$).

4.2   Approximation Strategy

Although Theorem 2 can reduce distance computation for a certain number of pairs of patterns, the effectiveness of this single strategy may not be sufficient. Therefore, we further exploit an approximation strategy which substantially improves the computation efficiency by slightly sacrificing the compression rate.

Recall that, in Fig. 3, only if the co-location distance between the current representative pattern (e.g., F_6) and its child prevalent sub-pattern (e.g., F_5) is smaller than ϵ, we may use the optimization strategy to infer the distance between F_6 and F_4 (F_3). Otherwise, we have to compute the distance between F_6 and F_4 (F_3), which is expensive since we have to re-gain the table instance of F_4 (F_3). Therefore, we consider the following approximation strategy.

If a representative pattern F_r cannot ϵ-cover its prevalent child sub-pattern F, we skip considering whether F_r ϵ-covers any descendant sub-pattern of F.

For example, in Fig. 3, if the co-location distance between F_6 and F_5 is greater than ϵ, all F_4, F_3 and F_2 will not be included in $set(F_6)$.

Figure 4 provides two examples to illustrate the influence of the approximation strategy. In Fig. 4 (a), let's assume the set of PCP that need to be summarized are F_5', F_4 and F_3, where F_3 is a child sub-pattern of F_4, which is a child sub-pattern of F_5. F_5' is a sibling pattern of F_5 (e.g., ABC and ABD). The exact coverage information shows that F_6 ϵ-covers F_4. However, since F_6 does not ϵ-cover F_5, F_4 is removed from $set(F_6)$ according to the approximation strategy. If using the greedy algorithm to find RCPs, the final number of RCPs found from the exact cover sets will be 2, which is the same as the final number of RCPs found from the approximate cover sets. It indicates that the approximation strategy does not incur any difference to the final number of RCPs under this situation.

In contrast, Fig. 4 (b) shows an example where this approximation strategy will result in difference in the final number of RCPs. In this example, suppose the set of PCP are F_4 and F_4'. The complete coverage information shows that F_6 ϵ-covers both F_4 and F_4'. Since F_6 does not ϵ-cover F_5 or F_5', F_4 and F_4' are removed from $set(F_6)$ in the approximate cover sets. Consequently, the final number of RCPs found from the exact cover sets is 1 while the final number found from the approximate cover sets is 2.

In general, we have the following lemma, implying that the final number of RCPs generated from the incomplete cover sets, produced by the approximation strategy, will be no smaller than the final number of RCPs generated from the complete cover sets.

Lemma 1. *Let $\mathcal{P}$ be a set of representative patterns with non-empty cover sets. That is, $\mathcal{P} \subseteq PCP^*$ and $\forall P \in \mathcal{P}$, $|set(P)| > 0$. Let $\mathcal{P}'$ be a set of representative patterns with non-empty cover sets found using the approximation strategy. Then we have (1) $\mathcal{P}' \subseteq \mathcal{P}$ and $\forall P \in \mathcal{P}'$, $|set(P)| \geq |set'(P)|$, where $set'(P)$ represents the cover set generated by the approximation strategy. (2) let $\mathcal{R}$ and $\mathcal{R}'$ be the minimum sets of RCPs generated from $\mathcal{P}$ and $\mathcal{P}'$, respectively, i.e., $\mathcal{R} \subseteq \mathcal{P}$ and $\mathcal{R}' \subseteq \mathcal{P}'$, we have $|\mathcal{R}| \leq |\mathcal{R}'|$.*

The proof is given in [8]. We will investigate the efficiency improvement gained by this approximate strategy and the incurred loss of compression rate in Sect. 5.

The *gen_cover_set()* Function. Integrating the optimization strategy and the approximation strategy discussed above, we present the gen_cover_set() function in Algorithm 3. Given the input representative pattern F_r, Algorithm 3 visits its sub-patterns using a depth-first search. Line 1 finds all child prevalent co-location patterns of the current pattern F. Lines 2–4 implement the optimization strategy, when the co-location distance between F_r and a sub-pattern can be inferred to be smaller than ϵ. Otherwise, we have to compute the co-location distance (line 7). If the co-location distance is smaller than ϵ, we check further descendent sub-patterns (lines 9–10). If not, the depth-first search can be stopped according to the approximation strategy.

5   Experimental Study

We have conducted comprehensive experiments to evaluate the proposed algorithms from multiple perspectives on both synthetic and real data sets. All algorithms are implemented in Python 2.7. All experiments are run on a PC with Intel Core Xeon 2.9 GHz CPU and 8 GB memory.

5.1   Experiments on Synthetic Data

Our synthetic data generation methodology is similar to the one used in [5] for co-location mining. In our experiment, three synthetic data sets are generated. SynData_1 is a sparse data set with 37 features and 29, 496 events. In contrast, SynData_2 has 52 features and 291, 520 events in total, with a larger number of events per feature. SynData_3 is the most dense dataset, containing 525 features and 424, 400 events. The default values of the parameters *minpi* and ϵ are 0.4 and 0.2, respectively.

Compression Rate. We first evaluate the compression rate achieved by representative co-location pattern (RCP) mining, in comparison with closed co-location pattern (CCP) mining and maximal co-location pattern (MCP) mining. Specifically, we define *compression rate* as $(1 - \frac{N^*}{N_{PCP}}) \times 100\,\%$, where N^* equals to the number of compressed patterns and N_{PCP} refers to the number of prevalent co-location patterns (PCP).

Besides comparing with CCP and MCP, we also conduct experiments to investigate the compression rate of RCP without relaxation (RCP-NoRelax). As discussed in Sect. 2.3, we may either generate ϵ-clusters from the spatial data and return the prevalent centroid patterns as representatives, or relax the restriction to allow representative patterns to be prevalent w.r.t. $(1 - \epsilon) * minpi$ in order to achieve higher compression rate.

Figure 5 shows the compression rates of MCP, CCP, RCP, and RCP-NoRelax on the three synthetic data sets by varying the parameters *minpi* and ϵ respectively. Overall, it can be observed that CCP has the lowest compression rate,

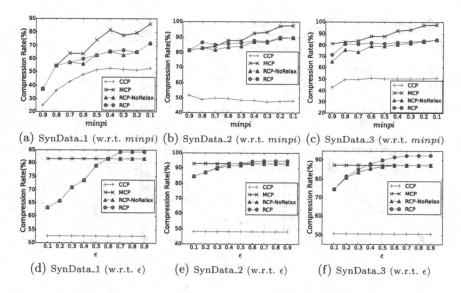

Fig. 5. Compression rate tests on synthetic data sets.

while RCP achieves a higher compression rate than RCP-NoRelax. Regarding the comparison between RCP and MCP, we observe that MCP has a higher compression rate when ϵ is fixed at 0.2 (Figs. 5a, 5b and 5c). However, as ϵ is getting larger, RCP's compression rate prevails (Figs. 5d, 5e and 5f). That is, by relaxing the condition on the co-location distance threshold ϵ, RCP can achieve a compression rate which is even higher than that of MCP. This is due to the definition of RCP, while the best compression rate of RCP-NoRelax is bounded by that of MCP.

Moreover, it can be observed that RCP obtains a high compression rate on a dense data set. For example, when $\epsilon = 0.2$, the best compression rate of RCP on SynData_1 is 71.9 % (Fig. 5a) while it is 89.9 % and 85.0 % on SynData_2 and SynData_3, respectively (Figs. 5b and 5c). This is because a representative pattern tends to cover more patterns on a dense data set. We also observe that when the prevalent threshold *minpi* gets smaller, which means more co-location patterns are generated, the compression rate of RCP is higher. When the requirement on preserving the prevalence information is relaxed (i.e., when the co-location distance threshold ϵ is increased), the compression rate of RCP also improves, which is consistent with the definition of ϵ-cover relationship.

RCPFast vs. RCPMS. In the following, we conduct experiments to compare the two proposed algorithms from different perspectives.

Computation Efficiency. We compare the overall efficiency of the *RCPFast* algorithm and the *RCPMS* algorithm, both implemented with respective optimization strategies. In particular, we also implement a variation of *RCPMS*, called *RCPMS-NA*, which uses the optimization strategy in Subsect. 4.1 only.

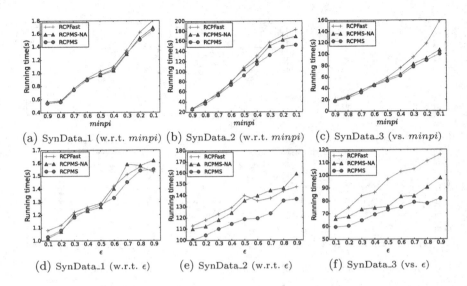

(a) SynData_1 (w.r.t. *minpi*) (b) SynData_2 (w.r.t. *minpi*) (c) SynData_3 (vs. *minpi*)

(d) SynData_1 (w.r.t. ϵ)   (e) SynData_2 (w.r.t. ϵ)   (f) SynData_3 (vs. ϵ)

Fig. 6. Performance tests with *minpi* and ϵ on synthetic data sets.

Hence, by comparing *RCPMS* and *RCPMS-NA*, we can study the effectiveness of the approximation strategy in Subsect. 4.2.

Figure 6 presents the running time on three synthetic data sets with respect to the variation of *minpi* and ϵ respectively. It can be observed that *RCPMS* outperforms *RCPFast* in all situations. Comparing the results on the three datasets, we note that the performance advantage of *RCPMS* is not as obvious on sparse data (SynData_1) as on dense data (SynData_2 and SynData_3). This is because when the data is sparse, the size of table instance of a co-location is small, resulting in a short time for computing co-location distance. Consequently, even if *RCPMS* reduces more number of co-location distance computation, the effect of computation saving of *RCPMS* is not obvious. The reasons for *RCPMS* being more efficient on the two dense data sets are different. For SynData_2, the data is dense in terms of the number of co-location instances, which leads to larger size of table instances and longer time to compute co-location distances. Specifically, the time of co-location distance computation is around 60ms on SynData_2 while it is 2.3 ms and 3.5 ms on SynData_1 and SynData_3, respectively. Therefore, by reducing a few more number of co-location distance computation, *RCPMS* can show efficiency improvement clearly. SynData_3 is dense in terms of the number of co-locations. For this type of dense data, *RCPMS* demonstrates efficiency advantage by directly reducing the number of co-location distance computation.

By comparing *RCPMS* and *RCPMS-NA*, it is obvious, especially on SynData_2 and SynData_3, that the approximation strategy contributes significantly to the efficiency of the *RCPMS* algorithm.

Reduction of Co-location Distance Computation. The optimization strategies devised form both *RCPFast* and *RCPMS* aim to skip some co-location distance computation. To investigate the effectiveness of these strategies more thoroughly, we conduct experiments to record the number of co-location distance computations for *RCPFast*, *RCPMS* and *RCPMS-NA*, compared with the original number (baseline). Figure 7 presents the results by varying *minpi* and ϵ respectively. It can be observed that all algorithms involves fewer co-location distance computations than the baseline does and the number of computations in *RCPMS* is the smallest. In addition, by comparing *RCPFast* against the baseline, we notice that *Theorem* 2 does reduce the number of co-location distance computations. However, when the reduced number is not great enough (e.g., Figs. 7a, 7b, and 7c). This is because it costs extra time for *RCPFast* to find all patterns that can be skipped according to *Theorem* 2.

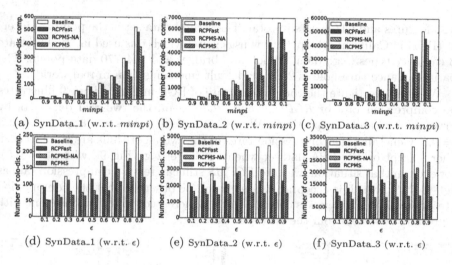

(a) SynData_1 (w.r.t. *minpi*) (b) SynData_2 (w.r.t. *minpi*) (c) SynData_3 (w.r.t. *minpi*)

(d) SynData_1 (w.r.t. ϵ)   (e) SynData_2 (w.r.t. ϵ)   (f) SynData_3 (w.r.t. ϵ)

Fig. 7. Co-location distance computation analysis on synthetic data sets.

Compression Rate. Although both Figs. 6 and 7 show that the approximation strategy significantly improves the efficiency of *RCPMS*, more RCPs will be discovered by *RCPMS* than *RCPFast*. Hence, we further carry out experiments to evaluate how many more RCPs will be produced by *RCPMS*. We present the results using *compression rate difference*, which is calculated as $\frac{N_M - N_F}{N_{PCP}} \times$ 100 %, where N_M and N_F refer to the numbers of patterns output by *RCPMS* and *RCPFast*, respectively. The results in Fig. 8 show that the compression rate difference is less than 5 % on all three datasets, regardless of the variation of parameters. Hence, *RCPMS* effectively improves the computation efficiency by sacrificing the compression rate very slightly.

5.2   Experiments on Real Data

Two real-world data sets are used in our experiments. The first one is an environmental dataset from the EPA databases (http://www.epa.gov/), which consists

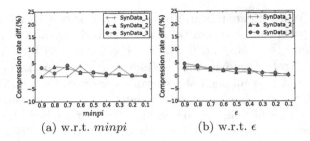

Fig. 8. Compression rate differences between *RCPMS* and *RCPFast* on synthetic data sets.

of 23 features and 647 events in total. The second data set is the points of interest (POI) in California (http://www.usgs.gov/) which was used in [7]. There are 63 category types (e.g., dam, school, and bridge) and 104, 770 data points. The spatial distance threshold is 2000 by default (meaning 2 km in real world).

We first investigate the compression rate of RCP mining. Figure 9 illustrates the compression rate of *RCPFast* and *RCPMS* on the two real data sets by varying *minpi* and ϵ respectively. We set the default values as *minpi* = 0.4 and ϵ = 0.2. Generally, the compression rates of the two algorithms are close to each other, except on the EPA dataset when ϵ is large (e.g., Fig. 9b where ϵ = 0.5). However, in that situation, we note *RCPMS* still can reach a compression rate as high as 75 %, which is acceptable. Also it can be observed that the compression rates increase when *minpi* is decreased or ϵ is increased, which is consistent with the results obtained from the synthetic data sets.

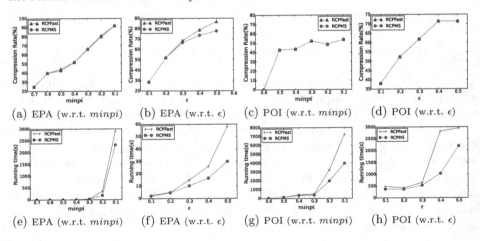

Fig. 9. Compression rate and performance on EPA and POI data sets.

Next, we study the efficiency of the proposed algorithms on real data sets. Figure 9 illustrates the running time of the two algorithms with respect to the variation of $minpi$ and ϵ respectively. It shows that $RCPMS$ outperforms RCP-$Fast$ on the two real datasets, especially when the data is getting dense (e.g., when $minpi$ is decreased) or the requirement of preserving prevalence information is relaxed (e.g., when ϵ is increased).

6   Related Works

The problem of prevalent co-location pattern mining was first introduced by Morimoto [12], where a *support* metric was defined as the number of instances of a co-location and was used to measure the prevalence of a co-location pattern. Shekhar and Huang [14] proposed to use *participation ratio* and *minimum participation index* as the interestingness measures that are more statistically meaningful. Various algorithms have been developed to mine prevalent co-location patterns based on the two measures, such as the Apriori-like algorithm [5], the partial-join algorithm [20] and the join-less algorithm [21].

Frequent pattern summarization has been studied extensively in traditional frequent itemset mining [1,6,13]. One common generalization is The distance-based approach [10,17]. It has been successfully applied in many applications such as summarizing uncertain data [2,9]. Although the framework achieves satisfactory compression rate, it cannot not be applied directly to summarize co-location patterns due to the lack of transaction concepts in co-location mining.

Some initial research efforts have been exerted to summarize prevalent co-location patterns. Mining maximal spatial co-location patterns from a large data set was studied in [11]. Wang et al. used an *order-clique-based* approach to identify table instances and mine maximal co-locations [16]. Closed co-location pattern has been studied by Yoo et al. [21]. To our knowledge, our work is the first distance-based approach to summarize co-location patterns using representative patterns, which preserves more prevalence information than maximal co-location patterns and enjoys higher compression rate than closed co-location patterns.

7   Conclusions

In this paper, we study the problem of summarizing spatial co-locations using representative patterns. A new measure is defined to appropriately quantify the prevalence distance between two co-location patterns. After formulating the problem of RCP mining, we propose two efficient algorithms for RCP mining: $RCPFast$ and $RCPMS$. $RCPFast$ adopts a *post-mining* framework while $RCPMS$ employs a novel *mine-and-summarize* paradigm to discover representative patterns. Experimental results show that RCP mining effectively summarizes prevalent co-location patterns, and $RCPMS$ significantly improves over $RCPFast$ on dense data sets by slightly sacrificing compression rate.

Acknowledgement. We thank the anonymous reviewers for their detailed suggestions for improving the paper. This work was supported, in part, by the Australian Research Council (ARC) Discovery Project under Grant No. DP140100545.

References

1. Calders, T., Goethals, B.: Mining all non-derivable frequent itemsets. In: Elomaa, T., Mannila, H., Toivonen, H. (eds.) PKDD 2002. LNCS (LNAI), vol. 2431, p. 74. Springer, Heidelberg (2002)
2. Chen, L., Liu, C., Zhang, C.: Mining Probabilistic Representative Frequent Patterns From Uncertain Data. In: SDM, pp. 73–81 (2013)
3. Cormen, T.H., Leiserson, C.E., Rivest, R.L., Stein, C.: Introduction to Algorithms, 2nd edn. MIT Press, Cambridge (2001)
4. Huang, Y., Pei, J., Xiong, H.: Mining co-location patterns with rare events from spatial data sets. GeoInformatica 10(3), 239–260 (2006)
5. Huang, Y., Shekhar, S., Xiong, H.: Discovering colocation patterns from spatial data sets: a general approach. IEEE Trans. Knowl. Data Eng. 16(12), 1472–1485 (2004)
6. Bayardo, Jr., R.J.: Efficiently mining long patterns from databases. In: SIGMOD Conference, pp. 85–93 (1998)
7. Li, F., Cheng, D., Hadjieleftheriou, M., Kollios, G., Teng, S.-H.: On trip planning queries in spatial databases. In: Medeiros, C.B., Egenhofer, M., Bertino, E. (eds.) SSTD 2005. LNCS, vol. 3633, pp. 273–290. Springer, Heidelberg (2005)
8. Liu, B., Chen, L., Liu, C., Zhang, C., Qiu, W.: RCP Mining: Towards the Summarization of Spatial Co-location Patterns. https://goo.gl/B0mwei
9. Liu, C., Chen, L., Zhang, C.: Summarizing probabilistic frequent patterns: a fast approach. In: SIGKDD, pp. 527–535 (2013)
10. Liu, G., Zhang, H., Wong, L.: Finding minimum representative pattern sets. In: KDD, pp. 51–59 (2012)
11. Modani, N., Dey, K.: Large maximal cliques enumeration in sparse graphs. In: CIKM, pp. 1377–1378 (2008)
12. Morimoto, Y.: Mining frequent neighboring class sets in spatial databases. In: KDD, pp. 353–358 (2001)
13. Pasquier, N., Bastide, Y., Taouil, R., Lakhal, L.: Discovering frequent closed itemsets for association rules. In: Beeri, C., Bruneman, P. (eds.) ICDT 1999. LNCS, vol. 1540, pp. 398–416. Springer, Heidelberg (1998)
14. Shekhar, S., Huang, Y.: Discovering spatial co-location patterns: a summary of results. In: Jensen, C.S., Schneider, M., Seeger, B., Tsotras, V.J. (eds.) SSTD 2001. LNCS, vol. 2121, pp. 236–256. Springer, Heidelberg (2001)
15. Wang, L., Zhou, L., Lu, J., Yip, J.: An order-clique-based approach for mining maximal co-locations. Inf. Sci. 179(19), 3370–3382 (2009)
16. Wang, S., Huang, Y., Wang, X.S.: Regional co-locations of arbitrary shapes. In: Nascimento, M.A., Sellis, T., Cheng, R., Sander, J., Zheng, Y., Kriegel, H.-P., Renz, M., Sengstock, C. (eds.) SSTD 2013. LNCS, vol. 8098, pp. 19–37. Springer, Heidelberg (2013)
17. Xin, D., Han, J., Yan, X., Cheng, H.: Mining compressed frequent-pattern sets. In: VLDB, pp. 709–720 (2005)
18. Yan, X., Cheng, H., Han, J., Xin, D.: Summarizing itemset patterns: a profile-based approach. In: KDD, pp. 314–323 (2005)
19. Yoo, J.S., Bow, M.: Mining Top-k closed co-location patterns. In: ICSDM, pp. 100–105 (2011)
20. Yoo, J.S., Shekhar, S.: A partial join approach for mining co-location patterns. In: GIS, pp. 241–249 (2004)

21. Yoo, J.S., Shekhar, S.: A joinless approach for mining spatial colocation patterns. IEEE Trans. Knowl. Data Eng. **18**(10), 1323–1337 (2006)
22. Zhang, X., Mamoulis, N., Cheung, D.W., Shou, Y.: Fast mining of spatial collocations. In: KDD, pp. 384–393 (2004)

Demonstrations

Pedestrian-Flow Analysis System for Improving Layout of Exhibitions

Akinori Asahara[1]([✉]), Nobuo Sato[1], and Masatsugu Nomiya[2]

[1] Hitachi Ltd., Center of Technology Innovation - Systems Engineering,
Tokyo, Japan
akinori.asahara.bq@hitachi.com
[2] Hitachi Ltd., Smart Information Systems Division, Information
and Telecommunication Systems Company, Tokyo, Japan

Abstract. A system for practical pedestrian-track analysis at an actual exhibition is demonstrated. Track data obtained at the exhibition was uploaded to a spatio-temporal database, and the key features of the technical exhibition were determined. New knowledge derived from these features was successfully applied to improve the layout of the next event.

1 Introduction

"Pedestrian tracking" gives crucial hints for improving plans of large facilities like shopping malls, airports, and exhibition halls. A population at a booth of an exhibition event, calculated from the tracks of visitors to that booth, can be taken as an example. Because the population is an effective indicator for evaluating the quality of the presentations at the booth, a time series of the population at the booth will show tendencies about the popularity of the booth. If the population at the booth quickly increases after a presentation at a neighboring booth, it is implied that the presentation encourages visitors' interests in the former booth. The correlation between the populations at the two booths thus implies that both exhibitions should be linked more closely. Similarly, many indicators derived from visitors' tracks show various criteria concerning the booths, which clarify the advantages and drawbacks of each booth.

Developed over the last few years, LIDAR (Light Detection and Ranging) systems for tracking pedestrians are available for obtaining the tracks of visitors to a large facility [1]. However, indicators for effectively improving planning are not established yet. One of the reasons for that is "lack of experimental data." A LIDAR system incurs high costs and uncertainness of usefulness; therefore, it has only been used for a few experiments. Under such a circumstance, the authors conducted an experiment on tracking pedestrians at two actual big technical exhibition events. Track data obtained at the event was imported to a spatio-temporal database, and the features key of the two events. New knowledge derived from these features was successfully applied for improving planning of the next event. A pedestrian-track analysis system that demonstrates improvements inferred by pedestrian tracking is described in the following.

C. Claramunt et al. (Eds.): SSTD 2015, LNCS 9239, pp. 473–477, 2015.
DOI: 10.1007/978-3-319-22363-6_25

2   Related Works

Analysis of the tracks, called "trajectory analysis," is used to extract information from stored trajectories by three methods: trajectory clustering [2,3] for retrieving similar trajectories from a database; extraction of a "representative path" (which is abstract information about a trajectory dataset [4]); and prediction of movement by various Markov-chain models [5].

Table 1. HIF2013 overview

Title	Contents
Event name	Hitachi Innovation Forum 2013 (HIF2013)
Place	Tokyo International Forum ($5000m^2$)
Date	2013/10/30 and 31 (two days)
Number of booths	210
Number of data records	6,801,655

3   Discovery: Hitachi Innovation Forum 2013

HIF2013 (Hitachi Innovation Forum 2013) is a private technical exhibition held in Tokyo, Japan. The details of the event are listed in Table 1, and the event layout is shown in Fig. 1(a). Tracks of visitors to HIF2013, obtained with 13 LIDAR sensors (drawn by white circles in the layout) over two days, are drawn as yellow lines in Fig. 1(b). Because the drawing is unreadable, numerical indices summarizing pedestrian movements (i.e., number of pedestrians in an area, hourly count of pedestrians walking through a corridor by time, and so on) are needed to extract hints for improving the plan of the next exhibition. Note that many indices for finding useful hints should be calculated by trial and error because useful indices are unknown before the calculation. The pedestrian tracks were thus uploaded to a spatial-relational database management system as very short tracks. The table schema of short tracks allows various summarizations without the need for programing, so many indices can be computed.

After many indices were computed, it was found that the sums of the durations of people's stays around booths (shown in Fig. 2) are concentrated at several booths. The cause of that concentration, that is, certain building structures obstructing flows of people, is shown in Fig. 3. Most visitors to HIF2013 walked in straight lines; however, their paths became curved if they met obstructions. People's flows thus bent toward specific sides of corridors, to which people's durations of stay around booths were concentrated. In conclusion, a new piece of knowledge was found; namely, the layout of a technical exhibition should be designed to bend the path of a pedestrian flow toward areas where their attention is desired.

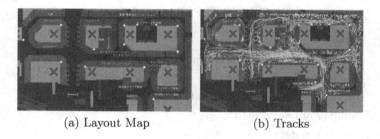

(a) Layout Map   (b) Tracks

Fig. 1. Pedestrian tracks at HIF2013

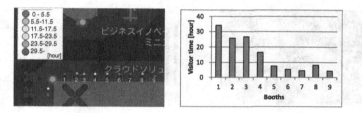

Fig. 2. Total duration around booths in HIF2013

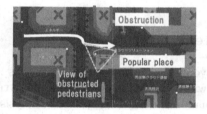

Fig. 3. Causal analysis of HIF2013

4   Improved: Hitachi Innovation Forum 2014

According to the analysis of HIF2013, the layout of a technical exhibition should be designed to bend the path of a pedestrian flow. This knowledge was thus applied to design the layout of next year's event, i.e., HIF2014 (Hitachi Innovation Forum 2014, the details of which are listed in Table 2). Many obstructions were set up at HIF2014 in order to equalize pedestrian flows to the booths, as shown in Fig. 4.

To clarify the improvements mentioned above, tracks of visitors to the event in 2014 were also obtained by 18 LIDAR sensors. Dispersion of the people's durations of stay at different booths are compared in Fig. 5. Standard deviations of people's duration of stay at booths at HIF2013 and that at HIF2014 are shown in Fig. 5(a). The normalized standard deviation for HIF2013 was over 1.4, while that for HIF2014 was 0.8. That is, dispersion was reduced by half. This result confirms that the knowledge gained by pedestrian tracking is useful for planning exhibitions.

Table 2. HIF2014 overview

Title	Contents
Event name	Hitachi Innovation Forum 2014 (HIF2014)
Place	Tokyo International Forum ($5000m^2$)
Date	2014/10/30, 31 (two days)
Number of booths	179
Number of data	9,126,297

Fig. 4. Layout of HIF2014

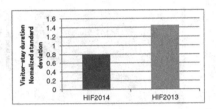

Fig. 5. Disparities among durations of stay at booths

5   Concluding Remarks

It is shown that the layout of a technical exhibition was improved by applying pedestrian tracking. It was demonstrated that pedestrian tracking with a LIDAR system is helpful for planning events. Trajectory analysis of the tracking data was carried out to calculate indices that would be helpful for knowledge extraction. That is, the knowledge extraction itself was completed manually. However, most managers of facilities cannot handle such difficult analysis, so automatic knowledge extraction must be developed as future work.

Demonstration

The system used for analysis of pedestrian flows at HIF2013 and HIF2014 will be exhibited as the demonstration. The dataset for HIF2013 and HIF2014 is uploaded to PostgreSQL (PostGIS). Many indices related to the event can be calculated by the system. GIS (Geographical Information System) will visualize the summary of track data. Examples of such visualization are given in Figs. 6 and 7. Density of people by colors on the map is shown in Fig. 6 (a). Averaged direction of people's movements (expressed with arrows) is shown in Fig. 6 (b). The durations of stay of visitors to booths at HIF2013 and HIF2014 are compared in Fig. 7 in terms of height of cylinders. The process for improving the event layout will be introduced by using these visualizations at the demo session.

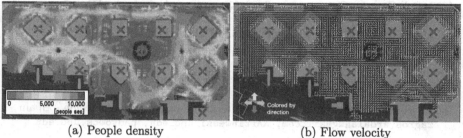

(a) People density   (b) Flow velocity

Fig. 6. Distributed indicators of HIF2014

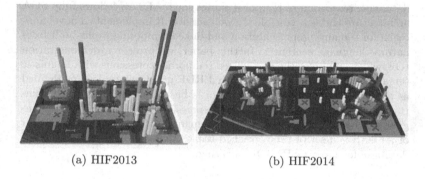

(a) HIF2013   (b) HIF2014

Fig. 7. Comparison of durations of stay at HIF2013 and HIF2014

References

1. Hitachi Information and Telecomunication Engineering, Ltd., Inc. LaserRadarvisionII (in Japanese). http://www.hitachi-ite.co.jp/products/lr/
2. Chudova, D., Gaffney, S., Mjolsness, E., Smyth, P.: Translation-invariant mixture models for curve clustering. In: Proceedings of the Ninth ACM SIGKDD International Conference on Knowledge Discovery and Data Mining, pp. 79–88. ACM, New York (2003)
3. Elias, F., Kostas, G., Yannis, T.: Index-based most similar trajectory search. In: Proceedings of IEEE 23rd International Conference on Data Engineering, pp. 816–825 (2007)
4. Lee, J.G., Han, J., Whang, K.Y.: Trajectory clustering: a partition-and-group framework. In: Proceedings of the 2007 ACM SIGMOD International Conference on Management of Data, pp. 593–604. ACM, New York (2007)
5. Asahara, A., Maruyama, K., Shibasaki, R.: A mixed autoregressive hidden-markov-chain model applied to people's movements. In: Proceedings of the 20th International Conference on Advances in Geographic Information Systems. SIGSPATIAL 2012, pp. 414–417. ACM (2012)

AETAS: A System for Semanticizing Temporal Expressions from Unstructured Contents

Zagros Ardalan, Carme Martín[(✉)], and Lluís Padró

TALP Research Center, Universitat Politècnica de Catalunya, Barcelona, Spain
zagros.ardalan@gmail.com, martin@essi.upc.edu, padro@cs.upc.edu

Abstract. AETAS is an online tool for converting text into RDF linked data with resolution of temporal expressions. AETAS follows fully SOA architecture and is accessible via web-service. It implements a novel approach for semantic representation and linked temporal graphs built from natural language sentences. In this paper, we present a demonstration tool, which combines the normalized temporal expressions with linguistic semantic frames and creates a linked RDF graph where time is defined as an individual dimension. The tool is based on SUTime which identifies and normalizes the temporal expressions and on FreeLing, a linguistic processor which extracts the semantics of sentences. The output of AETAS is a set of time-enriched triples that can be stored in a RDF database for later τ-SPARQL querying.

1 Introduction

Web archives already hold together more than 534 billion files —a phenomenon that many refer to as big data [1]. As more and more information becomes available year after year, its variety and richness in terms of temporal aspects becomes more manifest. A recent research area aimed at incorporating temporal aspects in modern information retrieval systems is temporal information retrieval (TIR) [2].

The performance of temporal event extraction from these open resources would dramatically increase if we were able to use relations and links between existing resources. The term Linked Data (LD) stands for a new paradigm of representing information on the Web in a way that enables the global integration of data and information in order to achieve unprecedented search and querying capabilities. The formalism underlying this "Web of Linked Data" is the Resource Description Framework (RDF) which encodes structured information as a directed labeled graph. Hence, in order to publish information as Linked Data, an appropriate graph-based representation has to be defined and created [3].

Processing of text documents in terms of the extraction and normalization of temporal expressions and of relations between events is very important for several NLP tasks requiring a deep understanding of language such as question answering or document summarization. Due to this fact, there has been significant research in temporal annotation of text documents. Research work on

© Springer International Publishing Switzerland 2015
C. Claramunt et al. (Eds.): SSTD 2015, LNCS 9239, pp. 478–483, 2015.
DOI: 10.1007/978-3-319-22363-6_26

fully utilizing the temporal information embedded in the text of documents for exploration and search purposes is very recent. The work by Alonso et al. [4] presents an approach for extracting temporal information and how it can be used for clustering search results. Schilder et al. [5] presented semantic tagging for temporal expressions on news articles. They used their system to be part of an experimental multi-document summarization system while covers indexical and vague temporal expressions. Meanwhile an interesting visualizing system proposed by Marcus et al. [6] extracts and visualizes the events from micro-blogs and tweets which are already being used for social science and augmented media experiences. FRED, an RDFizer system from natural language text has been proposed by Draicchio et al. [7]. They represent the natural language into RDF framework which is connected to DBPedia knowledge source without considering the temporal expressions.

Some work has been done in order to convert documents into RDF taking into account temporal expressions. Rula et al. [8] found out that the availability of temporal information describing the history and the temporal validity of statements and graphs is still very limited. Meanwhile Batsakis et al. [9] proposed a new framework to handle Spatial-Temporal information in OWL 2.0. To the best of our knowledge, there is still no formal study of temporality issues in RDF graphs and RDF query languages.

In this paper, we introduce AETAS, an end-to-end system which retrieves plain text data from web and blog news and represents and stores them in RDF, focusing on its temporal dimension. On top of the system, a querying layer has been deployed in order to allow users access by time to the extracted linked data.

The rest of the paper is structured as follows: Sect. 2 outlines the architecture of AETAS system for temporal expressions extraction and linguistic annotation, which relies on SUTime [10] and FreeLing [11]. Section 3 describes the nature of the demonstration.

2   AETAS Architecture

AETAS as a Latin word with meaning Time and Era, is the name of our system. AETAS is fully designed and implemented in service-oriented approach in order to get the benefits of SOA (Service Oriented Architecture) systems. AETAS has four main components which are responsible for different functionalities, as shown in Fig. 1. Two components have been implemented in-house: Temporal Mapper and RDFizer (see Sect. 2.2). The other two components are SUTime and FreeLing which are integrated as external components, and described in Sect. 2.1.

The process starts with a raw text document with its reference date, obtained from some web site such as news agencies or blogs. The document is fed to SUTime component, which returns annotated expressions. Then the Mapper component receives the document, and after some preprocessing, runs the document through FreeLing language processing web-service. Then, Mapper combines the linguistic information provided by FreeLing and the temporal expressions

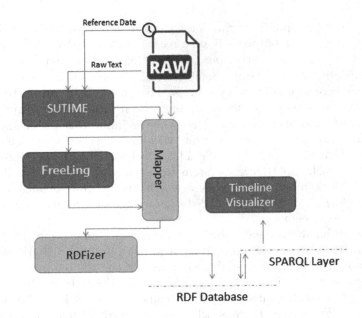

Fig. 1. AETAS Architecture.

identified by SUTime. The output of Mapper is piped to RDFizer, which creates triples and linked information based on the semantic representation made by FreeLing, and stores them into a BrightStarDB[1] native RDF database.

2.1   External Components

SUTime component [10] from Stanford NLP system is responsible for recognizing and extracting temporal expressions and normalize them in TIMEX3 standard. SUTime is one of best currently available temporal taggers, obtaining 90.32 F-score in TempEval-3 evaluation on 2013. Other top-performing systems in this shared task were HeidelTime and ClearTK (F-score 90.30 and 90.23 respectively). See the shared task summary paper [12] for details and references. SUTime offers three main features: (1) Extraction of temporal expressions from text, (2) Representation of temporal expressions as objects convenient to handle programmatically, and (3) Resolution of temporal expressions with respect to a reference date (e.g. the document date). SUTime is integrated in AETAS as the first processing step. It receives a raw text paragraph with a reference date and outputs a set of objects representing found temporal expressions with their normalized time and their temporal type.

    FreeLing is an open-source library providing language analysis services for a variety of languages. FreeLing is integrated in TextServer[2] NLP cloud platform,

---

[1] http://brigtstardb.com.

[2] http://textserver.cs.upc.edu/textserver.

which allows SOA access and an easy integration in AETAS. In our system, FreeLing is responsible for tokenization, lemmatization, morphological analysis, PoS tagging, named entity recognition, word sense disambiguation, dependency parsing, semantic role labeling, and co reference resolution, providing as output a semantic graph representing the main events described in the document and the involved actors in each of them. FreeLing web-service is considered as a black-box in AETAS, and we just call the service with appropriately formatted raw text document. Then, the Mapper component will extend the semantic representation of the document with the temporal expressions identified by SUTime, building a time-aware representation of the document meaning.

2.2   Internal Components

The Temporal Mapper component is the core module in AETAS process. Its main mission is to synchronize the outputs from SUTime and FreeLing to provide the final temporally enriched semantic representation.

After mapping and processing the resulting objects, the temporal semantic representation of the document is sent to the RDFizer component, which will convert the graph into RDF triples.

The RDFizer component in AETAS is responsible for creating RDF triples based on semantic representation of context and temporal information. Current approaches for coding temporal information consider it as additional data inside the data model. Therefore, temporal information is implicit in the data and difficult to access by programs. In AETAS system, time has been proposed as an additional semantic dimension of data. Therefore, it needs to be regarded as an element of the meta model instead of being just part of the data model.

RDFizer applies the foundations established by Gutierrez et al. [13], implementing an RDF-compatible syntax for temporal data. Then, these RDF triples are stored modeling time as a dimensional discrete value, as in most of the approaches dealing with temporal entities. Since RDFizer tries to roll up the time dimension as much as possible, another dimension can be added to the graph, connecting via an "associated with" edge the time point with any nodes related to it. This approach would increase the performance of information retrieval in SPARQL layer when the user tries to drill down into temporal data.

3   Demonstration

Since AETAS is just a RESTful web service, it does not have any graphical interface. Thus, a simple client web page[3] has been devised to demonstrate the capabilities of AETAS, where the user can input raw textual data and see how AETAS semanticizes the text and stores the result in a RDF database.

The demonstration page allows the user to turn on or off each component of the system, showing how each component affects the process flow. Finally, the

---

[3] http://xorrai.cs.upc.edu/aetas.

demo page allows the user to enter a τ-SPARQL query and obtain a timeline visualization of the events extracted from the input text. The timeline depiction is generated using the *Timeline* widget[4].

AETAS currently runs in two servers: FreeLing components are executed via TextServer platform, which runs on a 1,000+ core HPC cluster. SUTime, Mapper, and RDFizer run on a dedicated server at goDaddy web hosting[5].

Regarding execution resources, the most time-consuming step is the construction of the semantic graph performed by FreeLing (which may take over 30 s for an average news item of about 500 words). Nevertheless, since this occurs at indexation time, it is executed in batch mode and does not affect the querying performance. The τ-SPARQL querying of the RDF database and the Timeline generation are performed in a few seconds.

The demonstration will use a RDF database pre-filled with information extracted from news items on a particular event extracted from online newspapers. A τ-SPARQL query will be executed and the timeline results presented. Then, a new document reporting some related event not included in the database will be processed, and the same query will be issued again, obtaining the previous timeline extended with the newly extracted event.

Acknowledgements. This research has been partially funded by the Spanish Government via the SKATER project (TIN2012-38584-C06-01).

References

1. Costa, M., Couto, F., Silva, M.: Learning temporal-dependent ranking models. In: Proceedings of the 37th International ACM SIGIR Conference on Research and Development in Information Retrieval. SIGIR 2014, pp. 757–766. ACM, New York, NY, USA (2014)
2. Alonso, O., Baeza-yates, R., Strötgen, J., Gertz, M.: Temporal information retrieval: challenges and opportunities. In: 1st Temporal Web Analytics Workshop, pp. 1–8, Hyderabad, India (2011)
3. Augenstein, I., Padó, S., Rudolph, S.: LODifier: generating linked data from unstructured text. In: Simperl, E., Cimiano, P., Polleres, A., Corcho, O., Presutti, V. (eds.) ESWC 2012. LNCS, vol. 7295, pp. 210–224. Springer, Heidelberg (2012)
4. Alonso, O., Gertz, M., Baeza-Yates, R.: Clustering and exploring search results using timeline constructions. In: Proceedings of the 18th ACM Conference on Information and Knowledge Management. CIKM 2009, pp. 97–106. ACM, New York, NY, USA (2009)
5. Schilder, F., Habel, C.: From temporal expressions to temporal information: semantic tagging of news messages. In: Proceedings of the Workshop on Temporal and Spatial Information Processing. vol. 13 of TASIP 2001, pp. 9:1–9:8, Stroudsburg, PA, USA (2001)

---

[4] http://www.simile-widgets.org/timeline.

[5] http://www.godaddy.com.

6. Marcus, A., Bernstein, M.S., Badar, O., Karger, D.R., Madden, S., Miller, R.C.: Twitinfo: aggregating and visualizing microblogs for event exploration. In: Proceedings of the International Conference on Human Factors in Computing Systems, CHI 2011, May 7–12, 2011, Vancouver, BC, Canada, pp. 227–236 (2011)
7. Draicchio, F., Gangemi, A., Presutti, V., Nuzzolese, A.G.: FRED: from natural language text to RDF and OWL in one click. In: Cimiano, P., Fernández, M., Lopez, V., Schlobach, S., Völker, J. (eds.) ESWC 2013. LNCS, vol. 7955, pp. 263–267. Springer, Heidelberg (2013)
8. Stadtmüller, S., Maurino, A., Rula, A., Palmonari, M., Harth, A.: On the diversity and availability of temporal information in linked open data. In: Cudré-Mauroux, P., Heflin, J., Sirin, E., Tudorache, T., Euzenat, J., Hauswirth, M., Parreira, J.X., Hendler, J., Schreiber, G., Bernstein, A., Blomqvist, E. (eds.) ISWC 2012, Part I. LNCS, vol. 7649, pp. 492–507. Springer, Heidelberg (2012)
9. Batsakis, S., Petrakis, E.G.M.: SOWL: a framework for handling spatio-temporal information in OWL 2.0. In: Bassiliades, N., Governatori, G., Paschke, A. (eds.) RuleML 2011 - Europe. LNCS, vol. 6826, pp. 242–249. Springer, Heidelberg (2011)
10. Chang, A.X., Manning, C.: SUTime: a library for recognizing and normalizing time expressions. In: Proceedings of the Eight International Conference on Language Resources and Evaluation (LREC 2012), Istanbul, Turkey (2012)
11. Padró, L., Stanilovsky, E.: Freeling 3.0: towards wider multilinguality. In: Proceedings of the Eight International Conference on Language Resources and Evaluation (LREC 2012), Istanbul, Turkey (2012)
12. UzZaman, N., Llorens, H., Derczynski, L., Allen, J., Verhagen, M., Pustejovsky, J.: SemEval-2013 task 1: TempEval-3: evaluating time expressions,events, and temporal relations. In: Proceedings of the Seventh International Workshop on SemanticEvaluation (SemEval 2013). Second Joint Conference on Lexical andComputational Semantics (*SEM), pp. 1–9, Atlanta, Georgia, USA, June 2013
13. Gutiérrez, C., Hurtado, C.A., Mendelzon, A.O.: Formal aspects of querying RDF databases. In: Proceedings of the First International Workshop on Semantic Web and Databases (SWDB 2003), Co-located with VLDB-2003, pp. 293–307, Berlin, Germany (2003)

SCHAS: A Visual Evaluation Framework for Mobile Data Analysis of Individual Exposure to Environmental Risk Factors

Shayma Alkobaisi[1], Wan D. Bae[2]($\boxtimes$), and Sada Narayanappa[3]

[1] College of Information Technology, United Arab Emirates University,
Al Ain, UAE
shayma.alkobaisi@uaeu.ac.ae
[2] Mathematics, Statistics and Computer Science, University of Wisconsin-Stout,
Menomonie, USA
baew@uwstout.edu
[3] Microsoft, New York, USA
sada.narayanappa@microsoft.com

Abstract. Exposure to environmental risk factors as well as weather conditions are known to have negative effects on health. Until recently, there was little a society could do for an individual at risk, other than provide general warnings when the concentration of pollutants or weather conditions deviate from the norm. Similarly, the assessment of individuals' exposure over time has been confined to population and geographic averages, rather than individualized estimates. Recent advances in sensors and mobile technology have enabled real-time measurements of environmental variables and, at the same time, provided information about the spatio-temporal behavior of individuals. This can dramatically change the way health and wellness are assessed as well as how care and treatment are delivered. This paper presents a system framework called "Smart and Connected Health Alert System (SCHAS)" for individual-level environmental exposure in an attempt to better understand the relationships among exposures, symptoms and human health conditions. We demonstrate user interface, data acquisition and visual evaluation tools for large mobile sensor data analysis.

1 Motivation

Air pollution is one of the most important environmental determinants of health [11]. In fact, pollution-related respiratory chronic conditions, such as asthma and chronic obstructive pulmonary disease are estimated to affect a significant portion of people world-wide [1,6]. Understanding the relationship between health

---

S. Alkobaisi and W.D. Bae—This material is based upon works supported in part by the Information and Communication Technology Fund of United Arab Emirates under award number 21T042 and in part by the National Science Foundation under award number CNIC-1338378.

C. Claramunt et al. (Eds.): SSTD 2015, LNCS 9239, pp. 484–490, 2015.
DOI: 10.1007/978-3-319-22363-6_27

states and air pollution is a key prerequisite for optimizing care as well as for guiding societal interventions (pollution limits).

To simplify the relationship between stochastic, spatio-temporal sequences of pollutant concentration and their physiological consequences, researchers have recently began to entertain the notion of "exposome" [15,19]. Most recently, exposome has been defined as, "the cumulative measure of environmental influences and associated biological responses throughout the lifespan, including exposures from the environment, diet, behavior, and endogenous processes" [13].

Over the past years, a variety of exposome studies have been conducted in various disciplines [1,2]. While they have uncovered many important relationships between environment and human health, the assessment of individuals' exposure over time has been confined to population and geographic averages [14]. However, drawing conclusions on individual health effects based on the correlation of aggregated data can lead to biases in environmental health studies [12]. Recent advances in mobile sensor, communication and computation opened new opportunities to investigate the relationship between pollution, human behaviors and health outcomes as well as optimized interventions [16,17]. In particular, with appropriate sensing, big data analytics and model-based inference, it is becoming possible to guide individuals to minimize their exposure to dangerous pollutants, administer prophylactic treatments, etc.

Despite isolated commercial successes [2,5], considerable computing challenges exist in estimation of spatial, time-varying distribution of environmental risk factor concentration combined with human spatial behaviors. This task is extraordinarily time-consuming and resource-intensive. One cannot realistically sample too frequently in order to keep data amounts reasonable and sensors long lasting. No information of the individuals location is known other than these sampled values, and this results in uncertainty of individuals positions [7,18,20]. In addition, a given mobile sensor can likely measure only a subset of environmental variables. The system often needs to use publicly available environmental data that are less frequently updated and of lower resolution. This results in uncertainty of individuals exposures. Measuring these data is also associated with spatial and temporal uncertainties because of the approximations and interpolations used in modeling [10]. Estimating exposure depends on estimating position in road networks and approximation of environmental data, where the number of possible paths grows exponentially in the time between position reports. However, existing methods do not provide effective modeling for uncertainty of individual exposure. To scale the system up, often repeated steps (such as the spatial multi-join of relevant environmental variables) must be efficiently precalculated. Finally, behavior of individuals has a large impact on exposure estimates, and such behavior is not known a priori, but rather built iteratively from previous estimates of exposure and individuals responses.

The authors have introduced an initial framework of a health monitoring system designed to capture exposome for asthma patients [8,9]. This paper proposes a refined framework for exposure, estimation, evaluation and prediction. We present our research initiated on the following objectives: (1) design data models and define queries for the domain of environmental exposome, (2) develop efficient methods for data acquisition, (3) design user interface and visual analysis tools for exposure estimation, and (4) integrate the Hadoop file system and PostGres database for big data management.

2   System Overview

2.1   Data Models

We focus on three data components: individuals' moving trajectories, environmental data, and individuals' health conditions. The trajectory data is a record of device id (id) and a sequence of position values consisting of time (t), longitude (p_x), latitude (p_y), altitude (p_z) and velocity (p_v). If environmental sensors are equipped, n environmental sensor data ($e_1, e_2, .., e_n$), which measure the environmental elements (e.g., humidity), are collected as well. The numbers and types of environmental sensors could vary from one individual to another depending on the targeted environment factors of interest and the nature of the disease to be monitored. The real-time data collected via Android mobile devices are sampled with a pre-determined time interval, and transferred to the main system. The system also retrieves publicly available environmental data through web services. Sensors for individuals' biological signals, such as location and time of inhaler uses, peak flow meters and medication intake in the case of an asthma patient, are integrated to the system.

2.2   Queries

The following set of representative queries are examples how the proposed system can be used for exposure measurement:

$Q1$: How much time did patient A spend in a region with temperature 40 Celsius and 80 % humidity? First the regions need to be defined in 2-dimensional map based on the environmental values for each time stamp. A given geographical area is subdivided into several regions (Voronoi cells) using the locations of weather stations. Join operations on two datasets of temperature and humidity result in a heat map. Join operation of the patient's trajectory results in a set of points as vertices that intersect the environmental data in a set of edges. The exposure time to a particular environmental data value is proportional to the ratio of edge (an edge that crosses the regions where temperature 40 Celsius and humidity 80 %) and the length of the trajectory.

$Q2$: Return all individuals exposed to environmental variables X, Y, Z, during time period D - their exposure exceeds given thresholds T_x, T_y, T_z. This query corresponds to one of the common filtering queries: find individuals' daily exposure time to greater than 2500 ppm of CO_2. Exposure to CO_2 above 2500 ppm is considered as "adverse health effects expected", and the maximum allowed concentration within a 8 hour working period is 5000 ppm.

$Q3$: Return all environmental conditions within distance R of the position of an individual. $Q3$ is a common type of a select query based on spatio-temporal proximity. This query demonstrates the need for spatial operators and helps to assess environmental exposure by integrating spatio-temporal domain knowledge. Supposing that an incident of asthma episode (associated to an individual P) is notified with the location and time, we want to find all environmental conditions near the location at a reported time that might have triggered the attack.

2.3 System Design

We propose a system framework called "Smart and Connected Health Alert System (SCHAS)". The system requires consideration of both the nature of environmental exposures and their changes over time. It is unrealistic to measure every individual exposure continuously. Hence, data is sampled and this results in exposure uncertainty. The underlying individuals' moving trajectory uncertainty models attempt to incorporate weather and other environmental conditions to characterize and predict the path distribution, and estimate individuals' exposure to environmental triggers in time. The system uses k-shortest paths based on Dijkstra algorithm for path distribution but more sophisticated behavioral models can be applied to characterize and predict the expected mobility. Our proposed system models and their relationships are shown in Fig. 1 (a). The system consists of the following four main components: (1) user interface (UI), (2) data acquisition, (3) analytical processing engine, and (4) big data management. The system also implements security subsystem to ensure that the data and views are accessible to validated users. The SCHAS uses the standard module approach for the high-level architecture. Overview of the system is demonstrated in Fig. 1 (b).

The following technologies are used to implement the SCHAS: The system uses Linux Redhat 6.0 operating system; Java and Python are as main programming languages; Javascript is used for most of the client facing systems; Junit for writing test scripts; Jlog for logging. The SCHAS implements data repository using Hadoop file system and PostGres database. Raw data collected over

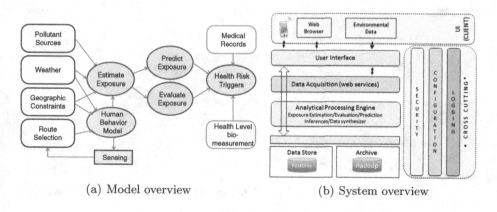

(a) Model overview   (b) System overview

Fig. 1. System design

time through mobile devices is stored in the Hadoop file system. These data are processed and extracted information is stored in the PostGres database tables. Third party data visualization tool such as OpenLayers [3] is used to create maps and integrate geographical datasets. The client codes that invoke Representational State Transfer (REST) API to pull results from its data repository are written in Javascript/AJAX, and are integrated with OpenLayers for the user interface. The Apache Tomcat is used as an application server. The UI layer connects to application server via JSP and using REST API. Our proposed system is evaluated in a pilot study of exposure effects on asthma. This pilot study provides a prototype in testing our proposed mathematical models, optimization methods and computing algorithms.

3   Demonstration

In our demonstration we show the UI module that interacts with Web browsers and mobile applications. The main functionalities of the UI are (1) to display individuals' data, such as moving trajectories (location/time) on maps, (2) to overlay environmental data, and (3) to allow users (including patients, doctors and medical staff, administrators, and researchers) to load the data, view the data, and select interesting data.

We present data acquisition modules that collect data through Bluetooth sensor equipped mobile devices and Web services: (1) GPS/GSM module for individuals' moving trajectories, (2) a sensor module for environmental data such as temperature, humidity, ozone, and nitrogen dioxide. Publicly available environmental data downloaded from Openweathermap [4] is also integrated into the system through web services, (3) sensor modules for biological signals of patients, such as peak flow meters and inhaler use time/location, and (4) user interface using mobile devices for individuals' current health status such as symptoms (e.g., users post that they are feeling difficulty in breathing or chest wheezing)

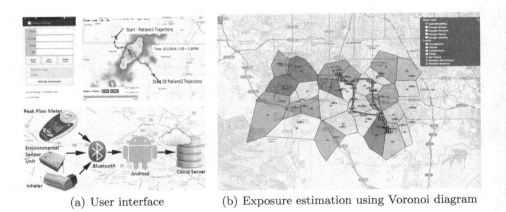

(a) User interface   (b) Exposure estimation using Voronoi diagram

Fig. 2. Visual evaluation framework

and medication use. The system implements sets of interfaces on mobile devices and examples are shown in Fig. 2 (a). Android mobile application is available at https://play.google.com/store/apps/details?id=org.geospaces.schas01. A prototype of user interfaces for individual exposure analysis is shown in Fig. 2 (b) and can be found at http://geospaces.org/SCHAS/html/maps/Openlayers3.html.

References

1. Centers for disease control and prevention. http://www.cdc.gov/asthma. Accessed June 2015
2. Icedot. https://icedot.org/. Accessed June 2015
3. Open layer. http://openlayers.org/. Accessed June 2015
4. Openweather map. http://openweathermap.org/. Accessed June 2015
5. Propeller health. http://www.propellerhealth.com/. Accessed June 2015
6. Trends in asthma morbidity and mortality. American Lung Association Epidemiology and Statistics and Health Education Division, September 2012
7. Alkobaisi, S., Vojtěchovský, P., Bae, W.D., Kim, S.H., Leutenegger, S.T.: The truncated tornado in tmbb: a spatiotemporal uncertainty model for moving objects. In: Bhowmick, S.S., Küng, J., Wagner, R. (eds.) DEXA 2008. LNCS, vol. 5181, pp. 33–40. Springer, Heidelberg (2008)
8. Bae, W.D., Alkobaisi, S., Narayanappa, S., Liu, C.C.: A real-time health monitoring system for evaluating environmental exposures. Int. J. Softw. **8**(4), 791–801 (2013)
9. Bae, W.D., Narayanappa, S., Alkobaisi, S., Bae, K.Y.: Mobis: a distributed paradigm of mobile sensor data analytics for evaluating environmental exposures. In: 10th ACM SIGSPATIAL Geographic Information Systems MobiGIS, pp. 93–96 (2012)
10. Gerharz, L., Pebesma, E.: Using geostatistical simulation to disaggregate air quality model results for individual exposure estimation on gps tracks. Stoch. Environ. Res. Risk Assess. **27**(1), 223–234 (2013)
11. Gibson, J.M., Brammer, A., Davidson, C., Folley, T., Launay, F., Thomsen, J.: Environmental Burden of Disease Assessment. Environmental Science and Technology Library, vol. 24. Springer, Netherlands (2013)
12. Gotway, C.A., Young, L.J.: Combining incompatible spatial data. J. Am. Stat. Assoc. **97**(458), 632–648 (2002)
13. Miller, G.W., Jones, D.P.: The nature of nature: refining the definition of the exposome. Toxicol. Sci. **137**(1), 1–2 (2014)
14. Newhouse, C.P., Levetin, E.: Correlation of environmental factors with asthma and rhinitis symptoms in tulsa, ok. Ann. Assoc. Am. Geogr. **92**(3), 356–366 (2004)
15. Peters, A., Hoek, G., Katsouyanni, K.: Understanding the link between environmental exposures and health: does the exposome promise too much? J. Epidemiol. Community Health **66**(2), 103–105 (2012)
16. Schappert, S., Rechtsteiner, E.: Ambulatory medical care utilization estimates for 2007. Vital Health Stat. Ser. 13, Data Nat. Health Surv. **169**, 1–38 (2011)
17. Seto, E.Y., Giani, A., Shia, V., Wang, C., Yan, P., Yang, A.Y., Jerrett, M., Bajcsy, R.: A wireless body sensor network for the prevention and management of asthma. In: IEEE International Symposium on Industrial Embedded Systems, pp. 120–123 (2009)

18. Trajcevski, G., Wolfson, O., Hinrichs, K., Chamberlain, S.: Managing uncertainty in moving objects databases. ACM Trans. Database Syst. **29**(3), 463–507 (2004)
19. Wild, C.P.: Exposome: from concept to utility. J. Epidemiol. **41**(1), 24–32 (2012)
20. Zheng, K., Trajcevski, G., Zhou, X., Scheuermann, P.: Probabilistic range queries for uncertain trajectories on road networks. In: Proceedings of the 14th International Conference on Extending Database Technology, pp. 283–294. ACM (2011)

Distributed SECONDO: A Highly Available and Scalable System for Spatial Data Processing

Jan Kristof Nidzwetzki$^{(\boxtimes)}$ and Ralf Hartmut Güting

Faculty of Mathematics and Computer Science, FernUniversität Hagen,
58084 Hagen, Germany
{jan.nidzwetzki,rhg}@fernuni-hagen.de

Abstract. Cassandra is a highly available and scalable data store but it provides only limited capabilities for data analyses. However, database management systems (DBMS) provide a lot of functions to analyze data but most of them scale poorly. In this paper, a novel method is proposed to couple Cassandra with a DBMS. The result is a highly available and scalable system that provides all the functions from the DBMS in a distributed manner. Cassandra is used as a data store and the DBMS SECONDO is used as a query processing engine. SECONDO is an extensible DBMS, it provides various data models, e.g. models for spatial data and moving objects data. With DISTRIBUTED SECONDO functions like spatial joins can be performed distributed and parallelized on many computers.

1 Introduction

SECONDO [1] is an extensible DBMS developed at FernUniversität Hagen. The system is designed with a focus on supporting spatial and spatio-temporal data management.

DISTRIBUTED SECONDO is a distributed version of SECONDO. The system is developed to handle huge amounts of data and provides mechanisms for analyzing data in a distributed and parallel way. Cassandra is used as a data storage and SECONDO is used as a query processing engine. DISTRIBUTED SECONDO allows adding new computers easily to handle more data or to analyze bigger amounts of data. Moreover, the removal of systems is supported. This allows reducing the amount of computers when the resources are no longer required.

The system offers all the functions implemented in SECONDO in a distributed and scalable way. Cassandra and SECONDO are loosely coupled with each other; unmodified versions of both components are used in DISTRIBUTED SECONDO. All the details about data distribution, parallel query processing and fault tolerance are encapsulated in one software component. That ensures that almost all of the functions and data models implemented in SECONDO can be used in a parallel manner, without changing the implementation.

With PARALLEL SECONDO [2] another SECONDO based prototype for distributed data processing does exist. PARALLEL SECONDO couples Hadoop [3] with SECONDO to achieve scalability and data distribution. In contrast to DISTRIBUTED SECONDO, PARALLEL SECONDO does not focus on data updates and its architecture contains a master node, which is a single point of failure.

© Springer International Publishing Switzerland 2015
C. Claramunt et al. (Eds.): SSTD 2015, LNCS 9239, pp. 491–496, 2015.
DOI: 10.1007/978-3-319-22363-6_28

BerlinMOD [4] is a benchmark for spatio-temporal database management systems. The benchmark generates trips of moving vehicles within Berlin. The data generated by BerlinMOD is used in this demonstration to simulate moving vehicles. In a real world scenario, every observed vehicle is equipped with a GPS receiver and sends out position updates every few seconds. These position updates are handled and analyzed by DISTRIBUTED SECONDO.

2   System Overview

DISTRIBUTED SECONDO is a distributed system, consisting of three different node types: *storage nodes* (SNs), *query processing nodes* (QPNs) and *management nodes* (MNs).

SNs are responsible to store data and they are running Cassandra [5]. The QPNs do the data processing and run SECONDO as a query processing engine. The MNs are running SECONDO too; they are used to import and export data and to specify the queries to be executed.

Query Execution: The usage of DISTRIBUTED SECONDO usually consists of two steps: first of all, data is loaded, converted to tuples and stored on the SNs. This is done by SECONDO instances running on the MNs. In the second step, the data is analyzed. For this purpose, the user creates a *global execution plan* (GEP) and submits it to DISTRIBUTED SECONDO.

The GEP is an ordered set of single queries. These queries are formulated in the SECONDO executable language. Usually, a single query contains three operations: (*i*) fetching data from the SNs, (*ii*) processing the data and (*iii*) storing the result on the SNs.

Work Units: For the parallel and distributed execution of queries, it is required, that the input data is partitioned into small work units. The structure of the logical ring of Cassandra is used to achieve this goal.

The key element of Cassandra is a distributed hash table (DHT), organized as a logical ring. The logical ring consists of numbers, called tokens. Tokens are ordered sequentially around the ring. The token with the highest number is connected back to the token with the lowest number.

Upon initialization, every Cassandra system obtains one or more tokens. The system is placed at the position according to its token. The range between two adjacent Cassandra nodes is called token range. A node is responsible for the token range between their token and the token of its predecessor node. The data stored in a token range will be called *work unit* in this paper.

Partitioning Data: Some functions, e.g. joins, can be executed more effectively if the work units are partitioned considering the structure of the data.

Example: A join of the relations A and B should be executed. The join operator reads a work unit of relation A. After that, the operator needs to find all corresponding tuples in relation B.

If the work units are created without considering the structure of the data, all work units of the relation B have to be read. If the work units are partitioned considering the join attribute, only the work unit containing the join attribute has to be read.

For partitioning spatial data in SECONDO, a grid is used [6]. The number of the cell determines the position where the data is stored within the logical ring of Cassandra. An example how spatial data is stored in DISTRIBUTED SECONDO is shown in Fig. 1.

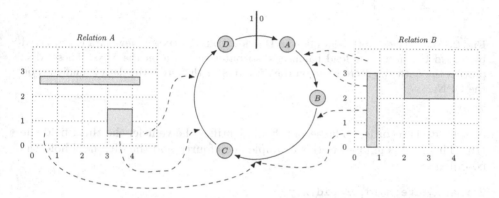

Fig. 1. Two relations with spatial data stored in cassandra. The spatial data is overlapped with a grid. The content of corresponding cells in both relations are stored at the same position in cassandra.

3   Demonstration

Two different demonstrations are presented to show the usage of DISTRIBUTED SECONDO. The first one handles GPS coordinate updates of moving vehicles and stores them. The second one performs a spatial join. An overview of the data flow and the components of DISTRIBUTED SECONDO are shown in Fig. 2.

3.1   Scalable Processing of GPS Updates

In the first demonstration, BerlinModPlayer is used as a data generator. The tool reads the trips of vehicles generated by BerlinMod, extracts GPS coordinates from the trips and writes them to a TCP socket. In a real world scenario, such data can be generated by vehicles, which drive through the streets of a city and send GPS updates every few seconds to a central system.

Position updates are represented as lines. Every line consists of five fields, each separated by a comma. The first field contains the time when the GPS

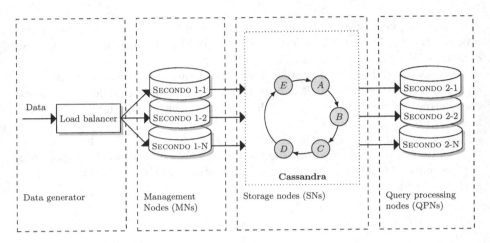

Fig. 2. An overview of the data flow in DISTRIBUTED SECONDO. New data is processed by a load balancer. The load balancer distributes the data on the MNs. These MNs process the data and store them on the SNs. In a further step, the data is analyzed on the QPNs.

coordinate is recorded. The second field identifies the vehicle, the third field the trip. This field is ignored in this example. The fourth and the fifth field hold the coordinates.

```
Format:␣date;moid;tripid;x;y
Example:␣2007-05-28␣06:00:00.001,122,17724,13.43745,52.42229
```

DISTRIBUTED SECONDO also includes a TCP load balancer. By using the load balancer, it is possible to distribute the stream of GPS coordinates on more than one MN. The SECONDO instances running on the MNs read the data from a network socket, parse the lines, convert them into tuples and store them onto the storage nodes. In the demonstration, it is measured how long the system takes to store 1 000 000 position updates. To show the scalability of the system, the demonstration is performed with a varying number of storage nodes.

Result: As illustrated by Fig. 3 the import of the data can be processed faster with additional storage nodes. Processing 1 000 000 GPS coordinates takes about 200 s using one storage node. Using six storage nodes, the processing can be done in about 110 s.

3.2 Performing a Distributed Spatial Join

In this demonstration, a parallel and distributed spatial join is performed. On the SNs two relations are stored. The relations describe the geographic information of the German state North Rhine Westphalia (NRW). Both relations are created from data fetched from the *Open Street Map* [7] project. The relation *Roads*

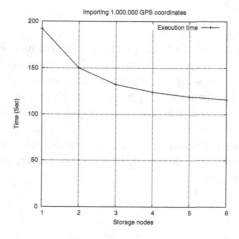

Fig. 3. Importing 1 000 000 GPS coordinates into DISTRIBUTED SECONDO.

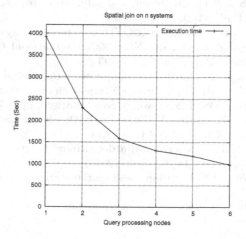

Fig. 4. A distributed spatial join on n systems.

contains the roads of NRW, the relation *Forests* all forests of NRW. The spatial data of both relations are partitioned using the same grid.

A spatial join is performed to find all roads that lead through a forest. For the calculation of the spatial join, each QPN fetches the content of one or more token ranges from the SNs, calculates the join and writes the result back to the SNs. To show the scalability of DISTRIBUTED SECONDO, the demonstration is performed with a varying number of query processing nodes.

Result: As shown in Fig. 4, the spatial join can be performed faster when the number of systems increases. On one system, the join requires an execution time of about 4 000 s, on six systems, the join can be executed in about 1 000 s.

4   Conclusion

In this paper, two use cases for DISTRIBUTED SECONDO have been demonstrated. In the first use case, it could be shown that the system is capable of handling more parallel GPS updates by adding additional storage nodes. In the second use case, it could be shown that DISTRIBUTED SECONDO could compute the result of a spatial join faster, when additional query processing nodes are added to the system.

Topics such as fault tolerance, data distribution and the creation of work units were only superficially addressed. In a further paper, the technical aspects will be discussed more precisely.

References

1. Güting, R.H., Behr, T., Düntgen, C.: SECONDO: a platform for moving objects database research and for publishing and integrating research implementations. IEEE Data Eng. Bull. **33**(2), 56–63 (2010)
2. Lu, J., Güting, R.H.: Parallel SECONDO: a practical system for large-scale processing of moving objects. In: IEEE 30th International Conference on Data Engineering, Chicago, ICDE, pp. 1190–1193 (2014)
3. Dean, J., Ghemawat, S.: MapReduce: simplified data processing on large clusters. In: Proceedings of the 6th Conference on Symposium on Operating Systems Design and Implementation, OSDI 2004, vol. 6, San Francisco, pp, 137–150 (2004)
4. Düntgen, C., Behr, T., Güting, R.H.: BerlinMOD: a benchmark for moving object databases. VLDB J. **18**(6), 1335–1368 (2009)
5. Lakshman, A., Malik, P.: Cassandra: a decentralized structured storage system. SIGOPS Oper. Syst. Rev. **44**(2), 35–40 (2010)
6. Patel, J.M., DeWitt, D.J.: Partition based spatial-merge join. SIGMOD Rec. **25**(2), 259–270 (1996)
7. Open Street Map. http://www.openstreetmap.org

EasyEV: Monitoring and Querying System for Electric Vehicle Fleets Using Smart Car Data

Gregor Jossé[✉], Matthias Schubert, and Ludwig Zellner

Institute for Informatics, Ludwig-Maximilians-University Munich,
Oettingenstr. 67, 80538 Munich, Germany
{josse,schubert,zellner}@dbs.ifi.lmu.de

Abstract. Electric vehicles (EVs) have great potential as a modern mobility concept. Electricity already relies on a broad infrastructure and is available anywhere in developed countries. Furthermore, EVs are emmission-free which makes them the preferable form of individual transportation in urban areas where air pollution is often alarmingly high. However, operating EVs has several drawbacks compared to common combustion engine cars. The range of most EVs is rarely above 150 km, and when running out of energy, recharging an EV usually takes up to several hours. In order to benefit from the advantages of EVs without being afflicted with the disadvantages, it is advisable to rely on the support from smart systems for trip and charge planning. In the project Shared E-Fleet, the shared use of a fleet of electric cars by a heterogeneous group of drivers is examined. In the presented demo, we introduce a spatio-temporal query system which was developed to support drivers and fleet managers alike. For the driver, the system provides assistance to keep in range of charging stations and provides routing alternatives to a specified destination. For the fleet manager, the system incorporates real-time information to identify possible delays or battery drainages and thereby detect deviations from the fleet schedule to allow for early rescheduling.

1 Introduction

With increasing air pollution, limited fossil fuel supply, and growing pressure to politically enforce reduction of CO_2 emissions, it seems, the days of common combustion engine cars are numbered. The number of electric vehicles (EVs) is constantly growing, especially in urban areas where air pollution is an increasing health threat. However, EVs have two major drawbacks which make them significantly less flexible than combustion engine cars. The first limitation is the range of EVs, which varies mostly from 100 to 150 km, excluding some exceptions like the Tesla Roadster which – according to the manufacturer – travels up to 400 km per charge. In addition, the energy consumption and therefore the range of an EV is dependent on the driving style but also on factors like heating, air conditioning, sound system, and headlights. The second limitation are the rather long recharging times. In contrast to refueling a car, a full recharge can take several hours.

© Springer International Publishing Switzerland 2015
C. Claramunt et al. (Eds.): SSTD 2015, LNCS 9239, pp. 497–502, 2015.
DOI: 10.1007/978-3-319-22363-6_29

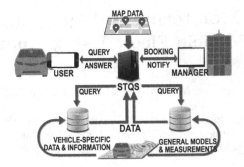

Fig. 1. System architecture of EasyEV. The manager is informed about anomalies concerning his fleet. The user can query the system, his on-board unit provides information to the system.

Despite these drawbacks, EVs have great potential to fulfill the mobility needs of urban populations. This is mainly because the average trip in a city is less than 30 km long and cars are actually parked most of the time. These two aspects can be taken advantage of. Even so,when in a setting where EVs are shared among several people, for instance by the employees of companies residing in the same office building. The business model of a shared fleet of EVs is examined in the project Shared E-Fleet. One goal of this project is to examine how spatio-temporal information systems can counteract the drawbacks and how the use of real-time information can improve the efficiency of a fleet of EVs. For instance, by monitoring the vehicles and keeping track of their remaining range limits, an automated booking system can inform the subsequent user of an expected delay or – even better – assign the user a different vehicle. Conversely, if the booking schedule is kept stable, charging times of vehicles can be optimized, yielding low electricity costs and reliably plannable charging processes.

In this demonstration we present EasyEV, our prototype for supporting drivers and fleet managers alike with spatio-temporal information services. EasyEV is currently employed in pilot projects in three different German cities involving seven BMW i3. Each car is equipped with an on-board transmission unit providing data like position, remaining range, charging status, and active electric devices. By evaluating this data, we are able to detect and predict anomalies, in order to inform the fleet manager about critical situations like expected belated returns or expected superabundances. In addition to these so-called monitoring features, EasyEV also supports the driver with multiple functionalities, such as providing directions to the nearest charging station,visually informing about reachable destinations, and computing alternative paths.

2   System Architecture

EasyEV runs as a platform-independent system which is connected to the fleet management and to the driver via app as well as to a real-time car database and an aggregated sensor information database. While the driver queries the system directly, the fleet management registers its vehicles and bookings with the system and is informed if anomalies occur. All connections are realized as platform-independent web service interfaces. Thus, the features of EasyEV are easily reused in a system requiring similar functionalities. At the heart of EasyEV lies a spatio-temporal query system (STQS) which answers directly to queries from users (routing features) and indirectly executes all recurrent tasks (monitoring features). For the STQS to answer queries instantly, the road network index has to be kept in memory. Hence, the STQS is modeled as a standalone server entity, while all communication with the server is handled by the aforementioned web services. The architecture is illustrated in Fig. 1.

The STQS relies on OpenStreetMap[1] (OSM) data for all road networks which we model as multicriterion (multiattribute) graphs. In a multicriterion graph, the vertices correspond to crossings, dead ends, etc., and the edges represent directed road segments connecting the vertices. A cost function maps every edge onto its cost vector. Each component of the cost vector corresponds to a particular cost dimension of the respective edge, e.g., distance, travel time, and energy consumption.

For the pilot projects, specific on-board units have been developed which collect data directly from the car computer as well as from additional built-in sensors, most prominently GPS and temperature sensors. The vehicle-specific data is fed into a database (see Fig. 1), while the environment-related data is aggregated and used to train time-dependent models reflecting outside influences on city-scale traffic, such as congestion or icy roads. We rely on external service providers feeding this kind of information into a data storage from where the STQS can query current influence factors. In order to be independent of the number of pilot test vehicles used during the demo presentation, we also simulate driving behavior according to the recorded trajectories. This enables us to replay actual historic driving data as well as display real-time movement enriched with simulated trajectories and bookings, ensuring a high frequency of queries and tasks.

In addition to the STQS, EasyEV offers a GUI to visualize trajectories, traffic-influencing factors, and query results. OSM data, is easily displayed using the open source library Leaflet(See Footnote 1) combined with the map design tool Mapbox (See Footnote 1). The GUI is browser-based and thereby platform-independent. It is dependent on the STQS-specific interfaces but can run on an entirely different machine. Hence, the architecture of EasyEV follows the model-view-controller principle, which facilitates portability and reusability.

3   Demo Features

We distinguish between routing features and monitoring features. The former are especially interesting from a user perspective, while the latter are particularly interesting from a fleet manager (operator) perspective. Note that in the pilot

---

[1] www.openstreetmap.org, www.leafletjs.com, www.mapbox.com.

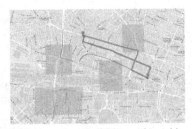

(a) Paths respecting (blue path) and ignoring (purple path) sensor-detected traffic delay (red road segments) and road ice (rectangular blue opaque regions).

(b) Respective range limits of two vehicles, illustrated as isochrone regions. The inner region is a more conservative range estimation than the outer region.

Fig. 2. Visualizations as displayed by the framework (Color figure online).

projects, monitoring and routing features are separated, as users do not have access to the trajectories of other users, while the monitoring system does not need access to all routing features.

Let us first present the routing features. As an elementary functionality, EasyEV supports shortest path searches given start and target nodes as well as w.r.t. predefined combinations of cost criteria. This has proven useful, as some users prefer a single path over a set of alternatives. However, EasyEV supports state-of-the-art algorithms for the computation of two sets of alternative paths, which we present without going into detail. The first set is known as the route or path skyline or the set of pareto optimal paths [1,3]. More recently, another concept of optimality was introduced in [5], called the linear path skyline. While the conventional path consists of all paths with cost vectors optimal under some monotone cost function, the result set of the linear path skyline consists of all paths with cost vectors optimal under a linear cost function. This is a restriction, limiting the number of results returned to the user and allowing for iterative result generation. EasyEV supports efficient methods for the computation of both sets, following [5,6]. In all the above cases, a result list is displayed which the user can browse.

As mentioned before, EasyEV does not only rely on static edge costs, constituted by the underlying road network, but also incorporates sensor information, as derived from data acquired by mobile (e.g., the vehicles themselves) and stationary sensors (e.g. traffic loop sensors). In our demo we focus on the influence factors road ice and traffic delay, however noting that, given a wider array of data providers, other factors could easily be included. These factors either affect specific roads (traffic delay) or whole areas (road ice). Both variants are displayed on the map. Given this kind of data, EasyEV computes paths avoiding affected roads or regions, depending on the duration of the delay or on the severity of the road icing, as depicted in Fig. 2(a).

In addition to sensor data, EasyEV employs static meta-information. Incorporating locations of public charging stations, EasyEV allows for ad hoc queries in order to intercharge when on the road, e.g., during a customer meeting. For the demo, this information is retrieved from providers like Open Charge Map or Plugshare[2]. Relying on probabilistic algorithms, similar to those presented in [2], EasyEV is able to provide paths along a number of charging stations, in order to maximize the probability that one of these stations is currently vacant.

Now, let us turn to the monitoring functionalities which serve the purpose of real-time as well as retrospective fleet analysis. In addition to the recorded trajectories, we also simulate bookings and corresponding trajectories in order to ensure high data density (during the demonstration). Trajectory simulation is done by computing (some pareto-optimal) path and sending positions along this path according to some randomly drawn travel time from a distribution around the speed limit. All trajectories, real-time, historic, and simulated, are retrieved from the vehicle database via web service. Most pilot test vehicles send their data every minute, some vehicles send their data every ten seconds. In order to clean the GPS signals of measurement errors and to be able to draw a valid trajectory between any two data points, we apply a map matching algorithm similar to [4]. Hence, we can display the roaming cars either in real-time or for defined past time spans at variable speeds. In both cases, we can display notifications having been or being sent by our monitoring system.

There are four types of notifications distributed to the fleet manager by the STQS. These notifications stem from the use cases covered in the project and will be displayed as pop-ups during the demo. The first one informs the manager of *illegal movement*, meaning that a registered veicle is moving without a valid booking. If a vehicle is not moved throughout the whole period of a booking, the manager receives a *no show* notification, as this might result in different billing. The other two notifications only occur during driving (with a valid booking). First, if a return within the booking period is unlikely, the fleet manager is informed of an expected *delay*. This is the case, if even the fastest path from the vehicle's current location to the booking-specific return station is longer than the remaining booking period. Second, if even the shortest path from the current location to the return station is longer than the currently remaining range of the vehicle, the fleet manager (and the driver) is informed of an expected *malfunction*.

In addition to the geo-positions at the respective timestamps, various other information is collected by the on-board unit and fed into the database. For example, the total number of kilometers driven, whether the passenger seat is occupied, whether wiper or lights are on, the current energy consumption, and the remaining distance. EasyEV displays all this information when hovering over the geo-position of an illustrated vehicle. Of all the data collected, the remaining distance is of particular interest, seeing as range limit exceedance results in vehicle deficiency and often significant rescheduling efforts. Hence, we incorporate the feature of displaying isochrone diagrams around pilot test EVs, as displayed in Fig. 2(b), where the inner isochrones are a more conservative estimation, and the outer isochrones depict the ideally reachable area.

---

[2] www.openchargemap.org, www.plugshare.com.

4   Conclusions

In this demo, we present EasyEV, a querying and monitoring system for fleets of EVs, which is being used in a pilot project investigating the potential of smart systems employed in optimization of EV fleets. EasyEV offers features for fleet managers and drivers alike. All computations are handled by a spatio-temporal query system, while the GUI is browser-based and the communication is handled by web services. Hence, EasyEV ensures high functionality and at the same time great flexibility and reusability.

References

1. Ehrgott, M., Gandibleux, X.: A survey and annotated bibliography of multiobjective combinatorial optimization. OR-Spektrum **22**, 425–460 (2000)
2. Jossé, G., Schmid, K.A., Schubert, M.: Probabilistic resource route queries with reappearance. In: EDBT 15, pp. 445–456 (2015)
3. Kriegel, H.P., Renz, M., Schubert, M.: Route skyline queries: a multi-preference path planning approach. ICDE 10, pp. 261–272 (2010)
4. Newson, P., Krumm, J.: Hidden markov map matching through noise and sparseness. In: ACM SIGSPATIAL GIS 09, pp. 336–343 (2009)
5. Shekelyan, M., Jossé, G., Schubert, M.: Linear path skylines in multicriteria networks. In: ICDE 15, pp. 459–470 (2015)
6. Shekelyan, M., Jossé, G., Schubert, M.: Paretoprep: efficient lower bounds for path skylines and fast path computation. In: SSTD 15 (2015)

TwitterViz: Visualizing and Exploring the Twittersphere

Christodoulos Efstathiades[1,2]([✉]), Helias Antoniou[2], Dimitrios Skoutas[1],
and Yannis Vassiliou[2]

[1] Institute for the Management of Information Systems Research Center "Athena",
Athens, Greece
cefstathiades@dblab.ece.ntua.gr, dskoutas@imis.athena-innovation.gr
[2] Knowledge and Database Systems Laboratory,
National Technical University of Athens, Athens, Greece
hantoniou@dblab.ece.ntua.gr, yv@cs.ntua.gr

Abstract. Micro-blogging platforms and social networks are a rich source of spatio-temporal information, which, together with additional information that can be mined from the social network's structure, makes them extremely valuable for monitoring users' opinions, sentiments and behavior, and, consequently, making more timely and effective decisions. In this demo, we present *TwitterViz*, a complete solution for the visualization and analysis of spatio-temporal Twitter data, in combination with the analysis of the Twitter graph, by leveraging the use of a popular graph database and using state of the art visualization tools that aim at providing insights to the non-expert user.

1 Introduction

Micro-blogging platforms, especially Twitter, have become a very popular communication tool, where millions of users share opinions on different aspects of everyday life. Twitter reports over 100 million active users and 500 million tweets exchanged every day. This huge amount of data and the fact that they are offered publicly in real time make their management and analysis challenging.

Many research studies have been conducted in order to determine whether Twitter can actually give insights as to how people behave. Such studies have focused on analyzing a variety of spatio-temporal phenomena (e.g. [4,7]), as well as topics, sentiments and social interactions (e.g. [5,6]). Typically, these works focus on specific problems and examine specific parts of the Twittersphere.

A recent survey of approaches for Twitter analytics can be found in [3]. It identifies the need for integrated solutions that provide a unified framework to be used by researchers and practitioners across disciplines, and it suggests the support of the following components for this purpose: (a) a focused crawler to allow for configuration by the user, (b) a pre-processor for the processing of tweets based on specific needs, (c) a defined data model that allows the efficient execution of complex queries, (d) the support of a query language and (e) the informative spatial as well as graph visualization.

© Springer International Publishing Switzerland 2015
C. Claramunt et al. (Eds.): SSTD 2015, LNCS 9239, pp. 503–507, 2015.
DOI: 10.1007/978-3-319-22363-6_30

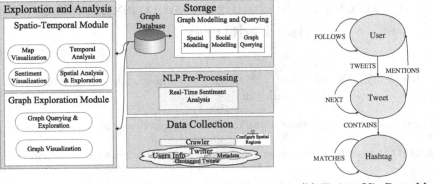

(a) TwitterViz architecture   (b) TwitterViz Data Model

Fig. 1. TwitterViz Architecture and Data Model for querying the Twittersphere

TwitterViz supports all of these components. The crawler allows spatial configuration in order to focus on specific geographic areas. The pre-processor uses natural language processing (NLP) in real time and the processed information is stored in a graph database with a defined data model as well as the support of a powerful graph query language. In addition, the visualization of the data both based on spatio-temporal characteristics as well as graph characteristics renders *TwitterViz* a complete solution for the management of Twitter data in order to provide useful analytics. *TwitterViz* provides a framework to the non-expert user for exploring and analyzing the Twittersphere using simple, unobtrusive yet powerful tools.

2   The TwitterViz System

TwitterViz comprises a modular pipeline that supports data collection, storage, analysis and visualization of Twitter data. Figure 1(a) shows the architecture of the system. We briefly describe each module below.

Data Collection. The *data collection* module supports the crawling of tweets. Tweets are collected from specific geographic regions based on users' preferences along with information about "followers" relationships.

Sentiment Analysis. NLP for sentiment analysis is conducted and each tweet is then tagged with a score denoting its sentiment. The pre-processing uses the *AlchemyAPI*[1] tool which is also used in various research works (e.g. [5]).

Storage. The *storage module* consists of a *Neo4j*[2] graph database, which naturally fits the overtly relationship-centered domain of social networks. The defined data model, depicted in Fig. 1(b), enables the Twitter graph construction using a

---

[1] http://www.alchemyapi.com.

[2] http://www.neo4j.org/.

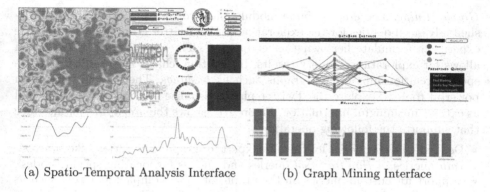

(a) Spatio-Temporal Analysis Interface   (b) Graph Mining Interface

Fig. 2. *TwitterViz* User Interface

variety of relationships, rendering it a complete model for the support of complex queries.

Visualization and Analysis. The user interface of the *TwitterViz* framework, shown in Fig. 2, consists of two main views: (a) the *Spatio-Temporal Analysis View* and (b) the *Graph Analysis View*. Both can be used simultaneously, independently or in combination with each other. In future versions, we intend to provide additionally an API that would allow third-party applications to reuse the results of the analysis creating custom visualizations for specific needs.

Spatio-Temporal Analysis. The user is given a variety of tools for spatio-temporal analysis and exploration of Twitter data, both on the map as well as a variety of charts. Spatial indexing is used to speedup range queries focusing on tweets in specific areas. The user interface allows for temporal visualization and analysis of tweets, as well as the simulation of the temporal evolution of tweets created during a specific time window. All tools can be used in combination with each other in order to reach to useful conclusions. Figure 2(a) shows the *spatio-temporal analysis* interface, which supports the following operations:

- Range queries on the map to visualize tweets from specific areas.
- Visualization of sentiment on tweets on the map in speficic geographic areas using a defined visual syntax. The user can investigate how the sentiments change in specific areas as well as how they change in time, by also applying other restrictions based on the social network's structure.
- Visualization of a user's followers' tweets on the map, combining information from the graph.
- Visualization and study of the temporal evolution of tweets in user-defined time windows.
- Analysis of the spatio-temporal distribution of tweets.
- Presentation of a variety of real-time statistics on the streaming data.

All of the supported operations can be combined with each other using restrictions, allowing a powerful spatio-temporal and social analysis based on the advanced visualization offered by *TwitterViz*.

Graph Mining. The *graph mining* module offers to the non-expert user off-the-shelf advanced queries for the exploration of the Twitter graph. In addition, the expert can formulate her own *Cypher* queries[3]. The visualization of the graph allows for exploration of query results. In addition, charts are used to visualize specific relationships on the graphs such as *hashtags*. The user is thus given powerful tools to explore the Twittersphere, visualize and analyze data as well as extract meaningful information. Figure 2(b) shows the *graph mining* interface that supports the following operations:

- Defined *Cypher* queries for the non-expert for graph exploration, like *shortest-path* queries, *n-Hop traversal* queries and queries for locating specific nodes.
- Support for custom queries on the graph for more complex analysis, such as pattern matching queries on the graph. Custom queries can be formulated easily and can combine the tweets' geo-social characteristics. Such example queries could be:
  - Find the tweets with hashtag "parthenon" that are within $0.5Km$ from the Athens historical center (substituting x,y with actual coordinates).

    ```
    START n=node:tweetWKT('withinDistance:[x,y, 0.5]')
    MATCH (n)-[]-(h:HashTag)
    WHERE h.Hashtag='parthenon'
    RETURN n,h
    ```

  - Find the *top-10* users in New York based on how many other users "follow" them.

    ```
    MATCH (n:User)<-[:FOLLOWS]-(m:User), (n)-[]->(t:Tweet)
    WHERE t.Region='NewYork' WITH n,count(m) AS total
    RETURN n ORDER BY total DESC limit 10
    ```

- Visualization of the results for all of the queries on the graph.
- Presentation of a variety of statistics for real-time graph analysis.

3 Demonstration

A use-case scenario that benefits from the use of *TwitterViz* is the following: We want to examine whether users who tweet from the same spatial neighborhood and who use the same *hashtags* in their tweets are close in the followers graph. This kind of scenario can be used to verify the results of a geo-social query as in [1] or a kRNN query as in [2]. We pick a very popular hashtag from an area in NYC *job* and we use TwitterViz to choose two random users who used *job* in their tweets. *TwitterViz* visualizes the resulting subgraph, showing that $u1$ and $u2$ are three hops apart. Figure 3 shows how we use *TwitterViz* spatial and graph analysis capabilities in order to gain insights.

For the demonstration at the SSTD 2015 conference, we intend to show-case the full capabilities of *TwitterViz* for spatio-temporal and graph analysis and visualization. Scenarios such as the above, that leverage the use of spatio-temporal exploration and sentiment analysis, and combine the findings with graph exploration in order to reach to useful results will be showcased. A current prototype of *TwitterViz* is available online[4] among with a video demonstration.

---

[3] *Cypher* is the graph query language used in *Neo4j*.

[4] https://web.imis.athena-innovation.gr/redmine/projects/twittervizdemo.

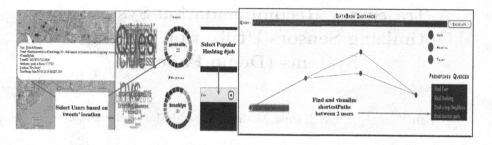

Fig. 3. Use Case: Find paths between co-tagged users

Acknowledgements. This work is supported by the EU/Greece funded KRIPIS Action: MEDA Project.

References

1. Armenatzoglou, N., Papadopoulos, S., Papadias, D.: A general framework for geo-social query processing. Proc. VLDB Endow. **6**(10), 913–924 (2013)
2. Efstathiades, C., Pfoser, D.: User-contributed relevance and nearest neighbor queries. In: Nascimento, M.A., Sellis, T., Cheng, R., Sander, J., Zheng, Y., Kriegel, H.-P., Renz, M., Sengstock, C. (eds.) SSTD 2013. LNCS, vol. 8098, pp. 312–329. Springer, Heidelberg (2013)
3. Goonetilleke, O., Sellis, T., Zhang, X., Sathe, S.: Twitter analytics: a big data management perspective. SIGKDD Explor. Newsl. **16**(1), 11–20 (2014)
4. Kling, F., Pozdnoukhov, A.: When a city tells a story: urban topic analysis. In Proceedings of the 20th International Conference on Advances in GIS, SIGSPATIAL 2012, pp. 482–485. ACM, New York, NY, USA (2012)
5. Quercia, D., Capra, L., Crowcroft, J.: The social world of twitter: topics, geography, and emotions. In: Breslin, J.G., Ellison, N.B., Shanahan, J.G., Tufekci, Z. (eds.) ICWSM. The AAAI Press, New York (2012)
6. Quercia, D., Ellis, J., Capra, L., Crowcroft, J.: Tracking "gross community happiness" from tweets. In: Proceedings of the ACM 2012 Conference on Computer Supported Cooperative Work, CSCW 2012, pp. 965–968. ACM, New York, NY, USA (2012)
7. Sengstock, C., Gertz, M., Abdelhaq, H., Flatow, F.: Reliable spatio-temporal signal extraction and exploration from human activity records. In: Nascimento, M.A., Sellis, T., Cheng, R., Sander, J., Zheng, Y., Kriegel, H.-P., Renz, M., Sengstock, C. (eds.) SSTD 2013. LNCS, vol. 8098, pp. 484–489. Springer, Heidelberg (2013)

A Trajectory Recommendation System via Optimizing Sensors Utilization in Airborne Systems (Demo Paper)

San Yeung[1], Sanjay Kumar Madria[1(✉)], and Mark Linderman[2]

[1] Computer Science Department, Missouri University of Science and Technology,
Rolla, MO 65409–0350, USA
{syq3b,madrias}@mst.edu
[2] Information Directorate, AFRL, Rome, New York 13440, USA
mark.linderman@us.af.mil

Abstract. Airborne sensory system is equipped on piloted or remotely-piloted aerial vehicles to collect and transmit imagery data back to the ground users. In traditional approaches where pilots to satisfy spatio-temporal tasks via image capturing, the pilot is required to manually decide an alternative trajectory to satisfy as many tasks as possible while maintaining a low deviation cost due to fuel constraint. Additionally, various constraints on tasks and original flight trajectory must be satisfied as well, such as temporal and Quality of Service constraints. We show a demo of a trajectory recommendation framework consists of two approaches to generate an optimized trajectory with the above goals by increasing sensor utilization via task aggregation and scheduling. We demonstrate a trajectory recommendation system that accepts user inputs and outputs visualization of intermediate processes and final trajectory.

1 Introduction

Airborne sensory system is equipped on piloted or remotely-piloted aerial vehicles to collect and transmit imagery data back to ground users. Requests for the airborne sensors are made up of one or more taskpoints containing spatio-temporal information, and each request could be sent to the system by the pilot or by ground operators directly. The goal is to seek a new trajectory such that it satisfies as many requests as possible while also satisfying the time constraint on each of the waypoints on the original trajectory. However, each taskpoint has additional temporal and QoS (Quality of Service) constraints that need to be satisfied as well. For real world trajectories, the number of waypoints is mission dependent (an acceptable range could be five to ten points), but the number of taskpoints could be hundreds or thousands. Currently, the pilot has the sole responsibilities of analyzing taskpoints and scheduling the final route. However, it is difficult for the pilot to manually analysis all of the taskpoints and associated constraints to produce an optimized trajectory. Therefore, an automatic trajectory recommendation system is necessary to perform quick analysis

© Springer International Publishing Switzerland 2015
C. Claramunt et al. (Eds.): SSTD 2015, LNCS 9239, pp. 508–513, 2015.
DOI: 10.1007/978-3-319-22363-6_31

on taskpoints and to generate an optimal trajectory considering all of the constraints.

The problem is challenging in three ways: (1) maximize taskpoints satisfaction, (2) satisfy original waypoints constraints, and (3) minimize trajectory deviation. Given a set of taskpoints, an ideal trajectory would satisfy all of them, but that is not realistic because each taskpoint has a hard temporal constraint, causing only portion of taskpoints could be satisfied. The QoS constraint is a critical real-world constraint and it requires that the sensor has to be at the required altitude level or lower when the image is taken for a taskpoint in order to obtain the desired image resolution (lower altitude means better image quality). This adds difficulty to satisfy maximum taskpoints because traveling from one altitude to another induces extra distance and time cost. Moreover, the pilot has to arrive at each original waypoint on or before the waypoint time constraint. This is critical because there might be additional mission tasks assigned to the pilot at the original waypoints. Lastly, the deviation cost between the original and new trajectory should not be too large due to the limited fuel supply on the aircraft. It means that we need to reach for a balance between number of taskpoints to satisfy and deviation cost.

In order to develop an efficient trajectory recommendation system, we aim to come up with a novel approach by optimizing airborne sensor utilization. Most existing works focus on generating cost-effective trajectory in a constrained environment, but not in the context of optimizing sensor utilization for trajectory recommendation [1–3]. The problem is real and in this demo paper, we present a trajectory recommendation framework by optimizing sensor utilization with the aim to satisfy as many task requests as possible considering the accompanying temporal as well as QoS constraints with low deviation from the original flight trajectory. We assume the fully implemented system to be integrated on the aircraft. Requests are issued dynamically and they are handled by the system fitted in the aircraft. The system is capable of generating the recommended trajectory on the fly, given that it has complete information on requests and original route with their corresponding constraints.

This demo paper proposed a trajectory recommendation framework consists of two approaches— (1) _Foot-Print Clustering Approach_ (**FPCA**) and (2) _Swath-Width Clustering Approach_ (**SWCA**). They are based on optimizing sensor utilization using the concepts of sensor _footprint diameter_ (maximum diameter of the area the sensor camera can cover) and _swath width_ (strip of surface area when the sensor takes an airborne image). The proposed solutions is novel by enhancing some basic works in clustering to integrate constraints in imagery sensor and flight path. FPCA partitions original flight trajectory using footprint diameter to generate the final trajectory while SWCA, an improved solution from FPCA, uses the idea of swath-width boundary followed by an optimization scheduling method to compute the optimized trajectory.

The rest of the paper is organized as follows. Section 2 gives a system overview with problem definition and approaches discussions. Section 3 presents the system demonstration. Finally, Sect. 4 concludes the paper and outlines future research directions and recommendations.

2   System Overview

The process of trajectory recommendation involves two critical steps: **Task Aggregation** and **Task Scheduling**. Task aggregation refers to the analysis of taskpoints distribution to identify dense regions that could be aggregated in order to satisfy multiple requests simultaneously. Task scheduling generates a new trajectory by selecting a set of clusters which met the constraint requirements. The new trajectory should have the following characteristics: (i) it contains original and new waypoints and their time constraints are satisfied, (ii) it should satisfy as many taskpoints as possible along with their constraints, and (iii) it should maintain a low deviation cost because fuel supply is limited.

2.1   Inputs

The inputs to the system are the original flight trajectory and sensor requests. The original trajectory is a set sequence of flight path consists of ordering of waypoints: $TR = \{(w_1, t_1), (w_2, t_2), ...(w_m, t_m)\}$ where $w_i's$ are waypoints, each consists of (longitude, latitude, altitude), and $t_i's$ are time constraints at those waypoints where $t_1 < t_2 < ... < t_m$. The second input is requests sent to the sensor from the pilot or from ground users. Each request is made up of one or more taskpoints: $REQ = \{tp_1, tp_2, ..., tp_n\}$ where each tp_i is a taskpoint consists of (longitude, latitude, priority, QoS, time). The priority of a taskpoint is expressed as *weight* in the domain of (*high, medium, low*). Intuitively, a taskpoint k with high priority means that it is able to satisfy multiple requests, thus it is deemed as a preferred point over those with a lower priority weight. In the case of having to select between k_i and k_j due to constraints, decision could be made based on priority weight. QoS constraint is the required altitude that the aircraft has to be at or lower in order to satisfy the imagery resolution of k. Time is the temporal constraint of a taskpoint of when the image needs to be taken by and be sent back. A taskpoint satisfied beyond the required time would be equivalent as unmet.

2.2   Benefits and Costs Measure

Recall that the two main objectives of the framework are to maximize the number of taskpoints satisfied and keeping a low deviation on the new trajectory. They conflict with each other; thus, we developed a *utility metrics* to balance the benefits and costs relationship that will help us in the process of taskpoints selection.

The utility metrics for individual taskpoint is defined as: $U_{tp_i} = \frac{pw}{D}$ where U_{tp_i} is the utility metrics for taskpoint i, pw is the priority weight of the taskpoint, and D is the distance (deviation) measured from i to the closest point on the original trajectory. It represents relationship such as that a taskpoint with low priority weight and distance cost could be preferred then one with high priority weight and high distance cost. This metrics will be used as the fundamental cluster selection criteria and will help determine the result of the final flight trajectory, avoiding it to incur high deviation cost.

2.3   Request Aggregation

The two proposed approaches, FPCA and SWCA, both consists of request aggregation and task scheduling phases. Performing request aggregation allows us to
group taskpoints based on spatial proximity. From the aggregation result, it
helps us to determine locations of new waypoints because dense clusters imply
groups of interests that we would like to capture. FPCA utilizes the idea of footprint diameter and applies it towards clustering. Figure 1a shows that it begins
by partitioning the original trajectory into equivalent sections using a system-
defined footprint diameter value, and then cluster taskpoints into corresponding
sections. This way, we can satisfy all taskpoints of a cluster by only satisfying
the cluster center. An additional step is needed to handle the QoS constraints
of taskpoints by further generating *QoS clusters* in each section (represented by
the colored rectangles in the figure). The purpose is to identify at which altitude
level could satisfy the most taskpoints.

The second approach, SWCA, first prunes taskpoints that are located outside
of *swath width boundaries*, and then employs a density-based clustering method
to cluster remaining taskpoints. The boundary threshold is calculated using the
following formula: $Boundary_{TH} = \sqrt{(\frac{1}{2}SW + Dev_{max})^2 + (\frac{1}{2}DM + Dev_{max})^2}$,
where SW and DM represents swath-width and footprint diameter, respectively.
Figure 1b shows boundary zones generated per each waypoint on the original
trajectory. The Dev_{max} value is user-defined to control the maximum deviation
of the new trajectory. By providing this value, the system could reduce the
number of taskpoints in clustering.

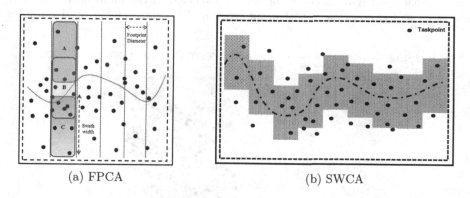

(a) FPCA   (b) SWCA

Fig. 1. Request aggregation and scheduling of proposed approaches

2.4   Task Scheduling

FPCA generates the final flight trajectory by using QoS clusters in each footprint
section. The main idea is to compute a sum utility metrics for each QoS cluster.
The scheduling algorithm then generates a subsection-trajectory in each section

by choosing the one with the highest weight, and ensuring that time constraints are satisfied with the original waypoint and other new waypoints already in the sub-trajectory. Clusters with higher weights would be considered first, then the algorithm will satisfy remaining lower weight clusters if time constraints allow. When a cluster cannot be inserted due to time constraint violation, some of the taskpoints may be removed from it to mitigate the violation and be re-inserted only if its updated utility weight remains to be competitive. The complete new trajectory is formed by merging all sub-trajectories together.

SWCA is inspired from the micro-and-macro clustering schema proposed in [4] and the solution is based on generating *representative trajectory*, tra_{rep}, for each cluster as a result from the density-based clustering. SWCA uses the sub-trajectory technique described in FPCA to generate representative trajectory. One advantage is that computing tra_{rep} allows us to assign utility metric for each cluster. Then, the scheduling algorithm can formed the new trajectory by selecting higher weight clusters first, and then satisfy remaining clusters as many as possible. Similar with FPCA, the algorithm tries to mitigate time constraint violation by updating the tra_{rep} of a cluster. The second advantage of SWCA is that during the scheduling process, it has knowledge of clusters weight in a macro-scale. On the other hand, FPCA has a weakness of detecting imbalance. Experiments had shown that SWCA has better performance than FPCA.

3   Demonstration

The goal of this demonstration is to visualize the final trajectories output by the two approaches. Figure 2 shows the system overview of the main user interface. The user will choose one set of input from multiple trajectories and requests

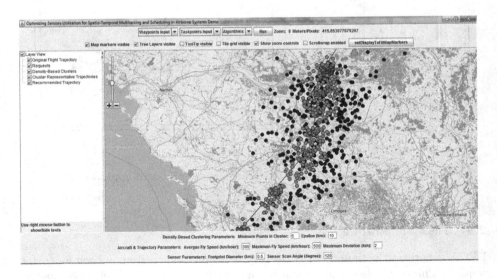

Fig. 2. System main GUI

sample. Then, he will choose which approach to run after confirming all system parameters, which consists of three components. The user can control parameters for clustering, *epsilon* and minimum number of points in a cluster, that will directly affect the clustering result. The user can control parameters of average and maximum flying speed that will impact time feasibility check, and the maximum deviation allowed from the original flight path. The user can also control the sensor footprint diameter and scan angle parameters, which will affect the swath width value.

In addition, user can selectively choose the display layers to visualize results. By selecting different layer to be visible, user is able to see the route of the original and final flight trajectory, original taskpoints distribution, clustering results and corresponding respective trajectories. Extra deviation traveled and the number of satisfied taskpoints will also be displayed.

4 Conclusions and Future Research

In this demo paper, we presented an optimization framework for sensory system utilization with respect to request aggregation and scheduling in presence of request QoS and time constraints. The goal is to generate a new trajectory with minimum deviation from the original trajectory while being able to satisfy maximum number of taskpoints. The first approach, FPCA, is designed using footprint diameter sections to generate QoS clusters. The second approach, SWCA, improves upon FPCA and utilizes the idea of representative trajectories in each cluster to perform the scheduling. The demonstration shows that users are able to submit parameters to each approach and select layers of view to visualize the final output. For future work, we would like to consider rotational sensor such that the swath width zones no longer only have vertical movements, and to adapt our framework to handle dynamic requests.

References

1. Babaei, A., Mortazavi, M.: Fast trajectory planning based on in-flight waypoints for unmanned aerial vehicles. Aircr. Eng. Aerosp. Technol. **82**(2), 107–115 (2010)
2. B. Luders.: Robust trajectory planning for unmanned aerial vehicles in uncertain environments. Ph.D. thesis. Massachusetts Institute of Technology (2008)
3. De Filippis, L., Guglieri, G., Quagliotti, F.: Path planning strategies for UAVS in 3D environments. J. Intell. Rob. Syst. **65**(1–4), 247–264 (2012)
4. Li, Z., Lee, J.-G., Li, X., Han, J.: Incremental clustering for trajectories. In: Kitagawa, H., Ishikawa, Y., Li, Q., Watanabe, C. (eds.) DASFAA 2010. LNCS, vol. 5982, pp. 32–46. Springer, Heidelberg (2010)

Tourismo: A User-Preference Tourist Trip Search Engine

Gregor Jossé[1]([✉]), Klaus Arthur Schmid[1], Andreas Züfle[1], Georgios Skoumas[2],
Matthias Schubert[1], and Dieter Pfoser[3]

[1] Ludwig-Maximilians-Universität München, Munich, Germany
{josse,schmid,zuefle,schubert}@dbs.ifi.lmu.de
[2] National Technical University of Athens, Athens, Greece
gskoumas@dblab.ece.ntua.gr
[3] George Mason University, Fairfax, USA
dpfoser@gmu.edu

Abstract. In this demonstration we re-visit the problem of finding an
optimal route from location A to B. Currently, navigation systems compute shortest, fastest, most economic routes or any combination thereof.
More often than not users want to consider "soft" qualitative metrics
such as popularity, scenic value, and general appeal of a route. Routing
algorithms have not (yet) been able to appreciate, measure, and evaluate such qualitative measures. Given the emergence of user-generated
content, data exists that records user preference. This work exploits user-
generated data, including image data, text data and trajectory data, to
estimate the attractiveness of parts of the spatial network in relation to
a particular user. We enrich the spatial network dataset by quantitative scores reflecting qualitative attractiveness. These scores are derived
from a user-specific self-assessment ("On vacation I am interested in:
family entertainment, cultural activities, exotic food") and the selection
of a respective subset of existing POIs. Using the enriched network, our
demonstrator allows to perform a bicriterion optimal path search, which
optimizes both travel time as well as the attractiveness of the route.
Users will be able to choose from a whole skyline of alternative routes
based on their preference. A chosen route will also be illustrated using
user-generated data, such as images, textual narrative, and trajectories,
i.e., data that showcase attractiveness and hopefully lead to a perfect
trip.

1 Introduction

Nowadays, social networks are a great source of rich geo-spatial data. Almost
every social network allows users to incorporate geo-social features into their data
stream. The different features include, amongst others, geo-tagged pictures (e.g.,
Flickr), geo-descriptive text (e.g., travel blogs), and tracked movement (e.g., runners' trajectories). For this demo, we rely on all these kinds of user-generated
data to define attractiveness on a real world road network. Our aim is to reflect
human fondness according to the crowd by using qualitative information and

© Springer International Publishing Switzerland 2015
C. Claramunt et al. (Eds.): SSTD 2015, LNCS 9239, pp. 514–519, 2015.
DOI: 10.1007/978-3-319-22363-6_32

making it measurable. We present Tourismo, a tourist search engine, which computes attractive paths along points of interest (POIs), tailored to the interest of the user issuing the query. Based on this enriched spatial network, which has information about the attractiveness of locations, we aim at answering *attractive path queries*. Currently, navigation systems, i.e., machines, perform this task for us, computing routes such as the shortest route, the fastest route, the most economic route [1], or some combination of such quantitative measure on a spatial network [2]. In all of these cases, the employed algorithms optimize cost measures inherent in the underlying road network. What is rarely reflected, however, is user preference on subjective measures, such as attractiveness and interestingness of a route. Often users are willing to take a suboptimal detour, a deviation from quantitative optimality (shortest, fastest, etc.), in order to improve the quality of their route. In order to see more attractions, for instance, a tourist may be willing to take a moderate detour from a fast, but not very attractive, highway.

How can we measure a subjective concept of "quality"? How to measure attractive, scenic, recreative routes? As machines are not (yet) capable to reflect this concept, we rely on the crowd to answer this question, i.e., we propose to use crowdsourced data to estimate the attractiveness of an area. Relying on different datasets, image data (from Flickr[1]), textual narratives (from travel blogs), and trajectory data (from Endomondo (see Footnote 1)), we investigate the applicability of different data sources as cost measures for the underlying road network. More precisely, we enrich the road network by quantitative scores of qualitative statements as follows:

- areas having a large density of Flickr images indicate a particularly attractive area, increasing the attractiveness score;
- locations mentioned in the positive context of travel blogs increase attractiveness scores;
- routes commonly used by other users are also considered more attractive.

Furthermore, we incorporate meta-information from OpenStreetMap (see Footnote 1) (OSM), in order to categorize POIs and, using the aforementioned popoularity score, propose routes according to the user's preferences and the fondness of the crowd. Tourismo presents solutions to enrich the underlying road network using the data sources. We show an initial approach to map these *attractiveness scores* to cost measures correlated with travel time, allowing to apply existing routing algorithms which aim at minimizing edge-labeled cost metrics. We apply algorithms for pareto-optimal route search similar to [3,4], to find paths which are optimal w.r.t. the popularity scores. Our framework allows to specify origin and destination, computes and displays the skyline of pareto-optimal paths. Furthermore, the reasons for attractiveness of each path are illustrated: Flickr images along the way, travel blog entries mentioning locations on the way, and historical trajectories which share the same route. Our demonstrator is an extension of [5] it has three major features: First, we

---

[1] www.flickr.com, www.endomondo.com, www.openstreetmap.org.

incorporate new, route search algorithms which enable higher dimensional cost spaces at the same reducing computation time. Second, the demonstration allows to specify the interest of a user, thus returning routes that contain POIs which are of particular interest to the user. Third, this version considers a third type of data to enrich the underlying road network with attractiveness information: In addition to geotagged images, and texts containing geospatial references, we also learn attractiveness from an existing base of historic trajectory data.

2   State of the Art

Recently, a lot of interesting research has been done in the context of finding scenic, interesting or popular routes. The first set of related work focuses on providing paths which are easier to memorize, describe, and follow. For example, the authors of [6,7] try to tackle the problem by introducing cost criteria that allow for a trade-off between minimizing the length of a path while also minimizing the complexity in terms of instructions or turns along the path. Furthermore, an existing research direction covers the problem of defining tourist routes, which maximize the subset of a set of pre-defined POIs which can be visited in a tourist tour that has a time-constraint [8,9]. In these works, the set of interesting POIs is given, and the main conceptual contribution of is to automatically extract interesting locations, as well as a quantitative estimate of the popularity of this location from a variety of data sources. Another research direction, which is not necessarily restricted to touristic routing but lacks the aspect of qualitative measures, are the Trip Planning Query and variants thereof [10,11].

The approach most similar to the one presented in this work is [12], which proposes a method for computing beautiful paths, as the authors phrase it. However, in order to quantify quality, the authors rely on explicit statements about the beauty of specific locations, obtained from a crowd-sourcing platform which collects user opinions on photos of specific locations. In contrast, we propose to mine this kind of information from existing crowd-sourced data, which does not require any monetary investment to aquire. Thus our approach has the crucial advantage that it is scalable as the used data is already available globally available, while having local expert users rate photos one by one can hardly be extended to a global scale.

Another important research direction is the *stitching* existing trajectories in order to obtain new trajectories which guarantee that each sub-trajectory is used by other users, and is thus, "popular" following the definition [13] of Chen et al. This, however, only reflects a notion common usage, not taking into account, why a specific sub-trajectory has been favored. For instance, when mining trajectories of commuters, the fastest path is most likely to be chosen by most users. Hence, we propose mining trajectories specific to recreational use and merging this information with the attractiveness scores we derive from other user-generated data sources.

3 Features

The main feature of this demonstrator is the estimation of attractiveness from text, image, and trajectory data. Details covering text and image data can be found in [5]. In this section, we briefly describe how we enrich the underlying road network using historical trajectory data. For our demonstrator, we use trajectories of walkers, runners and bikers that have uploaded their workouts to Endomondo. Our dataset contains eight million trajectories, which are located all around the world, but have a strong regional focus in Northern Europe. To match each of the GPS trajectories, we apply state-of-the-art map matching techniques, similar to those presented in [14]. In a first step, we perform a basic enrichment: For each edge e of the spatial network, we count the number $tra(e)$ of historical trajectories that contain this edge. This count can be used as an indication of attractiveness of the nodes delimiting the edge, following the assumption that runners are, in average, more likely to choose a particularly nice running trail. We assign vertices in a road network attractiveness scores derived from different datasets. We refer to the score of a vertex v derived from Flickr image data as $im(v)$, to the score derived from travel blog text data as $txt(v)$, and to the score derived from trajectory data as $tra(v)$. In contrast to [5], where the different data source scores were combined into one cost measure, we now propose to diversified measures. For each edge $e = (u, v)$ and each of the scores $f \in \{$im,txt,tra$\}$ we define:

$$p_f(e) = tt(e) \cdot \phi^{C_f(f(u)+f(v))}$$

where $tt(e)$ denotes the travel time alonge e, C_f denotes a scaling parameter dependent only on the data source, and $\phi \in (0, 1)$ is a scaling factor for the influence of the respective attractiveness score. Hence, we obtain three travel time correlated cost measures reflecting notions of attractiveness according to the different data sources. Consequently, we may query the enriched road network, computing pareto-optimal paths as presented in [3,4] w.r.t. to the introduced cost measures.

Additionally, Tourismo features category-specific path queries. If the user chooses to specify his personal touristic interests, they can choose one or more options from a list containing outdoor activities, cultural sightseeing, culinary interest, and more. In order to provide paths which fulfill these requirements, we mine the OSM meta-information. Thanks to a very active community, the data contains well-tended information about POIs, that is named, categorized, and subcategorized. For instance, the categories "food" and "tourist" contain sub-categories "bar", "restaurant", "fastfood" and "monument", "museum", "archeological", respectively. Mapping these categories onto the preferences, we filter POIs which correspond to the particular interest of the user. When querying a route with a specific set of interests, the user is provided a number of pareto-optimal paths, guiding him along POIs tailored to his preference.

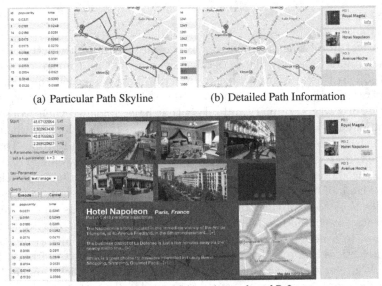

(a) Particular Path Skyline   (b) Detailed Path Information

(c) Detailed information about selected PoI

Fig. 1. Functionality of the presented framework.

4   Framework Description

The demonstrated framework allows users to validate that the notions of attractiveness defined in this paper indeed coincide with the general intuition. The result paths returned to the user yield competitive solutions in terms of travel time while passing POIs perceived as significant, appealing, and/or recognizable. Using OSM as a road network, our demonstrator visualizes a map relying on Google Maps. Upon selecting an origin and a destination location on the map, the user is presented with the skyline view as shown in Fig. 1(a). In this view, the path skyline is presented to the user. For each such route, the corresponding cost values are shown in a table in the lower left corner of Fig. 1(a). Using this table, the user can browse the choise and select a desired routeyielding the route view shown in Fig. 1(b). For the selected route A, this view shows the most "popular" points of interest on A. Once a point of interest is selected, the sources of popularity of this POI are displayed, as shown in Fig. 1(c). For this purpose, Fig. 1(c) shows all the pictures relevant for the selected POI, i.e., the set of images having a sufficiently low distance. The bottom-left corner shows all travel blog entries where this entry was mentioned in a positive context. Finally, the lower left corner shows a heatmap derived from all trajectories that share the same trajectory. During the demonstration, users will be able to specify start and target locations (also, if desired, specific categories of interest) and compute different sets of skyline paths.

Acknowledgements. This research has received funding from the Shared E-Fleet project (in the IKTII program) by the BMWi (grant no. 01ME12107), from the DFG (grant no. RE 266/5-1), from the DAAD supported by the BMBF (grant no. 57052426). Dieter Pfoser has been partially supported by NGA NURI (grant no. HM02101410004).

References

1. Andersen, O., Jensen, C.S., Torp, K., Yang, B.: Ecotour: reducing the environmental footprint of vehicles using eco-routes. In: MDM, pp. 338–340 (2013)
2. Graf, F., Kriegel, H.-P., Renz, M., Schubert, M.: MARiO: multi-attribute routing in open street map. In: Pfoser, D., Tao, Y., Mouratidis, K., Nascimento, M.A., Mokbel, M., Shekhar, S., Huang, Y. (eds.) SSTD 2011. LNCS, vol. 6849, pp. 486–490. Springer, Heidelberg (2011)
3. Shekelyan, M., Jossé, G., Schubert, M.: Linear path skylines in multicriteria networks. In: ICDE 2015, pp. 459–470 (2015)
4. Shekelyan, M., Jossé, G., Schubert, M.: Paretoprep: efficient lower bounds for path skylines and fast path computation. In: SSTD 2015 (2015)
5. Jossé, G., Franzke, M., Skoumas, G., Züfle, A., Nascimento, M.A., Renz, M.: A framework for computation of popular paths from crowdsourced data. In: ICDE, pp. 1428–1431 (2015)
6. Sacharidis, D., Bouros, P.: Routing directions: keeping it fast and simple. In: ACM SIGSPATIAL GIS, pp. 164–173 (2013)
7. Westphal, M., Renz, J.: Evaluating and minimizing ambiguities in qualitative route instructions. In: ACM SIGSPATIAL GIS, pp. 171–180 (2011)
8. Garcia, A., Arbelaitz, O., Linaza, M.T., Vansteenwegen, P., Souffriau, W.: Personalized tourist route generation. In: Daniel, F., Facca, F.M. (eds.) ICWE 2010. LNCS, vol. 6385, pp. 486–497. Springer, Heidelberg (2010)
9. Gavalas, D., Konstantopoulos, C., Mastakas, K., Pantziou, G.: A survey on algorithmic approaches for solving tourist trip design problems. J. Heuristics **20**, 291–328 (2014)
10. Kanza, Y., Safra, E., Sagiv, Y., Doytsher, Y.: Heuristic algorithms for route-search queries over geographical data. In: ACM SIGSPATIAL GIS, p. 11 (2008)
11. Chen, H., Ku, W.S., Sun, M.T., Zimmermann, R.: The partial sequenced route query with traveling rules in road networks. Geoinformatica **15**, 541–569 (2011)
12. Quercia, D., Schifanella, R., Aiello, L.M.: The shortest path to happiness: Recommending beautiful, quiet, and happy routes in the city. In: CoRR (abs/1407.1031) (2014)
13. Chen, Z., Shen, H.T., Zhou, X.: Discovering popular routes from trajectories. In: ICDE, pp. 900–911 (2011)
14. Newson, P., Krumm, J.: Hidden markov map matching through noise and sparseness. In: ACM SIGSPATIAL GIS, pp. 336–343 (2009)

Author Index